Sustainability Sciences in Asia and Africa

Agroecosystem Dynamics and Sustainable Practices

Series Editors

Kamel A. Abd-Elsalam, Plant Pathology Research Institute, Agricultural Research Center, Giza, Egypt

Olubukola O. Babalola, Faculty of Natural & Agricultural Sciences, North-West University, Mmabatho, South Africa

Sustainability as a scientific endeavor requires broad understanding of the interconnections in our global environment. It is therefore positioned at the crossroads of research in natural sciences, technology and human sciences. This book series supports the global efforts by providing timely coverage of the progress, opportunities, and challenges towards achieving sustainable development goals (SDGs) in Asia and Africa. It fosters the research in transdisciplinary academic fields spanning across agriculture and global food security, health and diseases, access to clean water and affordable energy, climate resilience and effects of global climate change on biodiversity, water and land. This series is planned to narrate the stories of scientific development and progression towards SDGs from the regions of Asia and Africa. The overall aim of the series is to provide a comprehensive publication platform and act as a knowledge engine in the growth of sustainability sciences with a special focus on developing nations. The series constitute of four sub-series each focusing on a specific theme closely linked to a set of SDGs. Each sub-series is leaders and experts in agricultural sciences, environmental sciences, biomedicine and medical sciences respectively. Books in this series are useful for environmentalists, biotechnologists, plant scientists, medical practitioners, conservation biologists, policy makers, NGOs and scientific bodies working in the areas of sustainability sciences in the developing regions.

This book series publishes research associated with the prevailing opportunities and challenges faced by agroecosystems in Asia and Africa. It also explores the future scope of agroecological techniques and offers guidance on sustainable management of these systems. The series discusses important areas such as sustainable agricultural production, soil health, water resource management, biodiversity conservation, animal health and productivity, insect management strategies, genome editing and climate change mitigation. Further, it includes the challenges faced by African agriculture in arid and semi-arid regions, particularly in response to climate change. It highlights issues such as water shortages, soil degradation, desertification, and the need for sustainable farming techniques.

Sustainable agroecosystem maintenance entails strategies that improve overall environmental health and economic viability while limiting negative environmental impacts. The series takes a multidisciplinary approach to investigate creative ways for boosting sustainable agriculture, increasing food security, and improving lives for people in global south. This book series acts as a global publishing platform to collate and disseminate knowledge on fostering resilient and productive agroecosystems, thus enhancing accessibility, and fostering informed decision-making.

Limenew Abate Worku • Rakesh Kumar Bachheti
Abdel Nasser B. Singab • Azamal Husen
Editors

Plant-Based Solutions for Sustainable Agriculture and Healthcare

Springer

ARC

Editors
Limenew Abate Worku
Department of Chemistry, College of Natural and Computational Science
Debre Tabor University
Debre Tabor, Ethiopia

Rakesh Kumar Bachheti
Department of Industrial Chemistry
Addis Ababa Science and Technological University
Addis Ababa, Ethiopia

Abdel Nasser B. Singab
Department of Pharmacognosy
Faculty of Pharmacy
Ain Shams University
Cairo, Egypt

Azamal Husen
Wolaita Sodo University
Wolaita, Ethiopia

ISSN 2730-6771 ISSN 2730-678X (electronic)
Sustainability Sciences in Asia and Africa
ISSN 3059-247X ISSN 3059-2488 (electronic)
Agroecosystem Dynamics and Sustainable Practices
ISBN 978-981-95-8909-8 ISBN 978-981-95-8910-4 (eBook)
https://doi.org/10.1007/978-981-95-8910-4

This Springer imprint is published by the registered company Springer Nature Singapore Pte Ltd.
The registered company address is: 152 Beach Road, #21-01/04 Gateway East, Singapore 189721, Singapore

Preface

Plants are the basis of life on planet Earth. Right from the primeval times of human civilization to the more complex systems of the modern world, plants have continued to be the backbone of human survival, the support system of the environment, and the driving fuels of economic development. Plants are the sources of food, medicines, shelter, fuels, fibers, and a vast array of environment-related services. In addition to these, plants have continued to influence cultures, beliefs, knowledge, and practices across various civilizations. In the modern era characterized by unprecedented levels of environmental destruction, climate change, food security threats, public health crises, resource depletion, and much more, the revival of interest in plant-based solutions is evidence not only to the advancements in science, technology, and innovation but also to the urgent needs of the global community. The twenty-first century poses the global community with interrelated challenges cutting across multiple disciplines. The challenges include steadily growing populations, greater agricultural needs, overreliance on chemical pesticides and fertilizers, declining soil quality, destruction of biodiversity, environmental destruction, and steadily escalating levels of chronic illnesses. While conventional strategies have played critical roles in the past in driving human progress, they are increasingly found to be insufficient, even counterproductive in contemporary times. Against this backdrop, the role of plants based on tradition, innovation, and science stands out as the most sustainable, resilient, and environmentally friendly responses to these interrelated challenges.

The notion of plant-based solutions goes far beyond culinary choices and alternative medicine. The use of plants is a vast spectrum of applications such as sustainable agriculture, rehabilitation of environment and ecology, healthcare and medicine, biomaterials and material science, and blue bioeconomy, among which plants are classified as natural ecosystem engineers with capabilities for reinstating degraded environments, sustaining ecosystem services including biogeochemistry processes, ameliorating soil quality, and combating pollution. Moreover, plants are a valuable source of bioactive compounds and drugs that are used for healthcare and sustainable economic development. This edited book has been designed and conceptualized as an all-encompassing and interdisciplinary sourcebook wherein various facets and aspects pertaining to the role and significance of plant contributions for mitigation and tackling various contemporary global challenges have been compiled and brought together. This book combines basic concepts and interpretations, historical and scientific developments, and latest innovations to provide a complete and holistic outlook on plant-driven strategies for sustainability and sustainable developments. This source book combines agriculture and food security with environment and ecology to recognize the pivotal role and significance of plant contributions.

The early chapters lay down the conceptual pillars of plant-based solutions. In the introductory chapter, the broad contours and implications of plant-based solutions in the contemporary world are introduced. The chapter places plant-based solutions within the context of sustainability and systems thinking. The early chapters conclude with an examination of traditional plant-based solutions that have long supported human civilizations. Furthermore, traditional knowledge systems and traditions of agriculture and herbal medicine are identified within these chapters as the rich reservoirs of sustainable solutions. These solutions serve to inform and guide the development of innovative solutions that are adapted to the specific needs and requirements of the contemporary world and its societies. Agriculture is one of those sectors that is most affected by plant-based innovations. The present-day agriculture faces two challenges: it has to increase its production and achieve it without damaging the environment. Overdependence on chemical fertilizers, pesticides, and herbicides has resulted in degradation of soil, water pollution, destruction of microbial life, as well as toxicity to plants, animals, and human life. Plant-based biofertilizers and biopesticides offer a hope to achieve soil enrichment, plant growth, as well as effective pest control without damaging the environment. This book has various contributions on plant-based innovations, specifically biofertilizers, biopesticides, their working, uses, and importance. A related aspect to sustainable agriculture is the relationship between plants and the improvement of soil health. Soil can be described as a living entity; soil fertility relates to the interactions among plants, microbes, organic matter, and environmental conditions. The involvement of plants can be noted through root growth, organic matter addition from plant residues, or plant biochemically mediated processes. This book shall discuss related topics on the use of plants to improve soil condition, increase nutrient availability, soil water holding capacity, and overall yield enhancement. Even more relevant is the significance of the knowledge on the relationship between plants and soil to ensure sustainable agriculture. Weed management is another major agricultural challenge conventionally solved with chemical herbicides, having well-documented hazards to both the environment and human health. The process of allelopathy—whereby plants release biochemical compounds into the environment that affect the growth and development of other plants around them—offers a natural alternative for managing weeds in a nonchemically intensive manner. Allelopathy is discussed in a separate chapter in this book that deals with plant interactions, allelochemicals, and their potential applications in integrated weed management systems. Using natural plant interactions, chemical inputs can be minimized without sacrificing crop yields or disturbing ecological balance. Beyond agricultural landscapes, plants play a transformative role in remedying environmental pollution and ecosystem degradation. The reasons for this are industrialization, the expansion of urban areas, and intensive land use practices that have contaminated soil, water, and air almost everywhere. Phytoremediation is generally described as the use of plants to remove, stabilize, or detoxify pollutants, emerging as a cost-effective and ecologically safe strategy for the restoration of environments. This book provides an in-depth discussion of the approaches involved in phytoremediation, including phytoextraction, phytostabilization, phytodegradation, and rhizosphere-mediated remediation. Recent advances and practical examples are used to illustrate how plants can be used in the rehabilitation of polluted sites, ecological restoration, and sustainable land management.

The integration of plant science and healthcare represents a central theme of this book. For thousands of years, medicinal plants have been the main source of health care for the majority of the world's population. Even today, a significant percentage of modern medicines is derived from plant constituents or is inspired by natural products. The sections specifically focused on medicinal plants and bioactive plant-derived compounds show the relevance of traditional medicine while covering recent advances in phytochemistry, pharmacology, and biomedical research; thus, these articles emphasize disease prevention, therapeutic management, and holistic healthcare systems via plant-based compounds. In the last few decades, increasing awareness has been directed toward nutrition regarding health maintenance and prevention of chronic diseases. Nutraceuticals, which have been defined as active food principles derived from natural sources with health-promoting properties, represent a rapidly growing field between nutrition and medicine. Plant-based nutraceuticals provide antioxidants, vitamins, minerals, polyphenols, and other bioactive components that facilitate metabolic health, immune function, and general well-being. This book investigates the scientific basis, functional properties, and health benefits of plant-derived nutraceuticals, emphasizing their importance in preventive healthcare and wellness-oriented lifestyles. Aromatic and essential oil–producing plants constitute another important dimension of plant-based healthcare. Plants producing such oils synthesize bioactive volatile compounds that possess antimicrobial, anti-inflammatory, antioxidant, and neuroactive properties. Essential oils are being increasingly used in complementary medicine, aromatherapy, mental health support, and wellness industries. The relevant chapter examines the contribution of aromatic plants to physical and mental well-being, drawing attention to both traditional applications and contemporary scientific evidence.

Plants are driving innovation in the most advanced areas of modern technology. New opportunities opened up by advances in nanotechnology have shown great potential in healthcare, diagnostics, and targeted drug delivery systems using plant-derived materials. Plant-mediated nanomaterials offer advantages, including biocompatibility, sustainability, and reduced toxicity when compared with many synthetic alternatives. This volume covers recent research on the synthesis of plant-based nanoparticles and applications in advanced healthcare, showing how biological systems can inspire innovative technological solutions. The development of plant-based bioplastics is a very relevant solution to the problem of plastic pollution all around the globe. Traditional plastics made from fossil feedstock are a source of serious concern for various environments because of their nonbiodegradable nature and other hazards created with a long-lasting effect on nature. The use of plant-based polymers and biodegradable materials is a green alternative for sustainable development with no adverse effect on nature because all such processes follow green chemistry principles. The chapters on plant-based materials and a circular bioeconomy show how plants can be a source of renewable material for waste management and sustainable industrial development. The circular bioeconomy combines agricultural production, industrial operations, and sustainable use of the environment in ways that make best use of biology-based resources. Plants are central to this paradigm of thinking because of their renewable contributions to the production of food, energy, material resources, and medicines. This book examines the role of plant-based solutions in achieving circular economy solutions that reduce waste while promoting economic resilience despite leaving small environmental footprints.

Although plant-based technologies have such great potentials, they need to be applied with caution and attention to ethical and environmental concerns. Some of these concerns include biodiversity conservation, intellectual property rights, and benefit sharing. All these concerns need to be considered carefully and taken into account when developing and using plant-based technologies. This book will scrutinize these concerns. The final parts of the book cover newer developments in plant biotechnology, which have been used in support of many sustainable applications mentioned throughout the book. Genetic modification, genome editing, tissue culture technology, and omic technologies have improved capabilities in plants, from improving traits, making plants tolerant, increasing production of desired molecules, to addressing particular requirements. When used in conjunction with ecological principles, there exists tremendous potential in using plant biotechnology for achieving greater sustainability. The book in hand has been planned to act as a ready reckoner for researchers, academicians, students, policymakers, and industry professionals. The book has been conceptualized to promote interdisciplinary collaboration, innovation, and evidence-driven decision-making in recognition of the multifaceted role of plants in the development of sustainable systems. The book demonstrates the incredible potential of plants in transforming the future toward a healthier and more sustainable dimension. The book will act as a solution-provider in this context. Overall, the aim of this book, therefore, is to encourage a greater appreciation for plants not simply for their exploitable qualities but for their inherent rôle in a sustainable future. Plants are very important for ecological equilibria, for food security and sustainability, and for human health and wellness and are, therefore, at the heart of a sustainable future for humanity.

Debre Tabor, Ethiopia — Limenew Abate Worku
Addis Ababa, Ethiopia — Rakesh Kumar Bachheti
Cairo, Egypt — Abdel Nasser B. Singab
Wolaita, Ethiopia — Azamal Husen

Contents

1 **Introduction to Plant-Based Solutions** 1
Limenew Abate Worku, Marie Yayinie, Teshale Adane, Tadesse Bassie Gelaw, Archana Bachheti, and Rakesh Kumar Bachheti
1.1 Overview of Plant-Based Solutions 2
1.2 The Rationale Behind Plant-Based Solutions 4
1.2.1 Environmental Concerns (Climate Change, Biodiversity Loss, Deforestation) 4
1.2.2 Health Benefits and Nutritional Aspects 4
1.2.3 Economic and Social Drivers 5
1.2.4 Ethical Considerations (Animal Welfare, Sustainability) 6
1.3 Plant-Based Diets and Food Products 6
1.3.1 Plant-Based Proteins 7
1.3.2 Plant-Based Carbohydrates 8
1.3.3 Plant-Based Lipids 8
1.4 Plant-Derived Materials 9
1.4.1 Natural Fibers 9
1.4.1.1 Cotton 10
1.4.1.2 Kapok 11
1.4.1.3 Flax 11
1.4.1.4 Ramie Fiber 12
1.4.1.5 Jute 12
1.4.1.6 Kenaf 12
1.4.1.7 Coir Fiber 13
1.4.1.8 Sisal 14
1.4.1.9 Bamboo 14
1.4.1.10 Wood 15
1.4.1.11 Banana 15
1.4.2 Biopolymers from Plant 16
1.4.3 Plant Oils and Resins 17
1.4.4 Natural Rubber 18
1.4.5 Plant-Based Medicines and Natural Remedies 18

1.4.6 Agricultural Innovations ... 22
1.5 Applications of Plant-Based Solutions ... 22
1.5.1 Crop Diversification in Smallholder Farms ... 22
1.5.2 Agroforestry Systems for Climate Resilience ... 24
1.5.3 Urban and Peri-Urban Agriculture ... 24
1.5.4 Restoration of Degraded Lands ... 25
1.6 Challenges and Limitations ... 26
1.7 Future Perspectives ... 27
1.8 Conclusion ... 28
References ... 28

2 Historical and Traditional Knowledge of Plant-Based Practices ... 41
Abdel Nasser B. Singab, Ahmed M. Elissawy, and Fadia S. Youssef
2.1 Introduction ... 42
2.2 Tribal Medical Practices ... 43
2.2.1 Main Aspects of Tribal Medical Practices ... 44
2.3 Ancient Traditional Systems of Medicine ... 44
2.3.1 Ancient Egyptian Medicine ... 44
2.3.2 Traditional Chinese Medicine (TCM) ... 46
2.3.3 Ayurvedic Medicine ... 48
2.3.4 Greco-Roman, Islamic, and Unani Medicine ... 50
2.4 Conclusion ... 52
References ... 54

3 Plant-Derived Biofertilizers and Biopesticides for Sustainable Agriculture ... 59
Esraa A. Elhawary, Dina M. Bahgat, Shaza H. Aly, Rania F. Abou El-Ezz, Heba A. S. El-Nashar, Mariam I. Gamal El-Din, and Abdel Nasser B. Singab
3.1 Introduction ... 60
3.2 Plant-Derived Biofertilizers ... 62
3.2.1 Examples of Plant-Derived Biofertilizers ... 63
3.2.2 Mechanisms of Action of Plant-Derived Biofertilizers ... 64
3.3 Plant-Derived Biopesticides ... 65
3.4 Impact of Plant-Derived Biofertilizers and Biopesticides on Sustainable Agriculture ... 68
3.4.1 Benefits of Plant-Derived Biofertilizers ... 69
3.4.2 Benefits of Plant-Derived Biopesticides ... 69
3.5 Challenges and Limitations of Natural Biofertilizers and Biopesticides ... 70
3.6 Recent Advances and Formulations for Plant-Derived Biofertilizers and Biopesticides ... 72
3.7 Conclusion and Future Perspectives ... 74
References ... 74

4 Role of Plants in Enhancing Soil Health and Crop Productivity 81
Noureddine Chaachouay, Ouafae Benkhnigue, and Lahcen Zidane
4.1 Introduction 82
4.2 Soil Health: Definition and Key Indicators 83
4.3 Mechanisms by Which Plants Improve Soil Health 84
4.4 Plant-Based Strategies for Soil Health Improvement 86
4.5 Impact of Soil Health on Crop Productivity 87
4.6 Case Studies and Examples 89
4.7 Challenges and Future Perspectives 90
4.8 Conclusion 91
References 91

5 Allelopathy: Natural Weed Management Using Plant Interaction 97
Deepti Singh Vashishth, Shazia Akhtar, Pooja Saklani, and Rinky Yadav
5.1 Introduction 98
5.2 Role of Allelopathic Plant Interaction in Weed Management 100
5.3 Allelopathic Plant Interaction Between Weed Plants and Other Weed Plants 100
5.4 Allelopathic Plant Interaction Between Crop Plants and Weeds 101
5.5 Allelopathic Plant Interaction Between Medicinal and Aromatic Plants and Weeds 104
5.6 Weed and Medicinal Plants Competition 108
5.7 Weed Autopathic Interactions (Autotoxicity) 111
5.8 Conclusion 112
References 112

6 Phytoremediation: Using Plants to Address Environmental Challenges 119
Marie Yayinie, Limenew Abate Worku, Teshale Adane, and Tadesse Bassie Gelaw
6.1 Introduction 120
6.1.1 Definition and Overview of Phytoremediation 120
6.1.2 Importance of Phytoremediation in Modern Environmental Management 121
6.1.3 Scope and Objectives of the Chapter 123
6.2 Background and Context 123
6.2.1 Overview of Environmental Contaminants Affecting Soils and Waters 123
6.3 Principles of Phytoremediation 125
6.3.1 Mechanisms of Phytoremediation 125
6.3.2 Phytoextraction 126
6.3.3 Phytostabilisation 127
6.3.4 Phytodegradation 128
6.3.5 Rhizofiltration 128

6.4 Plant Species Used in Phytoremediation 129
6.4.1 Plant Characteristics Required for Effective Phytoremediation 129
6.4.2 Selection Criteria for Phytoremediator Plants in Various Ecosystems 129
6.4.3 Hyperaccumulator Plants and Their Traits 132
6.4.4 Site Assessment and Preparation 133
6.4.5 Monitoring and Evaluating Remediation Progress 134
6.5 Enhancing the Proficiency of Phytoremediation 135
6.6 Case Studies and Real-World Applications 136
6.7 Environmental, Economic, and Social Benefits of Phytoremediation 137
6.8 Challenges and Future Perspectives of Phytoremediation 137
6.9 Conclusion 138
References 139

7 Medicinal Plants: Bridging Traditional Medicine and Modern Healthcare 145
Aneela Bashir, Ayman M. Al-Qaaneh, Asfandyar, Tahira Younis, Ali Raza Ishaq, Abdel Nasser B. Singab, and Heba A. S. El-Nashar
7.1 Introduction 146
7.2 Medicinal Plants: A House of Secondary Metabolites 147
7.3 Role of Secondary Metabolites in Diabetes Mellitus (DM) 148
7.3.1 Antioxidant Activity 149
7.3.2 Insulin Sensitization 149
7.3.3 Inhibition of Carbohydrate-Hydrolyzing Enzymes 149
7.3.4 Anti-Inflammatory Effects 150
7.3.5 Protection Against Diabetic Complications 150
7.3.6 Pancreatic β-Cell Protection and Regeneration 150
7.4 Plant-Based Phenolic Compounds 151
7.4.1 Functions of Phenolic Compounds 151
7.4.1.1 Defense Mechanisms 152
7.4.1.2 Structural Integrity 152
7.4.1.3 Ecological Interactions 152
7.4.1.4 Allelopathy 153
7.4.2 Plant-Derived Secondary Metabolites with Antidiabetic Activity 153
7.5 Microorganisms and Secondary Metabolites 153
7.5.1 Role of Intestinal Flora Metabolites in Treating Diabetic Nephropathy (DN) 159
7.6 Conclusion and Future Remarks 159
References 160

8 Bioactive Compounds from Plants for Disease Prevention and Management 165
Iriny M. Ayoub, Nehal Ibrahim, Yosra A. El Khodary, and Abdel Nasser B. Singab
8.1 Introduction 166
8.2 Classification of Bioactive Compounds 166
8.2.1 Terpenes or Terpenoids 167
8.2.1.1 Biosynthetic Pathways 168
8.2.1.2 Terpenoids Activity in Plants 171
8.2.1.3 Terpenoids Repellent Properties 171
8.2.1.4 Terpenoids Activity in Human Health 172
8.2.2 Alkaloids 175
8.2.2.1 Classification of Alkaloids 176
8.2.2.2 Alkaloids Biological Activities 177
8.2.3 Phenolic Compounds 187
8.2.3.1 Biological Activities of Polyphenols 188
8.2.3.2 Classification of Plant Polyphenols 190
8.3 Isolation Techniques of Bioactive Compounds 202
8.4 Mechanisms of Action of Bioactive Compounds from Plants for Disease Prevention and Management 204
8.4.1 Antioxidant Activity 204
8.4.2 Anti-inflammatory Activity 207
8.4.3 Antimicrobial Activity 210
8.4.4 Antidiabetic and Hypolipidemic Activities 214
8.4.5 Anticancer Activity 216
8.5 Case Studies 218
8.5.1 Curcumin 218
8.5.2 Resveratrol 219
8.5.3 Quercetin 219
8.5.4 Comparative Insights 220
8.6 Challenges and Future Prospects 220
8.6.1 Low Bioavailability and Stability 220
8.6.2 Extraction and Standardization of Plant Extracts 221
8.6.3 Clinical Validation and Regulatory Concerns 221
8.6.4 Future Research and Innovation Directions 221
8.6.5 Conclusion 222
References 222

9 Nutraceuticals: Plant-Based Nutrition for Health and Wellness 235
Salma Sameh, Ashaimaa Y. Moussa, Nada M. Mostafa, and Abdel Nasser B. Singab
9.1 Introduction 236
9.2 A Natural Approach to Brain Health 237
9.2.1 Monoamine Hypothesis 237
9.2.2 Neuroinflammation and Oxidative Stress 238

9.2.3 HPA Axis, Neurotrophic Factors, and Neurogenesis 238
9.2.4 Excitotoxicity 239
9.3 Therapeutic Effects of Herbal Extracts and Natural Compounds Obtained from Terrestrial and Marine Origins Against Psychiatric and Neurological Disorders 239
9.3.1 *Hypericum perforatum* 240
9.3.2 *Curcuma longa* 246
9.3.3 *Crocus sativus* 248
9.3.4 *Panax ginseng* 249
9.3.5 *Ginkgo biloba* 250
9.3.6 *Withania somnifera* 251
9.3.7 *Rhodiola rosea* 252
9.3.8 Brown Algae 253
9.3.9 Marine Sponge 253
9.4 Challenges, Safety, and Regulatory Considerations 254
9.4.1 Standardization and Quality Control 254
9.4.2 Pharmacokinetics and Bioavailability 255
9.4.3 Drug Interactions and Side Effects 255
9.4.4 Regulatory Landscape 255
9.5 Future Research 256
9.6 Conclusion 256
References 257

10 Innovations in Plant-Based Nanomaterials for Healthcare Applications 263
Heba A. S. El-Nashar, Ayman M. Al-Qaaneh, Eyad Ayman, Abdallah M. Ali, Omar Nabil, Hamza Elkoshairy, Omar Ibrahim, Mohammed Hamdy, and Abdel Nasser B. Singab
10.1 Introduction 264
10.2 Synthesis of Plant-Based Nanoparticles 266
10.3 Potential Therapeutic Effects of Plant-Based Nanoparticles 267
10.3.1 Neuroprotective Effects 267
10.3.2 Antimicrobial Effects 268
10.3.3 Antioxidant Effects 269
10.3.4 Antiparasitic Effects 270
10.3.5 Anticancer Effects 271
10.3.6 Wound Healing Effects 273
10.4 Role of Plant-Based Nanoparticles in Diagnostic Applications 274
10.5 Role of Plant-Based Nanoparticles in the Drug Delivery Systems 279
10.6 Conclusion and Future Remarks 280
References 281

11 Aromatic and Essential Oil Plants in Physical and Mental Well-Being . . . 289
Orchid A. Mahmoud, Jumana Haider Al-Anssari, Naglaa S. Ashmawy, Nada M. Mostafa, Omayma A. Eldahshan, and Abdel Nasser B. Singab
11.1 Introduction . . . 290
11.2 Methodology . . . 290
11.3 Methods of Extraction of Essential Oils . . . 291
11.4 Introduction to Essential Oils and Physical Well-Being . . . 291
11.4.1 Overview of Aromatherapy's Role in Physical Wellness . . . 291
11.5 Mechanisms of Action of Essential Oils Related to Physical Well-Being . . . 292
11.5.1 Anti-inflammatory Mechanisms of Essential Oils . . . 292
11.5.1.1 Changing Cytokines . . . 292
11.5.1.2 Inhibition of Key Enzymes and Transcription Factors . . . 293
11.5.1.3 Cellular-Level Effects . . . 293
11.5.1.4 Chemical Synergy . . . 294
11.5.2 Antimicrobial Mechanisms of Essential Oils . . . 294
11.5.2.1 Disruption of Microbial Cell Membranes . . . 294
11.5.2.2 Penetration and Action Within Gram-Negative Bacteria . . . 295
11.5.2.3 Targeting Intracellular Structures and Enzymes . . . 296
11.5.2.4 Induction of Oxidative Stress . . . 296
11.5.2.5 Inhibition of Quorum Sensing and Biofilm Formation . . . 296
11.5.2.6 Efflux Pump Inhibition and Resistance Modulation . . . 296
11.5.2.7 Synergistic Effects with Antibiotics . . . 297
11.5.2.8 Nano-Encapsulation and Antimicrobial Potency . . . 297
11.5.2.9 Broad-Spectrum Activity . . . 297
11.5.3 Analgesic Mechanisms of Essential Oils . . . 297
11.5.3.1 Modulation of Inflammatory Pathways . . . 298
11.5.3.2 Biphasic Inhibition in the Formalin Test . . . 298
11.5.3.3 Opioid-Like Activity . . . 298
11.5.3.4 Multimodal Mechanisms: Transient Potential Vanilloid 1 (TRPV1), Adrenergic, and More . . . 299
11.5.3.5 Peripheral Action and Ion Channel Modulation . . . 299
11.5.3.6 Immune-Modulatory Mechanisms of Essential Oils . . . 299
11.5.3.7 Activation of Innate Immunity . . . 299
11.5.3.8 Cytokine Regulation and Th1/Th2 Modulation . . . 300

11.6 Key Therapeutic Essential Oils Supporting Physical Health..... 301
11.6.1 *Nigella sativa* (Black Cumin) 301
11.6.2 *Thymus vulgaris* (Thyme) 301
11.6.3 *Chamaecyparis obtusa* (Japanese Cypress)........... 303
11.6.4 *Myrtus communis* (Myrtle) 303
11.6.5 *Elettaria cardamomum* (Cardamom)................. 303
11.6.6 *Pogostemon cablin* (Patchouli) 304
11.6.7 *Origanum vulgare* (Oregano) 304
11.6.8 *Cinnamomum zeylanicum* (Cinnamon) 304
11.6.9 *Lavandula officinalis* (Lavender)................... 305
11.6.10 *Salvia officinalis* (Sage).......................... 305
11.6.11 *Foeniculum vulgare* (Fennel)....................... 305
11.6.12 *Mentha piperita* (Peppermint) 305
11.6.13 *Curcuma longa* (Turmeric) 306
11.6.14 Eucalyptus 306
11.6.15 *Zingiber officinale* (Ginger)....................... 307
11.6.16 *Syzygium aromaticum* (Clove)..................... 307
11.7 Therapeutic Applications of Natural Agents 307
11.7.1 Pain Relief Uses................................... 307
11.7.2 Immune System Modulation Uses................... 308
11.7.3 Antimicrobial Uses 309
11.7.4 Respiratory Health Uses 309
11.7.5 Skin and Dental Health Uses....................... 310
11.7.5.1 Dental Health Applications 310
11.7.5.2 Skin Health Applications.................. 310
11.8 Clinical and Pre-clinical Evidence on the Therapeutic Effects of Essential Oils... 311
11.9 Introduction to Essential Oils and Mental Well-Being 312
11.10 Mechanisms of Action of Essential Oils Related to Mental Well-Being... 312
11.11 Key Therapeutic Essential Oils Supporting Mental Health...... 313
11.11.1 Lavender (*Lavandula* sp.) 313
11.11.2 Bergamot (*Citrus bergamia*) 315
11.11.3 Roman Chamomile (*Anthemis nobilis*) 316
11.11.4 German Chamomile (*Matricaria chamomilla*) 316
11.11.5 Lemongrass (*Cymbopogon citratus*) 316
11.11.6 Frankincense (*Boswellia* sp.)..................... 317
11.11.7 Rose (*Rosa damascena*)........................ 317
11.11.8 Rosemary (*Rosmarinus officinalis*) 317
11.11.9 Agarwood (*Aquilaria* sp.) 318
11.11.10 Valerian (*Valeriana* Spp.)........................ 318
11.12 Challenges and Future Considerations..................... 318
11.13 Conclusion and Future Remarks 319
References... 319

12 Plant-Based Bioplastics and Sustainable Materials 327
Tadesse Bassie Gelaw, Limenew Abate Worku, Marie Yayinie, Teshale Adane, Archana Bachheti, and Rakesh Kumar Bachheti
12.1 Introduction 328
12.1.1 Definition of Bioplastics 328
12.1.2 Overview of Bioplastics and Their Classification (Biodegradable Versus Non-Biodegradable) 328
12.1.3 Importance of Sustainable Materials 328
12.1.4 Environmental Impact of Traditional Plastics and the Need for Sustainable Alternatives 329
12.1.5 Scope of the Chapter 329
12.2 Overview of Plant-Based Bioplastics 330
12.2.1 Sources of Plant-Based Bioplastics 330
12.2.2 Types of Plant-Based Bioplastics 330
12.2.3 Starch-Based Bioplastics 330
12.2.4 Polylactic Acid (PLA) 331
12.2.5 Polyhydroxyalkanoates (PHA) 331
12.2.6 Cellulose-Based Bioplastics 332
12.3 Production Processes 332
12.3.1 Extraction and Processing of Raw Materials 332
12.3.2 Manufacturing Techniques 333
12.3.2.1 Extrusion 333
12.3.2.2 Injection Moulding 334
12.3.2.3 Blow Moulding 334
12.3.3 Mechanical Properties 335
12.3.4 Biodegradability and Environmental Impact 336
12.3.5 Analysis of Degradation Rates and Environmental Benefits 337
12.4 Applications of Plant-Based Bioplastics 338
12.4.1 Packaging Materials 338
12.4.2 Agricultural Uses 339
12.4.3 Consumer Products 339
12.4.4 Other Applications 340
12.5 Environmental Impact and Sustainability 340
12.5.1 Biodegradability and Compostability 341
12.5.2 Life Cycle Assessment of Bioplastics 341
12.5.3 Comparison with Fossil Fuel-Based Plastics 343
12.6 Challenges and Limitations 343
12.6.1 Economic Factors 343
12.6.2 Performance Issues 344
12.6.3 Public Perception and Market Acceptance 344
12.7 Future Trends and Innovations 345
12.7.1 Advances in Bioplastic Technology 345
12.7.2 Potential for Circular Economy Integration 345
12.8 Conclusion 346
References 346

13 Circular Bioeconomy: Plants as Renewable Resources 351
Teshale Adane, Limenew Abate Worku, Marie Yayinie, Tadesse Bassie Gelaw, Sheetal Tyagi, Rakesh Kumar Bachheti, and Archana Bachheti
13.1 Introduction 352
13.2 Conceptual Framework of Circular Bioeconomy 354
13.3 Plant Biomass as a Renewable Resource.................... 355
13.4 Technologies for Plant Biomass Conversion 356
13.5 Applications of Plant Biomass-Based Renewable Resources 359
13.6 Circularity Strategies in Plant Biomass-Based Bioeconomy..... 360
13.7 Environmental and Socioeconomic Impacts 361
13.8 Policy and Regulatory Frameworks........................ 363
13.9 Future Perspectives and Research Directions 363
13.10 Conclusions 365
References.. 365

14 Ethical and Environmental Challenges in Plant-Based Solutions ... 373
L. Srimathi Priya, Manikandan Ayyar, V. P. Santhi, K. Indhumathi, A. Dinesh, N. Subasree, K. Radhakrishnan, and A. Muthukrishnaraj
14.1 Introduction 374
14.2 Ethical Considerations.................................. 375
14.2.1 Moral Imperatives in Shifting to Plant-Based Systems 376
14.2.2 Biopiracy and Intellectual Property Rights 378
14.2.3 Indigenous Knowledge and Benefit Sharing 380
14.2.4 Equity and Accessibility of Plant-Based Technologies.... 381
14.2.5 Animal Ethics Versus Ecosystem Integrity 384
14.3 Environmental Impacts 385
14.3.1 Land Use Change and Biodiversity Loss.............. 386
14.3.2 Water Footprint and Agrochemical Use............... 388
14.3.3 Carbon Sequestration Potential Versus Emission Sources.. 390
14.3.4 Monoculture Practices and Soil Degradation 391
14.4 Socio-economic and Cultural Implications 392
14.4.1 Food Security and Agricultural Transition............. 392
14.4.2 Impact on Traditional Farming and Rural Livelihoods ... 394
14.4.3 Cultural Acceptance and Dietary Shifts............... 396
14.4.4 Ethical Supply Chains in Plant-Based Industries........ 398
14.5 Regulatory and Governance Challenges 400
14.5.1 International Frameworks and Protocols.............. 400
14.5.2 National Policies on Sustainable Agriculture and Ethics 403
14.5.3 Certification and Labeling Issues (e.g., Organic, Vegan, Non-GMO)....................................... 404
14.5.4 Corporate Accountability and Greenwashing 405
14.6 Case Studies... 407

14.6.1 Soy Cultivation in the Amazon Basin 407
14.6.2 Palm Oil Substitutes and Their Trade-Offs 408
14.6.3 Ethical Challenges in Plant-Based Meat Alternatives 408
14.6.4 Bioplastics and Bioenergy Crops: A Double-Edged Sword 411
14.7 Innovations and Ethical Frameworks for Sustainability 413
14.7.1 Integrating Circular Economy in Plant-Based Systems. . . 413
14.7.2 Ethical Design and Life Cycle Assessment Tools 414
14.7.3 Role of AI and Data Transparency in Ethical Plant-Based Solutions 416
14.7.4 Community-Led and Participatory Approaches. 417
14.8 Future Outlook and Recommendations 418
14.8.1 Balancing Ethical, Environmental, and Economic Goals . . . 418
14.8.2 Policy Recommendations for Ethical Transitions 420
14.8.3 Stakeholder Collaboration and Global Responsibility. . . . 422
14.9 Conclusion 424
References. 425

15 Advances in Plant Biotechnology for Sustainable Applications 429
L. Srimathi Priya, K. Radhakrishnan, V. P. Santhi, K. Indhumathi, A. Dinesh, N. Subasree, Manikandan Ayyar, and A. Muthukrishnaraj
15.1 Introduction 430
15.2 Genetic Engineering and Genome Editing 432
15.2.1 CRISPR/Cas and New Breeding Technologies 433
15.2.2 Transgenic Crops for Abiotic and Biotic Stress Resistance 434
15.2.3 Nutritional Biofortification Through Genetic Engineering 436
15.2.4 Regulatory and Ethical Considerations of GM Plants 437
15.3 Plant Tissue Culture and Micropropagation 439
15.3.1 Principles and Techniques 441
15.3.2 Conservation of Endangered Plant Species 442
15.3.3 Mass Propagation for Commercial Applications 443
15.3.4 In Vitro Stress Tolerance Screening. 445
15.4 Molecular Marker-Assisted Breeding 447
15.4.1 Types of Molecular Markers (SSR, SNP, AFLP, etc.) 447
15.4.2 Marker-Assisted Selection for Crop Improvement 448
15.4.3 Genomic Selection and Predictive Breeding 449
15.4.4 Applications in Biodiversity Conservation 450
15.5 Plant–Microbe Interactions and Biofertilizers. 451
15.5.1 Rhizosphere Engineering for Enhanced Growth 453
15.5.2 Plant Growth-Promoting Rhizobacteria (PGPR). 454
15.5.3 Mycorrhizal Associations for Nutrient Efficiency 455
15.5.4 Biocontrol Agents for Sustainable Pest Management 456
15.6 Secondary Metabolite Production and Phytochemicals. 459

15.6.1 Metabolic Engineering for Enhanced Compound Yield.... 459
15.6.2 Production of Pharmaceuticals and Nutraceuticals 460
15.6.3 Plant Cell and Hairy Root Cultures 462
15.6.3.1 Plant Cell Culture Technology.............. 462
15.6.3.2 Hairy Root Culture Systems 462
15.6.4 Industrial Applications and Market Potential 463
15.7 Phytoremediation and Environmental Applications............ 465
15.7.1 Engineered Plants for Heavy Metal and Pollutant Removal.. 465
15.7.2 Role of Biotechnology in Soil and Water Cleanup 466
15.7.3 Sustainable Urban Landscaping and Green Infrastructure 467
15.7.4 Case Studies in Environmental Biotechnology 468
15.8 Bioenergy and Biomaterials.............................. 470
15.8.1 Genetic Modification for Enhanced Biomass Yield...... 470
15.8.2 Algae- and Plant-Based Biofuel Production 471
15.8.3 Lignocellulosic Feedstocks for Bioplastics 473
15.8.4 Sustainability Metrics for Plant-Based Bioeconomy..... 474
15.9 Digital and Precision Tools in Plant Biotechnology 476
15.9.1 Integration of AI and Big Data in Plant Research 477
15.9.2 Remote Sensing, IoT, and Phenotyping Platforms 479
15.9.3 Bioinformatics and Systems Biology 480
15.9.4 Predictive Modeling for Sustainable Crop Performance 481
15.10 Challenges and Future Perspectives....................... 482
15.10.1 Public Perception and Policy Barriers 482
15.10.2 Climate Change Adaptation through Biotech Solutions .. 484
15.10.3 Bridging Research and Field-Level Application 485
15.10.4 Roadmap for Responsible Innovation 486
15.11 Conclusion.. 487
References.. 488

Editors and Contributors

About the Editors

Limenew Abate Worku obtained his Ph.D. in Organic Chemistry from Addis Ababa Science and Technology University, Ethiopia, completing his doctoral studies between September 2020 and March 2023. He earned his M.Sc. degree in Organic Chemistry from Bahir Dar University in 2014, following a B.Sc. degree in Chemistry from the same institution in 2006. In addition to his academic qualifications, he obtained a Certificate of Higher Diploma (CHD) as a certified professional teacher educator from Debre Tabor University in 2018. He has over 18 years of national and international teaching and research experience. He has held key academic leadership roles, including serving as Research, Community Service, and Technology Transfer Coordinator in the College of Natural and Computational Sciences, as well as Laboratory Coordinator in the same college. His professional experience reflects a strong commitment to academic leadership, research management, and institutional development. His research interests focus on natural products for industrial applications, green synthesis of nanoparticles and their applications, lignocellulosic materials, phytochemistry, and pulp and paper technology. He has established a strong research profile, with more than 58 peer-reviewed research publications in the areas of natural product chemistry and nanotechnology. In addition, he has contributed over 15 book chapters published by internationally recognized publishers, including Springer, Elsevier, Taylor & Francis, and Nova Publishers. His scholarly work demonstrates a deep passion for natural product research and a strong commitment to sustainable and environmentally friendly chemical technologies.

Rakesh Kumar Bachheti obtained his M.Sc. degree in Organic Chemistry from Hemwati Nandan Bahuguna (HNB) University, Garhwal, in 1998. Later, in 2001, he underwent a one-year Post Graduate Diploma in Pulp and Paper Technology from Forest Research Institute, Dehradun, India. He earned his Ph.D. in Organic Chemistry from Kumaun University, Nainital, in 2007. He has over 22 years of national and international research and teaching experience. He also serves as an adjunct faculty member at various national and international universities. His research interests include natural products for industrial applications, green synthesis of nanoparticles, and pulp and paper technology. He has a strong research profile, having presented his work at several national and international conferences, including in Malaysia, Thailand, and India. He has an h-index of 29 (Scopus), over 180 research publications, and 2 published patents. He also serves as an external examiner for Ph.D. thesis evaluations. As a dedicated academic mentor, he has successfully guided 40 M.Sc. and 8 Ph.D. students to completion. He has served as an Editor and Editorial Board Member for several international journals and has edited 14 books. He is also recognized for his editorial work, having edited 8 books. He has also contributed 41 book chapters published by renowned publishers such as Elsevier, Springer, CRC Press, Taylor & Francis, among others. He has received several accolades, including being recognized among the World's Top 2% Scientists by Stanford University and Elsevier for the years 2025 and 2024.

Abdel Nasser B. Singab is Professor Emeritus of Pharmacognosy since 2020. He got his bachelor's degree in Pharmacy from Zagazig University (1983), M.Sc. in Pharmacognosy from Al-Azhar University (1990), and Ph.D. in Pharmacognosy through Channel system (Egypt/Japan) Meiji Pharmaceutical Sciences University, Al-Azhar University (1996). He underwent 3 months postdoctoral fellowship at Kanazawa University (August–October 1999). He worked as a technical expert for teaching and research at the Faculty of Pharmacy, Addis Ababa University (2004–2006). Dr. Nasser has over 40 years of national and international experience in research and teaching experience, Aids Ababa University, Ethiopia Tripoli University (Libya), Ain Shams University, and Al-Azhar University (Egypt). He worked as Head of Pharmacognosy Department, Vice-Dean for Postgraduate Studies, Vice-Dean for Environmental Affairs, elected Dean of Faculty of Pharmacy, Vice-President for Postgraduate Studies Ain Shams University from 1999 to 2020; consultant of the Minister of Health and Population 2015–2018; and expert in pharmaceutical companies (Multinational and Egyptian). In 2025 as head of the pharmacy studies sector, supreme council of universities, Ministry of Higher Education. Egypt. Consultant to three Private Universities, Advisor & consultant of EDA (Egyptian Drug Authority), Member of the Board of Egyptian Medical Syndicates Union. Supervisor of Egyptian Syndicate of Pharmacy since 2019 Chairman of the Promotions Committee for Professors and Assistant Professors (Pharmacognosy). Member of the Board of Directors of the Egyptian Health Council, Chairman of the Center of Drug Discovery Research and Development Ain Shams University. International publication more than 250, h-index 41. Attended 210 Conferences, scientific forum and workshops (Academic & Professional). Supervision on more than 200 theses (master's and Ph.D.). 25 scientific chapters, 3 Arabic books on herbal medicine and the history of pharmacy,12 National and international projects (PI), His subjects of interest are pharmacognosy, ethnobotany, traditional medicine, pharmaceutical biotechnology, basic and hyphenated techniques of chromatography and spectroscopy and research interest is searching of bioactive compounds with enzyme inhibitors, anti-cancer, anti-platelet aggregations, anti-inflammatory agents, anti-diabetic, molluscicide and hepatoprotective agents from natural products resources (medicinal plants, endophytes, fungi, bacteria, marine organisms). Invention of methodology of extraction, chromatographic isolation, and purification of lead compounds as well-advanced methods of structural elucidation of natural compounds, expectation of bioactivity of organic compounds through literature, docking etc. Expert in medicinal plants cultivation and transfer from area to area worldwide. Top 2% Scientists by "Stanford University from 2020 to 2025". State Award 2021, (Golden Medal), University Award 2022. Founder of the center of drug discovery research &development, Medicinal Plant Research Station and founding Editor of the journal *Archives of Pharmaceutical Sciences*.

Azamal Husen is an eminent academic and researcher whose career spans over two decades of outstanding contributions to teaching, research, and scientific leadership. His academic journey has been marked by distinction and international engagement, having served as Full Professor and Head of the Department of Biology at the University of Gondar, Ethiopia, and Foreign Delegate at Wolaita Sodo University, Ethiopia. In India, he is currently an Adjunct Professor at Graphic Era University, and has previously served as Professor at Sankalchand Patel University and Visiting Faculty at several reputed Indian institutions, including the Forest Research Institute, Dehradun. His research has been supported by several prestigious national and international organizations, such as the World Bank, National Agricultural Technology Project, Indian Council of Agricultural Research, Indian Council of Forestry Research and Education, and the Japan Bank for International Cooperation. A prolific scholar, he has authored a vast number of research papers, books, and book chapters published by leading international publishers. His commitment to advancing global science is reflected in his editorial and peer-review work for numerous high-impact international journals. His expertise has been widely acknowledged through invitations to evaluate book proposals for leading global publishers such as Springer, Taylor & Francis, and Elsevier, as well as to review research projects for prestigious international funding agencies, including the Czech Science Foundation (Czechia) and the National Research, Development and Innovation Office (Hungary). He is a Fellow of the Plantae Group under the American Society of Plant Biologists, and a member of several prestigious scientific organizations, including the International Society of Root Research, the Asian Council of Science Editors, and the International Natural Product Sciences Taskforce. Throughout his distinguished career, he has been a keynote and invited speaker at numerous international conferences and symposia. He has played a significant role in academic development, contributing to curriculum design, faculty training, and educational module development across undergraduate, postgraduate, and doctoral programs. His outstanding achievements in teaching, research, and community service earned him special recognition from the University of Gondar. In the realm of scientific publishing, he serves as Editor-in-Chief of the *American Journal of Plant Physiology* and as Series Editor for several globally recognized academic collections, including *Exploring Medicinal Plants*, *Plant Chemistry, Plant Biology, Sustainability, and Climate Change*, *Smart Nanomaterials Technology*, and the *Handbook Series on Nanotechnology and Its Applications.* Consistently recognized among the World's Top 2% Scientists (2020 onwards) by Stanford University, he has also served as Guest Editor for several influential special issues, including *Advances in Plant Metabolomics* (*Plant Physiology and Biochemistry*), *Exploring Nanotechnology in Medicinal Plants* (*Journal of Herbal Medicine*), *Integrating Nanotechnology with Plant Compounds* (*Frontiers in Plant Science*), and so on. Through his scholarly excellence, editorial leadership, and global collaborations, he continues to advance cutting-edge research at the intersection of plant science, nanotechnology, and sustainability, establishing himself as a leading figure in contemporary biological and environmental research.

Contributors

Rania F. Abou El-Ezz Department of Pharmacognosy, Faculty of Pharmacy, Misr International University, Cairo, Egypt

Teshale Adane Department of Chemical Engineering, Gafat Institute of Technology, Debre Tabor University, Debre Tabor, Ethiopia

Shazia Akhtar Department of Environmental Science, Graphic Era (Deemed to be University), Dehradun, Uttarakhand, India

Jumana Haider Al-Anssari Department of Pharmaceutical Sciences, College of Pharmacy, Dubai Medical University, Dubai, United Arab Emirates

Abdallah M. Ali Faculty of Pharmacy, Ain Shams University, Cairo, Egypt

Ayman M. Al-Qaaneh Faculty of Allied Medical Sciences, Al-Balqa Applied University (BAU), Al-Salt, Jordan

Shaza H. Aly Department of Pharmacognosy, Faculty of Pharmacy, Badr University in Cairo, Cairo, Badr City, Egypt

Asfandyar State Key Laboratory of Biocatalysis and Enzyme Engineering, Engineering Research Center for Microbial Cell Factory of Hubei Province, Environmental Microbial Technology Center of Hubei Province, College of Life Sciences, Hubei University, Wuhan, China

Naglaa S. Ashmawy Department of Pharmacognosy, Faculty of Pharmacy, Ain Shams University, Cairo, Egypt

Department of Pharmaceutical Sciences, College of Pharmacy, Dubai Medical University, Dubai, United Arab Emirates

Eyad Ayman Faculty of Pharmacy, Ain Shams University, Cairo, Egypt

Iriny M. Ayoub Department of Pharmacognosy, Faculty of Pharmacy, Ain Shams University, Cairo, Egypt

Manikandan Ayyar Department of Chemistry, Karpagam Academy of Higher Education (Deemed to be University), Coimbatore, Tamil Nadu, India

Centre for Material Chemistry, Karpagam Academy of Higher Education (Deemed to be University), Coimbatore, Tamil Nadu, India

Archana Bachheti Department of Environmental Science, Graphic Era (Deemed to be University), Dehradun, Uttarakhand, India

Rakesh Kumar Bachheti Centre of Molecular Medicine and Diagnostics (COMMD), Saveetha Dental College & Hospitals, Saveetha Institute of Medical and Technical Sciences, Saveetha University, Chennai, Tamil Nadu, India

Dina M. Bahgat Department of Pharmacognosy, Faculty of Pharmacy, Ain Shams University, Cairo, Egypt

Aneela Bashir Key Laboratory of Biorheological Science and Technology of Ministry of Education, Bioengineering College, Chongqing University, Chongqing, China

Ouafae Benkhnigue Department of Botany and Plant Ecology, Scientific Institute, Mohammed V University, Rabat, Morocco

Noureddine Chaachouay Agri-Food and Health Laboratory (AFHL), ESEF of Berrechid; Hassan First University, Settat, Morocco

A. Dinesh Department of Chemistry, K. Ramakrishnan College of Engineering (Autonomous), Affiliated to Anna University, Samayapuram, Trichy, Tamil Nadu, India

Yosra A. El Khodary Department of Pharmacognosy, Faculty of Pharmacy, Ain Shams University, Cairo, Egypt

Omayma A. Eldahshan Department of Pharmacognosy, Faculty of Pharmacy, Ain Shams University, Cairo, Egypt

Center for Drug Discovery Research and Development, Faculty of Pharmacy, Ain Shams University, Cairo, Egypt

Esraa A. Elhawary Department of Pharmacognosy, Faculty of Pharmacy, Ain Shams University, Cairo, Egypt

Ahmed M. Elissawy Department of Pharmacognosy, Faculty of Pharmacy, Ain Shams University, Cairo, Egypt

Center of Drug Discovery Research and Development, Ain Shams University, Cairo, Egypt

Hamza Elkoshairy Faculty of Pharmacy, Ain Shams University, Cairo, Egypt

Heba A. S. El-Nashar Department of Pharmacognosy, Faculty of Pharmacy, Ain Shams University, Cairo, Egypt

Mariam I. Gamal El-Din Department of Pharmacognosy, Faculty of Pharmacy, Ain Shams University, Cairo, Egypt

Quadram Institute Bioscience, Norwich Research Park, Norfolk, UK

Tadesse Bassie Gelaw Department of Chemistry, College of Natural and Computational Science, Debre Tabor University, Debre Tabor, Ethiopia

Mohammed Hamdy Faculty of Pharmacy, Ain Shams University, Cairo, Egypt

Nehal Ibrahim Department of Pharmacognosy, Faculty of Pharmacy, Ain Shams University, Cairo, Egypt

Omar Ibrahim Faculty of Pharmacy, Ain Shams University, Cairo, Egypt

K. Indhumathi Krishi Vigyan Kendra, Tamil Nadu Agricultural University, Papparapatty, Dharmapuri, Tamil Nadu, India

Ali Raza Ishaq State Key Laboratory of Biocatalysis and Enzyme Engineering, Engineering Research Center for Microbial Cell Factory of Hubei Province, Environmental Microbial Technology Center of Hubei Province, College of Life Sciences, Hubei University, Wuhan, China

Orchid A. Mahmoud Department of Pharmacognosy, Faculty of Pharmacy, Ain Shams University, Cairo, Egypt

Nada M. Mostafa Department of Pharmacognosy, Faculty of Pharmacy, Ain Shams University, Cairo, Egypt

Ashaimaa Y. Moussa Department of Pharmacognosy, Faculty of Pharmacy, Ain Shams University, Cairo, Egypt

A. Muthukrishnaraj Amrita School of Artificial Intelligence, Amrita Vishwa Vidyapeetham, Coimbatore, Tamil Nadu, India

Omar Nabil Faculty of Pharmacy, Ain Shams University, Cairo, Egypt

L. Srimathi Priya Department of Natural Resource Management, Horticultural College and Research Institute (HC & RI), Tamil Nadu Agricultural University (TNAU), Periyakulam, Theni, Tamil Nadu, India

K. Radhakrishnan Department of Chemistry, Karpagam Academy of Higher Education (Deemed to be University), Coimbatore, Tamil Nadu, India

Centre for Material Science, Karpagam Academy of Higher Education (Deemed to be University), Coimbatore, Tamil Nadu, India

Pooja Saklani Department of Environmental Science, Graphic Era (Deemed to be University), Dehradun, Uttarakhand, India

Salma Sameh Department of Pharmacognosy, Faculty of Pharmacy, Ain Shams University, Cairo, Egypt

V. P. Santhi Anbil Dharmalingam Agricultural College and Research Institute (AD AC &RI), Tamil Nadu Agricultural University (TNAU), Trichy, Tamil Nadu, India

Abdel Nasser B. Singab Department of Pharmacognosy, Faculty of Pharmacy, Ain Shams University, Cairo, Egypt

Center for Drug Discovery Research and Development, Faculty of Pharmacy, Ain Shams University, Cairo, Egypt

N. Subasree Department of Chemistry, School of Basic Science, Vels Institute of Science, Technology and Advanced Studies (VISTAS), Pallavaram, Tamil Nadu, India

Sheetal Tyagi Department of Chemistry, School of Basic and Applied Sciences, Shri Guru Ram Rai University, Dehradun, Uttarakhand, India

Deepti Singh Vashishth Department of Environmental Science, Graphic Era (Deemed to be University), Dehradun, Uttarakhand, India

Limenew Abate Worku Department of Chemistry, College of Natural and Computational Science, Debre Tabor University, Debre Tabor, Ethiopia

Rinky Yadav Department of Environmental Science, Alpine Group of Institutions, Dehradun, Uttarakhand, India

Marie Yayinie Department of Chemistry, College of Natural and Computational Science, Debre Tabor University, Debre Tabor, Ethiopia

Tahira Younis Department of Biochemistry and Biotechnology, The Women University of Multan, Punjab, Pakistan

Fadia S. Youssef Department of Pharmacognosy, Faculty of Pharmacy, Ain Shams University, Cairo, Egypt

Lahcen Zidane Plant, Animal Productions and Agro-industry Laboratory, Department of Biology, Faculty of Sciences, Ibn Tofail University, Kenitra, Morocco

Chapter 1
Introduction to Plant-Based Solutions

Limenew Abate Worku, Marie Yayinie, Teshale Adane, Tadesse Bassie Gelaw, Archana Bachheti, and Rakesh Kumar Bachheti

Abstract In addition to addressing pressing global issues of resource management, sustainability, and health, the use of plants and plant-derived products can enhance food, agricultural materials, and medicine. It examines the rationale behind these solutions, which are driven by environmental imperatives such as preventing biodiversity loss and deforestation, reducing greenhouse gas emissions (up to 70% lower than those produced by animal-based systems), and offering health benefits such as lowered risks of chronic diseases and enhanced mental health through diets high in nutrients. Plant-derived materials such as natural fibers, biopolymers, oils, resins, and natural rubber are also important due to their biodegradability and industrial applications. Plant-based medicines, rooted in traditional systems, offer readily available remedies containing bioactive substances shown to be effective in treating a variety of ailments. Agricultural innovations with practical applications that increase ecosystem services, productivity, and resilience include crop diversification, agroforestry, urban farming, land restoration, and sustainable livestock integration. Despite challenges such as technological barriers, market reluctance, financial constraints, and potential ecological risks, future prospects are promising, driven by developments in gene editing, AI-driven agriculture, and policy support aligned with circular economies and the UN Sustainable Development Goals. All things

L. A. Worku · M. Yayinie (✉) · T. B. Gelaw
Department of Chemistry, College of Natural and Computational Science, Debre Tabor University, Debre Tabor, Ethiopia

T. Adane
Department of Chemical Engineering, Gafat Institute of Technology, Debre Tabor University, Debre Tabor, Ethiopia

A. Bachheti
Department of Environmental Science, Graphic Era (Deemed to be University), Dehradun, Uttarakhand, India

R. K. Bachheti
Centre of Molecular Medicine and Diagnostics (COMMD), Saveetha Dental College & Hospitals, Saveetha Institute of Medical and Technical Sciences, Saveetha University, Chennai, Tamil Nadu, India

L. A. Worku et al. (eds.), *Plant-Based Solutions for Sustainable Agriculture and Healthcare*, Sustainability Sciences in Asia and Africa,
https://doi.org/10.1007/978-981-95-8910-4_1

considered, plant-based treatments support a transition to fair, low-impact systems that may prevent significant environmental harm while advancing human welfare and economic growth. The aim of the chapter is to provide an overview of the role of plants and plant-derived products in offering sustainable, health-promoting, and eco-friendly solutions for global challenges.

Keywords Plant-based solutions · Application · Sustainable agriculture · Plant-derived materials · Plant-based diets · Biopolymers

1.1 Overview of Plant-Based Solutions

Using plants and plant-derived materials to address societal health and environmental problems is known as a plant-based solution (Prasad et al. 2025). There are many uses for plant-based solutions, such as sustainable agriculture, plant-based diets, bio-based products, and natural medicines. The versatility and potential of plants are also emphasized in the food industry, packaging, textiles, and pharmaceuticals (Krzywonos and Piwowar-Sulej 2022). Plant-based meals are predicted to grow from US$29.4 billion in 2020 to US$162 billion by 2030 a 19.4% increase (Nyhan et al. 2023). Additionally, studies reveal that, compared to conventional Western diets, plant-based diets use 30% less water, 50% less land, and produce 70% fewer greenhouse gases (Aleksandrowicz et al. 2016). These solutions are intended to support the health of ecosystems and communities in addition to boosting sustainability, cutting back on the use of fossil fuels and animal products, and creating healthier ecosystems (Selvakumar et al. 2025).

The potential of plant-based therapies to greatly enhance global health outcomes is another factor contributing to their growing popularity. According to McMacken and Shah (2017), plant-based diets are linked to decreased risks of long-term conditions like heart disease and type 2 diabetes. Hemler and Hu (2019b) suggest that a plant-based diet could avert up to 11 million preventable deaths annually. Furthermore, traditional plastics and synthetic fibers are being rapidly replaced by plant-based products. Between 2020 and 2027, the bioplastics market is expected to grow by 12.5%, from US$4.3 to 10 billion (Golden et al. 2024). Plant-based therapies not only reduce negative environmental impacts but also enhance public health and stimulate economic growth through sustainable innovation (Mešić et al. 2024).

Humans have used plants for food, medicine, and resources for thousands of years (Schaal 2019; Husen 2021, 2022, 2023a, b, 2024a, b). Early plant-based treatments that increased survival and well-being included farmers' methods and herbal remedies (Kevany et al. 2018). The emphasis has shifted toward synthetic and animal-based products due to industrialization and technological advancements, frequently at the expense of environmental health. Due to health issues, resource depletion, and climate change, plant-based methods are growing in popularity (Feigin et al. 2023). Modern inventions such as bioengineered crops and plant-based meat substitutes meet current demands by combining cutting-edge science and conventional wisdom (Mešić et al. 2024).

In recent decades, plant-based solutions have gained popularity due to the development of innovative products and environmentally friendly methods made possible by scientific and technological advancements (Mešić et al. 2024). Crop yields, nutrient profiles, and pest resistance have all improved as a result of biotechnology and precision agriculture, addressing concerns about food security while reducing environmental impact (Hayat et al. 2025). The development of plant-based meat substitutes that closely resemble the flavor and texture of animal products, such as those made from mycoprotein and pea protein, has given consumers more options (Kurek et al. 2022). The development of plant-based meat substitutes that closely resemble the flavor and texture of animal products, such as those made from mycoprotein and pea protein, has given consumers more options (De Bakker and Dagevos 2012; Nielsen et al. 2019). Furthermore, international agreements and policy initiatives focused on sustainability and climate change have highlighted the importance of integrating plant-based solutions into global plans. When taken as a whole, these developments demonstrate how innovation and tradition interact dynamically, making plant-based solutions a crucial component of a sustainable future (Mešić et al. 2024).

As society grapples with escalating problems such as resource scarcity, public health, and climate change, plant-based solutions are becoming increasingly crucial in the modern world (Sadhukhan et al. 2020). With an estimated 10 billion people on the planet by 2050, sustainable resource management and food production are essential (McKenzie and Williams 2015). Plant-based solutions offer a comprehensive approach to addressing some of the most significant global issues by leveraging the power of plants in a range of industries. These solutions support sustainable diets by encouraging the consumption of plant-based meals, which are both healthier and substantially less damaging to the environment than animal-based products (Almeaiweed et al. 2025). While environmental remediation techniques employ plants to restore contaminated soils and damaged ecosystems, innovations in plant-based meat and dairy substitutes offer consumers sustainable and healthful options. The resilience of food systems is enhanced by sustainable agricultural practices such as agroecology and the conservation of agricultural biodiversity. A circular economy is also supported by the development of bio-based industrial products that lessen reliance on fossil fuels (Kozicka et al. 2023). Significant reductions in greenhouse gas emissions reduce land and water use, less water pollution, and support biodiversity preservation, among other environmental benefits (Smith et al. 2013). Because plant-based diets eliminate the need for antibiotics in meat production, they improve mental health, reduce the incidence of chronic illnesses, and increase food security (Santo et al. 2020). Despite these benefits, widespread adoption is still hindered by affordability and economic viability concerns. If plant-based solutions are to fulfill their potential in creating a sustainable and healthy future, these obstacles must be addressed through creativity, legislative support, and education (Tartaruga et al. 2024).

1.2 The Rationale Behind Plant-Based Solutions

1.2.1 Environmental Concerns (Climate Change, Biodiversity Loss, Deforestation)

The adoption of plant-based solutions is mainly motivated by environmental concerns. The Food and Agriculture Organization (FAO) estimates that agriculture, especially cattle production, accounts for 14.5% of global greenhouse gas emissions (Twine 2021). Large areas of forest are cleared for pastureland or the production of animal feed crops, which significantly contributes to habitat damage, deforestation, and biodiversity loss. By lowering carbon footprints, conserving water, and safeguarding natural habitats, switching to plant-based systems can greatly lessen these impacts (Ghazoul and Medina 2024). For instance, research indicates that adopting a plant-based diet could cut global agricultural land use by as much as 50%, safeguarding ecosystems and slowing climate change (Poore and Nemecek 2018).

In addition to lowering greenhouse gas emissions, plant-based approaches are crucial for addressing a number of environmental problems, including soil degradation and water scarcity (Smith et al. 2013). Intensive livestock farming often leads to soil erosion, nutrient depletion, and pollution from manure runoff, all of which can harm aquatic ecosystems and reduce agricultural productivity (Menzi et al. 2010). On the other hand, plant-based farming systems, especially those that employ crop diversification and regenerative techniques, promote soil health and enhance carbon sequestration (Sher et al. 2024). Furthermore, the production of plant-based foods typically uses much less freshwater. For instance, it can take up to 15,000 L of water to produce 1 kg of beef, but many plant-based proteins require only a small fraction of that amount (Pimentel and Pimentel 2003). By using plant-based solutions, we can reduce the strain on freshwater supplies and encourage more sustainable water management, which is essential given the increasing global water stress (Mekonnen and Hoekstra 2012; Lal 2020).

1.2.2 Health Benefits and Nutritional Aspects

Plant-based solutions provide substantial health benefits because diets rich in fruits, vegetables, legumes, nuts, and whole grains are linked to lower risks of chronic conditions such as heart disease, type 2 diabetes, obesity, and certain cancers. Saturated fats and cholesterol are often lower in plant-based diets, but vitamins, antioxidants, and dietary fiber are more abundant. According to research published in The Lancet, adopting a plant-based diet could prevent millions of premature deaths worldwide (Willett et al. 2019). A plant-based diet may enhance both digestive and mental health. Maintaining a balanced intake of vital nutrients such as iron, vitamin B12, and omega-3 fatty acids is crucial, and this can be done with supplements or careful dietary planning (Craig et al. 2021).

Plant-based diets have been shown to improve weight control and metabolic health while reducing the risk of chronic diseases. According to research, plant-based eaters generally have lower body mass indices (BMIs) and a lower incidence of metabolic syndrome, a group of conditions that increase the risk of heart disease and stroke (Vajdi et al. 2023). Plant-based diets with a high fiber content improve feelings of fullness and control blood sugar levels, both of which are beneficial for managing and preventing type 2 diabetes. Furthermore, plant-based diets are high in phytonutrients, bioactive compounds derived from plants with antioxidant and anti-inflammatory properties that support cellular health in general and help prevent disease (Satija and Hu 2018).

Another area where plant-based nutrition may be beneficial is mental wellness. Diets high in fruits, vegetables, and whole grains may be associated with fewer cases of anxiety and depression, according to recent research (Haghighatdoost et al. 2023). Dietary decisions and the makeup of the gut microbiota impact the gut-brain axis, a complex network of communication between the brain and the gastrointestinal tract (Maiuolo et al. 2021). Plant-based diets may improve mood and cognitive function by promoting a diverse and healthy gut microbiome (Medawar et al. 2019). Even though more research is required to understand these pathways, the potential advantages for mental health fully contribute to the growing appeal of plant-based diets (Fehér et al. 2020).

Despite these benefits, it is crucial to address any nutritional deficiencies in plant-based diets to guarantee optimal health. Nutrients like vitamin B12, which is primarily found in animal products, must be considered through fortified foods or supplements (Akinwumi et al. 2025). Since non-heme iron from plants is absorbed less efficiently than heme iron from meat, eating meals high in vitamin C, in addition to iron-rich plant sources, can improve absorption (Kolarš et al. 2025). Similarly, omega-3 fatty acids, which are essential for heart and brain health, can be found in flaxseeds, chia seeds, walnuts, and supplements made from algae. With careful planning and comprehension, plant-based diets can meet all nutritional requirements at various stages of life (Rodriguez-Leyva et al. 2010).

1.2.3 Economic and Social Drivers

Social and economic factors also play a fundamental role in the growing interest in plant-based solutions. Due to consumer demand for ethical, environmentally friendly, and health-conscious products, the market for plant-based foods is rapidly growing (Sharma et al. 2024). This expansion generates jobs in the manufacturing, food technology, and agricultural sectors, among other new economic opportunities (Rais et al. 2013). Compared to animal agriculture, plant-based solutions can increase food security from a social perspective by enabling more efficient use of resources, thereby allowing more food to be produced per unit of land and water (Mešić et al. 2024). Additionally, plant-based meals are frequently more accessible and less expensive, which contributes to a decrease in hunger and malnutrition. This change is also consistent with broader social movements that promote environmental justice, animal welfare, and fair access to healthful food (Kraak and Aschemann-Witzel 2024).

1.2.4 Ethical Considerations (Animal Welfare, Sustainability)

Ethics must be taken into account when promoting plant-based solutions, especially when it comes to sustainability and animal welfare. Common industrial animal farming practices such as confinement, cruel treatment, and mass killing severely impair animal welfare. In an effort to reduce animal suffering, a large number of consumers and advocacy groups support plant-based alternatives. Sustainability ethics also emphasize the significance of protecting ecosystems and natural resources for the benefit of future generations. Plant-based solutions minimize environmental harm, increase biodiversity, and use fewer resources. This ethical viewpoint promotes a holistic approach to resource and food production that respects the environment and living things (Chiu and Lin 2009).

1.3 Plant-Based Diets and Food Products

The potential health benefits, environmental sustainability, and ethical implications of plant-based diets have garnered a lot of attention lately. These diets emphasize consuming mostly plant-based foods, including fruits, vegetables, grains, nuts, seeds, and legumes, while limiting or avoiding animal products. Because plant-based diets are high in fiber, vitamins, antioxidants, and healthy fats, they may reduce the risk of chronic conditions such as diabetes, heart disease, and certain cancers (Kent et al. 2022). When compared to diets heavy in animal products, a plant-based diet can also lower greenhouse gas emissions, water use, and land degradation (Aleksandrowicz et al. 2016). The foods associated with plant-based diets have changed significantly, from traditional fruits and vegetables to a variety of cutting-edge substitutes that aim to mimic the flavor, texture, and nutritional content of animal-derived foods. These include plant-based meat substitutes made from soy or pea protein, or mycoprotein; dairy substitutes like almond, oats, and soy milk; and egg substitutes made from algae or legumes (Alcorta et al. 2021). In addition to assisting individuals in switching to plant-based diets, the manufacturing of these goods attracts flexitarians and environmentally conscious consumers who wish to consume fewer animal products. As the market for plant-based meals grows, continued research and technological developments are enhancing the nutritional benefits and sensory attributes of these products, making plant-based diets more accessible and pleasurable for a wider range of individuals (Viroli et al. 2023).

1.3.1 Plant-Based Proteins

Plant proteins are widely used in plant-based products due to their many functional properties including their capacity to thicken, gel, emulsify, foam, and retain fluids (Loveday 2020). A significant source of essential amino acids is found in proteins. These proteins can be obtained from a wide range of botanical sources, each with distinct qualities, such as soybeans, peas, faba beans, mung beans, lentils, algae, and microalgae (McClements and Grossmann 2021). The majority of plant proteins are globular and often form complex multimers composed of various protein types linked by chemical or physical bonds (Fukushima 1991). The biological source of these proteins, as well as any modifications to their natural states and associations during separation and purification, determines their functional capacity. One of the biggest issues facing the plant-based food industry is the lack of plant proteins with reliable functional properties (McClements and Grossmann 2021).

Plant-based proteins are the primary component of meat substitutes. The primary ingredients are mushrooms, wheat gluten, and soy (Kurek et al. 2022). Soy is preferred because of its higher nutritional content when producing protein-dense products like tofu, which is made by coagulating and compressing soy curds into a solid block (Guan et al. 2021). Soy flour, the least processed soy product, is utilized to make soy protein isolate (90% protein), soy protein concentrate (70% protein), and soy texturized vegetable protein (Lusas and Riaz 1995). A popular technique for producing soy meat substitutes, such as texturized vegetable protein, is extrusion, which enables a variety of final product sizes and shapes (Zhang et al. 2021). Wheat gluten, also known as seitan, is extracted from wheat flour during the starch extraction process and is highly valued for its capacity to leaven, bind, and form dough. Wheat gluten gives products made with it a meat-like texture due to its chewy and stretchy properties. Additionally, these products include mushrooms to make them more chewy (Alcorta et al. 2021). Proteins from legumes such as peas, lentils, lupines, or chickpeas have been used to create meat substitutes. Of them, pea protein is undoubtedly the most promising. Additionally, oilseed proteins from rapeseed and canola can function as structural agents when heated, enhancing textures that resemble meat (Kurek et al. 2022). Mycoprotein, a protein-rich material that can be made for human consumption, is derived from the mycelium that the fungus *Fusarium venenatum* forms during fermentation (Cheriaparambil and Grossmann 2025). Mycoprotein is combined with water, a small amount of egg albumen, and a small amount of roasted barley malt extract, alternatively, to give it a savory flavor. A natural flavoring is added. The filamentous nature of mycoprotein gives this product a texture that resembles flesh (Finnigan 2011).

1.3.2 Plant-Based Carbohydrates

Sugars, oligosaccharides, and polysaccharides are examples of carbohydrates that can be used as functional ingredients in the production of animal product analogs (McClements and Grossmann 2021). Depending on their origin and isolation method, plant-based carbohydrates exhibit distinct molecular, physicochemical, functional, and biological properties. A variety of functional qualities, including sweetness, viscosity, gelling, emulsification, structure formation, stability, and moisture retention, can be added to plant-based foods using these ingredients (Williams and Phillips 2021). Carbohydrates affect human nutrition and health because they can be either fermentable or non-fermentable, as well as digestible or indigestible (like fiber). Choosing carbohydrate ingredients that provide the required quality and nutritional qualities in the finished product is therefore essential. Through phase separation and interacting mechanisms, polysaccharides are commonly used in conjunction with proteins to produce desired texture and sensory properties in plant-based foods (Dekkers et al. 2018).

1.3.3 Plant-Based Lipids

A variety of lipid-rich botanical sources, including algae, canola, coconut, cocoa, corn, flaxseed, olive, palm, safflower, soybean, and sunflower, can be used to extract plant-based fats and oils (Zhang et al. 2025). Triacylglycerols' ability to form a three-dimensional network of fat crystals is essential in many applications because it contributes desirable textural properties, such as the fluidity of butter and spreads (Marangoni et al. 2020). Additionally, the functionality of many food products depends on changes in solid fat content with temperature. When attempting to replicate the properties of animal fats with plant-based fats, this factor is especially crucial (Marangoni et al. 2020). The melting point of fats rises as the number of carbon atoms in fatty acid chains increases or the number of double bonds decreases. The spreadability of butter, the foamability of whipped cream, the meltability of cheese, and the hardness of ice cream are just a few of the desired quality attributes of products made from animals that are impacted by the crystallization properties of lipids (Yang et al. 2024). Because of this, it is often necessary to replicate the crystallization properties of animal fats using plant-derived lipids. This approach can be difficult because plant-derived fats often contain more unsaturated fatty acids than animal fats, making them more fluid at room temperature (Hu and McClements 2022). Hydrogenating these lipids to raise their saturation levels is one way to address this issue, but it may have adverse nutritional effects. Because of this, food producers often incorporate naturally occurring high-melting plant-based fats, such as cocoa butter and coconut oil, into their products. High levels of saturation in these fats, however, may have detrimental health effects such as an elevated risk of heart disease (Hemler and Hu 2019a).

The fatty acid composition of plant-based fats and oils significantly influences their nutritional properties and oxidative stability. Fats and oils high in polyunsaturated fatty acids (PUFAs), especially omega-3 fatty acids (found in flaxseed or algae oils), are thought to have health benefits, such as reducing the risk of heart and brain disorders (Saini and Keum 2018; Shahidi and Ambigaipalan 2018). These fats may replace fish oils, which are high in omega-3 polyunsaturated fatty acids (PUFAs). However, more research is needed to validate these claims through meta-analyses and randomized clinical trials. However, it is imperative to prevent these PUFAs from oxidizing during processing and storage because this can lead to the production of hazardous reaction byproducts and disagreeable off-flavors (Arab-Tehrany et al. 2012; Nogueira et al. 2019). This can be achieved in a number of ways, including controlling light, oxygen, and temperature; reducing pro-oxidant pollution; introducing antioxidants; using chelating agents; or using structural strategies. Using these techniques will be essential to creating the next wave of plant-based foods with improved nutritional value (McClements and Decker 2018; Jacobsen 2015; Jacobsen et al. 2013).

1.4 Plant-Derived Materials

Plant-derived materials include fibers, oils, resins, biopolymers, and other natural substances derived from different plant parts (Mohammadinejad et al. 2016). Because they are renewable, biodegradable, and versatile plant-derived materials, they have been utilized for centuries in a variety of industries, including construction, packaging, and textiles. Cellulose fibers from cotton and flax, lignin from wood, and natural rubber from rubber tree latex are typical examples (Tripathy and Mishra 2016). Due to growing environmental concerns and the need for sustainable substitutes for synthetic materials, plant-derived materials are increasingly being researched as environmentally acceptable alternatives across a range of applications, helping reduce reliance on fossil fuels and plastic pollution (Rajendran et al. 2025).

1.4.1 Natural Fibers

Natural fibers, as the name implies, are derived from natural sources such as plants, animals, or minerals (Thapliyal et al. 2023; Worku et al. 2022). Plant-based fibers include cotton and kapok, which come from seeds; sisal, banana, and pineapple fibers, obtained from leaves; and bast fibers like jute, flax, hemp, kenaf, and ramie (Fig. 1.1), which are extracted from the stems of plants (Eyupoglu 2020). Among cellulose fibers widely used in the textile industry are cotton, linen, flax, hemp, and jute, as indicated in the following sub-section (Shuvo 2020).

Fig. 1.1 Some common sources of natural fibers

1.4.1.1 Cotton

Cotton, belonging to the botanical genus Gossypium and classified within the subtribe Hibisceae of the Malvaceae family, produces fibers that grow around its seeds and are composed of pure cellulose (Oh et al. 2024). The unique arrangement of cellulose in cotton fibers imparts both strength and excellent absorbent properties. Structurally, these fibers consist of multiple layers that resemble the coils of a spring (Radoor et al. 2022).

Sports socks, sneakers, bandanas, handkerchiefs, scarves, stoles, hats, caps, cotton canvas satchels, tote bags, and tapestry baggage are just a few of the many fashion items made from cotton, a common basic raw material (Dhir 2022). It is suitable for a wide range of apparel and accessories due to its fundamental characteristics, which include softness, breathability, moisture absorption, and temperature control (Salleh et al. 2021). Because of their natural roughness and moisture-absorbing properties, cotton fibers are incredibly breathable and feel comfortable against the skin. These qualities make cotton especially popular for applications requiring rapid water absorption such as beach jackets and terry-cloth clothing (Dhir 2022).

Additionally, cotton is a multipurpose, low-maintenance, hypoallergenic, static-free, and pill-resistant fabric. It is a durable choice because it maintains its original hue and texture over time. Cotton is a year-round material that performs well in both warm and cold climates due to its characteristics. Cotton fabric possesses a multitude of chemical and physical properties. Its composition and structure significantly influence its strength, durability, and dye-absorbing ability (Kumar et al. 2022).

The primary component of cotton fibers is cellulose, which accounts for 88.0–96.5% of their mass. Cotton's high cellulose content is responsible for its basic physical properties (Nickerson 1940). The chemical composition of cotton fibers is generally as follows: 88.0–96.5% cellulose, 1.0–1.9% proteins, 0.7–1.2% pectin, 0.4–1.0% waxes, 0.7–1.6% inorganic materials, 0.5–1.0% organic acids and 0.1–1.0% sugars and dyes (Liu 2018).

1.4.1.2 Kapok

Kapok is typically grown in tropical climates and is a member of the Bombacaceae family. The fiber encloses the kapok seeds. The cellulose fiber is hydrophobic, lightweight, and has a yellowish or light brown hue. In addition to being used as a biofuel and reinforcement, kapok fiber is best used as an adsorbent, oil absorber, and buoyant material (Dhir 2022).

1.4.1.3 Flax

Flax is a crystalline cellulosic fiber with a diameter of 12–16 μm and a length of 90 cm. Canada, the Netherlands, Belgium, and France are the primary countries where flax is grown. The fiber source is the stem of the *L. usitatissimum* plant. Retting and burning are the two methods used to remove the fiber and change its characteristics. Fibers separate due to pectin degradation induced by enzymatic treatment during the retting process. In addition to being utilized in furniture, home textiles, and interior décor products, flax fiber is widely used to produce linen (Dhir 2022).

Flax, sometimes referred to as linseed or common flax, is a plant that is a member of the genus *Linum usitatissimum* and the family Linaceae, which has 300 species and 13 genera (Diederichsen and Richards 2003). In contrast to its sister constituent, linseed, which has short, thick stems with more branches, flax is grown as a taller crop with thinner stems and fewer branches. One of the earliest cultivated plants to grow in temperate climates, flax is a self-pollinating crop (Goudenhooft et al. 2019). There has always been high demand for the fibers and seeds of this plant. Flax fibers' exceptional qualities make them a significant raw resource for textiles. These fibers are categorized as bast (stem) fibers, along with jute, ramie, and hemp. 70% cellulose and 30% non-cellulose materials, such as water-soluble compounds, mineral salts, hemicellulose, lignin, pectin, waxes, and lipids, make up flax fibers (Pandey 2007).

1.4.1.4 Ramie Fiber

In China, Japan, and Malaysia, ramie fiber—which is derived from the stem of herbaceous perennial plants—is extensively grown. Ramie is one of the strongest, longest, and fastest-growing natural bast fibers, reaching heights of 1–2 m. Similar to linen, flax is used to make ramie fiber. Applications for this fiber are numerous and include clothing items like cardigans and sweaters, upholstery, gas mantles, fishing nets, cars, furniture, building supplies, pulp, paper, agrochemicals, and composites (Dhir 2022).

1.4.1.5 Jute

Jute is typically grown in Asian nations, including China, Bangladesh, India, and Myanmar (Sharna and Kamruzzaman 2020). Jute usually takes 4 months to reach a height of 15–20 cm. About 4 months after cultivation, the fibers are gathered (Maity et al. 2012). The fibers must undergo a retting process, whether chemical or biological, to remove the pectin between the wood core and the bast. By submerging the bundled stalks in water for around 20 days, biological retting makes it easier to separate the fibers by extracting pectin from the space between the bast and the wood core. Chemical retting, on the other hand, uses materials such as $N_2H_8C_2O_4$ and Na_2SO_3. The fibers are then allowed to dry after this procedure (Dhir 2022).

Jute fiber is made from plants in the genus Corchorus of the family Malvaceae. Jute is a lignocellulosic fiber that is composed of both wood and textiles. It falls under the category of bast fiber, which is made from the plant's skin or bast (Rahman and Rahman 2024). Jute fiber is composed of cellulose (64.4%), hemicellulose (12%), pectin (0.2%), lignin (11.8%), water soluble (1.1%), wax (0.5%), and water (10%) (Ali et al. 2018). Several cells make up jute fiber. These cells consist of cellulose-based, crystalline microfibrils connected to a complete layer by amorphous lignin and hemicellulose. When many cellulose and lignin/hemicellulose layers adhere to one primary cell wall and three secondary cell walls, a composite with multiple layers is created. These cell walls have different compositions, particularly the cellulose-to-lignin/hemicellulose ratio and the arrangement of the cellulose microfibrils (Thomas et al. 2011).

1.4.1.6 Kenaf

Kenaf (*Hibiscus cannabinus* L.) is an annual spring crop cultivated for a long time (4000 BC). It started in Africa, spread throughout Asia in the 1900s (first in India and later in China), and then spread to the northern and central United States in the 1940s. Kenaf belongs to the Malvaceae family, specifically the section Furcaria. It is closely related to cotton, okra, hollyhock, and roselle. An industrial crop with great potential for growth in tropical or subtropical regions is kenaf. It is a source of raw material for fiber-based industries (Xu et al. 2020). After being harvested, kenaf plants are processed by pulping the entire stalk with a mechanical fiber separator. Following extraction, the fibers are separated from non-fibrous components such as wax, pectin, and other substances using either chemical or bacterial treatments. The fiber has special qualities like stiffness, strength, toughness, and outstanding pesticide resistance, in addition to being environmentally friendly and biodegradable. In addition to textiles, kenaf fibers are mainly used to make paper, rope, cords, and storage bags. In addition to furniture, buildings, packaging, and the automotive industry, kenaf fiber is now used in composites (Dhir 2022).

1.4.1.7 Coir Fiber

The husk of the coconut fruit is used to make coir, one of the thickest natural fibers, which is mostly grown in tropical countries like Malaysia, India, Sri Lanka, Indonesia, and the Philippines. Compared with other natural fibers, coir fiber has a higher microfibrillar angle, a higher lignin concentration, and lower levels of cellulose and hemicellulose. These characteristics provide several benefits, including strength, robustness, weather resistance, and high elongation at break. Because of its unique properties, coir fiber can be used for a variety of purposes, including upholstery, ropes, mats, beds, construction, agriculture, brushes, and more (Everitt et al. 2013; Smole et al. 2013; Thyavihalli Girijappa et al. 2019). Coir fiber is produced from the fruits of the coconut tree (*Cocos nucifera*). Husk describes the outer layer of the fruit. The outer skin of the coconut husk is smooth and impermeable with a fibrous area. Fibro-vascular coir bundles embedded in non-fibrous parenchymatous connective tissue comprise the mesocarp.

Coir is a natural fiber made from the husk or outer shell of young palms or ripe fruits (*Cocos nucifera*). These fibers can be extracted and used to create a range of products, including reinforced concrete, cement fiber boards that take the place of wood, geotextile textiles and screens, and reinforced polymeric composites (Hao et al. 2018). Coconut fiber is widely available, reasonably priced, and safe for the environment. A naturally occurring polymer called cellulose is also present. Coconut fiber nanocellulose can be used in many ways because it offers greater functionality and a smaller surface area than cellulose (Poornachandhra et al. 2023). Several applications for coconut fiber-derived nanocellulose have been documented. Because of its strength, biocompatibility, and photonic properties, it can be used for a wide range of applications, such as enhancing mechanical and thermal properties, eliminating heavy metals, fabric tents, mechanically reinforced polymer composites, packaging, and optoelectronics. However, the use of coconut nanocellulose is limited by the energy and chemicals needed to separate lignin and hemicellulose from the fiber. To mitigate these challenges, low-energy methods, affordable, environmentally friendly reagents, and economical sources of value-added compounds should all be employed (Abate et al. 2022).

Coir is used to make a variety of products, including floor mats, yardage, doormats, rugs, and geotextiles (Kavitha 2015). As a result, the coir industry is now the largest employer in many coastal regions of India, employing nearly 400,000 people

Fig. 1.2 Coconut tree (*Cocos nucifera*) (**a**) and coir fiber (**b**)

in fiber processing, coir yarn spinning, and coir fabric manufacturing (Bordoloi 2020). The entire coastal belt is utilized for spinning and processing. Coir has been combined with materials like sisal, banana, and others at the fiber stage to increase the final product's strength and pliability (Fig. 1.2).

1.4.1.8 Sisal

Sisal fiber is obtained from the leaves of the *Agave sisalana* plant (Sahu and Gupta 2017). In India, there are four varieties of sisal plants such as Sisalana, Vergross, Istle, and Natale. Each variety produces different amounts of fiber. The leaves of the first two varieties yield more fiber than those of the other two. The plant's age and origin affect its fiber composition. Moisture (87.25%), fiber (4%), cuticle (0.75%), and miscellaneous dry matter (8%) make up the leaf's chemical composition. The fiber has broad, blunt ends and spans in length from 60 to 120 mm. Typically angular in shape, cells are 500–6000 μm long and 5–40 μm wide, with a maximum aspect ratio (fiber length/diameter) of 150. The parenchyma contains calcium oxalate crystals, often accompanied by tiny globules. The xylem, ribbon, and mechanical fibers make up the sisal leaf (Anandjiwala and John 2010; Akram Khan et al. 2011; Fednand et al. 2022).

1.4.1.9 Bamboo

Bamboo is the world's most adaptable and rapidly growing plant. Because of its environmental, social, and economic benefits, it has become indispensable in the daily lives of many people in tropical areas (Emamverdian et al. 2020). Bamboo belongs to the Poaceae family of grasses, specifically to the subfamily Bambusoideae.

Fig. 1.3 Bamboo plants from eastern part of Ethiopia (Pawe woreda)

There are now no native bamboo species in Europe, and the majority of native bamboo is found in Asia, America, and Africa (Ahmad et al. 2021). Over 1500 species and about 123 genera of bamboo plants have been identified worldwide. A number of variables, including soil type, rainfall, temperature, and altitude, affect bamboo's natural distribution in tropical and subtropical regions. Bamboo's altitudinal distribution ranges from sea level to 4300 m, while its latitudinal range is between 47° and 50° 300′ N. As a result, in both temperate and tropical regions, bamboo is mainly associated with mesic to wet forest types (Ahmad et al. 2021). Cellulose, hemicellulose, and lignin are the main chemical constituents of bamboo (Worku et al. 2024). More than half of the total is composed of cellulose and hemicellulose, which are found in bamboo as holocellulose (Rusch et al. 2023; Maulana et al. 2020) (Fig. 1.3).

1.4.1.10 Wood

Wood originates from trees and shrubs. It is a heterogeneous, hygroscopic, cellular, and anisotropic solid substance made up of cellulose (40–50%) and hemicellulose (15–25%), which are bonded by lignin (15–30%). It is typically categorized as either softwood (pine) or hardwood (oak). Softwood species consist primarily of wood cells of a single type, known as tracheids, resulting in a more uniform structure than that of hardwoods. Softwood lacks vessels (pores), whereas hardwoods are generally filled with vessels (Pournou 2020).

1.4.1.11 Banana

The banana is classified under the genus Musa and the family Musaceae. Although commonly referred to as a tree, this plant is technically a large herb, characterized by a trunk or stem made up of overlapping leaf bases that encase it. Each stem produces fruit only once, after which it is succeeded by new suckers that subsequently flower, bear fruit, and perish (Nelson et al. 2006). The banana plant's trunk yields fibers used to make products such as mats, bags, and bins. Only four Musa varieties—Sentuluvan, Aethalpalal, Rasagatali, and Palayankottai—produce fiber, even though many are grown for their fruit (Balda et al. 2021). While the other three varieties yield white fiber, the Senthulavan variety is distinguished by its red color and produces a reddish-brown fiber. Southern India, particularly Kanyakumari, Kerala, Tamil Nadu, and Tirunelveli, is where these varieties are primarily grown (Pradheep et al. 2021). The fiber creates a distinctive and elegant fabric that is smooth, silky, lightweight, adaptable, and fun to work with when combined with silk and polyester (Hu et al. 2020). Banana fiber is used for ornamental items like lamp shades, wall hangings, table mats, ladies' purses, flower vases, and kids' caps (Bhoj 2022).

1.4.2 Biopolymers from Plant

Biopolymers, which are naturally occurring macromolecules derived from plants such as cellulose, lignin, and starch, are essential for many industrial applications and for plant structure (Singh et al. 2023). Cellulose, the most prevalent organic polymer on Earth and the main structural element of plant cell walls, gives them their strength and stiffness. Its densely packed linear chains of glucose units create microfibrils that increase the mechanical stability of plants. Lignin is a complex aromatic polymer that serves as a binding agent inside the cell wall, making it more rigid and resistant to microbial degradation (Shaghaleh et al. 2018). Conversely, starch is the primary polysaccharide that plants use to store energy. It is composed of amylose and amylopectin molecules, which can be easily broken down to release glucose when needed (MacNeill et al. 2017).

Because of their renewable and biodegradable qualities, biopolymers are essential in the fields of agriculture and environmental management. Cellulose- and lignin-rich plant leftovers significantly increase soil organic matter, improving soil structure, water retention, and nutrient cycling (Valle et al. 2024). By fostering microbial populations that aid in breakdown and nutrient mineralization, the addition of these biopolymers to the soil eventually increases soil fertility and crop yield (Harisha et al. 2023). Unlike synthetic polymers, starch-based products are also used to produce biodegradable films and coatings for agricultural applications, such as controlled-release fertilizers and seed coatings, thereby reducing environmental contamination (Azeem 2025).

In addition to their natural applications, biopolymers have garnered significant interest from various industries as eco-friendly substitutes for chemicals and petroleum-based plastics (Rajendran et al. 2025). While lignin is currently being studied as a sustainable source for bio-based adhesives, resins, and carbon fibers, cellulose is widely used in the production of paper, textiles, and biofuels (Sharma et al. 2025). Starch is utilized in the food, pharmaceutical, and packaging industries due to its biodegradability and ability to form films. By transforming plant biomass

Fig. 1.4 Isolated lignin (**a**) and cellulose (**b**) from bamboo plants

into practical, environmentally friendly products, the increasing focus on biopolymers is in line with international initiatives to lessen dependency on fossil fuels, cut greenhouse gas emissions, and advance the ideas of a circular economy (Dhalsamant et al. 2025) (Fig. 1.4).

1.4.3 Plant Oils and Resins

Plant oils and resins are significant natural compounds derived from various plant parts, such as seeds, fruits, bark, and leaves. Plant oils, which are often high in fatty acids, have several functions for the plant, such as storing energy and protecting it from environmental stress (Rahim et al. 2023). These oils are widely used in the food, cosmetic, pharmaceutical, and industrial industries due to their nutritional and chemical properties. For example, oils such as castor, coconut, and olive oil are highly prized for their numerous applications and health benefits (Abdalla et al. 2024). On the other hand, as a defense against pests and illnesses, plants—mostly conifers—produce complex mixtures of volatile and non-volatile substances known as resins. Because of their antibacterial and aromatic properties, they have long been used in varnishes, adhesives, incense, and pharmaceutical compositions (López-Goldar et al. 2020).

By offering natural alternatives to synthetic chemicals, plant oils and resins play a significant role in advancing sustainable practices in the domains of agriculture and environmental management (Höfer and Bigorra 2007). Due to their fungicidal and pesticidal properties, some plant oils are useful biopesticides that reduce reliance on dangerous synthetic pesticides (Gupta et al. 2023). By increasing the amount of organic matter and microbial activity in the soil, resins can improve soil health. Furthermore, more research is being done on the potential of both oils and resins for the manufacturing of biodegradable materials and bio-based products, which promote environmentally friendly advancements. Because of their renewable and versatile properties, plant oils and resins are essential to the development of green industrial technologies and sustainable agriculture (Kumar et al. 2025b).

1.4.4 Natural Rubber

Natural rubber, a biopolymer known for its waterproof, elastic, and long-lasting properties, is mostly derived from the latex sap of the rubber tree (Kurian and Mathew 2011). Tires, gloves, seals, and many other products are made from natural rubber in a variety of industries, including manufacturing, healthcare, and automotive (Kurian and Mathew 2011). Because natural rubber is renewable and biodegradable, it is more environmentally friendly than petroleum-based synthetic rubber. Rubber production byproducts can be used as organic supplements to increase soil fertility, and planting rubber trees helps conserve soil and sequester carbon. Natural rubber management is crucial to preserving a balance between economic gains, environmental preservation, and social well-being in rubber-growing regions (Seid et al. 2022).

1.4.5 Plant-Based Medicines and Natural Remedies

Many cultures have used natural remedies and plant-based drugs for thousands of years to treat a range of ailments and enhance overall health (Chaachouay and Zidane 2024). These drugs contain bioactive compounds, such as phenolics, terpenoids, flavonoids, and alkaloids, and are derived from a range of plant parts, including leaves, roots, bark, flowers, and seeds (Roy et al. 2022). Among other therapeutic properties, these compounds have analgesic, antibacterial, antioxidant, and anti-inflammatory properties. Current scientific research has demonstrated the safety and efficacy of many plant-based therapies used in traditional medical systems such as Ayurveda, Traditional Chinese Medicine, and Indigenous healing techniques (Anand et al. 2019).

Recently, plant-based medicines have seen a renaissance in interest due to their potential as innovative pharmaceuticals and their relatively lower side effects (Najmi et al. 2022). Many modern drugs, including aspirin, quinine, and paclitaxel, were developed from plant-derived compounds. Natural therapies are a vital part of basic healthcare due to their accessibility and affordability, especially in remote and resource-constrained areas (Chaachouay and Zidane 2024). To protect biodiversity and guarantee the continuous availability of these priceless resources, sustainable medicinal plant collection and cultivation are essential. In order to improve medication development and advance holistic medical practices worldwide, conventional wisdom must be combined with cutting-edge scientific methods (Woldearegay and Teso 2022).

According to Wachtel-Galor and Benzie (2012), traditional medicine is most frequently used because it is more cost-effective, more closely aligns with the patient's ideology, allays worries about the adverse effects of chemical (synthetic) medications, fulfills a desire for more individualized healthcare, and increases public access to health information. According to Wachtel-Galor and Benzie (2012), herbal remedies are primarily used to promote health and treat chronic conditions rather than life-threatening ones. However, when conventional medicine fails to treat a disease like advanced cancer or newly emerging infectious diseases, the use of traditional remedies rises. Additionally, many people believe that traditional medicines are safe, natural, and non-toxic. This is not always the case, particularly when herbs are taken with other herbs, over-the-counter medications, or prescription drugs (Canter and Ernst 2004; Cohen and Ernst 2010).

Whether or not people have physical or financial access to allopathic medicine, traditional medicine is a thriving worldwide commercial enterprise and offers a crucial health care service regardless of why an individual uses it (Conboy et al. 2007; Evans 2008). According to estimates, the United States spent US$13.7 billion on alternative therapy in 1990. By 1997, this had doubled, and herbal remedies were expanding more quickly than any other alternative therapy (Engebretson 2002). An estimated $80 million, $1 billion, and $22.3 billion are spent on traditional medicine annually in Australia, Canada, and the United Kingdom, respectively. According to Wachtel-Galor and Benzie (2012), these numbers show that traditional medicine, including herbal medicine, is incorporated into many health care systems and is taught to doctors in many developed countries.

It is impossible to overlook the overall commercial value of the ethnobotanical market. For instance, in 1995, the annual retail sales of herbal products in the United States were estimated to be US$5.1 billion, and the total turnover of nonprescription-bound herbal medicines in pharmacies was equivalent to nearly 30% of the total turnover of nonprescription-bound medicines in Germany. The Indian herbal industry uses about 960 plant species, of which 178 are high-volume, exceeding 100 metric tons annually. Herbal medicine is widely practiced in India (Sahoo et al. 2010). According to Eisenberg et al. (1998), the total value of herbal medicine produced in China in 1995 was 17.6 billion Chinese yuan, or roughly $2.5 billion USD. This pattern has persisted, and in 2003–2004 Western Europe's yearly revenues hit US$5 billion (De Smet 2005). According to Wachtel-Galor and Benzie (2012), sales of herbal products in China reached US$14 billion in 2005, while sales of herbal medicines in Brazil reached US$160 million in 2007. According to Tilburt and Kaptchuk (2008), the annual global market for these products is estimated to be close to US$60 billion.

Herbs are currently used to treat a wide range of illnesses and conditions, including depression, inflammation, prostate issues, cardiovascular disease, and immune system support, among others (Yatoo et al. 2018). Traditional herbal remedies were a major part of China's 2003 strategy to contain and treat severe acute respiratory syndrome (SARS), and in Africa, the African flower has been used for decades to treat HIV-related wasting symptoms (De Smet 2005; Tilburt and Kaptchuk 2008). In most developed nations, essential oils, herbal extracts, and herbal teas are sold in pharmacies alongside conventional medications. Herbal medicines are also very popular in Europe, with Germany and France leading the continent in over-the-counter sales (De Smet 2005).

Herb teas, syrups, essential oils, ointments, salves, rubs, capsules, and tablets that contain a ground or powdered form of a raw herb or its dried extract are examples of how herbs and plants can be processed and consumed (Umaru et al. 2020). Alcoholic extracts (tinctures), vinegars (acetic acid extracts), hot water extracts (tisanes), long-term boiled extracts (typically from roots or bark; decoctions), and cold infusions of plants (macerates) are examples of plant and herb extracts that differ in the solvent used, temperature, and duration of extraction. According to Wachtel-Galor and Benzie (2012), there is no standardization, and the components of a product or herbal extract are likely to differ significantly between batches and producers.

Plants are abundant in many different substances. Aromatic compounds, the majority of which are phenols or their oxygen-substituted derivatives like tannins, are among the numerous secondary metabolites (Thirumurugan et al. 2018). Many of these substances possess antioxidant qualities. In addition to using plant components directly as therapeutic agents, ethnobotanicals are crucial for pharmacological research and drug development because they can be used as models for pharmacologically active compounds or as starting materials for drug synthesis (Li and Vederas 2009). Opium extracted from the seeds and pods of the poppy Papaver somniferum was used to make morphine, the first pharmacologically active pure compound, about 200 years ago. This finding demonstrated that regardless of the source or age of the material, medications derived from plants can be refined and given in precise dosages (Hartmann 2007). The discovery of penicillin improved this strategy (Li and Vederas 2009). As a result of this ongoing trend, products derived from plants and natural sources (such as fungi and marine microorganisms) or their analogs have made significant contributions to today's commercial drug preparations. Examples include antibiotics (*penicillin*, *erythromycin*); digoxin, a cardiac stimulant obtained from foxglove (*Digitalis purpurea*); salicylic acid, an aspirin precursor obtained from willow bark (Salix spp.); reserpine an antipsychotic and antihypertensive medication derived from Rauwolfia species; lipid-lowering drugs and antimalarials like quinine from Cinchona bark; *lovastatin* from a fungus (Rishton 2008; Schmidt et al. 2008).

Additionally, over 60% of cancer treatments currently on the market or in clinical trials are derived from natural products. Over 70% of the 177 cancer treatments approved globally are based on natural products or mimetics, many of which have been enhanced through combinatorial chemistry. Plant-based cancer treatments include combretastatin, derived from the South African bush willow; paclitaxel, isolated from the Pacific yew tree; and camptothecin, derived from the Chinese happy tree, *Camptotheca acuminata*, and used to prepare irinotecan and topotecan (Brower 2008). Additionally, it is estimated that 121 active compounds derived from plants are currently in use, accounting for around 25% of all prescribed medications worldwide (Sahoo et al. 2010). The United States approved 13 natural product-derived medications between 2005 and 2007. Over 100 medications derived from natural products are undergoing clinical trials (Li and Vederas 2009), and 11% of the 252 medications on the World Health Organization's (WHO) list of essential medicines are solely derived from plants (Sahoo et al. 2010) (Table 1.1).

Table 1.1 Well-known medicinal plants used for medicine

Medicinal plants	Plant part	phytochemicals	Medicinal use	Reference
Aloe vera	Leaves	Anthraquinones, polysaccharides	Wound healing, anti-inflammatory	Lefsih et al. (2018)
Azadirachta indica	Leaves, bark	Azadirachtin, nimbin	Antimicrobial, antimalarial	Joshi and Prabhakar (2021)
Curcuma longa	Rhizome	Curcumin	Anti-inflammatory, antioxidant	Ramsewak et al. (2000)
Ocimum sanctum	Leaves	Eugenol, ursolic acid	Stress relief, immune booster	Chatterjee and Sarkar (2025)
Zingiber officinale	Rhizome	Gingerol, shogaol	Digestive aid, anti-nausea	Haniadka et al. (2012)
Echinacea	Roots, leaves	Alkamides, caffeic acid derivatives	Immune stimulant	Ahmadi (2024)
Ginkgo biloba	Leaves	Flavonoids, terpenoids	Cognitive enhancement, circulation	Liebgott et al. (2000)
Withania somnifera	Roots	Withanolides	Adaptogen, anti-stress	Della Porta et al. (2023)
Lavandula angustifolia	Flowers	Linalool, linalyl acetate	Anxiety relief, sleep aid	Manzoor et al. (2025)
Evodia rutaecarpa	Fruit, leaves	Quinolone alkaloids	Antibacterial (against Helicobacter pylori)	Hamasaki et al. (2000)
Boesenbergia pandurata	Rhizomes	4-Hydroxypanduratin A	Antimutagenic	Trakoontivakorn et al. (2001)
Silybum marianum	Seeds	Silymarin	Liver protection, antioxidant	Kmail (2024)
Matricaria chamomilla	Flowers	Apigenin, chamazulene	Anti-inflammatory, digestive aid	Melnyk et al. (2024)
Hypericum perforatum	Flowers	Hypericin, hyperforin	Antidepressant	Wurglics and Schubert-Zsilavecz (2006)
Valeriana officinalis	Roots	Valerenic acid, valepotriates	Sleep aid, anxiety relief	Nandhini et al. (2018)
Panax ginseng	Roots	Ginsenosides	Energy booster, immune support	Kang and Min (2012)
Cinnamomum verum	Bark	Cinnamaldehyde, eugenol	Antimicrobial, blood sugar regulation	Pathak and Sharma (2021)
Allium sativum	Bulb	Allicin	Antimicrobial, cardiovascular health	Kendler (1987)

1.4.6 Agricultural Innovations

Leading agricultural advances that support resilience and sustainability in farming systems are plant-based solutions. Crop diversification improves soil health and lowers pest and disease outbreaks (Mihrete and Mihretu 2025). Agroforestry, or the integration of trees and shrubs with crops and/or livestock, contributes to sustainable agriculture by providing several ecological benefits, such as carbon sequestration, improved water retention, and habitat for beneficial organisms. Combining these techniques improves soil fertility, productivity, and climate change mitigation over time, making them essential instruments for advancing sustainable farming globally (Kumar et al. 2023).

In addition to their ecological benefits, plant-based solutions such as agroforestry and crop diversification offer farmers significant financial benefits. Farmers can spread financial risk and improve income stability by growing a variety of crops and incorporating trees to lessen reliance on a single commodity (Mihrete and Mihretu 2025). According to Dwivedi et al. (2017), to meet consumers' increasing demand for plant-based, sustainably produced foods, various cropping systems can create new markets for specialty and niche products. Additionally, agroforestry systems often produce other goods, such as fruits, nuts, wood, and medicinal plants, which can serve as supplementary sources of income. According to Keprate et al. (2024b), smallholder farmers and rural communities depend on this economic resilience to enhance their standard of living and lessen poverty.

These agricultural innovations are also essential for improving nutrition and food security. To prevent malnutrition and micronutrient deficiencies, crop diversification provides a variety of nutrient-rich foods, ensuring a more varied diet (Ali and Bhattacharjee 2023). This is made possible by agroforestry systems that preserve genetic diversity by safeguarding native tree species and establishing microclimates that encourage the growth of a variety of crops. Additionally, even in challenging weather conditions, these practices enhance soil health and water management, resulting in more consistent yields (Lelamo 2021). The development of resilient and sustainable food production systems that can meet the demands of an expanding population increasingly depends on incorporating plant-based solutions into farming practices as climate change continues to pose challenges to global food systems (Said et al. 2025) (Fig. 1.5).

1.5 Applications of Plant-Based Solutions

1.5.1 Crop Diversification in Smallholder Farms

Crop diversification is implemented by smallholder farmers who cultivate a range of crop species on their land. By using complementary nutrients, this strategy increases soil fertility while lowering susceptibility to pests, diseases, and market swings

Fig. 1.5 Agricultural innovations (agroforestry systems contribute)

Fig. 1.6 Crop diversification in smallholder farms

(Shaffril et al. 2024). For instance, intercropping legumes and cereals improves nitrogen fixation and lessens the need for artificial fertilizers.

According to research crop variety can significantly increase farm resilience and productivity. A study by Lin (2011) found that diversified cropping systems that protect against environmental stresses, such as pest outbreaks and drought, improve overall production stability. Additionally, diversified farms tend to have higher levels of biodiversity, which supports natural pest predators and beneficial insects and lessens the need for chemical pesticides (Letourneau 2012). Crop diversity also improves soil health. According to a meta-analysis by Cong et al. (2018), intercropping systems increase soil organic matter and microbial biomass, which are critical for nutrient cycling and soil structure. In crop rotations and intercropping systems, legumes enhance biological nitrogen fixation, supplying up to 50% of the nitrogen needed by subsequent cereal crops (Peoples et al. 2017). Diversified farming techniques provide smallholder farmers with additional revenue streams while reducing the risks associated with market price fluctuations. For instance, by selling various crops at different times of year, diversified farms can smooth income flows (Snapp et al. 2019) (Fig. 1.6).

1.5.2 Agroforestry Systems for Climate Resilience

Agroforestry creates multipurpose landscapes by combining trees, crops, and/or livestock. Home gardens, silvopasture, and alley cropping are examples of applications. These systems increase resilience to climate unpredictability by improving carbon sequestration, water retention, and shade, which reduce heat stress on crops and animals (Kehinde et al. 2024). Additionally, by anchoring the soil, agroforestry techniques reduce erosion and increase soil fertility through improved nutrient cycling, thereby enhancing pollination and pest control (Fahad et al. 2022). Agroforestry enhances climate resilience by reducing runoff and providing windbreaks against extreme weather events. By increasing farm productivity and diversifying revenue streams, these technologies can help farmers adjust financially to climate variability (Keprate et al. 2024a).

At least 156 million metric tons of carbon dioxide equivalent could be sequestered by agroforestry techniques on 110.9 million eligible acres in the US each year, potentially eliminating emissions from more than 33.9 million cars (Udawatta et al. 2022). By absorbing excess soil nutrients, trees in these systems reduce the need for nitrogen fertilizers and lower nitrous oxide emissions, a dangerous greenhouse gas that is 300 times more potent than CO_2 (Ussiri and Lal 2012). A USDA study claims that by enhancing soil quality and providing pest and weather protection, agroforestry can boost crop yields by as much as 56%. Growing wild fruits and timber can also diversify its sources of income (Dissanayaka et al. 2024).

According to research from the USDA National Agroforestry Center, silvopasture is essential for mitigating livestock heat stress. Farms such as Minnesota's Early Boots Farm have reported increased productivity following the planting of more than 5000 trees, the expansion of grazing areas, and the implementation of rotational grazing with funding from the Environmental Quality Incentives Program (Smith et al. 2022). The Conservation Reserve Enhancement Program has restored riparian forest buffers along more than 925 miles of streams in Washington State. Six million trees have been planted to improve salmon habitats, fortify banks, and filter runoff—all of which have improved water quality and flood resilience (Roni et al. 2018).

1.5.3 Urban and Peri-Urban Agriculture

Urban and peri-urban agriculture promotes sustainable food systems in densely populated areas by employing plant-based techniques to convert constrained areas into productive green zones (Righini et al. 2023). While vertical farming uses stacked hydroponic or aeroponic systems for year-round production, rooftop gardens cultivate fruits, vegetables, and herbs on vacant building surfaces (Singh 2023). Urban and peri-urban agriculture employs plant-based techniques to convert small areas into productive green spaces, thereby supporting sustainable food systems.

While vertical farming uses stacked hydroponic or aeroponic systems for year-round production, rooftop gardens cultivate fruits, vegetables, and herbs on unused building surfaces (Gunapala et al. 2025). In addition to fostering social cohesion, mental health benefits, and educational opportunities, these strategies help peri-urban communities—those on the periphery of cities—in preserving agricultural history and bridging rural-urban divides. Microclimates that provide shade and evapotranspiration can reduce city temperatures by 1–5 °C during heatwaves (Debie 2025).

Urban and peri-urban agriculture could supply up to 20% of the world's vegetable production and 15% of fruits, potentially feeding 800 million urban residents by 2050 and reducing global food transport emissions by 10–20%, according to a 2022 Food and Agriculture Organization (FAO) study (Teng 2024). Rooftop gardens in New York City retain 50–75% of stormwater, reducing urban flooding and easing strain on sewer systems. Up to 0.7 kg of CO_2 can be absorbed annually by one square meter of vegetated roof (Wang Jr 2023). The 2021 Nature Sustainability assessment claims that vertical farming innovations use 90–95% less water and produce 10–100 times more per unit area than typical field agriculture (Vatistas et al. 2022). For instance, AeroFarms in New Jersey demonstrates scalability for climate-resilient food production by using zero pesticides to produce the equivalent of two acres of farmland on just a tenth of an acre (Wightman and Woodbury 2020).

According to a 2019 University of Melbourne study conducted in Melbourne, Australia, urban gardens support 20–30% more pollinator species than nearby lawns, enhancing pollination services worth millions of dollars in crop value (Pauli et al. 2021). In major areas like Los Angeles, urban green spaces like gardens may remove up to 1000 tons of air pollutants annually, according to the U.S. Environmental Protection Agency (EPA) (Nowak and Heisler 2010). This is because plants act as organic filters for ozone and sulfur dioxide. Low-income urban residents who participate in community gardens consume 25% more fruits and vegetables, which reduces obesity rates and diet-related health disparities by boosting access to fresh food, according to social impact research published in the American Journal of Public Health in 2020 (Hume et al. 2022).

1.5.4 *Restoration of Degraded Lands*

On degraded soils, plant-based strategies such as growing cover crops and nitrogen-fixing trees can improve soil structure, boost organic matter, and reduce erosion (Singh et al. 2024). Over time, agroforestry and diverse cropping systems boost biodiversity, restore ecosystems, and increase land production. Numerous studies have demonstrated the ecological and financial advantages of plant-based restoration techniques, which have proven highly effective in restoring degraded areas. According to a 2021 World Resources Institute (WRI) report on global land restoration, programs such as the African Union's Great Green Wall, which has planted over 100 million trees and shrubs in 11 countries, have restored 20 million hectares of degraded Sahelian land since 2007. This has increased soil organic carbon by 20–50% through nitrogen-fixing species such as Acacia senegal and Faidherbia albida, increased biodiversity by 30–40%, and improved groundwater recharge to combat desertification (Mansourian and Berrahmouni 2021). Cover cropping improved soil structure on 15 million acres of degraded farmland in the United States between 2015 and 2020, reducing erosion by up to 90% and increasing water infiltration by 50%, resulting in crop yield gains of 10–25% in subsequent seasons (Chen et al. 2022). For example, a study conducted in the Corn Belt found that diversified rotations with legumes restored microbial diversity and sequestered 0.5–1 ton of CO_2 per hectare annually (Evans 2022). Agroforestry trials in Brazil's Atlantic Forest restored 1.2 million hectares of deforested pastures, boosting land productivity by 40% through integrated tree-crop systems that supported over 200 native species and enhanced nutrient cycling, according to a 2019 Science Advances article. This demonstrates that these techniques not only stop degradation but also provide ecosystem services worth $1–2 billion a year, including improved water quality, fruits, and timber (Gupta et al. 2020) (Table 1.2).

1.6 Challenges and Limitations

Plant-based solutions offer exciting opportunities for sustainable agriculture and climate resilience, but several complex challenges hinder their widespread adoption. Examples of technological and production issues that are exacerbated by developing regions' limited access to high-tech equipment include the need for complex breeding techniques to produce resilient crop varieties suitable for a range of climates, as well as scalable developments in precision farming and soil management to maximize yields without depleting resources (Tilman et al. 2011). Additional challenges include consumer behavior and market acceptance. Only 20–30% of consumers worldwide regularly choose plant-based alternatives, which slows demand-driven innovation (Mamia 2023). Despite increased awareness, many consumers are reluctant to switch from well-known animal-based products due to taste preferences, cultural customs, and misconceptions about nutritional completeness (Bryant et al. 2019). Low-income farmers in the Global South are disproportionately affected by inadequate subsidies, trade policies that support conventional agriculture, a lack of incentives for carbon credits or ecosystem services, and initial setup costs for agroforestry or urban farming systems, all of which make implementation even more challenging (Pretty et al. 2018). Finally, but just as importantly, possible ecological effects need to be managed appropriately. These include the potential for monoculture growth to lead to biodiversity loss, excessive water use in intensive vertical farming, or the inadvertent spread of invasive species from non-native plant introductions (Tilman et al. 2017). This highlights the importance of comprehensive context-specific strategies to reduce these adverse effects and maximize beneficial ones.

Table 1.2 Applications of Plant-Based Solutions in Sustainable Agriculture and Ecosystem Resilience

Application	Description	Use	Reference
Crop Diversification in Smallholder Farms	Cultivating multiple crop species and varieties within fields, e.g., intercropping legumes with cereals for nitrogen fixation.	Reduces vulnerability to pests, diseases, and market fluctuations; improves soil fertility, yield stability, and biodiversity, and provides multiple income sources.	Mihrete and Mihretu (2025); Shaffril et al. (2020)
Agroforestry Systems for Climate Resilience	Integrating trees and shrubs with crops/ livestock in systems like alley cropping, silvopasture, and home gardens.	Enhances carbon sequestration, soil fertility, water retention, erosion control, biodiversity, and resilience to climate variability; diversifies income and boosts yields.	Vinodhini et al. (2023)
Urban and Peri-Urban Agriculture	Rooftop gardens, vertical farming, and community gardens using diverse plant species in urban/ peri-urban spaces.	Boosts local food production, reduces food miles/emissions, improves air quality/ microclimates, promotes biodiversity, social cohesion, and food security.	Biasi and Brunori (2023); De Neergaard et al. (2009)
Restoration of Degraded Lands	Planting nitrogen-fixing trees/cover crops and using agroforestry/ diversified cropping to rehabilitate ecosystems.	Restores soil structure/organic matter, prevents erosion, enhances biodiversity/ productivity, nutrient cycling, and groundwater recharge; provides economic ecosystem services.	Gupta et al. (2020)
Sustainable Livestock Integration	Incorporating fodder trees/shrubs into cropping systems for nutritious livestock feed and nutrient cycling.	Supports circular nutrient flows, improves animal health, reduces external feed reliance/GHG emissions, and enhances farm resilience across diverse environments.	Kumar et al. (2025a)

1.7 Future Perspectives

Future developments have the potential to completely transform global sustainability, including CRISPR-Cas9 gene editing to produce nutrient-dense, drought-resistant crops, AI-powered precision agriculture for optimal resource use, and advancements in plant-based meat and dairy substitutes that mimic animal products with 90% lower environmental footprints. By 2035, 60% of the protein market will be made up of plant-based products, according to a McKinsey report from 2023. These innovations easily align with the circular economy's tenets by promoting closed-loop systems that reduce inputs and outputs, such as urban farming, waste-to-compost models, and agroforestry nutrient recycling. Additionally, they directly support the United Nations Sustainable Development Goals (SDGs), particularly SDG 2 (Zero Hunger) through increased food security, SDG 13 (Climate Action) through carbon sequestration, and SDG 15 (Life on Land) through ecosystem restoration. If scaled globally in line with IPCC scenarios, they could prevent 10–20 gigatons of CO2 emissions annually. Implementing supportive frameworks, such as increased subsidies for regenerative practices, carbon pricing to incentivize emissions reductions, and international agreements under the Paris Accord, will be necessary to realize this vision. Education campaigns can influence consumer behavior and empower farmers through community training programs, fostering widespread acceptance and increased research funding, such as the EU's Horizon Europe, which allocates €1 billion to sustainable supply chains, ensuring equitable, inclusive progress toward a future powered by plants.

1.8 Conclusion

In conclusion, plant-based solutions stand out as a crucial element of sustainable development because they address the interconnected issues of resource scarcity, health concerns, and climate change by fusing conventional knowledge with contemporary innovation. From the environmental justification of reducing emissions and safeguarding biodiversity to the health benefits of preventing illness and the ethical considerations of animal welfare, these approaches demonstrate the remarkable adaptability of plants as renewable resources. The development of plant-based diets, proteins, lipids, and materials demonstrates how they can enhance food security, restore degraded lands, and create jobs when combined with agricultural techniques like agroforestry and urban farming. Industry estimates that growth will exceed US$162 billion by 2030, along with significant environmental benefits such as improved soil health and reduced water use, which support this. Even though there are still problems with technology, consumer acceptance laws, and the environment, these issues can be addressed through focused research, instruction, and frameworks that prioritize justice, especially for smallholders in the Global South. As new advancements in biotechnology and circular economies align with global objectives such as the SDGs, plant-based solutions not only promise a more resilient and healthy world but also enable communities to create a future where sustainability and prosperity can coexist. Adopting this paradigm shift is essential to humanity's long-term well-being and necessitates concerted efforts to implement these revolutionary strategies on a global scale.

References

Abate L, Bachheti A, Bachheti RK, Husen A, Pandey D, Megra MB (2022) Coir fiber-based cellulose, nanocellulose, and their cutting-edge applications. In: Coir fiber and its composites. Elsevier, pp 309–331

Abdalla S, Aroua MK, Gew LT (2024) A comprehensive review of plant-based cosmetic oils (virgin coconut oil, olive oil, argan oil, and jojoba oil): chemical and biological properties and their cosmeceutical applications. ACS Omega 9(44):44019–44032

Ahmad Z, Upadhyay A, Ding Y, Emamverdian A, Shahzad A (2021) Bamboo: origin, habitat, distributions and global prospective. In: Biotechnological advances in bamboo: the "Green Gold" on the earth. Springer, pp 1–31
Ahmadi F (2024) Phytochemistry, mechanisms, and preclinical studies of Echinacea extracts in modulating immune responses to bacterial and viral infections: a comprehensive review. Antibiotics 13(10):947
Akinwumi FE, Akinyemi AO, Akangbe B, Odeniran OM, Sehkar J, Olajimbiti CO, Ajayi OH (2025) Risk of osteoporosis and anemia in plant-based diets: a systematic review of nutritional deficiencies and clinical implications. Cureus 17(7):e88461
Akram Khan M, Guru S, Padmakaran P, Mishra D, Mudgal M, Dhakad S (2011) Characterisation studies and impact of chemical treatment on mechanical properties of sisal fiber. Compos Interfaces 18(6):527–541
Alcorta A, Porta A, Tárrega A, Alvarez MD, Vaquero MP (2021) Foods for plant-based diets: challenges and innovations. Foods 10(2):293
Aleksandrowicz L, Green R, Joy EJ, Smith P, Haines A (2016) The impacts of dietary change on greenhouse gas emissions, land use, water use, and health: a systematic review. PLoS One 11(11):e0165797
Ali A, Bhattacharjee B (2023) Nutrition security, constraints, and agro-diversification strategies of neglected and underutilized crops to fight global hidden hunger. Front Nutr 10:1144439
Ali A, Shaker K, Nawab Y, Jabbar M, Hussain T, Militky J, Baheti V (2018) Hydrophobic treatment of natural fibers and their composites—a review. J Ind Text 47(8):2153–2183
Almeaiweed NH, Aloud SS, Alotaibi KD, Al Watban MA, Alrobaish WS, Alorf MS (2025) Sustainable management approaches to heavy metal pollution in arid soils using soil amendments and plant-based remediation. Sustainability 17(16):7558
Anand U, Jacobo-Herrera N, Altemimi A, Lakhssassi N (2019) A comprehensive review on medicinal plants as antimicrobial therapeutics: potential avenues of biocompatible drug discovery. Metabolites 9(11):258
Anandjiwala RD, John M (2010) Sisal-cultivation, processing and products. In: Industrial applications of natural fibres: structure, properties and technical applications. Wiley, pp 181–195
Arab-Tehrany E, Jacquot M, Gaiani C, Imran M, Desobry S, Linder M (2012) Beneficial effects and oxidative stability of omega-3 long-chain polyunsaturated fatty acids. Trends Food Sci Technol 25(1):24–33
Azeem B (2025) Stimuli-responsive starch-based biopolymer coatings for smart and sustainable fertilizers. Gels 11(9):681
Balda S, Sharma A, Capalash N, Sharma P (2021) Banana fibre: a natural and sustainable bioresource for eco friendly applications. Clean Technol Environ Policy 23(5):1389–1401
Bhoj R (2022) An intensification of sustainable eco-friendly sisal fiber crafts in healthcare industry. Cardiometry 23:310–318
Biasi R, Brunori E (2023) Agrobiodiversity-based landscape design in urban areas. Plants 12(24):4121
Bordoloi S (2020) The coir industry and its social relations of production. In: The political economy of uneven rural development: case of the nonfarm sector in Kerala, India. Springer, pp 97–144
Brower V (2008) Back to nature: extinction of medicinal plants threatens drug discovery. J Natl Cancer Inst 100:838
Bryant C, Szejda K, Parekh N, Deshpande V, Tse B (2019) A survey of consumer perceptions of plant-based and clean meat in the USA, India, and China. Front Sustain Food Syst 3:432863
Canter PH, Ernst E (2004) Herbal supplement use by persons aged over 50 years in Britain: frequently used herbs, concomitant use of herbs, nutritional supplements and prescription drugs, rate of informing doctors and potential for negative interactions. Drugs Aging 21(9):597–605
Chaachouay N, Zidane L (2024) Plant-derived natural products: a source for drug discovery and development. Drugs Drug Candidates 3(1):184–207
Chatterjee A, Sarkar B (2025) Polyphenols and terpenoids derived from Ocimum species as prospective hepatoprotective drug leads: a comprehensive mechanistic review. Phytochem Rev 24(2):2087–2129

Chen L, Rejesus RM, Aglasan S, Hagen SC, Salas W (2022) The impact of cover crops on soil erosion in the US Midwest. J Environ Manag 324:116168
Cheriaparambil R, Grossmann L (2025) Properties and cultivation of Fusarium spp. to produce mycoprotein as an alternative protein source. Sustain Food Proteins 3(1):e70002
Chiu TH, Lin C-L (2009) Ethical management of food systems: plant based diet as a holistic approach. Asia Pac J Clin Nutr 18(4):647–653
Cohen PA, Ernst E (2010) Safety of herbal supplements: a guide for cardiologists. Cardiovasc Ther 28(4):246–253
Conboy L, Kaptchuk TJ, Eisenberg DM, Gottlieb B, Acevedo-Garcia D (2007) The relationship between social factors and attitudes toward conventional and CAM practitioners. Complement Ther Clin Pract 13(3):146–157
Cong W, Meng J, Ying SC (2018) Impact of soil properties on the soil methane flux response to biochar addition: a meta-analysis. Environ Sci Process Impacts 20(9):1202–1209
Craig WJ, Mangels AR, Fresán U, Marsh K, Miles FL, Saunders AV, Haddad EH, Heskey CE, Johnston P, Larson-Meyer E (2021) The safe and effective use of plant-based diets with guidelines for health professionals. Nutrients 13(11):4144
De Bakker E, Dagevos H (2012) Reducing meat consumption in today's consumer society: questioning the citizen-consumer gap. J Agric Environ Ethics 25(6):877–894
De Neergaard A, Drescher AW, Kouamé C (2009) Urban and peri-urban agriculture in African cities. In: African indigenous vegetables in urban agriculture. Routledge, pp 35–64
De Smet PA (2005) Herbal medicine in Europe—relaxing regulatory standards. N Engl J Med 352(12):1176–1178
Debie E (2025) Synergistic integration of urban agriculture and green infrastructure to enhance urban sustainability in Bahir Dar, Ethiopia. City Environ Interact 27:100216
Dekkers BL, Boom RM, van der Goot AJ (2018) Structuring processes for meat analogues. Trends Food Sci Technol 81:25–36
Della Porta M, Maier JA, Cazzola R (2023) Effects of Withania somnifera on cortisol levels in stressed human subjects: a systematic review. Nutrients 15(24):5015
Dhalsamant K, Dalai A, Pattnaik F, Acharya B (2025) Biodegradable carbohydrate-based films for packaging agricultural products—a review. Polymers 17(10):1325
Dhir YJ (2022) Natural fibers: the sustainable alternatives for textile and non-textile applications. In: Natural fiber. IntechOpen
Diederichsen A, Richards K (2003) Cultivated flax and the genus Linum L.: taxonomy and germplasm conservation. In: Flax. CRC Press, pp 34–66
Dissanayaka D, Udumann S, Atapattu AJ (2024) Synergies between tree crops and ecosystems in tropical agroforestry. In: Agroforestry. Wiley, pp 49–87
Dwivedi SL, Van Bueren ETL, Ceccarelli S, Grando S, Upadhyaya HD, Ortiz R (2017) Diversifying food systems in the pursuit of sustainable food production and healthy diets. Trends Plant Sci 22(10):842–856
Eisenberg DM, Davis RB, Ettner SL, Appel S, Wilkey S, Van Rompay M, Kessler RC (1998) Trends in alternative medicine use in the United States, 1990–1997: results of a follow-up national survey. JAMA 280(18):1569–1575
Emamverdian A, Ding Y, Ranaei F, Ahmad Z (2020) Application of bamboo plants in nine aspects. Sci World J 2020(1):7284203
Engebretson J (2002) Culture and complementary therapies. Complement Ther Nurs Midwifery 8(4):177–184
Evans S (2008) Changing the knowledge base in Western herbal medicine. Soc Sci Med 67(12):2098–2106
Evans JF (2022) Carbon dioxide net ecosystem exchange altered by crop rotation diversification and use of cover crops. University of Guelph
Everitt NM, Aboulkhair NT, Clifford MJ (2013) Looking for links between natural fibres' structures and their physical properties. Paper read at Conference papers in science

Eyupoglu S (2020) Sustainable plant-based natural fibers. In: Sustainability in the textile and apparel industries: sourcing natural raw materials. Springer, pp 27–48

Fahad S, Chavan SB, Chichaghare AR, Uthappa AR, Kumar M, Kakade V, Pradhan A, Jinger D, Rawale G, Yadav DK (2022) Agroforestry systems for soil health improvement and maintenance. Sustainability 14(22):14877

Fednand C, Bigambo P, Mgani Q (2022) Modification of the mechanical and structural properties of sisal fiber for textile applications. J Nat Fibers 19(15):10834–10845

Fehér A, Gazdecki M, Véha M, Szakály M, Szakály Z (2020) A comprehensive review of the benefits of and the barriers to the switch to a plant-based diet. Sustainability 12(10):4136

Feigin SV, Wiebers DO, Lueddeke G, Morand S, Lee K, Knight A, Brainin M, Feigin VL, Whitfort A, Marcum J (2023) Proposed solutions to anthropogenic climate change: a systematic literature review and a new way forward. Heliyon 9(10):e20544

Finnigan T (2011) Mycoprotein: origins, production and properties. In: Handbook of food proteins. Woodhead Publishing, pp 335–352

Fukushima D (1991) Structures of plant storage proteins and their functions. Food Rev Int 7(3):353–381

Ghazoul J, Medina L (2024) Deforestation and land clearing. In: Encyclopedia of biodiversity. Elsevier, pp 3–605

Golden JS, Handfield RB, Daystar J, Pires S (2024) An economic impact analysis of the US biobased products industry: 2023 update. USDA

Goudenhooft C, Bourmaud A, Baley C (2019) Flax (Linum usitatissimum L.) fibers for composite reinforcement: exploring the link between plant growth, cell walls development, and fiber properties. Front Plant Sci 10:411

Guan X, Zhong X, Lu Y, Du X, Jia R, Li H, Zhang M (2021) Changes of soybean protein during tofu processing. Foods 10(7):1594

Gunapala R, Gangahagedara R, Wanasinghe W, Samaraweera AU, Gamage A, Rathnayaka C, Hameed Z, Baki ZA, Madhujith T, Merah O (2025) Urban agriculture: a strategic pathway to building resilience and ensuring sustainable food security in cities. Farm Syst 3:100150

Gupta SR, Dagar JC, Teketay D (2020) Agroforestry for rehabilitation of degraded landscapes: achieving livelihood and environmental security. In: Agroforestry for degraded landscapes: recent advances and emerging challenges, vol 1. Springer, pp 23–68

Gupta I, Singh R, Muthusamy S, Sharma M, Grewal K, Singh HP, Batish DR (2023) Plant essential oils as biopesticides: applications, mechanisms, innovations, and constraints. Plants 12(16):2916

Haghighatdoost F, Mahdavi A, Mohammadifard N, Hassannejad R, Najafi F, Farshidi H, Lotfizadeh M, Kazemi T, Karimi S, Roohafza H (2023) The relationship between a plant-based diet and mental health: evidence from a cross-sectional multicentric community trial (LIPOKAP study). PLoS One 18(5):e0284446

Hamasaki N, Ishii E, Tominaga K, Tezuka Y, Nagaoka T, Kadota S, Kuroki T, Yano I (2000) Highly selective antibacterial activity of novel alkyl quinolone alkaloids from a Chinese herbal medicine, Gosyuyu (Wu-Chu-Yu), against Helicobacter pylori in vitro. Microbiol Immunol 44(1):9–15

Haniadka R, Rajeev AG, Palatty PL, Arora R, Baliga MS (2012) Zingiber officinale (ginger) as an anti-emetic in cancer chemotherapy: a review. J Altern Complement Med 18(5):440–444

Hao L, Sapuan S, Hassan M, Sheltami R (2018) Natural fiber reinforced vinyl polymer composites. In: Natural fibre reinforced vinyl ester and vinyl polymer composites. Elsevier, pp 27–70

Harisha CB, Meena KK, Rane J, Halli HM, Manjanna BK, Patil BS, Chaudhary A, Naik V, Sorty AM (2023) Bacterial derived biopolymer to alleviate nutrient stress and yield enhancement in turmeric (Curcuma longa L.) by mediating physiology and rhizosphere microbes on poor soils of semi-arid tropics. Arch Agron Soil Sci 69(13):2645–2662

Hartmann T (2007) From waste products to ecochemicals: fifty years research of plant secondary metabolism. Phytochemistry 68(22–24):2831–2846

Hayat MK, Khan TY, Sial TN, Khan HA, Ashraf ZU, Sheraz M, Oumar B, Usman M, Nadeem M, Belizaire P (2025) The biotech and precision agriculture revolution: feeding the world sustainably. Policy Res J 3(3):293–316

Hemler EC, Hu FB (2019a) Plant-based diets for cardiovascular disease prevention: all plant foods are not created equal. Curr Atheroscler Rep 21(5):18

Hemler EC, Hu FB (2019b) Plant-based diets for personal, population, and planetary health. Adv Nutr 10:S275–S283

Höfer R, Bigorra J (2007) Green chemistry—a sustainable solution for industrial specialties applications. Green Chem 9(3):203–212

Hu X, McClements DJ (2022) Development of plant-based adipose tissue analogs: freeze-thaw and cooking stability of high internal phase emulsions and gelled emulsions. Foods 11(24):3996

Hu J, Jahid MA, Harish Kumar N, Harun V (2020) Fundamentals of the fibrous materials. In: Handbook of fibrous materials. Wiley, pp 1–36

Hume C, Grieger JA, Kalamkarian A, D'Onise K, Smithers LG (2022) Community gardens and their effects on diet, health, psychosocial and community outcomes: a systematic review. BMC Public Health 22(1):1247

Husen A (2021) Traditional herbal therapy for the human immune system. CRC Press, Boca Raton. https://doi.org/10.1201/9781003137955

Husen A (2022) Herbs, shrubs and trees of potential medicinal benefits. CRC Press, Boca Raton. https://doi.org/10.1201/9781003205067

Husen A (2023a) Exploring poisonous plants: medicinal values, toxicity response and other potential uses. CRC Press, Boca Raton. https://doi.org/10.1201/b23017

Husen A (2023b) Antidiabetic medicinal plants and herbal treatments. CRC Press, Boca Raton. https://doi.org/10.1201/b23347

Husen A (2024a) Antimalarial medicinal plants. Taylor & Francis Group, LLC. https://doi.org/10.1201/9781003378396

Husen A (2024b) Medicinal spices and condiment crops. Taylor & Francis Group, LLC. https://doi.org/10.1201/9781003387046

Jacobsen C (2015) Some strategies for the stabilization of long chain n-3 PUFA-enriched foods: a review. Eur J Lipid Sci Technol 117(11):1853–1866

Jacobsen C, Sørensen A-D, Nielsen NS (2013) Stabilization of omega-3 oils and enriched foods using antioxidants. In: Food enrichment with omega-3 fatty acids. Elsevier, pp 130–149

Joshi M, Prabhakar B (2021) Azadirachta indica (neem) in various infectious diseases. In: Infectious diseases. Bentham Science Publishers, pp 128–147

Kang S, Min H (2012) Ginseng, the "immunity boost": the effects of Panax ginseng on immune system. J Ginseng Res 36(4):354

Kavitha M (2015) Production process of coir and coir products. IMPACT: IJRBM 3(3):39–47

Kehinde OC, Ofuowoichoyama UG, Ogunsola MO, James AB (2024) Integrating trees into agricultural landscapes: benefits and challenges of agroforestry systems. Int Inst Acad Res Dev 10(7):94–121

Kendler BS (1987) Garlic (Allium sativum) and onion (Allium cepa): a review of their relationship to cardiovascular disease. Prev Med 16(5):670–685

Kent G, Kehoe L, Flynn A, Walton J (2022) Plant-based diets: a review of the definitions and nutritional role in the adult diet. Proc Nutr Soc 81(1):62–74

Keprate A, Bhardwaj D, Sharma P, Verma K, Abbas G, Sharma V, Sharma K, Janju S (2024a) Climate resilient agroforestry systems for sustainable land use and livelihood. In: Transforming agricultural management for a sustainable future: climate change and machine learning perspectives. Springer, pp 141–161

Keprate A, Sharma V, Bhatnagar S, Thakur R, Abbas G, Bhardwaj D, Sharma P (2024b) Economic studies in agroforestry for livelihood security. In: Agroforestry. Wiley, pp 443–479

Kevany KM, Baur G, Wang GC (2018) Shifting food systems: increasing well-being through plant-based approaches. Explore 14(6):435–442

Kmail A (2024) Mitigating digestive disorders: action mechanisms of Mediterranean herbal active compounds. Open Life Sci 19(1):20220857

Kolarš B, Mijatović Jovin V, Živanović N, Minaković I, Gvozdenović N, Dickov Kokeza I, Lesjak M (2025) Iron deficiency and iron deficiency anemia: a comprehensive overview of established and emerging concepts. Pharmaceuticals 18(8):1104

Kozicka M, Havlík P, Valin H, Wollenberg E, Deppermann A, Leclère D, Lauri P, Moses R, Boere E, Frank S (2023) Feeding climate and biodiversity goals with novel plant-based meat and milk alternatives. Nat Commun 14(1):5316

Kraak VI, Aschemann-Witzel J (2024) The future of plant-based diets: aligning healthy marketplace choices with equitable, resilient, and sustainable food systems. Annu Rev Public Health 45(1):253–275

Krzywonos M, Piwowar-Sulej K (2022) Plant-based innovations for the transition to sustainability: a bibliometric and in-depth content analysis. Foods 11(19):3137

Kumar P, Sai Ram C, Srivastava JP, Behura AK, Kumar A (2022) Synthesis of cotton fiber and its structure. In: Natural and synthetic fiber reinforced composites: synthesis, properties and applications. Wiley-VCH, pp 17–36

Kumar R, Veeraragavan M, Baral K, Saikanth D, Singh V, Upadhyay L, Raj S (2023) Agroforestry and its potential for sustainable land management and climate action: a review. Int J Environ Climate Chang 13(12):620–629

Kumar A, Vaishnav R, Deori C, Timung B, Roy S, Pradhan R, Prakash R, Bochalya RS (2025a) Silvopastoral systems in India for fodder production and livestock sustainability: a review. J Exp Agric Int 47(3):112–135

Kumar R, Farda B, Mignini A, Djebaili R, Koolman L, Paul A, Mondal S, Joel JM, Pandit A, Panneerselvam P (2025b) Microbial solutions in agriculture: enhancing soil health and resilience through bio-inoculants and bioremediation. Bacteria 4(3):28

Kurek MA, Onopiuk A, Pogorzelska-Nowicka E, Szpicer A, Zalewska M, Półtorak A (2022) Novel protein sources for applications in meat-alternative products—insight and challenges. Foods 11(7):957

Kurian T, Mathew N (2011) Natural rubber: production, properties and applications. In: Biopolymers: biomedical and environmental applications. Wiley, pp 403–436

Lal R (2020) Regenerative agriculture for food and climate. J Soil Water Conserv 75(5):123A–124A

Lefsih K, Iboukhoulef L, Petit E, Benouatas H, Pierre G, Delattre C (2018) Anti-inflammatory and antioxidant effect of a D-galactose-rich polysaccharide extracted from Aloe vera leaves. Adv Appl Chem Biochem 1:18–26

Lelamo LL (2021) A review on the indigenous multipurpose agroforestry tree species in Ethiopia: management, their productive and service roles and constraints. Heliyon 7(9):e07874

Letourneau DK (2012) Integrated pest management–outbreaks prevented, delayed, or facilitated? In: Insect outbreaks revisited. Wiley, pp 371–394

Li JW-H, Vederas JC (2009) Drug discovery and natural products: end of an era or an endless frontier? Science 325(5937):161–165

Liebgott T, Miollan M, Berchadsky Y, Drieu K, Culcasi M, Pietri S (2000) Complementary cardioprotective effects of flavonoid metabolites and terpenoid constituents of Ginkgo biloba extract (EGb 761) during ischemia and reperfusion. Basic Res Cardiol 95(5):368–377

Lin BB (2011) Resilience in agriculture through crop diversification: adaptive management for environmental change. Bioscience 61(3):183–193

Liu Y (2018) Chemical composition and characterization of cotton fibers. In: Cotton fiber: physics, chemistry and biology. Springer, pp 75–94

López-Goldar X, Lundborg L, Borg-Karlson AK, Zas R, Sampedro L (2020) Resin acids as inducible chemical defences of pine seedlings against chewing insects. PLoS One 15(5):e0232692

Loveday SM (2020) Plant protein ingredients with food functionality potential. Nutr Bull 45(3):321–327

Lusas EW, Riaz MN (1995) Soy protein products: processing and use. J Nutr 125:573S–580S

MacNeill GJ, Mehrpouyan S, Minow MA, Patterson JA, Tetlow IJ, Emes MJ (2017) Starch as a source, starch as a sink: the bifunctional role of starch in carbon allocation. J Exp Bot 68(16):4433–4453
Maity S, Chowdhury S, Datta AK (2012) Jute biology, diversity, cultivation, pest control, fiber production and genetics. In: Organic fertilisation, soil quality and human health. Springer, pp 227–262
Maiuolo J, Gliozzi M, Musolino V, Carresi C, Scarano F, Nucera S, Scicchitano M, Oppedisano F, Bosco F, Ruga S (2021) The contribution of gut microbiota–brain axis in the development of brain disorders. Front Neurosci 15:616883
Mamia P (2023) Along the clean label trend: identifying target customers of plant-based beverage snacks and their reasons against and for buying. Master's thesis
Mansourian S, Berrahmouni N (2021) Review of forest and landscape restoration in Africa 2021. Food & Agriculture Organization
Manzoor S, Rakha A, Rasheed H, Bhat ZF, Khan MSA, Abdi G, Aadil RM (2025) A comprehensive review on anxiolytic effect of Lavandula angustifolia Mill. in clinical studies. Food Sci Nutr 13(9):e70993
Marangoni AG, Van Duynhoven JP, Acevedo NC, Nicholson RA, Patel AR (2020) Advances in our understanding of the structure and functionality of edible fats and fat mimetics. Soft Matter 16(2):289–306
Maulana M, Marwanto M, Nawawi D, Nikmatin S, Febrianto F, Kim N (2020) Chemical components content of seven Indonesian bamboo species. IOP Conf Ser Mater Sci Eng 935:012028
McClements DJ, Decker E (2018) Interfacial antioxidants: a review of natural and synthetic emulsifiers and coemulsifiers that can inhibit lipid oxidation. J Agric Food Chem 66(1):20–35
McClements DJ, Grossmann L (2021) A brief review of the science behind the design of healthy and sustainable plant-based foods. NPJ Sci Food 5(1):17
McKenzie FC, Williams J (2015) Sustainable food production: constraints, challenges and choices by 2050. Food Secur 7(2):221–233
McMacken M, Shah S (2017) A plant-based diet for the prevention and treatment of type 2 diabetes. J Geriatr Cardiol 14(5):342
Medawar E, Huhn S, Villringer A, Veronica Witte A (2019) The effects of plant-based diets on the body and the brain: a systematic review. Transl Psychiatry 9(1):226
Mekonnen MM, Hoekstra AY (2012) A global assessment of the water footprint of farm animal products. Ecosystems 15(3):401–415
Melnyk N, Nyczka A, Piwowarski JP, Granica S (2024) Traditional use of chamomile flowers (Matricariae flos) in inflammatory-associated skin disorders. Prospects Pharm Sci 22(4):59–73
Menzi H, Oenema O, Burton C, Shipin O, Gerber P, Robinson T, Franceschini G (2010) Impacts of intensive livestock production and manure management on the environment. In: Livestock in a changing landscape, vol 1. Island Press, pp 139–163
Mešić A, Jurić M, Donsì F, Maslov Bandić L, Jurić S (2024) Advancing climate resilience: technological innovations in plant-based, alternative and sustainable food production systems. Discov Sustain 5(1):423
Mihrete TB, Mihretu FB (2025) Crop diversification for ensuring sustainable agriculture, risk management and food security. Global Chall 9(2):2400267
Mohammadinejad R, Karimi S, Iravani S, Varma RS (2016) Plant-derived nanostructures: types and applications. Green Chem 18(1):20–52
Najmi A, Javed SA, Al Bratty M, Alhazmi HA (2022) Modern approaches in the discovery and development of plant-based natural products and their analogues as potential therapeutic agents. Molecules 27(2):349
Nandhini S, Narayanan KB, Ilango K (2018) Valeriana officinalis: a review of its traditional uses, phytochemistry and pharmacology. Asian J Pharm Clin Res 11(1):36–41
Nelson SC, Ploetz RC, Kepler AK (2006) Musa species (banana and plantain). Species Profiles Pac Isl Agrofor 15(2):251–259

Nickerson R (1940) Cotton fibers constitution, structure, and mechanical properties. Ind Eng Chem 32(11):1454–1462

Nielsen AB, Thorsen-Meyer H-C, Belling K, Nielsen AP, Thomas CE, Chmura PJ, Lademann M, Moseley PL, Heimann M, Dybdahl L (2019) Survival prediction in intensive-care units based on aggregation of long-term disease history and acute physiology: a retrospective study of the Danish National Patient Registry and electronic patient records. Lancet Digital Health 1(2):e78–e89

Nogueira MS, Scolaro B, Milne GL, Castro IA (2019) Oxidation products from omega-3 and omega-6 fatty acids during a simulated shelf life of edible oils. LWT 101:113–122

Nowak D, Heisler G (2010) Air quality effects of urban trees and parks, Research series monograph. National Recreation and Parks Association Research Series Monograph, Ashburn, pp 1–44

Nyhan L, Sahin AW, Schmitz HH, Siegel JB, Arendt EK (2023) Brewers' spent grain: an unprecedented opportunity to develop sustainable plant-based nutrition ingredients addressing global malnutrition challenges. J Agric Food Chem 71(28):10543–10564

Oh E, Godoy Zúniga MM, Nguyen TB, Kim B-H, Trung Tien T, Suhr J (2024) Sustainable green composite materials in the next-generation mobility industry: review and prospective. Adv Compos Mater 33:1368

Pandey S (2007) Ramie fibre: part I. Chemical composition and chemical properties. A critical review of recent developments. Text Prog 39(1):1–66

Pathak R, Sharma H (2021) A review on medicinal uses of Cinnamomum verum (Cinnamon). J Drug Deliv Ther 11(6-S):161–166

Pauli N, Mouat C, Prendergast K, Chalmer L, Ramalho CE, Ligtermoet E (2021) The social and ecological values of native gardens along streets: a socio-ecological study in the suburbs of Perth. Clean Air and Urban Landscapes Hub

Peoples MB, Swan AD, Goward L, Kirkegaard JA, Hunt JR, Li GD, Schwenke GD, Herridge DF, Moodie M, Wilhelm N (2017) Soil mineral nitrogen benefits derived from legumes and comparisons of the apparent recovery of legume or fertiliser nitrogen by wheat. Soil Res 55(6):600–615

Pimentel D, Pimentel M (2003) Sustainability of meat-based and plant-based diets and the environment. Am J Clin Nutr 78(3):660S–663S

Poore J, Nemecek T (2018) Reducing food's environmental impacts through producers and consumers. Science 360(6392):987–992

Poornachandhra C, Jayabalakrishnan RM, Balasubramanian G, Lakshmanan A, Selvakumar S, Maheswari M, John JE (2023) Coconut husk fiber: a low-cost bioresource for the synthesis of high-value nanocellulose. Biointerface Res Appl Chem 13:504

Pournou A (2020) Wood anatomy, chemistry and physical properties. In: Biodeterioration of wooden cultural heritage: organisms and decay mechanisms in aquatic and terrestrial ecosystems. Springer, pp 1–41

Pradheep K, Joseph John K, Latha M, Suma A (2021) Status of crop plants of agricultural importance in Kerala state, India: an update. Genet Resour Crop Evol 68(5):1849–1873

Prasad S, Gupta E, Yadav S, Babulal KS, Mishra S (2025) Plant-based food industry: overview and trends. In: The future of plant protein: innovations, challenges, and opportunities. Springer, pp 73–97

Pretty J, Benton TG, Bharucha ZP, Dicks LV, Flora CB, Godfray HCJ, Goulson D, Hartley S, Lampkin N, Morris C (2018) Global assessment of agricultural system redesign for sustainable intensification. Nat Sustainability 1(8):441–446

Radoor S, Karayil J, Shivanna JM, Jayakumar A, Parameswaranpillai J, Siengchin S (2022) Cotton fibers, their composites and applications. In: Plant fibers, their composites, and applications. Elsevier, pp 379–390

Rahim MA, Ayub H, Sehrish A, Ambreen S, Khan FA, Itrat N, Nazir A, Shoukat A, Shoukat A, Ejaz A (2023) Essential components from plant source oils: a review on extraction, detection, identification, and quantification. Molecules 28(19):6881

Rahman MA, Rahman MN (2024) Climatic adaptation and sustainability of jute (Corchorus spp.). J Agric Sci Eng 6(2):80–95
Rais M, Acharya S, Sharma N (2013) Food processing industry in India: S&T capability, skills and employment opportunities. J Rural Dev 32:451–478
Rajendran S, Al-Samydai A, Palani G, Trilaksana H, Sathish T, Giri J, Saravanan R, Lalvani JIJ, Nasri F (2025) Replacement of petroleum based products with plant-based materials, green and sustainable energy—a review. Eng Rep 7(4):e70108
Ramsewak R, DeWitt D, Nair M (2000) Cytotoxicity, antioxidant and anti-inflammatory activities of curcumins I–III from Curcuma longa. Phytomedicine 7(4):303–308
Righini I, Stanghellini C, Hemming S, Graamans L, Marcelis LF (2023) Resources for plant-based food: estimating resource use to meet the requirements of urban and peri-urban diets. Food Energy Secur 12(3):e462
Rishton GM (2008) Natural products as a robust source of new drugs and drug leads: past successes and present day issues. Am J Cardiol 101(10):S43–S49
Rodriguez-Leyva D, Bassett CM, McCullough R, Pierce GN (2010) The cardiovascular effects of flaxseed and its omega-3 fatty acid, alpha-linolenic acid. Can J Cardiol 26(9):489–496
Roni P, Åberg U, Weber C (2018) A review of approaches for monitoring the effectiveness of regional river habitat restoration programs. N Am J Fish Manag 38(5):1170–1186
Roy A, Khan A, Ahmad I, Alghamdi S, Rajab BS, Babalghith AO, Alshahrani MY, Islam S, Islam MR (2022) Flavonoids a bioactive compound from medicinal plants and its therapeutic applications. Biomed Res Int 2022(1):5445291
Rusch F, Wastowski AD, de Lira TS, Moreira KCCSR, de Moraes Lúcio D (2023) Description of the component properties of species of bamboo: a review. Biomass Convers Biorefinery 13(3):2487–2495
Sadhukhan J, Dugmore TI, Matharu A, Martinez-Hernandez E, Aburto J, Rahman PK, Lynch J (2020) Perspectives on "game changer" global challenges for sustainable 21st century: plant-based diet, unavoidable food waste biorefining, and circular economy. Sustainability 12(5):1976
Sahoo N, Manchikanti P, Dey S (2010) Herbal drugs: standards and regulation. Fitoterapia 81(6):462–471
Sahu P, Gupta M (2017) Sisal (Agave sisalana) fibre and its polymer-based composites: a review on current developments. J Reinf Plast Compos 36(24):1759–1780
Said Z, Vigneshwaran P, Saboor S, Rauf A, Ahmad Z (2025) Climate and carbon policy pathways for sustainable food systems. Environ Sustain Indic 27:100730
Saini RK, Keum Y-S (2018) Omega-3 and omega-6 polyunsaturated fatty acids: dietary sources, metabolism, and significance—a review. Life Sci 203:255–267
Salleh KM, Armir NAZ, Mazlan NSN, Wang C, Zakaria S (2021) Cellulose and its derivatives in textiles: primitive application to current trend. In: Fundamentals of natural fibres and textiles. Elsevier, pp 33–63
Santo RE, Kim BF, Goldman SE, Dutkiewicz J, Biehl E, Bloem MW, Neff RA, Nachman KE (2020) Considering plant-based meat substitutes and cell-based meats: a public health and food systems perspective. Front Sustain Food Syst 4:569383
Satija A, Hu FB (2018) Plant-based diets and cardiovascular health. Trends Cardiovasc Med 28(7):437–441
Schaal B (2019) Plants and people: our shared history and future. Plants People Planet 1(1):14–19
Schmidt B, Ribnicky DM, Poulev A, Logendra S, Cefalu WT, Raskin I (2008) A natural history of botanical therapeutics. Metabolism 57:S3–S9
Seid D, Kassa H, Firdu D (2022) Impact of rubber tree plantations chronosequence on soil fertility and soil organic carbon stocks, Gurafarda District, Southwest Ethiopia
Selvakumar P, Seenivasan S, Patel S, Das A, Ravikumar S (2025) Environmental sustainability. In: Renewable energy and the economic welfare of society. IGI Global Scientific Publishing, pp 103–132

Shaffril HAM, Idris K, Sahharon H, Samah AA, Samah BA (2020) Adaptation towards climate change impacts among highland farmers in Malaysia. Environ Sci Pollut Res 27(20):25209–25219

Shaffril HAM, Samah AA, Samsuddin SF, Ahmad N, Tangang F, Sidique SFA, Rahman HA, Burhan NAS, Shah JA, Khalid NA (2024) Diversification of agriculture practices as a response to climate change impacts among farmers in low-income countries: a systematic literature review. Clim Serv 35:100508

Shaghaleh H, Xu X, Wang S (2018) Current progress in production of biopolymeric materials based on cellulose, cellulose nanofibers, and cellulose derivatives. RSC Adv 8(2):825–842

Shahidi F, Ambigaipalan P (2018) Omega-3 polyunsaturated fatty acids and their health benefits. Annu Rev Food Sci Technol 9(1):345–381

Sharma N, Yeasmen N, Dube L, Orsat V (2024) Rise of plant-based beverages: a consumer-driven perspective. Food Rev Int 40(10):3315–3341

Sharma A, Amin MM, Aziz MA, Siddiquee MN (2025) Production of high tensile strength bio-based carbon fibers: advances, challenges, and emerging applications. Chem Asian J 20:e00144

Sharna SC, Kamruzzaman M (2020) Comparative advantages of jute export in Bangladesh, China and India. Res Agric Livest Fish 7(2):183–190

Sher A, Li H, ullah A, Hamid Y, Nasir B, Zhang J (2024) Importance of regenerative agriculture: climate, soil health, biodiversity and its socioecological impact. Discov Sustain 5(1):462

Shuvo II (2020) Fibre attributes and mapping the cultivar influence of different industrial cellulosic crops (cotton, hemp, flax, and canola) on textile properties. Bioresour Bioprocess 7(1):51

Singh B (2023) New systems of vegetable production: protected cultivation, hydroponics, aeroponics, vertical, organic, microgreens. In: Vegetables for nutrition and entrepreneurship. Springer, pp 31–56

Singh NK, Baranwal J, Pati S, Barse B, Khan RH, Kumar A (2023) Application of plant products in the synthesis and functionalisation of biopolymers. Int J Biol Macromol 237:124174

Singh NR, Singh A, Devi NP, Kumar YB, Sangma RHC, Philanim W, Devi MP, Bhutia PL (2024) Agroforestry for soil health. In: Agroforestry. Wiley, pp 255–283

Smith P, Haberl H, Popp A, Erb K h, Lauk C, Harper R, Tubiello FN, de Siqueira Pinto A, Jafari M, Sohi S (2013) How much land-based greenhouse gas mitigation can be achieved without compromising food security and environmental goals? Glob Chang Biol 19(8):2285–2302

Smith MM, Bentrup G, Kellerman T, MacFarland K, Straight R, Ameyaw L, Stein S (2022) Silvopasture in the USA: a systematic review of natural resource professional and producer-reported benefits, challenges, and management activities. Agric Ecosyst Environ 326:107818

Smole MS, Hribernik S, Kleinschek KS, Kreže T (2013) Plant fibres for textile and technical applications. Adv Agrophys Res 10:52372

Snapp S, Bezner Kerr R, Ota V, Kane D, Shumba L, Dakishoni L (2019) Unpacking a crop diversity hotspot: farmer practice and preferences in Northern Malawi. Int J Agric Sustain 17(2):172–188

Tartaruga I, Sperotto F, Carvalho L (2024) Addressing inclusion, innovation, and sustainability challenges through the lens of economic geography: introducing the hierarchical regional innovation system. Geogr Sustainability 5(1):1–12

Teng P (2024) Urban agriculture and food security in development planning. Asian J Agric Rural Dev 21:61–76

Thapliyal D, Verma S, Sen P, Kumar R, Thakur A, Tiwari AK, Singh D, Verros GD, Arya RK (2023) Natural fibers composites: origin, importance, consumption pattern, and challenges. J Compos Sci 7(12):506

Thirumurugan D, Cholarajan A, Raja S, Vijayakumar R (2018) An introductory chapter: secondary metabolites. In: Secondary metabolites – sources and applications. IntechOpen, pp 1–13

Thomas S, Paul S, Pothan L, Deepa B (2011) Natural fibres: structure, properties and applications. In: Cellulose fibers: bio-and nano-polymer composites: green chemistry and technology. Springer, pp 3–42

Thyavihalli Girijappa YG, Mavinkere Rangappa S, Parameswaranpillai J, Siengchin S (2019) Natural fibers as sustainable and renewable resource for development of eco-friendly composites: a comprehensive review. Front Mater 6:226
Tilburt JC, Kaptchuk TJ (2008) Herbal medicine research and global health: an ethical analysis. Bull World Health Organ 86:594–599
Tilman D, Balzer C, Hill J, Befort BL (2011) Global food demand and the sustainable intensification of agriculture. Proc Natl Acad Sci 108(50):20260–20264
Tilman D, Clark M, Williams DR, Kimmel K, Polasky S, Packer C (2017) Future threats to biodiversity and pathways to their prevention. Nature 546(7656):73–81
Trakoontivakorn G, Nakahara K, Shinmoto H, Takenaka M, Onishi-Kameyama M, Ono H, Yoshida M, Nagata T, Tsushida T (2001) Structural analysis of a novel antimutagenic compound, 4-hydroxypanduratin A, and the antimutagenic activity of flavonoids in a Thai spice, fingerroot (Boesenbergia pandurata Schult.) against mutagenic heterocyclic amines. J Agric Food Chem 49(6):3046–3050
Tripathy DB, Mishra A (2016) Renewable plant-based raw materials for industry. Sustain Inorg Chem 353:156
Twine R (2021) Emissions from animal agriculture—16.5% is the new minimum figure. Sustainability 13(11):6276
Udawatta RP, Walter D, Jose S (2022) Carbon sequestration by forests and agroforests: a reality check for the United States. Carbon Footprint 2(1):2
Umaru IJ, Shuaibu SI, Adam RB, Habibu B, Umaru KI, Haruna D, David B (2020) Effect of herbal medicine and its biochemical implication. Int J Adv Biochem Res 4:46–57
Ussiri D, Lal R (2012) Soil emission of nitrous oxide and its mitigation. Springer Science & Business Media
Vajdi M, Karimi A, Tousi AZ, Hosseini B, Nikniaz Z, Farhangi MA (2023) Association between plant-based diets and metabolic syndrome in obese adults from Iran: a cross-sectional study. BMC Endocr Disord 23(1):109
Valle C, Voss M, Calcio Gaudino E, Forte C, Cravotto G, Tabasso S (2024) Harnessing agri-food waste as a source of biopolymers for agriculture. Appl Sci 14(10):4089
Vatistas C, Avgoustaki DD, Bartzanas T (2022) A systematic literature review on controlled-environment agriculture: how vertical farms and greenhouses can influence the sustainability and footprint of urban microclimate with local food production. Atmosphere 13(8):1258
Vinodhini S, Manibharathi S, Pavithra G, Sakthivel S (2023) Agroforestry: integrating trees into agricultural systems. Recent Approaches Agric 2:246–272
Viroli G, Kalmpourtzidou A, Cena H (2023) Exploring benefits and barriers of plant-based diets: health, environmental impact, food accessibility and acceptability. Nutrients 15(22):4723
Wachtel-Galor S, Benzie IF (2012) Herbal medicine: an introduction to its history, usage, regulation, current trends, and research needs. In: Herbal medicine: biomolecular and clinical aspects. CRC Press/Taylor & Francis
Wang J Jr (2023) Researching city-scale water resource improvement through rainwater: green roof in private realm. University of Victoria
Wightman JL, Woodbury PB (2020) New York Agriculture and Climate Change: key opportunities for mitigation, resilience, and adaptation. Final report on Carbon Farming Project for the New York State Department of Agriculture and Markets
Willett W, Rockström J, Loken B, Springmann M, Lang T, Vermeulen S, Garnett T, Tilman D, DeClerck F, Wood A (2019) Food in the Anthropocene: the EAT–Lancet Commission on healthy diets from sustainable food systems. Lancet 393(10170):447–492
Williams PA, Phillips GO (2021) Introduction to food hydrocolloids. In: Handbook of hydrocolloids. Elsevier, pp 3–26
Woldearegay M, Teso M (2022) Medicinal plants of Ethiopia: conservation, traditional knowledge, and sustainable use. Ecol Insights 4(1):1–18

Worku LA, Bachheti RK, Tadesse MG (2022) Isolation and characterization of natural cellulose from Oxytenanthera abyssinica (lowland Ethiopian bamboo) using alkali peroxide bleaching stages followed by aqueous chlorite in buffer solution. Int J Polym Sci 2022(1):5155552

Worku LA, Tadesse MG, Bachheti RK, Bachheti A, Husen A (2024) Synthesis of carboxylated cellulose nanocrystal/ZnO nanohybrids using Oxytenanthera abyssinica cellulose and zinc nitrate hexahydrate for radical scavenging, photocatalytic, and antibacterial activities. Int J Biol Macromol 267:131228

Wurglics M, Schubert-Zsilavecz M (2006) Hypericum perforatum: a "modern" herbal antidepressant: pharmacokinetics of active ingredients. Clin Pharmacokinet 45(5):449–468

Xu J, Tao A, Qi J, Wang Y (2020) Bast fibres: kenaf. In: Handbook of natural fibres. Elsevier, pp 71–92

Yang D, Lee Y-Y, Lu Y, Wang Y, Zhang Z (2024) Internal factors affecting the crystallization of the lipid system: triacylglycerol structure, composition, and minor components. Molecules 29(8):1847

Yatoo MI, Gopalakrishnan A, Saxena A, Parray OR, Tufani NA, Chakraborty S, Tiwari R, Dhama K, Iqbal HM (2018) Anti-inflammatory drugs and herbs with special emphasis on herbal medicines for countering inflammatory diseases and disorders – a review. Recent Patents Inflamm Allergy Drug Discov 12(1):39–58

Zhang T, Dou W, Zhang X, Zhao Y, Zhang Y, Jiang L, Sui X (2021) The development history and recent updates on soy protein-based meat alternatives. Trends Food Sci Technol 109:702–710

Zhang J, Zhang M, Chen K, Deng D (2025) Improvement strategies for fats and oils used in future food processing based on health orientation and sustainability: research progress, challenges and solutions. Crit Rev Food Sci Nutr 65(1):47–63

Chapter 2
Historical and Traditional Knowledge of Plant-Based Practices

Abdel Nasser B. Singab, Ahmed M. Elissawy, and Fadia S. Youssef

Abstract Over millennia, mankind has used plants as a primary source of healing for various ailments. Medicinal plants played a crucial role in various cultural, ecological, and economic aspects of life throughout the development of human civilization; moreover, different civilizations have developed diplomatic relationships based on medicinal plants. Practices in traditional medicine can be traced through various historical records, particularly from ancient civilizations such as the Egyptians, Chinese, and Indians. Greco-Roman and Islamic civilizations. These records not only reflect knowledge of healing plants but also describe their use, with the relevant philosophical and cultural background. Nowadays, amid the global call to "go back to nature" and the widespread resurgence of traditional medicinal practices, it is essential to provide an evidence-based rationale for using traditional medicinal plants. This book chapter provides an overview of traditional systems of medicine, shedding light on the plant-based practices associated with these systems, tracing their development alongside the relevant civilizations, and concluding with the recent development of herbal medicine, grounded in evidence and scientific evaluation of traditional practices.

Keywords Medicinal plants · Traditional medicine · Plant-based practices · Ancient Civilizations · Evidence-based

---

A. N. B. Singab (✉) · A. M. Elissawy
Department of Pharmacognosy, Faculty of Pharmacy, Ain Shams University, Cairo, Egypt

Center for Drug Discovery Research and Development, Faculty of Pharmacy, Ain Shams University, Cairo, Egypt
e-mail: abdelnassersingab@pharma.asu.edu.eg; vpr.nassersingab@asu.edu.eg

F. S. Youssef
Department of Pharmacognosy, Faculty of Pharmacy, Ain Shams University, Cairo, Egypt

L. A. Worku et al. (eds.), *Plant-Based Solutions for Sustainable Agriculture and Healthcare*, Sustainability Sciences in Asia and Africa,
https://doi.org/10.1007/978-981-95-8910-4_2

2.1 Introduction

Survival strategies involve securing the basic needs required for human life; these include food, water, shelter, and healing resources. The search for curative sources was not limited to mankind; animals, for example, utilize medicinal plants for healing purposes. This instinct is the main driving force for survival; in other words, the utilization of healing herbs in very ancient eras was not based on any rationale and was merely dependent on trial and error and observation (Huffman 2022; Stojanoski 1999; Petrovska 2012).

The use of herbal medicine is probably as old as the existence of humanity on earth; continuous archeological discoveries provide solid evidence for this fact. One of these discoveries revealed the presence of remains of medicinal plants dating back to 60,000 years in Neanderthal tombs in Iraq, including *Ephedra, Althea, and Achillea*. This evidence proves that the use of herbal medicines dates back even further than that time (Solecki 1975; Jamshidi-Kia and Lorigooini 2018; Bachheti et al. 2022).

Over time, mankind developed independent practices for healing using herbal remedies, bridged with metaphysical rationalization of disease and healing. This is evident in the presence of specialized practitioners who use healing herbs with rituals, spells, and practices that blend the medicinal value of herbal remedies with metaphysical beliefs (Petrovska 2012; Elendu 2024; Abate et al. 2022).

With the emergence of writing, different civilizations began to record their experiences related to disease and healing in medical recipes. One of the oldest records was discovered near Nagpur and dates to 5000 years ago, written in Sumerian. The record documents the systemic use of herbal remedies, various dosage forms, and complex medications. The record includes 12 recipes using about 250 herbs, including commonly known drugs such as poppy and henbane (Khan 2014; Bhajbhuje 2013).

The different ancient civilizations in Egypt, China, Iraq, Persia, and India established their medical systems based on the inherent cultures of each, resulting in characteristic systems that reflected cultural, religious, and political variations. Nevertheless, a common aspect across the different ancient systems was the reliance on medicinal plants as the primary healing component, though they were used with practices that underpinned different treatment philosophies (Halberstein 2005; Khan 2014; Elendu 2024).

With the development of human knowledge and the emergence of new civilizations, dramatic changes had occurred in medical practices, still relying mainly on herbal remedies, until the late eighteenth and beginning of the nineteenth century, with the advances in chemical sciences, when there was a trend towards the utilization of isolated chemical compounds, still mainly obtained from herbal extracts. By the end of the nineteenth century and the beginning of the twentieth century, with great advances in synthetic chemistry, pharmacological and biological sciences, a major shift towards the utilization of synthetic drugs in contrast to herbal remedies had occurred, although it is estimated that more than 50% of synthetic drugs are derived from or inspired by natural drugs (Zunic et al. 2017; Mathur and Hoskins 2017; Christensen 2021).

2.2 Tribal Medical Practices

From the earliest beginnings, current humans (*Homo sapiens*) have populated the Earth for about 250,000–300,000 years; however, written history and the estimated presence of civilization (agrarian civilization) date back only 5000–6000 years. For about 98% of human history, humans lived as hunters or gatherers in tribal societies (Hublin et al. 2017; Stearns and Langer 2001). The tribal societies still live until now around the world, and observation of medical practices among these tribes can give an insight into the medical practices present thousands of years ago; being closed societies, the tribes depend on the knowledge transmitted from generation to generation spanning tens of thousands of years. Of course, some of the ancient practices had been filtered out due to trial and error experiences; however, the fundamental aspects of herbalism are preserved (Shankar et al. 2012; Alum 2024; Bachheti et al. 2021).

Recent findings reported the discovery of an unknown Neanderthal caveman near Iraq who lived some 60,000 years ago. The findings confirmed the presence of pollen grains around the caveman belonging to eight different plants that flourish in this area, including yarrow, marshmallow, and ephedra, giving insights into the use of these medicinal plants for healing purposes at that time (Solecki 1975). Another preserved Neanderthal was found and examined, revealing the presence of various herbal drugs in his stomach (Shipley and Kindscher 2016).

Another interesting finding is the discovery of the mummified hunter Iceman in the Alps mountains, dating back to about 5500 years ago. The body was preserved excellently so that clear, deep examinations of the internal organs were possible. One of the interesting findings was the presence of birch polypore, bracken fern, and different types of bee pollen in his stomach concurrently with confirmed evidence of his infection with *Helicobacter pylori* and parasitic infestation in his gastrointestinal tract (GIT). Knowing that the found herbal drugs in his gut are related to the treatment of GIT disorders, such as diarrhea, bacterial and viral infections, and parasitic infestations, provides evidence that ancient tribes knew the disease and knew how to treat it. Moreover, the body of the Iceman was marked with ink tattoos in special areas that shadow the locations of the Chinese acupuncture, giving evidence of the application of acupuncture techniques at that time in a primitive way by applying ink tattoos (Maixner and Krause-Kyora 2016; Zink and Maixner 2019; Capasso 1998).

It is clear now that tribal medical practices were mainly practiced by a special group of people within the tribe who had the knowledge and experience for healing and had "spiritual" powers that enabled them to communicate with hidden powers or spirits and, in some cases, could communicate with the healing herbs. This knowledge and these powers were transmitted orally from generation to generation, and in many cases, the practitioner had to comply with a rigorous "training program." Close observation of tribal medicine nowadays reveals little difference from ancient tribal practices, as we mentioned earlier, reflecting the closed nature of tribes and their rigorous cultural preservation.

2.2.1 *Main Aspects of Tribal Medical Practices*

Despite intracultural variations, there are main aspects of tribal medical practices that outline the framework of these ancient practices (Mutatkar 2021; Elendu 2024; Shankar et al. 2012); the main aspects include:

Nomenclature: This process is essential to build trust bridges between the practitioner and the patient, giving precise names of the underlying conditions and the precise names of the healing herbs utilized. Names of the procedures usually imply the feeling of trust, at least psychologically revealing the knowledge of the practitioner and his ability to identify, diagnose, and treat the source of illness; nevertheless, this is actually different from our modern perception of illness, causes of disease, and healing.

The practitioner's personality includes the practitioner's holy image in the patient's eyes, as well as personal features such as wisdom, knowledge, caring, sincerity, and prestige.

Raising patient expectations: This feature has a significant impact on the psychological side in dealing with hopeless and desperate patients; using some words, procedures, unknown spells, and rattles, among others, is an essential part of healing practice in tribal medicine, along with building a reputation for the practitioner. For example, the patient may travel or recall certain practitioners hundreds of miles apart based on this reputation and patient expectations; astonishingly, this is usually associated with greater chances for healing.

Rigorous training: The process of learning and training is an essential component in building the expertise and reputation of the healer, usually over a period of several years. The learning process is entirely oral and primarily relies on observation and a deep connection with expert practitioners.

The major aspects of tribal medicine are not far from modern medical practices in outline; however, it differs significantly in its scientific, philosophical, and rational foundations. In the following sections, we will dive deeper into the traditional systems of medicine (Fig. 2.1), which shadowed the rise of agrarian civilizations, during which medical practices became more organized and systematic yet remained deeply connected to cultural, philosophical, and religious backgrounds.

2.3 Ancient Traditional Systems of Medicine

2.3.1 *Ancient Egyptian Medicine*

Egyptian civilization is considered one of the oldest civilizations, marked by the presence of a unified, strong national government established by Menes at 3100 B.C.E. with the beginning of the dynastic period, with the first dynasty (the Old Kingdom) starting at 2686 B.C.E. and lasting for about 500 years, marked by the

Fig. 2.1 Ancient traditional systems of medicine

building of great pyramids. Medical practices were an integral part of the ancient Egyptian civilization (3100–362 B.C.E.). The main characteristic of the medical practices is the merging between magico-religious practices and the empirical observation practices that apply logic and rationale; as a result, two types of practi tioners were present: the shamanic healers "*Sau*" and the physicians "swnw." In all cases, special spells and rituals were performed with the prescribed therapy (Kemp 2007; Nunn 2006; Metwaly et al. 2021).

As early as the first dynasty, there were specialized medical schools called "Peri-Ankh," or "the house of life," such as that established by the great scientist, philosopher, physician, and engineer Imhotep, who later became a symbolic god for healing after his death (Garland 1928). Most of the medical practices in ancient Egypt were discovered in specialized medical texts written on papyrus, "medical papyrus." To date, only a limited number of medical papyri have been discovered and translated; however, they contain a wealth of information about the Egyptian medical system. Some of these papyri refer to practices much older than their date of issue, such as the Ebres Papyrus, shedding light on medical practices dating back up to 3000 years B.C.E. (Leake 1953).

There are more than 40 medical papyri; however, eight important papyri address various medical conditions, some of which are specialized in particular branches of medicine. This includes *Kahun* (1900 B.C.E.), *Edwin-Smith* (1600 B.C.E.), *Ebers* (1550 B.C.E.), *Hearst* (1550 B.C.E.), *Erman* (1550 B.C.E.), *London* (1350 B.C.E.), and *Chester Beatty* (1200 B.C.E.) (Leake 1953; Pommerening 2012). The most comprehensive papyrus from a pharmaceutical perspective is the Ebers Papyrus, which describes over 800 herbal formulations and more than 400 herbal drugs targeting internal diseases, as well as ophthalmology, dermatology, gynecology, and orthopedics. It includes some surgical procedures (Hartsock and Halverson 2023). The Kahun papyrus, on the other hand, is mainly focused on gynecology/obstetrics, describing some procedures for the determination of fertility, such as the insertion of a garlic clove in the vagina of the woman; if she feels the taste of garlic on her tongue, then she is fertile. Another example includes the determination of the sex of the fetus by applying the urine of the pregnant woman on barley and wheat seedlings simultaneously, which grows first indicated the sex of the fetus (Smith 2011). The Edwin-Smith Papyrus, in contrast, specialized in standard surgical procedures, describing about 50 cases, along with diagnoses, prognoses, and treatment strategies (Stiefel et al. 2006). Other papyri, such as the Erman Papyrus, specialized in pediatrics, and the Beatty Papyrus in anorectal disorders. Inspection of these medical papyri reveals a coherent, consistent, and developing medical system over thousands of years (Leake 1953).

Diagnosis was based on observable signs and symptoms, such as pulse quality, skin color, consistency of the feces, urine frequency, sweating, and fever. In the clear cases such as wounds, delivery, constipation, diarrhea, the diagnosis relied mainly on logic and rationale "naturalistic mechanisms," whereas other mysterious cases, such as sudden fever caused by infection, were attributed to "supranatural mechanisms" such as anger of the gods and the demonic influence (Metwaly et al. 2021; David 2000; Abate et al. 2021). Also, ancient Egyptian physicians were renowned for their credibility, as they could describe the prognosis of the disease and inform the patient whether they could treat the condition (Leake 1953; David 2000). The therapeutic strategies relied mainly on herbal medicine, a balanced diet, and lifestyle modifications. The main goal was to restore balance and flow; most probably, this was inferred from the surrounding environment, where the river Nile represented the gift for living, and must be flowing continuously, and should be clean. Applying this to the human body, all the tracts inside the body should be free of blockages (Metwaly et al. 2021; Manniche 1989).

The Egyptians used a multitude of herbal remedies, many from the surrounding environment and some imported from nearby Nubia, Somalia, Ethiopia, and Syria. The Materia medica involves the use of plants such as garlic and onion, fennel, aloe, linseed, moringa, Willow bark, castor oil, and senna pods; resins such as myrrh and frankincense; gums such as acacia gum; vehicles such as honey and beer, and minerals such as gypsum and malachite (Metwaly et al. 2021). They not only prescribed these remedies but also described the proper dose and the toxic dose, invented special dosing formulas and equipment, and specified dosing intervals and the duration of treatment (Omar 2022; Rosso 2019; Scarborough 2006).

2.3.2 *Traditional Chinese Medicine (TCM)*

Medical practices in China have developed over the course of prehistoric eras. They had survived as the main medical system until the present, reflecting the preservation of ancient culture from generation to generation and the application of recent medical, biological, and chemical advances to validate the utility of different traditional practices, resulting in a comprehensive, coherent, and validated clinical system that relies mainly on the practice of herbal medicine. Chinese Herbal Medicine (CHM) is another term for traditional Chinese medical practices that rely solely on herbal drugs. TCM can be regarded as the most successful system of traditional medicine, having maintained its core framework for thousands of years (Gaur 2024).

The Shang Dynasty (1600 B.C.E.) is thought to be the first true Chinese dynasty marked by the development of Chinese writings, according to the beliefs of the Chinese at that time. The leading cause of illness was mainly magico-religious and represented by the offended relationship between the patient and one of his ancestors, and the main role of the healer/practitioner was to unveil this offended relationship through the "oracle." According to this procedure, a question was written on bones or tortoise shells and then burned exhaustively, followed by the observation of the resulting cracks and their thorough interpretation, followed by presenting a sacrifice to the ancestor. Of course, this was accompanied by the administration of certain medicinal herbs or other primitive medical interventions; however, the main reason for healing was thought to be the introduction of a sacrifice to the ancestor. This was clear through the preventive measures taken at that time to periodically present sacrifices to the ancestors and the division of the resources between the living and the dead (Unschuld 1985; Wang et al. 2024).

During the Zhou dynasty (1040 B.C.E.), the use of oracles from burned bones and tortoise shells had greatly declined and was replaced by the *I Ching*, "one of the oldest Chinese texts." The *I Ching* contained 64 hexagrams that had to be consulted by the practitioner to get the oracle and define the imbalance underlying the medical condition. The hexagrams were based mainly on the concept of *Yin and Yang*, the two opposing yet interdependent balances of life (Zeng 2023). The concepts of Yin and Yang represent the foundation of TCM, and they became clearly apparent in later periods of the Zhou Dynasty. The main cause of illness, from the Zhou point of view, was the influence of harmful demons; in contrast, the Shang dynasty regarded it as an offended ancestor soul. Medical practices involved the use of medicinal herbs by specialized priests concurrently with special spells to get rid of the harmful demons; later on, in the sixth century B.C.E., 13 points in the human body were identified as the "hearts of the demons," and puncturing these points with simultaneous spells for each illness was essential for healing (Reid 1996; Ma et al. 2021; Cai and Zhen 2003).

The Han Dynasty (206 B.C.E.–220 C.E.) established a systematic organization of earlier practices in Chinese medicine, laying the foundations for the Chinese medical system to date. One of the important texts marking the medical developments in this dynasty is "Yellow Emperor's Inner Classic," and "Treatise on Cold Damage and Miscellaneous Illnesses" consolidated the major philosophy of Traditional Chinese medicine, including the "*Qi*" as the essential life force or the essence of life that flows and transfers inside the body through meridians and the importance of keeping vital "*Qi*" in a good state through breathing exercises, a balanced diet, and sometimes medicinal herbs to ensure longevity and a good health state; Yin and Yang, the two balanced yet opposing forces, with Yin representing the negative, dark, and cold side, whereas Yang represents the positive, white, and hot side; and the five elements of life (wood, earth, fire, water, and metal) and their connection to the vital organs of the body. The causes of illness were described clearly through these texts as being mainly a deficiency of flowing energy "Qi" or disturbance in external forces "the six excesses" (cold, heat, fire, wind, dryness, and moisture) or internal forces marked by the different emotions of the patient, emphasizing the importance of psychological and environmental components (Curran 2008; Daidoji 2013; Bensky et al. 2004; Furth et al. 2021; Unschuld 1985).

Another important text issued in the Han Dynasty, "Divine Farmer's Classic of Materia Medica," systematically categorizes about 360 medicinal herbs and describes their pharmacological effects in terms of the five tastes (salty, acrid, bitter, sweet, and pungent) and the link between each taste with certain functions/organs, as well as the classification of these tastes into Yin and Yang and the interrelation with the five elements mentioned above (Zhao et al. 2018; Rogaski 2001).

The post-Han to Tang Dynasty (third–ninth century C.E.) was marked by the collection and organization of foundational works during the Han Dynasty and the development of a more robust and organized medical system; this era was marked by the emergence of the first state-sponsored Materia Medica, *Xinxiu Bencao*, written by Su Jing, expanding the pharmacopeia to include more than 850 medicinal herbs (Buell 2022). Another pioneer of TCM, Sun Simiao, was active during this era, compiling more than 5000 herbal formulations and is considered "the king of medicine" to this day (Rogaski 2001).

2.3.3 Ayurvedic Medicine

Ayurvedic medicine is a traditional system of medicine that originated in India between 2500 B.C. and 50 B.C. and was widely used in India and Nepal (Subhose et al. 2005). In literature, Ayurveda can be defined as "the science of life," in which the ancient Indian health care system sheds light on human beings and their diseases and claims that positive health is characterized by a well-balanced metabolism. Meanwhile, "science of longevity" is another definition of Ayurveda, as this system offers a comprehensive approach to a healthy, long life. Besides, its provided programs help revitalize the body through healthy, well-balanced nutrition and diet and offer many treatment strategies to alleviate well-known disorders, including food allergies, which are treated with advanced techniques (Pandey et al. 2013).

Meanwhile, Ayurveda is not magic but an interactive system that requires the full participation of patients, who must be responsible and self-empowered to achieve its goals in a friendly and successful manner (Mehl-Madrona 2007). Many therapies have evolved in Ayurvedic medicine for 2000 years or more, including herbal products, meditation, special diets, massage, enemas, laxatives, and essential oils. Moreover, its preparations depend on complex herbal products, metals, and mineral substances, in addition to the teaching and adoption of various surgical strategies, such as cataract surgery, lithotomy, rhinoplasty, sutures, and the extraction of foreign bodies (Leslie 1999; Mukhopādhyāya 1987).

At the beginning, Ayurveda divided medicine into two branches, surgeons and physicians, but later it was expanded to include eight branches, which characterized the physician's art, as referred to in Sanskrit works on Ayurveda such as Ashtanga Hridaya, Charaka Samhita, and Sushruta Samhita (Kapur 2015). These eight branches are Kayachikitsa, internal or general medicine of the body; Shalya Tantra, which includes surgery and foreign object extraction; Shalakya Tantra, which alle viates disorders invading upper body openings such as the nose, eyes, mouth, and ears; Bhutavidya, which is psychological and spiritual medicine; Kaumarabhratya, which includes pediatrics, obstetrics, and gynecology where it discusses baby care, either prenatal or postnatal, in addition to conception methods, diseases in childhood, and a child's sex and intelligence; Agada Tantra, which includes toxicology and also clarifies animal, minerals, and vegetables toxins and epidemics in addition to providing clues for toxicity recognition and their antidotes; Rasayana Tantra, which deals with geriatrics and provides tonics for elevating strength and lifespan; and finally, Vajikaranta Tantra, which deals with eugenics and aphrodisiacs and provides treatments for infertility problems and for enhancing viability and volume of semen as well as sexual pleasure and spiritual development (Kapur 2016).

Ayurvedic medicine postulated that the human body is composed mainly of *dhatus* (tissues), *malas* (waste products), and *dashas* (humoral biomaterials). *There* are seven dhatus: *rakta* (blood), *rasa* (chyle), *meda* (fat), *māmsa* (muscles), *majja* (marrow), *asthi* (bones), and *shukra* (semen). Meanwhile, *dashas* are postulated to be three: (1) *vata*, the air, which is equated with the nervous system by modern authors; (2) *kapha*, phlegm, or earth and water, which are equated with mucus by modern authors; and (3) *pitta*, the bile, or fire, which is equated with enzymes by modern authors. However, *malas* are composed of urine, sweat, and feces. Ayurveda emphasizes the importance of balance and equilibrium (*Samyata*) among *dhatus*, *malas*, and *dashas* for maintaining health and wellness. Basically, *Samyata* is achieved through various lifestyle factors, including *Ahara* (diet), *Manas* (mental wellbeing), and *Dinacharya* (daily routine) (Ali et al. 2025).

Plant-based preparations constitute 90% of ayurvedic remedies that are effective in healing. Meanwhile, the consumption of inaccurately identified substances was greatly threatened in traditional Ayurvedic texts. The substances used should be fully identified, including nomenclature, properties, and applications. Besides, it stated that the misuse of medicines can make useful medicinal plants harmful when improperly used, and vice versa (Rammanohar 2012; Husen 2023). Ayurvedic plant-based preparations were mainly obtained from fruits, leaves, seeds, roots, and barks such as cardamom and cinnamon. William Dymock and his colleagues summarized the well-known Ayurvedic herbal products, including their chemical composition, uses, toxicology, and microscopical constitution, as well as their reported stories, myths, and commercial relationships, in their book *Pharmacographia Indica* (Dymock et al. 1893).

Among these herbal formulations is Triphala, which is composed of three fruits: *Emblica officinalis* (*Amalaki*), *Terminalia bellerica* (*Bibhitaki*), and *Terminalia chebula* (*Haritaki*). It serves as a potent remedy for gastrointestinal tract disorders. Additionally, Triphala was found to be effective in appetite enhancement, hyperacidity reduction, and dental caries prevention, in addition to exerting antimutagenic, antibacterial, anti-inflammatory, adaptogenic, antioxidant, hypoglycemic, and radioprotective effects (Peterson et al. 2017). Furthermore, tulsi (*Ocimum tenuiflorum*) and ashwagandha (Withania somnifera) are widely used herbs in Ayurvedic medicine (Kumar et al. 2017; Mishra et al. 2000). Moreover, *Madya* (alcoholic beverages) was widely used in Ayurvedic medicine, as it is well known for regulating *dashas* by elevating *pitta* and reducing both *kapha* and *vatta.* Alcoholic beverages are categorized based on raw material and fermentation process into several classes: cereal-based, sugar-based, cereal-based with herbs, fruit-based, vinegar-fermented products, and tonic wines (Sekar 2007). Besides, alcoholic beverages were considered medicinal in Ayurveda, manifested by enhancing taste and digestion, as well as creating dryness and causing a purgative effect. *Madya* was supposed to act quickly, entering and clarifying small body cavities (McGovern 2019).

Ayurvedic medicine highly utilizes purified opium, and it is incorporated in eight of its preparations, causing a notable balance in *kapha* and *vata* doshas while elevating *pitta dosha.* It is highly effective in counteracting dysentery and diarrhea and in enhancing muscular and sexual capabilities. Regarding the brain, purified opium-containing Ayurvedic preparations exhibited pain-relieving and sedative effects (Chopra and Chopra 1955). Many Ayurvedic texts, such as *Sarngadhara Samhita* (1300–1400 C.E.) and the book *Yoga Ratnakara* (1700–1800 C.E., unknown author), mentioned the benefits of opium, where the former reported its effect as an aphrodisiac agent to retard male ejaculation. In contrast, the latter manifested its role in relieving diarrhea in its herbal-mineral preparation. In addition, *Bhaisajya Ratnavali* reported the use of purified opium together with camphor in the treatment of acute gastroenteritis, where opium's depressant effect on respiration is contrasted by camphor's respiratory stimulant effect (Chaturvedi et al. 1981).

Besides, *Cannabis indica* was also popular in Ayurveda and was listed in *Sarngadhara Samhita* for its potential to treat diarrhea, whereas in *Bhaisajya Ratnavali*, *Cannabis. Indica* was reported as a component in the aphrodisiac preparations (Gupta et al. 2020). Additionally, medicinal oils and tar were widely used in Ayurveda for bleeding stoppage, as they were supposed to exert their actions via blood vessel ligation, facilitating blood clotting, heat cauterization, and blood vessel constriction. Furthermore, massage with oils is widely used by ayurvedic practitioners, including multiple methods such as head massage, regular consumption, smearing, anointing, and oil pulling, or pouring oil on the patient's forehead via shirodhara (Cacho and Lum 2021; Ratan 2020).

2.3.4 *Greco-Roman, Islamic, and Unani Medicine*

Hippocrates (460–379 B.C.E.), or more precisely the Hippocratic school of medicine, established the early foundations of Western medicine until the emergence of modern medical practices by the beginning of the eighteenth century. It is not surprising that this Greek physician has been regarded as "the father of medicine"

(Kleisiaris et al. 2014). The early works of the great Greek physicians, starting with Hippocrates, also had a great impact on the development of Islamic medicine (Karenberg and Hort 1998). In brief, ancient Greek medical practices are based on the natural healing power of life and primarily depend on a balanced diet, a healthy environment, the psychological aspects of patients, and a healthy lifestyle (Kleisiaris et al. 2014). The works of Hippocrates demonstrated the filtration of magico-religious aspects of other traditional medical systems at the time and the shift towards a more rational and scientific understanding of medical conditions (Lloyd 1978). These practices were regarded as evolutionary at the time; however, the Hippocratic writings were less concerned with documenting the herbal medicines used and more with the overall healthcare system. Hippocrates thought of illness as an imbalance between the physical, emotional, and mental states of the body, and the practitioner should target the natural healing powers of the patients through the understanding of the "Four Humors" and restore the balance between these four humors, viz., the blood, yellow bile, black bile, and phlegm (Bujalkova et al. 2001; Kleisiaris et al. 2014).

Later, the following Greek physicians showed great interest in documenting herbal remedies, describing their names, morphology, uses, and doses. Starting with Theophrastus (327–285 B.C.E.), who described 500 medicinal plants in *Historia Plantarum* (Scarborough 1978; Einarson 1976). Dioscorides (first century C.E.) described more than 600 medicinal plants in his great five-volume work "*De Materia Medica*," benefiting from his journeys with the Roman army as a field physician, gathering information about different medicinal plants during his journeys at different localities (Scarborough 1996; Yarnell and Touwaide 2019).

Galen (129–216 C.E.), a genius physician, made tremendous changes in medical practice and education, and his works served as the basis of Western medicine for about 1500 years until the Western Renaissance (Hajar 2021). Galen was a pioneer in human anatomy, physiology, surgery, and therapeutics. In his work "*De Simplicibus*," Galen classified medicinal plants and described their relation to the humoral system established by Hippocrates; moreover, he categorized each class of medicinal plants into different grades according to the potency of their effects for each condition (Ventura 2017). Galenical was named in honor of Galen, who developed the use of medicinal plant extracts, for example, infusions, decoctions, and alcoholic extracts. Moreover, Galen relied mainly on polyherbal formulations prepared from various medicinal plant extracts rather than on simple dried herbs (Hajar 2021).

The term "Unani" is an Arabic translation of the word "Greek"; this term implies that the main framework of medical practices during the Islamic era was based on translations, refining, and modifications of the Greek, or more precisely, the Greco-Roman works (Alrawi and Fetters 2012). Actually, the term "Islamic" medicine includes two main branches. The first branch arises from traditional practices from the Bedouin people living in the deserts of the Middle East and North Africa that had been developed by the Muslims since the seventh century C.E., applying medicinal plants from the surrounding environment but greatly influenced by the magico-religious practices, for example, prayers from the holy book *the Quran*. Some prayers may be written on pieces of paper or leather and then macerated in water and drunk together with the administration of healing herbs. Also, “the Quran” provides some recommendations for certain foods such as olive, honey, and figs that promote health or certain behaviors such as prohibiting alcohol intake and promoting fasting periods; however, specific phrases from the prophet “Mohamed” (Hadeeth) had much information about specific medical conditions and their treatment using the available medicinal plants in their environment (Saad et al. 2005; Alrawi and Fetters 2012).

The second branch of “Islamic” medicine is what we call “Unani” medicine, which had been nourished mainly starting from the Abbasid dynasty, where the Abbasid Khalifahs, “kings,” had encouraged a massive wave of translation of the Greco-Roman, Indian, and Persian works in all fields such as basic sciences, chemistry, biology, philosophy, and medicine. The greater impact in the medical field was the translation of the great works of Greek physicians and philosophers, such as Hippocrates, Galen, Aristotle, and Socrates. Actually, this was a revival of these great works, recognizing that after the decline of the Greco-Roman empires, they had been neglected and replaced by theological practices in the western part of the world (Brentjes and Morrison 2010).

One of the great practitioners of Unani medicine was Abu Bakr El-Razi (Rhazes), who had included most of his medical and therapeutic experiences in the work “El-hawii fi al-Tebb,” or the “Comprehensive Book of Medicine,” and who gained a big reputation in the medical schools in the Middle East and Europe for centuries (Tibi 2006). Despite the great works and developments of the Muslim physicians and practitioners, no one had gained the reputation of the great physician Ibn Sina (Avicenna). Ibn Sina made significant advances in the art of medicine, and his great work, *Qanun al* Tibb (Canon of Medicine), was the basic medical text taught in Middle Eastern and European medical schools. He had described more than 750 medicinal plants and described their use as simple drugs and complex drugs (polyherbal); moreover, he had extensively relied on the rationale and scientific methods for testing the medicinal plants on animals and small groups of humans before the widespread use of the drugs (Amr and Tbakhi 2007).

The apothecaries (pharmacies) were also a cornerstone of Islamic Unani medicine, where significant developments occurred, including herbal drugs from various regions such as India, Persia, China, Europe, and North Africa; many of these herbs were introduced to the European market primarily through these apothecaries. Moreover, they were pioneers in the field of pharmaceutical formulations, using equipment such as alembics and mortar-and-pestle sets, as well as techniques such as distillation and volatile oil extraction (Skrbo and Masic 2017; Hajar 2013).

2.4 Conclusion

Herbal medicine consumption is probably as old as humanity on Earth, as continuous archaeological discoveries provide solid evidence for this. Close observation of tribal medicine nowadays reveals little difference from ancient tribal practices, with the main aspects of tribal medical practice focusing on nomenclature, the practitioner's personality, raising patient expectations, and rigorous training. The ancient civilizations of Egypt, China, Iraq, Persia, and India established their medical systems based on the distinct cultures of each, resulting in distinct systems that reflect cultural, religious, and political variations. Ancient traditional systems of medicine include ancient Egyptian medicine, traditional Chinese medicine (TCM), Ayurvedic medicine, Greco-Roman medicine, Islamic medicine, and Unani medicine. In ancient Egyptian medicine, most medical practices were documented in specialized medical texts written on papyrus, also known as "medical papyrus." There are eight important medical papyri that address various medical conditions; some are specialized in particular branches of medicine. The most comprehensive papyrus from a pharmaceutical perspective is the Ebers Papyrus, which describes over 800 herbal formulations and more than 400 herbal drugs targeting internal diseases. Meanwhile, traditional Chinese medicine (TCM) can be regarded as the most successful system of traditional medicine, having maintained its core framework for thousands of years. It is composed of several dynasties: the Shang, Zhou, and Han Dynasties. The post-Han to Tang Dynasty (3rd–ninth century C.E.) marked the collection and organization of foundational works from the Han Dynasty and the development of a more robust, organized medical system. Ayurvedic medicine is a traditional system of medicine originating in India and widely used there and in Nepal. Meanwhile, Ayurveda is not magic but an interactive system that requires the full participation of patients, who must be responsible and self empowered to achieve its goals in a friendly and successful manner. Plant-based preparations constitute 90% of ayurvedic remedies effective in healing, such as Triphala, Tulsi, Ashwagandha, purified opium, and *Cannabis indica*. Regarding Greco-Roman, Islamic, and Unani medicine, many prominent physicians made significant contributions, including Hippocrates, Galen, Abu Bakr El-Razi, and Ibn Sina. The works of Hippocrates demonstrated the filtration of magico-religious aspects of other traditional medical systems at the time and the shift towards a more rational and scientific understanding of medical conditions. Galen was a genius physician who made tremendous changes in medical practice and education, and his works served as the basis of Western medicine for about 1500 years until the Western Renaissance. One of the great practitioners of Unani medicine was Abu Bakr El-Razi (Rhazes), who had included most of his medical and therapeutic experiences in the work "El-hawii fi al-Tebb," or the "Comprehensive Book of Medicine," which gained a big reputation in the medical schools in the Middle East and Europe for centuries. However, the great Muslim physician Ibn Sina (Avicenna) achieved fame for his extensive contributions to medicine, and his great work *Qanun al* Tibb (Canon of Medicine) is the introductory medical text taught in Middle Eastern and European medical schools. He had described more than 750 medicinal plants and described their use as simple drugs and complex drugs (poly-herbal); moreover, he had extensively relied on the rationale. Nowadays, the great work performed by the traditional practitioners and physicians is being strengthened and consolidated by modern and advanced research.

References

Abate LA, Bachheti RK, Bachheti A, Husen MG, Pandey DP (2021) Potential role of forest-based plants in essential oil production: an approach to cosmetic and personal health care applications. In: Husen A, Bachheti RK, Bachheti A (eds) Non-timber forest products: food, healthcare and industrial applications. Springer, Cham, pp 1–18

Abate L, Tadesse MG, Bachheti A, Bachheti RK (2022) Traditional and phytochemical bases of herbs, shrubs, climbers, and trees from Ethiopia for their anticancer response. Biomed Res Int 2022:1589877

Ali M, Sharma RK, Sharma DC, Pareek P, Bano R (2025) Achieving equilibrium in Ayurveda: the role of Dosha, Dhatu, and Mala in promoting wellness. Int J Ayurveda Pharma Res 13:116–119

Alrawi SN, Fetters MD (2012) Traditional arabic & islamic medicine: a conceptual model for clinicians and researchers. Global J Health Sci 4(3):164–169

Alum EU (2024) The role of indigenous knowledge in advancing the therapeutic use of medicinal plants: challenges and opportunities. Plant Signal Behav 19(1):2439255

Amr SS, Tbakhi A (2007) Ibn Sina (Avicenna): the prince of physicians. Ann Saudi Med 27(2):134–135

Bachheti A, Deepti RK, Bachheti, and A. Husen. (2021) Medicinal plants and their pharmaceutical properties under adverse environmental conditions. In: Harsh environment and plant resilience. Springer, Cham, pp 457–502

Bachheti RK, Worku LA, Gonfa YH, Zebeaman M, Pandey DP, Bachheti A (2022) Prevention and treatment of cardiovascular diseases with plant phytochemicals: A review. Evid Based Complement Alternat Med 2022. https://doi.org/10.1155/2022/5741198

Bensky D, Clavey S, Stöger E (2004) Chinese herbal medicine: Materia medica, 3rd edn. Seattle, WA: Eastland Press

Bhajbhuje M (2013) Ethno-medicinal survey for wound healing plants from rural area of Nagpur. Int J Life Sci 1(2):93–109

Brentjes S, Morrison R (2010) The sciences in Islamic societies. In: Irwin R (ed) The new Cambridge history of Islam, volume 4: Islamic cultures and societies to the end of the eighteenth century. Cambridge University Press, Cambridge, pp 564–569

Buell D (2022) Food and dietary medicine in Chinese herbal literature and beyond. In: Lo V, Yang D, Stanley-Baker M (eds) Routledge handbook of Chinese medicine, pp 328–336; Abingdon: Routledge

Bujalkova M, Straka S, Jureckova A (2001) Hippocrates' humoral pathology in nowaday's reflections. Bratislavske lekarske listy 102(10):489–492

Cacho V, Lum E (2021) Integrative sleep medicine. Oxford: Oxford University Press

Cai J, Zhen Y (2003) Medicine in ancient China. In: Selin H (ed) Medicine across cultures. Science across cultures: the history of non-Western science, vol 3. Springer, Dordrecht

Capasso L (1998) 5300 years ago, the Ice Man used natural laxatives and antibiotics. Lancet 352(9143):1864

Chaturvedi G, Tiwari S, Rai N (1981) Medicinal use of opium and cannabis in medieval India. Indian J Hist Sci 16(1):31–35

Chopra RN, Chopra I (1955) Quasi-medical use of opium in India and its effects. Bull Narc 7(3–4):1–24

Christensen SB (2021) Natural products that changed society. Biomedicine 9(5):472

Curran J (2008) The Yellow Emperor's classic of internal medicine. Br Med J 336(7647):777

Daidoji K (2013) The adaptation of *the Treatise on Cold Damage* in eighteenth-century Japan: text, society, and readers. Asian Med (Leiden, Netherlands) 8(2):361–393

David AR (2000) Medical practices in ancient Egypt. London: KPI

Dymock W, Warden CJH, Warden CJH, Hooper D (1893) Pharmacographia Indica: A history of the principal drugs of vegetable origin, met with in British India.London: K. Paul, Trench, Trübner & Co

Einarson B (1976) The manuscripts of Theophrastus' Historia Plantarum. Class Philol 71(1):67–76

Elendu C (2024) The evolution of ancient healing practices: from shamanism to Hippocratic medicine: A review. Medicine (Baltimore) 103(28):e39005
Furth C, Hsu E, Chang C-F (eds) (2021) Thinking with cases: specialist knowledge in Chinese cultural history. Leiden: Brill
Garland LH (1928) The lure of medical history: Imhotep: Patron of physicians and deity of medicine in ancient Egypt. California West Med 29(2):96–99
Gaur R (2024) A brief history: traditional Chinese medicinal system, pharmacological research. Modern Chinese Medicine 10:100387
Gupta AK, Jain A, Roy P, Singh R (2020) Pharmacological evaluation of Cannabis indica for their aphrodisiac potential. Int J Ayurvedic Med 11:399
Hajar R (2013) The air of history part III: the Golden age in Arab Islamic Medicine an introduction. Heart Views 14(1):43–46
Hajar R (2021) Medicine from Galen to the present: A short history. Heart Views 22(4):307–308
Halberstein R (2005) Medicinal plants: historical and cross-cultural usage patterns. Ann Epidemol 15:686–699
Hartsock JA, Halverson CME (2023) Lost in translation: the history of the Ebers Papyrus and Dr. Carl H. von Klein. J Med Libr Assoc 111(4):844–851
Hublin JJ, Ben-Ncer A, Bailey SE, Freidline SE, Neubauer S, Skinner MM et al (2017) “New fossils from Jebel Irhoud, Morocco and the Pan-African origin of Homo Sapiens” (PDF). Nature 546(7657):289–292
Huffman MA (2022) Folklore, animal self-medication, and phytotherapy-something old, something new, something borrowed, some things true. Planta Med 88(3–04):187–199
Husen A (2023) Exploring poisonous plants: medicinal values, toxicity response and other potential uses. CRC Press, Boca Raton, pp 33487–32742. https://doi.org/10.1201/b23017
Jamshidi-Kia F, Lorigooini Z (2018) Amini-Khoei, H. Medicinal plants: past history and future perspective. J Herbmed Pharmacol 7:1–7
Kapur M (2015) Basic principles of ayurveda, psychological perspectives on childcare in Indian indigenous health systems. New Delhi: Springer; pp 15–29
Kapur M (2016) Psychological perspectives on childcare in Indian indigenous health systems. New Delhi: Springer
Karenberg A, Hort I (1998) Medieval descriptions and doctrines of stroke: preliminary analysis of select sources. Part I: the struggle for terms and theories - late antiquity and early Middle Ages. J Hist Neurosci 7(113 Pt 2):162–173
Kemp BJ (2007) Ancient Egypt: anatomy of a civilisation. London & New York: Routledge
Khan H (2014) Medicinal plants in light of history: recognized therapeutic modality. J Evid Based Complem Altern Med 19(3):216–219
Kleisiaris CF, Sfakianakis C, Papathanasiou IV (2014) Health care practices in ancient Greece: the Hippocratic ideal. J Med Ethics Hist Med 7:6
Kumar S, Dobos GJ, Rampp T (2017) The significance of Ayurvedic medicinal plants. J Evid Base Complem Altern Med 22(3):494–501
Leake CD (1953) The old Egyptian medical Papyri. Med J Aust 2:584
Leslie C (1999) The roots of Ayurveda: selections from Sanskrit medical writings. JSTOR, Medical Anthropology Quarterly, pp 384–386
Lloyd GER (1978) Hippocratic writings. Pelican, London
Ma D, Wang S, Shi Y, Ni S, Tang M, Xu A (2021) The development of traditional Chinese medicine. J Tradit Chinese Med Sci 8:1–9
Maixner F, Krause-Kyora B (2016) The 5300-year-old Helicobacter pylori genome of the Iceman. Science 351(6269):162–165
Manniche L (1989) An ancient Egyptian herbal. Austin, TX: University of Texas Press
Mathur S, Hoskins C (2017) Drug development: Lessons from nature. Biomed Rep 6(6):612–614
McGovern PE (2019) Alcoholic beverages as the universal medicine before synthetics. In: Chemistry's role in food production and sustainability: past and present. Washington, DC, USA; ACS Publications, pp 111–127

Mehl-Madrona L (2007) Narrative medicine: the use of history and story in the healing process. Rochester, Vermont; Simon and Schuster
Metwaly AM, Ghoneim MM, Eissa IH, Elsehemy IA, Mostafa AE, Hegazy MM, Afifi WM, Dou D (2021) Traditional ancient Egyptian medicine: A review. Saudi J Biol Sci 28(10):5823–5832
Mishra L-C, Singh BB, Dagenais S (2000) Scientific basis for the therapeutic use of Withania somnifera (ashwagandha): a review. Altern Med Rev 5(4):334–346
Mukhopādhyāya G (1987) The surgical instruments of the Hindus: with a comparative study of the surgical instruments of the Greek, Roman, Arab, and the modern European surgeons: 396 illustrations in 82 plates. *Meharchand Lachmandas*, New Delhi
Mutatkar RK (2021) Tribal health & culture: symbiotic model of one health. Indian J Med Res 153(3):327–332
Nunn JF (2006) Ancient Egyptian medicine. Norman, OK: University of Oklahoma Press
Omar S (2022) Pharmacology and drugs in ancient Egypt. Int J Adv Stud World Archaeol 5(2):152–170
Pandey M, Rastogi S, Rawat A, Medicine A (2013) Indian traditional ayurvedic system of medicine and nutritional supplementation. J Evidence-Based Complem 1:376327
Peterson CT, Denniston K, Chopra D (2017) Therapeutic uses of triphala in ayurvedic medicine. J Altern Complem Med 23(8):607–614
Petrovska BB (2012) Historical review of medicinal plants' usage. Pharmacogn Rev 6(11):1–5
Pommerening T (2012) Altägyptische Rezepte-Eine diachrone Betrachtung. Geschichte der Pharmazie 64:33–38
Rammanohar P (2012) Clinical evidence in the tradition of Avurveda: evidence-based practice in complementary and alternative medicine. Berlin, Heidelberg: Springer-Verlag
Ratan R (2020) Essential oils: your aromatherapy guide to Ayurvedic healing. Summer Hill, NSW: Rockpool Publishing Simon and Schuster
Reid D (1996) The Shambhala guide to traditional Chinese medicine. Shambhala Publications, Boston
Rogaski R (2001) Paul U. Unschuld, Medicine in China: historical artifacts and images. Munich, London, and New York: Prestel, 2000. 220 pp. with 140 full color and 171 black and white illustrations. East Asian Sci Technol Med 18(1):126–128
Rosso AM (2019) Antidotes and counter-poisons in the ancient world: onions (hdw) (Allium cepa L.) in Egypt, the preferred antitoxic for snake bites. J Am Res Center Egypt 55:173–187
Saad B, Azaizeh H, Said O (2005) Tradition and perspectives of Arab herbal medicine: a review. Evid Based Complement Alternat Med 2(4):475–479
Scarborough J (1978) Theophrastus on herbals and herbal remedies. J Hist Biol 11(2):353–385
Scarborough J (1996) Drugs and medicines in the roman world. Expedition 38(2):38–51
Scarborough J (2006) Pharmacy in antiquity. In: The Cambridge history of medicine. Cambridge University Press
Sekar S (2007) Traditional alcoholic beverages from Ayurveda and their role on human health. Indian J Tradit Knowl 6:144–149
Shankar R, Lavekar GS, Deb S, Sharma BK (2012) Traditional healing practice and folk medicines used by Mishing community of North East India. J Ayurveda Integr Med 3(3):124–129
Shipley GP, Kindscher K (2016) Evidence for the Paleoethnobotany of the Neanderthal: a review of the literature. Scientifica (Cairo) 2016:8927654
Skrbo A, Masic I (2017) Influence of Arabian pharmacy on diseases tretament during Ottoman's period in Bosnia and Herzegovina. Med Arch (Sarajevo, Bosnia and Herzegovina) 71(3):219–225
Smith L (2011) The Kahun gynaecological Papyrus: ancient Egyptian medicine. J Fam Plann Reprod Health Care 37(1):54–55
Solecki R (1975) Shanidar IV, a Neanderthal flower Burial in Northern Iraq. Science 190:880–881
Stearns PN, Langer WL (eds) (2001) The Encyclopedia of world history: ancient, medieval, and modern, chronologically arranged, 6th edn. Houghton Mifflin. ISBN 978-0-395-65237-4

Stojanoski N (1999) Development of health culture in Veles and its region from the past to the end of the 20th century. Society of Science and Art, Veles, pp 13–34
Stiefel M, Shaner A and Schaefer SD (2006) The Edwin Smith Papyrus: the birth of analytical thinking in medicine and otolaryngology. The laryngoscope, 116(2):182–188
Subhose V, Srinivas P, Narayana A (2005) Basic principles of pharmaceutical science in Ayurveda. J Indian Med Heritage 35(2):83–92
Tibi S (2006) Al-Razi and Islamic medicine in the 9th century. J R Soc Med 99(4):206–207
Unschuld PU (1985) Medicine in China: a history of ideas. University of California Press, Berkeley
Ventura I (2017) Classification systems and pharmacological theory in medieval collections of Materia Medica: A short history from the antiquity to the end of the 12th century. In: Pommerening T, Bisang W (eds) Classification from antiquity to modern times: sources, methods, and theories from an interdisciplinary perspective. De Gruyter, Berlin/Boston, pp 101–166
Wang P, Zhang K, Wang X, Han S, Liu Y, Wan J, Guan H, Kuang Z, Jin L, Bai X, Liu Y (2024) An open dataset for oracle bone character recognition and decipherment. Sci Data 11(1):976
Yarnell E, Touwaide A (2019) Accuracy of Dioscorides,' De materia medica (first century C.E.), regarding Diuretic activity of plants. J Altern Complem Med (New York, NY) 25(1):107–120
Zeng J (2023) The *I Ching* as a potential Jungian application: history and practice. J Anal Psychol 68:913–932
Zhao Z, Guo P, Brand E (2018) A concise classification of *bencao* (*materia medica*). Chin Med 13:18
Zink R, Maixner F (2019) The current situation of the Tyrolean Iceman. Gerontology 65(6):699–706
Zunic L, Skrbo A, Dobraca A (2017) Historical contribution of pharmaceutics to botany and pharmacognosy development. Mater Sociomed 29(4):291–300

Chapter 3
Plant-Derived Biofertilizers and Biopesticides for Sustainable Agriculture

Esraa A. Elhawary, Dina M. Bahgat, Shaza H. Aly, Rania F. Abou El-Ezz, Heba A. S. El-Nashar, Mariam I. Gamal El-Din, and Abdel Nasser B. Singab

Abstract Sustainable agriculture is one of the sustainability goals for many countries worldwide. Insects and herbivores represent a challenging problem for plant agriculture and crop production. Chemical-based pesticides have many drawbacks, mainly related to their environmental and human health hazards, the potential for target insects to develop resistance over time, and their unsustainability. Thus, plant-derived biopesticides and biofertilizers may provide a sustainable, safe, affordable, and natural resource for green agriculture. This book chapter focuses on the main natural resources for biopesticides and biofertilizers, their impact on agriculture, the challenges and limitations of these natural sources, and their possible formulations.

Keywords Sustainability · Biopesticide · Biofertilizer · Green agriculture · Biocontrol · Bioresources · Nutrient cycling · Metabolites · Pest resistance

E. A. Elhawary · D. M. Bahgat · H. A. S. El-Nashar
Department of Pharmacognosy, Faculty of Pharmacy, Ain Shams University, Cairo, Egypt

S. H. Aly
Department of Pharmacognosy, Faculty of Pharmacy, Badr University in Cairo, Cairo, Badr City, Egypt

R. F. Abou El-Ezz
Department of Pharmacognosy, Faculty of Pharmacy, Misr International University, Cairo, Egypt

M. I. Gamal El-Din
Department of Pharmacognosy, Faculty of Pharmacy, Ain Shams University, Cairo, Egypt

Quadram Institute Bioscience, Norwich Research Park, Norfolk, UK

A. N. B. Singab (✉)
Department of Pharmacognosy, Faculty of Pharmacy, Ain Shams University, Cairo, Egypt

Center for Drug Discovery Research and Development, Faculty of Pharmacy, Ain Shams University, Cairo, Egypt
e-mail: abdelnassersingab@pharma.asu.edu.eg; vpr.nassersingab@asu.edu.eg

L. A. Worku et al. (eds.), *Plant-Based Solutions for Sustainable Agriculture and Healthcare*, Sustainability Sciences in Asia and Africa,
https://doi.org/10.1007/978-981-95-8910-4_3

3.1 Introduction

The pursuit of food production to feed the world's ever-growing population has been a matter of concern from at least 300 AD to the present (Hezakiel et al. 2024; Chowdhury et al. 2025; Baky et al. 2024). The statistics pointed to abundant agro-production to meet global demand, with minimal input from most of the third world (Siyavush 2015; Haque et al. 2024). Economies without crop pests would have achieved greater global and regional growth (Abu-Hashem et al. 2024). Pests, including fungi, bacteria, nematodes, weeds, rodents, and insects, can trigger damage to crops and loss of their stored products (Oerke 2006). In 2018, the loss of potential yield due to pests reached 13%, with the main harm occurring in developing countries (Pandya 2018). It is necessary to use conventional fertilizers and pesticides deliberately in order to improve crop yields to an industrial level. In China, the United States, and the European Union, there has been an increase in the stringency and binding of legislation relating to the use of pesticides and the approval of plant derivatives allowed in biological control regulations (for example, clove essential oil from *Syzygium aromaticum*, terpenes derived from *Chenopodium ambrosioides* L., *Ginko biloba* L. fruit extract, *Psoralea corylifolia* L. seed extract, and *Tephrosia candida* DC).

The application of synthetic chemicals, or agrochemicals such as organochlorines, organophosphates, carbamates, or pyrethroids, is associated with serious implications, including poisoned food, pest resistance, environmental degradation (ozone depletion), and health challenges (Sanganyado et al. 2015; Carvalho 2017). The residues of synthetic pesticides such as toxaphene, 1,1,1-trichloro-2,2-bis(4-chlorophenyl) ethane (DDT), polychlorinated biphenyls (PCBs), and organophosphate compounds may cause health-related problems such as irritation, enzyme inhibition, allergic sensitization, oxidative damage, and neurotransmission inhibition (Lushchak et al. 2018). Further, the global estimate of pesticides has reached around $38 billion (Pan-Germany 2012). The manipulation of pesticides for fiber crops accounts for around 67%, fruits 50%, vegetables 46%, spices 43%, oilseeds 28%, and pulses 23% (Kumar et al. 2021). An annual report by the Ministry of Chemicals and Fertilizers in India (MoCF) reported that chemical pesticide production increased from 186,000 metric tons (MT) in 2014–2015 to 217,000 MT in 2018–2019 (Miresmailli and Isman 2014). These challenges have heightened the need for safe and effective alternative agents, particularly those of plant origin, as they are cheap, target-specific, less harmful to human health, biodegradable, environmentally friendly, and supportive of sustainable production.

For many years, plant-derived compounds and extracts have been studied for their ability to control insect pests (Souto et al. 2021). Plant-derived biopesticides are materials of natural origin with active components that control pests via mechanisms that are not toxic to the host, the environment, or individuals (Leahy et al. 2014). Mechanistically, plant extracts could counteract the feeding, oviposition, and mating behaviors of insect pests, thereby inhibiting pest breeding, growth, and development (Lengai and Muthomi 2018). Further, some essential oils serve as anti-feeding, insect repellants, and oviposition deterrents (Chaudhari et al. 2021). Likewise, they comprise active components with larvicidal, ovicidal, and insecticidal properties that can affect insect metamorphosis (Cruz-Castillo et al. 2023). The essential oils are effective pesticides in view of their insignificant persistence in the environment (eco-friendly), varied modes of action, little toxicity, high accessibility, renewable source, and lowest probabilities of resistance owing to numerous components that may interfere with many facets of pest physiology and biochemistry (Koul et al. 2008).

Plant-derived pesticides include bioprospecting from natural sources, in vitro and in vivo tests, cleaning, formulation design, and registration (Lengai and Muthomi 2018). They are obtained via screening a number of plants and their constituents through field trials. The most effective plant or compound can be identified, combined with carriers, surfactants, emulsifiers, and other components, and tracked by intensive in vitro studies and optimization tests to obtain an efficacious product (Altemimi et al. 2017). Pyrethrins, rotenones, azadirachtin, nicotine, and gedunin are examples of plant-derived pesticides that have already been used globally (Şengül et al. 2022).

The long-term use of agricultural materials, such as chemical pesticides and fertilizers, led to lower soil fertility, manifested by losses in biodiversity, water retention, and acidification, and by disruptions of biogeochemical cycles. These affected the soil health and plant yield (Harman et al. 2021). Owing to these constraints, the search for alternative biofertilizers, such as plant microbiomes, is important for developing sustainable crop production systems. Using biofertilizers can increase plant yields by about 25% and reduce inorganic N and P fertilizer use by about 25–50% and 25%, respectively (Uddin et al. 2014). In addition to improving soil microbial status, biofertilizers stimulate the natural soil microbiota, thereby influencing nutrient availability and decomposition of organic matter (Chaudhari et al. 2021). Among the plant microbiomes are the plant growth-promoting rhizobacteria (PGPR) that exhibited different benefits, including nutrients' dissolution (Rezvani et al. 2021), inhibition of plant diseases (Bektas and Kusek 2021), nitrogen fixation (Hara et al. 2024), and enhanced phytochemical yield (Rezvani et al. 2021). Plant--soil microbial interactions positively influence agriculture by increasing plant survival, nutritional status, soil fertility, and crop productivity (Chen et al. 2024; Husen 2025). For example, applying biofertilizers to apricots alters the composition and activity of soil microbes, thereby enhancing nutrient cycling efficiency (Baldi et al. 2021).

Recently, soil with a bacterial consortium was reported to display a positive influence on the nutrient uptake efficiency of plants, restoration of agricultural soil, pathogen defense, and stress conditions (Aguilar-Paredes et al. 2020).

This book chapter provides updated information on the application of medicinal plants and their derived bioactive components as eco-friendly biofertilizers and biopesticides for the management of pests that trigger deterioration of crops and their stored supplies. In addition, we discussed the different pesticidal modes of action regarding the pesticidal activity of plant products as well as their applications against a wide spectrum of storage insect pests. Further, we highlighted the challenges and limitations of natural biofertilizers and biopesticides. This chapter will help drug researchers to develop a new biofertilizer and pesticide formulation from natural sources.

3.2 Plant-Derived Biofertilizers

The global demand for sustainable alternatives to chemical fertilizers has fueled growing interest in biofertilizers, formulations that improve plant growth and productivity through biological processes such as nutrient solubilization, nitrogen fixation, or phytohormone stimulation. Traditionally, the term "biofertilizer" has been associated mainly with microbial inoculants (e.g., *Rhizobium*, *Azotobacter*, *Azospirillum*, and mycorrhizal fungi) that enhance plant nutrition by colonizing the rhizosphere. Over the past two decades, the definition of biofertilizers has significantly expanded with the development of plant-derived biofertilizers that utilize plant biomass and extracts as nutrient sources or as materials that promote beneficial microbial activity (Santos et al. 2024; Bachheti et al. 2022; Abate et al. 2022; Worku et al. 2022)

It is important to clarify how plant-derived biofertilizers differ from other closely related categories. Traditional organic fertilizers, such as manure, compost, and bone meal, mainly act as nutrient sources and generally do not have direct stimulatory effects on plant growth. Microbial biofertilizers, on the other hand, consist of

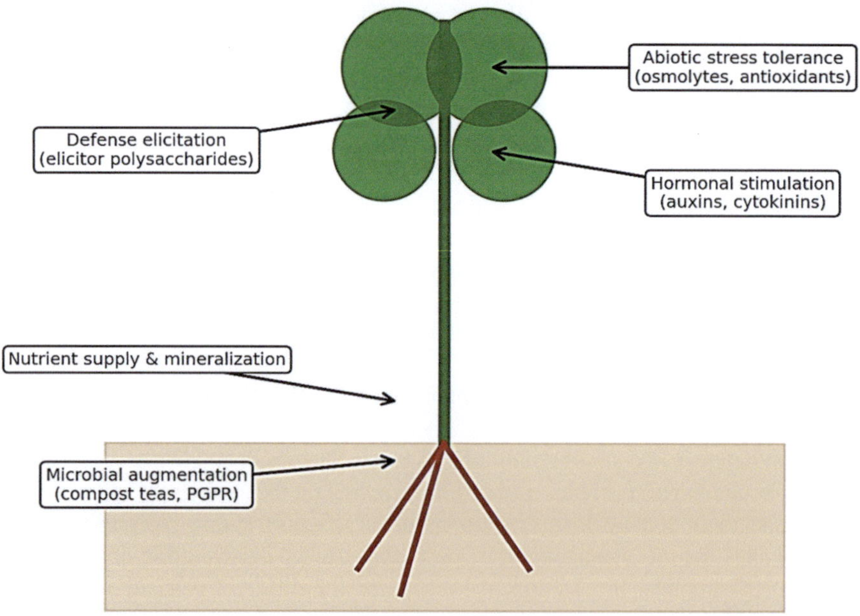

Fig. 3.1 The multifunctionality of plant-derived biofertilizers

living microorganisms, such as *Rhizobium* inoculants, whose action depends on their ability to colonize the rhizosphere and form symbiotic relationships with plant roots (Nygaard and Thorup-Kristensen 2011). Plant-derived biofertilizers fall between these two groups. They provide nutrients and contain plant-based bioactive molecules that can augment plant vigor or microbial activity. They can also act as carriers for beneficial microbes, although they do not necessarily contain living inoculants. Because of their multifunctional nature, plant-derived biofertilizers are increasingly regarded as valuable components of integrated nutrient management (INM) systems, where they complement both organic manures and microbial inoculants to improve soil fertility and crop yield (Kushwah et al. 2024).

Plant-derived biofertilizers are renewable and versatile. They align well with circular bioeconomy concepts by converting agricultural and food-processing residues into useful agricultural inputs, thus minimizing waste and reliance on synthetic products. Besides, they play an important role in restoring soil health. Their organic matter and secondary metabolites can enhance microbial activity and support beneficial communities in the rhizosphere. Unlike mineral fertilizers, which tend to release nutrients rapidly and sometimes unevenly, plant-derived fertilizers decompose gradually, providing a steady supply of nutrients that helps maintain soil fertility over time (Brahmaprakash and Sahu 2012).

3.2.1 Examples of Plant-Derived Biofertilizers

The range of plant-derived biofertilizers is extensive, including products such as seaweed extracts, fermented plant juices, compost and vermicompost derivatives, plant-based teas, green manures, and mixtures containing biochar. Each one differs in its chemical composition, mode of action, and agronomic applicability:

- *Seaweed extracts and macroalgal biostimulants*
- Extracts from brown and green macroalgae (e.g., *Ascophyllum nodosum*, *Laminaria* spp.) are manufactured by water extraction, enzymatic or chemical hydrolysis, and concentration. They contain polysaccharides (alginate, fucoidan, and laminarin), low-molecular-weight oligosaccharides, betaines, and hormone-like molecules, all reported to improve root growth, nutrient uptake, and stress tolerance. Commercial seaweed biostimulants are widely applied as foliar sprays or soil drenches (Sangha et al. 2014; Gautam et al. 2024).
- *Fermented plant extracts*
- Produced by controlled fermentation (often lactic or anaerobic) of green plant parts or fruits with sugar or carbohydrate sources. Fermentation increases soluble organics, small organic acids, and microbial metabolites; traditional systems such as Korean Natural Farming use Fermented Plant Juice (FPJ) to promote vegetative growth and Fermented Fruit Juice (FFJ) to enhance flowering and fruiting, typically applied as foliar feeds or soil amendments to boost plant vigor. Composition depends strongly on feedstock and fermentation parameters.

Compost and vermicompost teas serve as microbial inoculum carriers and nutrient suppliers (Suthar et al. 2017).

- *Compost, vermicompost, and compost teas*
- Mature composts and vermicomposts are bulk organic fertilizers rich in humic substances, slow-release N, and mineral fractions. Liquid extracts (aerated/non-aerated compost teas, vermiwash) deliver water-soluble nutrients and live microbial consortia. Tea preparations are highly variable but can supply plant growth-promoting rhizobacteria (PGPR), beneficial fungi, and metabolites (Abdel-Haleem et al. 2022).
- *Green manures and plant residues*
- Incorporation of cover crops/legumes supplies mineralizable N and labile C, improves soil structure, and fosters beneficial microbial shifts (including mycorrhizae and nitrogen-fixers). Residues are often combined with other plant-derived inputs to create integrated biofertilizer strategies (Singh et al. 2013).

3.2.2 Mechanisms of Action of Plant-Derived Biofertilizers

The action of plant-derived biofertilizers is multifaceted, operating through several interlinked mechanisms. One of the primary mechanisms by which plant-derived biofertilizers function is through the direct supply of nutrients and their gradual mineralization. During microbial decomposition and fermentation, complex organic compounds are broken down into simpler, soluble forms of nitrogen (N), phosphorus (P), and essential micronutrients (Figiel et al. 2025). These become more readily available to plants for uptake, thereby improving nutrient-use efficiency compared with raw plant residues (Singh et al. 2021). Compost and vermicompost, in particular, provide a dual benefit: they release certain nutrients rapidly to meet immediate plant demands while also contributing to long-term fertility through slow mineralization of more recalcitrant organic fractions (Figiel et al. 2025). This balance between immediate and sustained nutrient availability helps maintain soil fertility and reduces the need for synthetic inputs.

Another important mechanism involves hormonal and signaling modulation, often referred to as the biostimulant effect. Seaweed extracts and certain fermented plant products have been shown to contain, or to trigger the production of, hormone-like compounds resembling auxins, cytokinins, and brassinosteroids, as well as osmoprotectants such as betaines (Baltazar et al. 2021). These bioactive compounds play an important role in promoting root growth, strengthening sink capacity, and shaping plant responses to stress. Mechanistic studies have also shown that specific algal polysaccharides, including laminarin and alginate oligomers, can trigger plant defense pathways. By activating innate immunity, these molecules help plants withstand both abiotic and biotic stresses, offering benefits that go beyond simple nutrient supply (Poveda and Díez-Méndez 2023).

Another key mechanism involves enhancing and modulating soil microbial communities. Liquid formulations such as compost teas and vermicompost leachates contain a wide range of microorganisms—bacteria and fungi alike—that are introduced directly into the rhizosphere or phyllosphere. These beneficial microbes help suppress pathogens through competition, antibiosis, and the production of siderophores or lytic enzymes (Arosha and Sarvananda 2022). They also improve nutrient cycling by fixing nitrogen, solubilizing phosphorus, and generating growth--promoting metabolites. In some formulations, plant-derived carriers are co-applied with well-known plant growth promoting rhizobacteria (PGPR) to enhance these effects further (Parthasarathi et al. 2023). However, such combinations must be carefully tested, as acidic conditions or antimicrobial substances formed during fermentation can sometimes inhibit microbial survival.

Plant-derived biofertilizers also help mitigate abiotic stress—an increasingly important function in the context of climate change and limited natural resources. Components such as osmolytes, antioxidants, and signaling oligosaccharides found in seaweed extracts or fermented plant juices help plants tolerate drought, salinity, and temperature extremes (Johnson et al. 2024). These molecules stabilize cell membranes, protect proteins from damage, and activate antioxidant defense systems, thereby reducing oxidative stress. Meta-analyses indicate that the advantages of plant-based biofertilizers are especially evident under challenging or low-input conditions, where they help maintain yields and enhance crop resilience (Ouhaddou et al. 2023) (Fig. 3.1).

Plant-derived biofertilizers are expected to play a major role in climate-smart agriculture. They can reduce dependence on synthetic fertilizers, lower greenhouse gas emissions, improve soil organic carbon, and support sustainable crop produc tion. Despite these benefits, key challenges remain across standardization, regulatory approval, quality assurance, and large-scale production. Climate-smart agriculture reduces reliance on synthetic fertilizers, mitigates greenhouse gas emissions, improves soil organic carbon, and supports crop resilience.

3.3 Plant-Derived Biopesticides

Biopesticides are natural, biologically occurring compounds used to control various agricultural pests (Kumar et al. 2021). They are agents for pest management based on biochemicals derived from living microorganisms, insects, and plants (Souto et al. 2021). Types of biopesticides include biochemical pesticides, microbial pesticides, and plant-incorporated protectants. This section will focus on plant-derived biopesticides (PDBs) (Souto et al. 2021; Gupta et al. 2023).

Conventional agriculture's reliance on synthetic pesticides has led to environmental degradation, pest resistance, and risks to non-target species and human health (Acheuk et al. 2022). There is a need for plant-derived biopesticides, as they are assumed to be effective against various crop pests, readily biodegradable, available in large quantities, and cost-effective. They offer low persistence and residuality, preventing environmental pollution and minimizing adverse effects on living organisms. PDBs often exhibit high host specificity and are less prone to pest resistance (Souto et al. 2021; Acheuk et al. 2022).

Humans have used plants and their metabolites as pesticides against insects and microbes since the onset of agriculture (Souto et al. 2021). Botanical pesticides were applied in Europe and North America over 150 years ago (Pavela 2016), predating the discovery of major synthetic pesticide classes. Africa has centuries-long traditions of using plants for pest suppression (Pavela 2016).

Growing demand for organic products and alternative, eco-friendly approaches to substitute synthetic pesticides (Acheuk et al. 2022). Regulatory stringency in the use of chemical pesticides, environmental and safety considerations, and new technological achievements are boosting biopesticide consumption (Fusar and Fontefrancesco 2024).

PDBs are characterized by bioactive mixtures, extracts, or compounds from plant materials. Plant parts used for extraction include leaves, roots, stems, fruits, seeds, fruit peels, wood, bark, resin, and flowers (Isman 2006). Major classes of natural compounds with pesticidal activity include fatty acids, glycolipids, aromatic phenols, aldehydes, ketones, alcohols, terpenoids (e.g., monoterpenes, sesquiterpenes), flavonoids, alkaloids, limonoids, naphthoquinones, saccharides, polyolesthers, saponins, and sapogenins (Díaz and Rossini 2012; Aly et al. 2023a). Essential oils (EOs) and plant extracts are the botanical products most frequently used as biopesticides, acting through plant-synthesized molecules as part of their intrinsic defensive mechanisms (Tripathi et al. 2000; Aly et al. 2020). EOs are hydrophobic concentrates of volatile chemical compounds, including terpenes, alcohols, aldehydes, esters, fatty acids, ketones, phenols, as well as nitrogenous and sulfuric compounds (Aly et al. 2023b; Aly et al. 2025; Abdel-Hameed et al. 2024).

- *Mechanism of action*

Repel attacking insects, inhibit food ingestion, reduce growth at various developmental stages, inhibit egg-laying, or directly kill insects by inhibiting essential metabolic reactions. They can also affect insect behavior, physiology, morphology, and metabolism (Pavela 2016; Lucia and Guzmán 2021) (Fig. 3.2). Their modes of action include paralysis of insect nerve cells, disruption of muscle control, inhibition of pest growth and development by interfering with endocrine systems, repellence, protein denaturation, and inhibition of enzymatic or metabolic pathways vital to the pests (Kumar et al. 2021; Souto et al. 2021).

- *Examples*

Neem oil (from *Azadirachta indica*) contains azadirachtin, a tetranortriterpenoid known for its insecticidal activity and growth-regulating effects. It can impact multiple biological targets in cotton bollworms and affect insect metamorphosis (Ascher 1993).

Pyrethrins (from *Tanacetum cinerariifolium*) are potent insecticides and the most widely exploited bioinsecticide globally. They are natural mixtures, plentiful in flower heads. Pyrethrum represents about 80% of botanical insecticides worldwide (Jeran et al. 2021). It disrupts sodium and potassium ion exchange in insect nerve fibers, leading to immediate paralysis (Silver et al. 2014).

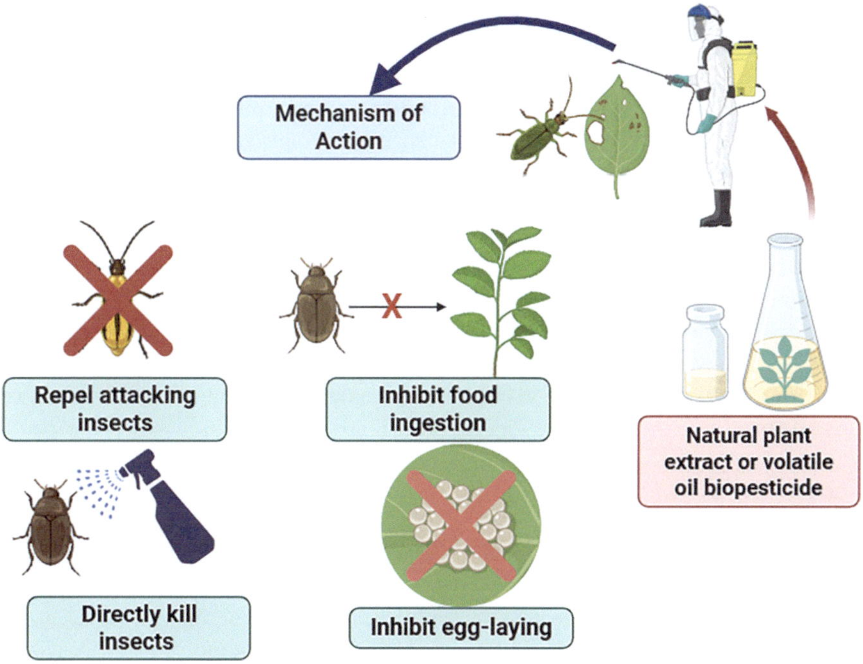

Fig. 3.2 Schematic representation of the mechanism of action of the plant-derived biopesticides

Several essential oils have been widely explored for their toxicity against various crop pests. Citronella oil (*Cymbopogon nardus*, *C. winterianus*) showed insecticidal effects against black cutworm (*Agrotis ipsilon*) and cotton leafworm (*Spodoptera littoralis*). Used as a repellent against dog ticks (*Amblyomma cajennense*) (Moustafa et al. 2021). The main components of Lemongrass oil (*Cymbopogon citratus*) are citral (geranial and neral), β-myrcene, and citronellal (Aly et al. 2025). Effective against black cutworm and tomato leafminer (*Tuta absoluta*) (Ngongang et al. 2022). Clove oil (*Syzygium aromaticum*) has eugenol as its main constituent. Included in commercial biopesticide formulations (El Gohary et al. 2021). Additionally, thyme oil (*Thymus vulgaris*) with its components like eugenol, thymol, and carvacrol shows larvicidal activity against seaside mosquito species (*Ochlerotatus caspius*) (Knio et al. 2008). Mint oil (*Mentha piperita*) contains geranial, geraniol, linalool, methyl cinnamate, and methyl eugenol, showing herbicidal effects. Menthol can augment the toxicity of carbamate pesticides (Kalemba and Synowiec 2019; Jankowska et al. 2019). *Rosmarinus officinalis* volatile oil has demonstrated insecticidal effectiveness against *T. confusum* and *A. obtectus* (Mohamed 2015). *Citrus sinensis* volatile oil, with D-limonene as its primary active compound, has shown toxicity to *Aedes aegypti* larvae and house flies (Amusan et al. 2005). In conclusion, plant-based biopesticides, such as essential oils and various plant extracts, could be a better, more environmentally friendly way to control pests, diseases, and weeds than synthetic agrochemicals. They exhibit several modes of action, including insecticidal, herbicidal, fungicidal, and nematicidal properties, often distinguished by improved target selectivity and biodegradability. PDBs, particularly those incorporating innovative technologies such as nanoencapsulation, exhibit enhanced efficacy, prolonged durability, and controlled release of their constituents, making them advantageous for sustainable agriculture. Ongoing investigations into their chemistry and mechanisms of action will enhance understanding of their potential. Future research directions include the development of new technologies for product manufacture, such as microencapsulation, and simplification of costly authorization requirements. Focus on detailed mechanisms of action, including synergistic interactions, and effects on non-target species, particularly beneficial insects. Optimizing cultivation processes and extraction procedures to ensure homogeneous chemical composition and consistent efficacy. Investing in field trials for bacterial VOCs and concentrated natural bacterial VOCs with suitable formulations.

3.4 Impact of Plant-Derived Biofertilizers and Biopesticides on Sustainable Agriculture

Currently, global agricultural systems rely on chemical pesticides and fertilizers to protect crops from multiple pests and transmitted diseases and to increase agronomic productivity to meet the incredibly increasing demand for food worldwide. However, their extensive and widespread use poses serious ecological hazards and health risks (Verardi et al. 2025).

Exposure to organic pesticides, whether through direct contact or ingestion of contaminated food and water, is harmful to livestock. It is the main cause behind serious health problems such as the development of allergies, respiratory disorders, hormonal imbalances causing reproductive system alterations, and, for sure, the increased rates of carcinogenicity incidence. In addition, the residues of non-biodegradable pesticides damage the soil and leach into underground water, causing contamination (Srivastav 2020; Rana et al. 2019).

At the same time, the use of chemical fertilizers is linked to many adverse ecological impacts, starting with the process of their synthesis and production, which accounts for about 2% of global gas emissions. The improper use leads to soil acidification, soil degradation, and disruption of the ionic balance. They corrupt natural microbiomes and thus suppress the activity of beneficial microbiota, leading to the accumulation of harmful substances and contamination of groundwater (Calabi-Floody et al. 2018; Nakachew et al. 2024).

Despite these major drawbacks, global pesticide consumption totaled 3.69 million metric tons in 2022. In 2023, the global market for crop protection products was estimated to be 77.85 billion USD (Statista). The global fertilizer market amounted to 190 billion USD in 2021, with a rough increase of 12% from 2020. It is forecast to reach about 240 billion by 2030, driven by the growing global population (Statista).

The current situation underscores the importance of incorporating biofertilizers and biopesticides into sustainable agricultural models to enhance food production while conserving natural resources, maintaining agroecosystems, and minimizing health risks.

3.4.1 Benefits of Plant-Derived Biofertilizers

Biofertilizers are biological formulations that improve plant nutrient uptake when added to the growing medium. They are categorized into two major classes: the organic residue-based and the microbial-based biofertilizers. The former one consists of farmyard manure, crop residues, and green manure (Abbey et al. 2019).

The application of biofertilizers enhances plant growth and performance either directly or indirectly. Direct benefits occur when plant biomass (organic residue) decomposes: it provides the soil with essential macro- and micronutrients, including nitrogen, phosphorus, potassium, and trace elements, thereby increasing plant productivity (Chaudhary et al. 2021). The indirect benefit of biofertilizers is due to their ability to stimulate the soil natural microbiota responsible for decomposition of organic matter and nutrient accessibility (Chaudhary et al. 2022). In addition, soil microbiota helps the plant defend itself against various pathogenic bacteria, fungi, and viruses. The symbiotic microbial population stimulates the plant release of natural growth-promoting substances such as auxins, cytokinins, and gibberellins, which improve root development and plant vigor (Mahmud et al. 2021).

3.4.2 Benefits of Plant-Derived Biopesticides

Biopesticides derived from a plant source exhibit multiple non-specific modes of action including intestinal disruption, neuromuscular toxicity, and metabolic inhibition. Thus, they reduce pest populations very efficiently (Dar et al. 2021). Despite this efficiency, they are safe for people, fish, birds, and bees. They support beneficial soil organisms, sparing the natural microbiota, which itself is considered a defense line against pests. The complexity of biopesticides' chemical structures and their ability to target multiple receptors make it harder for pests to develop resistance than with chemical pesticides. So, there is no need to increase the dose or use combinations as in conventional pesticides. Biodegradability is another beneficial characteristic of biopesticides, as they break down rapidly in soil, leaving minimal residues on crops and in the environment (Khursheed et al. 2022).

To wrap up, biofertilizers and biopesticides make a significant contribution to sustainable agriculture by promoting environmental protection and biodiversity conservation while causing minimal disruption to ecosystems. Economically, they have lower costs and can be locally produced.

3.5 Challenges and Limitations of Natural Biofertilizers and Biopesticides

The market for biopesticides is still in its infancy, despite proof that botanical pesticides are effective against a wide variety of agricultural pests. The commercialization of botanical pesticides relies heavily on the development of plants and on the availability of plant sources in sufficient quantities. Up to now, the source plants have either been cultivated for food, medicine, or other uses. Additionally, growing plants to produce botanical pesticides requires a lot of space, making it extremely competitive with food production for arable land. Furthermore, it does not appear easy to formulate biopesticides, as a single plant may have a variety of active ingredients with varying chemical characteristics. Furthermore, organic solvents are used in the pesticide extraction process, which pollutes the environment when disposed of (Siddiqui et al. 2025).

Furthermore, because of their high rate of biodegradability, biological insecticides have a limited shelf life. When it comes to microbial pesticide specificity, microorganisms can only control a subset of the pests present in an area. They might not be able to manage other pests that are present in treated regions, which could cause ongoing harm. Heat, UV radiation, and desiccation limit the effectiveness of microbial pesticides, making the delivery system a crucial microbiological component. Short shelf life is also caused by the need for specific formulations and storage techniques, which lead to costly development and production and to inconsistent field performance. In addition, the cooperation between community and business representatives is an important part of improving innovation, production, and marketing of environmentally friendly options (Ndolo et al. 2019).

To improve and broaden the applications and use of biopesticides, further advanced research and the formulation of promising candidates are advised. Furthermore, the availability of biopesticides and bioinsecticides for farmers in developing countries is a challenge. This can be improved by modifying regulatory processes to make it easier to register molecules with proven safety (Essiedu et al. 2020; Folina et al. 2025).

Several obstacles may render the widespread use of natural pesticides and insecticides. These obstacles could be explained by variations in the efficacy of these natural agents, attributed to changes in environmental conditions such as soil, water supply, pH, temperature, plant species, and others (Fig. 3.3).

- *Limited commercial penetration*: an example is biofertilizers of fungal origin; they are scarce in the market, although they have a potent effect compared to chemical agents (Ferreyra-Suarez et al. 2024).
- *Production and regulatory hurdles*: several obstacles may face large-scale production of natural biocontrol agents, including suitable production techniques, regulatory issues, and limitations imposed by environmental conditions (Ferreyra-Suarez et al. 2024).

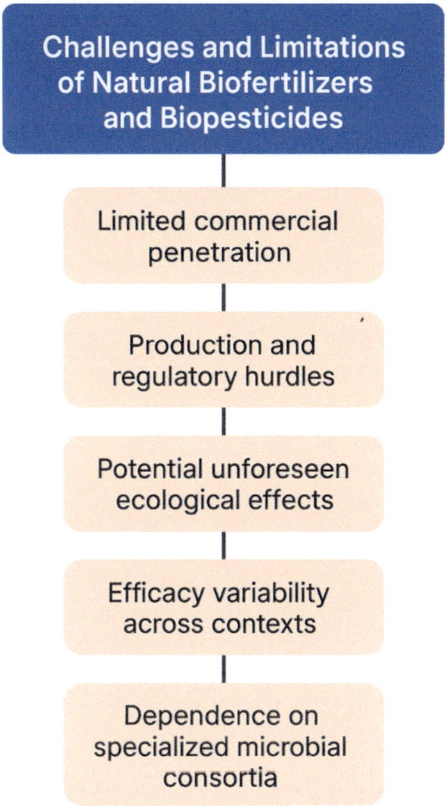

Fig. 3.3 Summary of the challenges and limitations for natural biofertilizers and biopesticides

- *Potential unforeseen ecological effects*: an in-depth study of the changing environmental factors related to biopesticide effects on non-target microbes in soil, which may alter these organisms' responses to other agents and affect soil enzymes.
- *Efficacy variability across contexts*: the selection of suitable plant strains, soil conditions, and plant species/variety is a cornerstone in the production of effective biocontrol agents (Abioye et al. 2024).
- *Dependence on specialized microbial consortia*: the synergistic use of certain microbial strains with biofertilizers and biopesticides may enhance their efficacy, provided the formulation, quality steps, and quality assurance of the entire process are considered (Folina et al. 2025).

3.6 Recent Advances and Formulations for Plant-Derived Biofertilizers and Biopesticides

Plant extracts are considered an important type of biopesticide; being derived from natural sources, they are also known as biochemicals. They comprise a broad part of biopesticides' classification, which also includes microbes, macro-organisms, and certain minerals. They are naturally derived or semi-synthesized chemical compounds with structures and functions similar to those of naturally occurring compounds.

Plant-based biopesticides have proven effective in controlling pests and diseases. Compared to synthetic pesticides, biopesticides have a more favorable environmental profile, with lower toxicity to non-target organisms and greater biodegradability.

Herbal extracts and volatile oils serve as naturally derived insecticidal agents obtained directly from plant origins, encompassing a broad spectrum of bioactive secondary metabolites (Magierowicz et al. 2020). These active compounds possess diverse modes of action, including repellence, attraction, antifeedant effects, disruption of respiratory processes, interference with host plant recognition, inhibition of oviposition, and suppression of adult insect emergence (Ali et al. 2017). Notable examples include neem oil and lemongrass oil. The insecticidal efficacy of such extracts is greatly affected by the choice of extraction solvent, which determines the spectrum and yield of active compounds. All plants produce a wide variety of bioactive constituents, including phenolic compounds, quinones, alkaloids, steroids, terpenoids, alcohols, and saponins (Mizubuti et al. 2007; Nefzi et al. 2016).

Plant extracts play a vital role in sustainable pest management, as they are increasingly recognized as complementary and viable alternatives to conventional chemical pesticides. Their application is particularly beneficial for short-term pest control, offering a safer, more targeted mode of action than synthetic counterparts (Mawcha et al. 2025b). From an environmental perspective, plant extracts and other biopesticides are generally considered ecologically benign due to their minimal residual activity, low toxicity to non-target organisms, and reduced risk to human health. Moreover, their use affords other advantages such as conservation of biodiversity, maintenance of soil health, and the reduction of post-harvest chemical contamination. In terms of classification, the International Biocontrol Manufacturers' Association (IBMA) identifies "natural products," including plant extracts, as one of the four principal categories of biocontrol agents. Similarly, the European Environment Agency defines biopesticides as products containing an active ingredient that is a microorganism (virus, fungus, or bacterium) or a natural compound of plant origin (Mawcha et al. 2025b).

Formulation is considered the most critical step in developing biopesticide agents derived from microorganisms and phytochemicals (Leggett et al. 2016), as it involves incorporating the active ingredient—whether a viable microorganism or a bioactive metabolite — into a matrix of functional and inert excipients (Grewal et al. 2005). Optimized formulations are essential for successful integration into pest management programs, as they enhance dispersion, attachment, persistence at target sites, and stability during storage, transport, and application. Biopesticide formulations generally fall into two main categories: dry and liquid. Dry types include dust powders, granules, seed dressers, and wettable powders (commonly used for *Bacillus* products and fungal spores), produced through methods such as spray drying, freeze drying, or air drying; whereas liquid formulations encompass water-based systems (suspension concentrates, suspo-emulsions, and capsule suspensions), as well as oil- and polymer-based products, which require the incorporation of inert ingredients such as stabilizers, stickers, surfactants, and antifreeze to conserve both the shelf life and efficacy.

An optimal formulation should extend the product's shelf life, ideally for at least 6 months and preferably up to 2 years under ambient conditions. Furthermore, formulations are designed to enable application through existing equipment, ensuring practicality for end users. Improving product stability, bioactivity, delivery, and persistence at the target site are additional key objectives, with strategies such as the inclusion of "stickers" to enhance dispersion and adhesion to foliage. Finally, successful formulations must prioritize user convenience and ensure compatibility with agricultural practices and application technologies, thereby maximizing their overall performance and adoption (Mawcha et al. 2025a).

The improvement in biopesticide technology reflects the convergence of multiple scientific innovations aimed at creating more structured, organized, and sustainable crop protection solutions by enhancing the efficacy, target specificity, and ecological safety, not only to align with the principles of sustainable agriculture but also to be able to meet the growing demand for environmentally compatible pest management strategies (Fenibo et al. 2022). Significant progress has been achieved through genetic engineering tools such as CRISPR-Cas9 and RNA interference (RNAi), synthetic biology, and nanotechnology, all of which have enabled the development of novel microbial and botanical products with high precision and reduced environmental impact. Advances in formulation science enhance the stability, bioavailability, and targeted release of these agents, while exploring synergistic interactions with other pest management approaches holds potential to reduce dependence on chemical pesticides. Application innovations, such as controlled droplet systems, variable-rate technology, intelligent spray systems, seed treatments, soil applications (including drenching, fumigation, and rhizosphere delivery), robotics, and aerial application, are equally important because they collectively increase the precision and effectiveness of biopesticide delivery. Similarly, emerging research highlights the promise of plant extracts and essential oils, alongside RNA-based products and nanotechnology-driven approaches, as rapidly advancing frontiers in sustainable crop protection. All these innovations emphasize the vital role of research in advancing biopesticides as viable, sustainable, and effective alternatives to synthetic pesticides (Kumar and Singh 2015).

3.7 Conclusion and Future Perspectives

Biopesticides and biofertilizers from plant origin may play an essential role in advancing environmentally friendly, sustainable agriculture. This may be achieved through nutrient cycling, improved plant quality, and pest control, thereby increasing crop yield. On the other hand, several hurdles stand in the way of the large-scale application of such bioagents, including variation in crop yields, unstable formulation techniques, regulatory issues, and low farmer awareness. Thus, guiding future improvements in the production and development of biofertilizers and biopesticides may involve overcoming such obstacles. Future measures may involve developing promising, stable, advanced formulations with controlled-release properties, using techniques such as oil-based and microbial carriers, and nanoencapsulation.

Furthermore, future perspectives should focus on regulating the strains used, limiting field variations, and negotiating regulatory affairs to facilitate the production of natural biocontrol agents. To improve product stability and market share, strong regulatory and quality assurance procedures, including standardization, certification, and monitoring, must be established. These measures could be combined with proper farmer education and training programs executed by the regulatory affairs. This may include crop selection, pest management, and awareness of environmental conditions. Crop resilience and carbon sequestration can be further increased under changing climate conditions by utilizing microbial and biostimulant inputs in combination with climate-smart integration. Finally, applying microbiome-specific solutions with advanced "omics" tools may enable tailoring bio-inputs to specific crops, soils, and regions, thereby improving their effectiveness and reliability.

References

Abate L, Tadesse MG, Bachheti A, Bachheti RK (2022) Traditional and phytochemical bases of herbs, shrubs, climbers, and trees from Ethiopia for their anticancer response. Biomed Res Int 2022. https://doi.org/10.1155/2022/1589877

Abbey L, Abbey J, Leke-Aladekoba A, Iheshiulo EM-A, Ijenyo M (2019) Biopesticides and biofertilizers: types, production, benefits, and utilization. In: Byproducts from agriculture fisheries: adding value for food, feed, pharma, fuels, pp 479–500. John Wiley & Sons, Hoboken, NJ, USA

Abdel-Haleem E-S, Hosney MF, Abeer B, Abdelrasheed KG (2022) Combined use of compost, compost tea, and vermicompost tea improves soil properties, and growth, yield, and quality of (*Allium cepa* L.). Notulae Botanicae Horti Agrobotanici Cluj-Napoca 50:12565–12565

Abdel-Hameed UK, Abualghaith AS, Aly SH, Soliman MM, Munshi LA, Mohammed SAA, Eldahshan OA, Abdelghffar EAR (2024) GC/MS analysis and protective effects of *Mentha longifolia* L. essential oil against Antituberculosis drug-induced organs toxicity in Wistar albino rats. Plants 13:3231

Abioye OM, Olasehinde DA, Abadunmi T (2024) The role of biofertilizers in sustainable agriculture: An eco-friendly alternative to conventional chemical fertilizers. Appl Sci Eng Prog 17:6883–6883

Abu-Hashem AA, Hakami O, El-Shazly M, El-Nashar HAS, Yousif MNM (2024) Caffeine and purine derivatives: a comprehensive review on the chemistry, biosynthetic pathways, synthesis-related reactions, biomedical prospectives and clinical applications. Chem Biodivers 21:e202400050

Acheuk F, Basiouni S, Shehata AA, Dick K, Hajri H, Lasram S, Yilmaz M, Emekci M, Tsiamis G, Spona-Friedl M (2022) Status and prospects of botanical biopesticides in Europe and Mediterranean countries. Biomolecules 12:311

Aguilar-Paredes A, Valdés G, Nuti M (2020) Ecosystem functions of microbial consortia in sustainable agriculture. Agronomy 10:1902

Ali AM, Mohamed DS, Shaurub EH, Elsayed AM, Studies Z (2017) Antifeedant activity and some biochemical effects of garlic and lemon essential oils on *Spodoptera littoralis* (Boisduval) (Lepidoptera: Noctuidae). J Entomol 5:1476–1482

Altemimi A, Lakhssassi N, Baharlouei A, Watson DG, Lightfoot DA (2017) Phytochemicals: extraction, isolation, and identification of bioactive compounds from plant extracts. Plants 6:42

Aly SH, Elissawy AM, Eldahshan OA, Elshanawany MA, Abdel Nasser B, Singab. (2020) Variability of the chemical composition of the essential oils of flowers and the alkaloid contents of leaves of *Sophora secundiflora* and *Sophora tomentosa*. J Essent Oil Bear Plants 23:442–452

Aly SH, Elissawy AM, Salah D, Alfuhaid NA, Zyaan OH, Mohamed HI, Singab ANB, Farag SM (2023a) Phytochemical investigation of three *Cystoseira* species and their larvicidal activity supported with in silico studies. Mar Drugs 21:117

Aly SH, Kandil NH, Hemdan RM, Kotb SS, Zaki SS, Abdelaziz OM, AbdelRazek MMM, Almahli H, El Hassab MA, Al-Rashood ST (2023b) GC/MS profiling of the essential oil and lipophilic extract of *Moricandia sinaica* Boiss. and evaluation of their cytotoxic and antioxidant activities. Molecules 28:2193

Aly SH, Uba AI, Nilofar N, Majrashi TA, El Hassab MA, Eldehna WM, Zengin G, Eldahshan OA (2025) Chemical composition and biological activity of lemongrass volatile oil and *n*-Hexane extract: GC/MS analysis, in vitro and molecular modelling studies. PLoS One 20:e0319147

Amusan AAS, Idowu AB, Arowolo FS (2005) Comparative toxicity effect of bush tea leaves (*Hyptis suaveolens*) and orange peel (*Citrus sinensis*) oil extract on larvae of the yellow fever mosquito Aedes aegypti. Tanzan J Health Res 7:174–178

Arosha KPL, Sarvananda L (2022) Vermicomposting tea a potential liquid biofertilizer. Front Life Sci Res 1:1–5

Ascher KRS (1993) Nonconventional insecticidal effects of pesticides available from the neem tree, *Azadirachta indica*. Arch Insect Biochem Physiol 22:433–449

Bachheti RK, Worku LA, Gonfa YH, Zebeaman M, Pandey DP, Bachheti A (2022) Prevention and treatment of cardiovascular diseases with plant phytochemicals: a review. Evid Based Complement Alternat Med 2022. https://doi.org/10.1155/2022/5741198

Baky MH, Elkenawy NM, El-Nashar HAS, Abib B, Farag MA (2024) Comparison of autoclaving and γ-radiation impact on four spices aroma profiles and microbial load using HS-SPME GC–MS and chemometric tools. Sci Rep 14:5752

Baldi E, Gioacchini P, Montecchio D, Mocali S, Antonielli L, Masoero G, Toselli M (2021) Effect of biofertilizers application on soil biodiversity and litter degradation in a commercial apricot orchard. Agronomy 11:1–18

Baltazar M, Correia S, Guinan KJ, Sujeeth N, Bragança R, Gonçalves B (2021) Recent advances in the molecular effects of biostimulants in plants: An overview. Biomolecules 11:1096

Bektas I, Kusek M (2021) Biological control of onion basal rot disease using phosphate solubilising rhizobacteria. Biocontrol Sci Tech 31:190–205

Brahmaprakash GP, Sahu PK (2012) Biofertilizers for sustainability. J Indian Inst Sci 92:37–62

Calabi-Floody M, Medina J, Rumpel C, Condron LM, Hernandez M, Dumont M, de La Luz M, Mora. (2018) Smart fertilizers as a strategy for sustainable agriculture. J Adv Agro 147:119–157

Carvalho FP (2017) Pesticides, environment, and food safety. Food Energy Secur 6:48–60

Chaudhari AK, Singh VK, Kedia A, Das S, Dubey NK (2021) Essential oils and their bioactive compounds as eco-friendly novel green pesticides for management of storage insect pests: prospects and retrospects. Environ Sci Pollut Res 28:18918–18940

Chaudhary A, Parveen H, Chaudhary P, Khatoon H, Bhatt P (2021) Rhizospheric microbes and their mechanism. In: Microbial technology for sustainable environment. Springer, Singapore
Chaudhary P, Singh S, Chaudhary A, Sharma A, Kumar G (2022) Overview of biofertilizers in crop production and stress management for sustainable agriculture. Front Plant Sci 13:930340
Chen Q, Song Y, An Y, Yinglin L, Zhong G (2024) Soil microorganisms: their role in enhancing crop nutrition and health. Diversity 16:734
Chowdhury R, Bhuia S, Rakib AI, Al Hasan S, Shill MC, El-Nashar HAS, El-Shazly M, Islam MT (2025) Gigantol, a promising natural drug for inflammation: a literature review and computational based study. Nat Prod Res 39:1241–1257
Cruz-Castillo AU, Rodriguez-Valdez LM, Correa-Basurto J, Nogueda-Torres B, Andrade-Ochoa S, Nevarez-Moorillon GV (2023) Terpenic constituents of essential oils with larvicidal activity against *Aedes aegypti*: a QSAR and docking molecular study. Molecules 28:2454
Dar SA, Wani SH, Mir SH, Showkat A, Dolkar T, Dawa T (2021) Biopesticides: mode of action, efficacy and scope in pest management. J Adv Res Biochem Pharmacol 4:1–8
Díaz M, Rossini C (2012) Bioactive natural products from Sapindaceae deterrent and toxic metabolites against insects. In: Insecticides–pest engineering, F. Perveen edn. InTech, Rijeka, pp 287–308
El Gohary E, Gohary E, Farag SM, El-Sayed AA, Khattab RR, Mahmoud DM (2021) Insecticidal activity and biochemical study of the clove oil (*Syzygium aromaticum*) nano-formulation on *Culex pipiens* l.(diptera: Culicidae). Egypt J Aquat Biol Fish 25:227–239
Essiedu JA, Adepoju FO, Ivantsova MN (2020) Benefits and limitations in using biopesticides: a review. In: AIP conference proceedings, 080002. AIP Publishing LLC, AIP Publishing LLC, Melville, NY, USA.
Fenibo EO, Ijoma GN, Matambo T (2022) Biopesticides in sustainable agriculture: current status and future prospects. In: New and future development in biopesticide research: biotechnological exploration, pp 1–53, Elsevier, Amsterdam,
Ferreyra-Suarez D, García-Depraect O, Castro-Muñoz R (2024) A review on fungal-based biopesticides and biofertilizers production. Ecotoxicol Environ Saf 283:116945
Figiel S, Rusek P, Ryszko U, Brodowska MS (2025) Microbially enhanced biofertilizers: technologies, mechanisms of action, and agricultural applications. Agronomy 15:1191
Folina A, Kakabouki I, Baginetas K, Bilalis D (2025) Integration of bioresources for sustainable development in organic farming: a comprehensive review. Resources 14:102
Fusar Poli E, Fontefrancesco MF (2024) Trends in the implementation of biopesticides in the Euro-Mediterranean region: a narrative literary review. Sustain Earth Rev 7:14
Gautam A, Chauhan A, Singh A, Mundepi S, Pant M, Husen A (2024) Use of seaweed extract-based biostimulants in plant growth, biochemical constituents, and productions. In: Husen A (ed) Biostimulants in plant protection and performance. Elsevier Inc, Cambridge, MA, pp 129–148. https://doi.org/10.1016/B978-0-443-15884-1.00022-1
Grewal PS, Ehlers R-U, David IS-I (2005) Nematodes as biocontrol agents. CABI, Wallingford, UK.
Gupta I, Singh R, Muthusamy S, Sharma M, Grewal K, Singh HP, Batish DR (2023) Plant essential oils as biopesticides: applications, mechanisms, innovations, and constraints. Plants 12:2916
Haque MF, El-Nashar HAS, Akbor MS, Alfaifi M, Bappi MH, Chowdhury AK, Hossain MK, El-Shazly M, Albayouk T, Na'il Saleh (2024) Anti-inflammatory activity of d-pinitol possibly through inhibiting COX-2 enzyme: in vivo and in silico studies. Front Chem 12:1366844
Hara S, Kakizaki K, Bamba M, Itakura M, Sugawara M, Suzuki A, Sasaki Y, Takeda M, Tago K, Ohbayashi T (2024) Does rhizobial inoculation change the microbial community in field soils? A comparison with agricultural Land-use changes. Microbes Environ 39:ME24006
Harman G, Khadka R, Doni F, Uphoff N (2021) Benefits to plant health and productivity from enhancing plant microbial symbionts. Front Plant Sci 11:610065
Hezakiel HE, Thampi M, Rebello S, Sheikhmoideen JM (2024) Biopesticides: a green approach towards agricultural pests. Appl Biochem Biotechnol 196:5533–5562
Husen A (2025) Agricultural crop improvement: plant and soil relationships. Taylor & Francis Group, LLC. https://doi.org/10.1201/9781032630366

Isman MB (2006) Botanical insecticides, deterrents, and repellents in modern agriculture and an increasingly regulated world. Annu Rev Entomol 51:45–66

Jankowska M, Lapied B, Jankowski W, Stankiewicz M (2019) The unusual action of essential oil component, menthol, in potentiating the effect of the carbamate insecticide, bendiocarb. Pestic Biochem Physiol 158:101–111

Jeran N, Grdiša M, Varga F, Šatović Z, Liber Z, Dabić D, Biošić M (2021) Pyrethrin from Dalmatian pyrethrum (*Tanacetum cinerariifolium* (Trevir.) Sch. Bip.): biosynthesis, biological activity, methods of extraction and determination. Phytochem Rev 20:875–905

Johnson R, Joel JM, Puthur JT (2024) Biostimulants: the futuristic sustainable approach for alleviating crop productivity and abiotic stress tolerance. J Plant Growth Regul 43:659–674

Kalemba D, Synowiec A (2019) Agrobiological interactions of essential oils of two menthol mints: Mentha piperita and Mentha arvensis. Molecules 25:59

Khursheed A, Rather MA, Jain V, Wani AR, Rasool S, Nazir R, Malik NA, Majid SA (2022) Plant based natural products as potential ecofriendly and safer biopesticides: a comprehensive overview of their advantages over conventional pesticides, limitations and regulatory aspects. Microb Pathog 173:105854

Knio KM, Usta J, Dagher S, Zournajian H, Kreydiyyeh S (2008) Larvicidal activity of essential oils extracted from commonly used herbs in Lebanon against the seaside mosquito, Ochlerotatus caspius. Bioresour Technol 99:763–768

Koul O, Walia S, Dhaliwal GS (2008) Essential oils as green pesticides: potential and constraints. Biopestic Int 4:63–84

Kumar S, Singh A (2015) Biopesticides: present status and the future prospects. J Fertil Pestic 6:1–2

Kumar J, Ramlal A, Mallick D, Mishra V (2021) An overview of some biopesticides and their importance in plant protection for commercial acceptance. In: Plants

Kushwah N, Billore V, Sharma OP, Singh D, Chauhan APS (2024) Integrated nutrient management for optimal plant health and crop yield. Plant Sci Arch 8:10–12

Leahy J, Mendelsohn M, Kough J, Jones R, Berckes N (2014) Biopesticide oversight and registration at the US Environmental Protection Agency. In: Biopesticides: state of the art and future opportunities, pp 3–18, ACS, Washington, DC.

Leggett MJ, Peter Setlow SA, Sattar, Maillard J-Y (2016) Assessing the activity of microbicides against bacterial spores: knowledge and pitfalls. J Appl Microbiol 120:1174–1180

Lengai GMW, Muthomi JW (2018) Biopesticides and their role in sustainable agricultural production. J Biosci Med 6:7–41

Lucia A, Guzmán E (2021) Emulsions containing essential oils, their components or volatile semiochemicals as promising tools for insect pest and pathogen management. Adv Colloid Interf Sci 287:102330

Lushchak VI, Matviishyn TM, Husak VV, Storey JM, Storey KB (2018) Pesticide toxicity: a mechanistic approach. EXCLI J 17:1101

Magierowicz K, Górska-Drabik E, Golan K, Protection. (2020) Effects of plant extracts and essential oils on the behavior of *Acrobasis advenella* (Zinck.) caterpillars and females. J Plant Dis 127:63–71

Mahmud AA, Upadhyay SK, Srivastava AK, Bhojiya AA (2021) Biofertilizers: a Nexus between soil fertility and crop productivity under abiotic stress. Curr Res Environ Sustain 3:100063

Mawcha KT, Malinga L, Muir D, Ge J, Ndolo D (2025a) Recent advances in biopesticide research and development with a focus on microbials. F1000Res 13:1071

Mawcha KT, Malinga L, Muir D, Ge J, Ndolo D (2025b) Recent advances in biopesticide research and development with a focus on microbials. J FRes 13:1071

Miresmailli S, Isman MB (2014) Botanical insecticides inspired by plant–herbivore chemical interactions. Trends Plant Sci 19:29–35

Mizubuti ESG, Júnior VL, Forbes GA (2007) Management of late blight with alternative products. J Pest Technol 1:106–116

Mohamed GS (2015) Toxicity of Basil (*Ocimum basilicum* L.) and Rosemary (*Rosmarinus officinali* L.) extracts on Tribolum confusum (DuVal)(Coleoptera: Teneberionidae). J Phytopathol Dis Manag 1:27–33

Moustafa MAM, Awad M, Amer A, Hassan NN, Ibrahim E-DS, Ali HM, Akrami M, Salem MZM (2021) Insecticidal activity of lemongrass essential oil as an eco-friendly agent against the black cutworm Agrotis ipsilon (Lepidoptera: Noctuidae). Insects 12:737

Nakachew K, Yigermal H, Assefa F, Gelaye Y, Ali S (2024) Review on enhancing the efficiency of fertilizer utilization: strategies for optimal nutrient management. Open Agric 9:20220356

Ndolo D, Njuguna E, Adetunji CO, Harbor C, Rowe A, Den Breeyen A, Sangeetha J, Singh G, Szewczyk B, Anjorin TS (2019) Research and development of biopesticides: challenges and prospects. Outlooks Pest Manag 30:267–276

Nefzi AJ, Jabnoun-Khiareddine H, Medimagh-Saïdana S, Haouala R et al (2016) Aydi Ben Abdallah R. J Microb Biochem Technol 8:144–150

Ngongang MD, Tchonang PE, Sameza ML, Mback MNLN, Lordon CD, Boyom FF (2022) Chemical constituents of essential oils from *Thymus vulgaris* and *Cymbopogon citratus* and their insecticidal potential against the tomato borer, *Tuta absoluta* (Lepidoptera: Gelechiidae). Int J Trop Insect Sci 42:31–43

Nygaard Sorensen J, Thorup-Kristensen K (2011) Plant-based fertilizers for organic vegetable production. J Plant Nutr Soil Sci 174:321–332

Oerke E-C (2006) Crop losses to pests. J Agric Sci 144:31–43

Ouhaddou R, Ech-Chatir L, Anli M, Ben-Laouane R, Boutasknit A, Meddich A (2023) Secondary metabolites, osmolytes and antioxidant activity as the main attributes enhanced by biostimulants for growth and resilience of lettuce to drought stress. Gesunde Pflanzen 75:1737–1753

Pandya IY (2018) Pesticides and their applications in agriculture. Asian J Appl Sci Technol 2:894–900

Parthasarathi R, Sivasakthivelan P, Arivukkarasu K, Akash K, Gnanasekaran A, Poonguzhali P, Vijayakumar N (2023) Biosurfactant-based biocontrol of agriculturally important plant pathogens. Eur Chem Bullet 12:3279

Pavela R (2016) History, presence and perspective of using plant extracts as commercial botanical insecticides and farm products for protection against insects–a review. Plant Prot Sci 52:229

Poveda J, Díez-Méndez A (2023) Use of elicitors from macroalgae and microalgae in the management of pests and diseases in agriculture. Phytoparasitica 51:667–701

Rana A, Sharma N, Tyagi M (2019) Impact of chemical pesticides *vs.* biopesticides on human health and environment. Int J All Res Writ 2:45–51

Rezvani Boroujeni, Samira MK, Asgharzadeh A, Baharlouei J (2021) Evaluating the potential of Halothiobacillus bacteria for sulfur oxidation and biomass production under saline soil. Geomicrobiol J 38:57–65

Sanganyado E, Munjanja B, Nyakubaya VT (2015) Biopesticide residue in water. In: Biopesticides handbook. CRC Press, Boca Raton, pp 93–117

Sangha JS, Kelloway S, Critchley AT, Prithiviraj B (2014) Seaweeds (macroalgae) and their extracts as contributors of plant productivity and quality: the current status of our understanding. Adv Bot Res 71:189–219

Santos F, Melkani S, Oliveira-Paiva C, Bini D, Pavuluri K, Gatiboni L, Mahmud A, Torres M, McLamore E, Bhadha JH (2024) Biofertilizer use in the United States: definition, regulation, and prospects. Appl Microbiol Biotechnol 108:511

Şengül Demirak, Meryem Ş, Canpolat E (2022) Plant-based bioinsecticides for mosquito control: impact on insecticide resistance and disease transmission. Insects 13:162

Siddiqui A, Ahmad S, Afaq U (2025) Innovations in biopesticides: the role of plant-based biopesticides in sustainable agriculture. J Biopest 18:18–35

Silver KS, Yuzhe D, Nomura Y, Oliveira EE, Salgado VL, Zhorov BS, Dong K (2014) Voltage-gated sodium channels as insecticide targets. Adv Insect Physiol 46:389–433

Singh S, Singh RJ, Kumar K, Singh B, Shukla L (2013) Biofertilizers and green manuring for sustainable agriculture. In: Modern technologies for sustainable agriculture, pp 129-150, New India Publishing Agency, New Delhi.

Singh D, Thapa S, Geat N, Mehriya ML, Rajawat MVS (2021) Biofertilizers: mechanisms and application. In: Biofertilizers. Elsevier, Amsterdam.

Siyavush MB (2015) Agro-production factors and marketing: determination of current situation and perspectives on the basic of coordination. Aust J Basic Appl Sci 9:1–7

Souto AL, Sylvestre M, Tölke ED, Tavares JF, Barbosa-Filho JM, Cebrián-Torrejón G (2021) Plant-derived pesticides as an alternative to pest management and sustainable agricultural production: prospects, applications and challenges. Molecules 26:4835

Srivastav AL (2020) Chemical fertilizers and pesticides: role in groundwater contamination. In: Agrochemicals detection, treatment and remediation. Elsevier, Amsterdam.

Statista Fertilizer industry worldwide—statistics & facts, https://www.statista.com/topics/8956/fertilizer-industry-worldwide/#topicOverview:. Accessed on August 2025

Suthar H, Hingurao K, Vaghashiya J, Parmar J (2017) Fermentation: a process for biofertilizer production. In: Microorganisms for Green Revolution: volume 1: microbes for sustainable crop production. Springer, Singapore.

Tripathi AK, Prajapati VPV, Aggarwal KK, Kumar SKS (2000) Effect of volatile oil constituents of *Mentha* species against the stored grain pests, *Callosobruchus maculatus* and *Tribolium castaneum*

Uddin M, Hussain S, Khan M, Masroor A, Hashmi N, Idrees M, Naeem M, Dar TA (2014) Use of N and P biofertilizers reduces inorganic phosphorus application and increases nutrient uptake, yield, and seed quality of chickpea. Turk J Agric For 38:47–54

Verardi A, Sangiorgio P, Mura BD, Moliterni S, Spagnoletta A, Dimatteo S, Bassi D, Cortimiglia C, Rebuzzi R, Palazzo S, Errico S (2025) *Tenebrio molitor* Frass: a cutting-edge biofertilizer for sustainable agriculture and advanced adsorbent precursor for environmental remediation. Agronomy 15:758

Worku LA, Bachheti RK, Tadesse MG (2022) Isolation and characterization of natural cellulose from *Oxytenanthera abyssinica* (Lowland Ethiopian Bamboo) using alkali peroxide bleaching stages followed by aqueous chlorite in buffer solution. Int J Polym Sci 2022. https://doi.org/10.1155/2022/5155552

Chapter 4
Role of Plants in Enhancing Soil Health and Crop Productivity

Noureddine Chaachouay, Ouafae Benkhnigue, and Lahcen Zidane

Abstract Plants play vital roles in enhancing soil health and sustaining agricultural productivity through several synergistic and interactive mechanisms. This chapter gives an overview of the current scientific literature related to the role of plants in soil ecosystem in terms of four major processes that contribute to the addition of organic matter by plant residues and roots exudates that stimulate microbial activity and humus formation; nutrient cycling by processes related to nitrogen fixation, phosphorus solubilization, and micronutrient mobilization; enhanced physical structure by root contribution; and enhanced soil biodiversity by microbiome--related interactions in the rhizosphere-plant interface. Plant-related scientific literature on cover crops, agroforestry, and green manure-related case studies in different corners of the world has conclusively revealed that specific plant related cover crops, agroforestry, and green manure are capable of playing important roles in sustaining water, nutrient, and carbon capture capabilities in addition to reducing the dependence on chemical inputs. Some plants, such as legumes, deep-rooted perennials, drought-tolerant millets, etc., have potential that can be harnessed to regenerate soils across different agro-ecologies. However, the major hindrance to its utilization on a larger scale is cost, as well as farmers' ignorance. Scientific development of newer technologies related to microbiological consortia, biochar, and precision agriculture will be of great help in boosting natural processes. Taken together, this chapter examines how strategically used plants can lead to healthier soils and more productive, sustainable agricultural practices. Moreover, it describes how

N. Chaachouay (✉)
Agri-Food and Health Laboratory (AFHL), ESEF of Berrechid; Hassan First University, Settat, Morocco

O. Benkhnigue
Department of Botany and Plant Ecology, Scientific Institute, Mohammed V University, Rabat, Morocco

L. Zidane
Plant, Animal Productions and Agro-industry Laboratory, Department of Biology, Faculty of Sciences, Ibn Tofail University, Kenitra, Morocco

L. A. Worku et al. (eds.), *Plant-Based Solutions for Sustainable Agriculture and Healthcare*, Sustainability Sciences in Asia and Africa,
https://doi.org/10.1007/978-981-95-8910-4_4

plants enhance soil fertility and structure, with the consequence that this should result in sustainable increases in crop yield. It explains the processes, such as nutrient cycling and soil organic matter accumulation, that enable this.

Keywords Cover Cropping · Microbial Ecology · Nutrient Cycling · Plant-Soil Interactions · Soil Health · Sustainable Agriculture

4.1 Introduction

The concept of soil health is considered a major aspect of sustainable production because it directly affects crop growth, sustainability, and food security (Parr et al. 1992). It supplies key nutrients, maintains water balance, and supports a multifaceted micro-community that sustains ecosystem functioning. Given that soil quality is becoming a security concern amid agricultural intensification and climate change, the world has shifted its focus to soil preservation and rehabilitation (Lal 2012). Plants can significantly contribute to this task by improving soil structure, increasing nutrient content, and promoting biological activity; thus, their absence is not supportive of a sustainable agricultural system (Tarafdar 2022).

Plant-soil interactions determine the interactions between the plant and soil ecosystems and include numerous processes that constitute a highly dynamic relationship (Husen 2024, 2025). Organic compounds released by root exudates trigger microbial growth, which ensures recycling of nutrients and decomposition of organic matter (Ma et al. 2022; Neumann 2007). These exudates also contribute to the physical cementation of soil particles, helping them aggregate and preventing erosion. In addition, root channels increase aeration and water uptake and enhance overall soil health (Bodner et al. 2021). Such interactions can be described as examples of mutualism between plants and soil, in which both benefit and continue to thrive together.

The most noticeable contributions of plants to soil health are the availability of organic matter from crop and animal litter and root degradation (Neumann 2007). When such materials break down, they also replenish the soil with humus, thereby enhancing its fertility, water-retention capacity, and structure. Socio-forestry and cover crop systems are effective at storing organic matter in the soil and at controlling nutrients and weeds (Singh et al. 2024). The continual incorporation of organic matter helps plants maintain long-term productivity and reduce reliance on synthetic fertilizers (Syers 1997).

In addition to their physical and chemical benefits, plants enhance soil biodiversity by supporting a wide variety of microorganisms, fungi, and earthworms (Sylvain and Wall 2011). Because they can increase the length of the root and improve nutrient uptake, mycorrhizal fungi are symbiotic with plant roots. The organic matter is further decomposed by earthworms and related bacteria, which release nutrients in usable forms for plants (Akhila and Entoori 2022). A wide array of soil microbiome-enhanced crop productivity is a reason for a more resilient agricultural system, characterized by increased disease resistance and tolerance to stress (Afridi et al. 2022).

Plants can enhance soil well-being in a practical way that supports sustainable farming. Methods such as crop rotation, intercropping, or conservation tillage take advantage of plant-soil relationships to conserve soil and minimize environmental damage (Rehman et al. 2025). By focusing on more environmentally friendly plant systems, farmers will be able to sustain their production without endangering the long-term sustainability of their businesses (Srivastav and Das 2025). This chapter examines how strategically used plants can result in healthier soils and more productive, sustainable agricultural activities. It also describes how plants enhance soil fertility and structure, with the consequence that this should result in sustainable increases in crop yield. It explains the processes, such as nutrient cycling and soil organic matter accumulation, that enable it.

4.2 Soil Health: Definition and Key Indicators

Soil health is generally perceived as the continued capacity of soil to support a living ecological community that nourishes plants, animals, and people and is physically, chemically, and biologically balanced (Lehmann et al. 2020). A good soil supports dense crop growth, controls water flow, recycles nutrients, and is resilient to damage (Rawal and Choudhary 2023). There are three main categories of indicators scientists use to measure soil health: physical, chemical, and biological. These characteristics reflect the soil's functionality and its ability to sustain agricultural productivity and environmental balance.

Examples of physical soil characteristics that serve as indicators of soil health include soil structure, porosity (and water retention), and their effects on root growth, penetration, aeration, and water availability (Usharani et al. 2019). Well--aggregated soils with stable crumb structure are resistant to erosion and enhance water infiltration, but poor structure leads to compaction, resulting in low productivity (Rajput et al. 2022). Porosity affects the soil's ability to facilitate gas exchange, since this is also necessary for plant respiration; this depends on the aggregate structure of the particles. In addition, the water-holding capacity plays a crucial role in plant drought tolerance, as it can retain water. These two physical properties underpin the soil's ability to support vegetation and alleviate environmental stresses (Abdul Khalil et al. 2015).

The fertility of the soil can be assessed using chemical indicators, such as nutrient content, pH, and organic matter, and predictions of future productivity can be made (Diacono and Montemurro 2011). Optimum levels of macronutrients (e.g., nitrogen, phosphorus, and potassium) stimulate optimal plant growth, and soil pH helps nutrients remain in a soluble state and also affects the microflora (Shrivastav et al. 2020). A significant component of healthy soils is the health of the organic matter, which enhances the CEC, WRC, and also microbial diversity. Microbial biomass, earthworm populations, and enzymatic activities can serve as biological indicators and provide insight into the living components of the soil (Alkorta et al. 2003). High levels of microbial activity can accelerate the decomposition of organic matter, and, through bioturbation, earthworms can improve soil texture. Through the enzymatic activities of phosphatase and dehydrogenase, there is some indication of active metabolic function, and prompt nutrient mineralization can also occur (Neemisha and Sharma 2022). Each indicator provides a moderate assessment of soil health, which ultimately leads to sustainable land use.

4.3 Mechanisms by Which Plants Improve Soil Health

Plants are key agents of soil-based health, functioning within a complex set of interdependent processes that enhance the physical, chemical, and biological well-being of soil systems (Tarafdar 2022; Abate et al. 2021). These natural processes are a self-perpetuating mechanism in which plants and resident microorganisms function synergistically. This interaction enables the cooperation of soils and plants to support primary soil activity and enhance overall soil functionality (Wang et al. 2024; Abate et al. 2022). The main processes are the assimilation of organic material, the circulation of nutrients, the improvement of the soil structure, and the amplification of microbial populations (Chaachouay et al. 2025a, b). All of these processes accumulate and maintain fertile, strong soils that sustain high plant growth and curb environmental deterioration. It will be critical to make informed decisions regarding the execution of agricultural processes that align with natural processes rather than being hostile to them.

Weed discontinuity is one of the main contributions of plants to soil health, as they continuously add organic matter. This is achieved through several processes, including the deposition of above-ground litter (leaves, stems), root turnover, and root exudation (Chaachouay et al. 2025a, b). Plants die; their remains, in the form of tissues, are broken down in the soil matrix to form a biological building block that eventually forms stable humus through a complex biochemical reaction (Witkamp 1971). Complex organic substances are decomposed into simpler substances through the presence of soil fauna (e.g., earthworms, arthropods) and microorganisms (bacteria, fungi). Among the consequences of its humus-forming process, and in particular its ability to reach slow carbon pools that can last decades or centuries, are the benefits of improved water retention, cation exchange capacity, and soil productivity (Gerke 2022). Root exudates (sugars, organic acids, and proteins) are the best substrates for root-zone microbes and form hotspots of biological activity that drive nutrient cycling (Ma et al. 2022).

Plants also play a major role in averaging the complex process of nutrient cycling in the soil, thereby helping maintain soil fertility. Leguminous plants can form a symbiotic union with nitrogen-fixing microorganisms, which convert atmospheric nitrogen into ammonium compounds, making them available to plants within specialized root nodules (Raza et al. 2020). The capacity of this type of biological nitrogen fixation to add significant amounts of nitrogen to agricultural systems means that dependence on synthetic fertilizers can be reduced. Besides nitrogen, plants maximize phosphorus uptake through mycorrhizal associations (Kebede 2021). These fungal networks spread much farther than the root system of the plant itself to gain access to phosphorus pools that would not otherwise be accessible to plants. There is a mutualistic relationship between the fungus and the plant, in which the fungus obtains carbohydrates in exchange for mobilizing phosphorus to the plant (Lanfranco et al. 2016). Root exudates chelate metal ions, and some root--associated microorganisms also influence the cycling of potassium and micronutrients (such as iron, zinc, and manganese) in plants. They offer more sustainable and efficient fertilization systems than conventional methods, as they minimize nutrient loss from leaching or volatilization and maintain soil chemical balance (More et al. 2020).

One effect of plant roots on soil structure is the way roots grow and interact biologically with the soil. The roots, as they dig into the ground, open channels that enhance aeration and water absorption while also disrupting compacted soil (Ogorek et al. 2025). Tap roots would go deep into the subsoil, while fibrous root systems would produce dense networks within the topsoil (Bodner et al. 2021). Besides structural amelioration, plant root systems before and after the erosion process make a significant contribution to the settlement of erosion (Degens 1997). The physical attributes of fine root systems include the ability to anchor the soil and provide vegetative cover over the ground that averts erosion by wind and water (Gyssels et al. 2005). In particular, perennials and cover crops help ensure the ground is always covered, thus preventing the harmful effects of bare ground or tilled ground. Plant-based systems are highly effective in long-term soil protection because they provide physical security and improve soil structure.

One of the most biologically active spheres in terrestrial ecosystems is the soil environment, which is in direct contact with plant roots. Plants actively produce root exudates to regulate their rhizosphere microbiome: beneficial microorganisms or pathogens may be attracted or repelled, respectively (Olanrewaju et al. 2019). Such selective enrichment produces heterogeneous communities of microbes, which carry out critical ecosystem processes, such as nutrient cycling, disease control, and organic matter breakdown. The species *Pseudomonas* and *Bacillus* can solubilize phosphorus, fix nitrogen, and produce plant growth hormones (Timofeeva et al. 2023). Mycorrhizal fungi promote root growth and enhance nutrient uptake in plants (Shao et al. 2018). Mutual exchange of carbon compounds, nutrients, and services creates a self-reinforcing relationship in which both parties are winners. The microbial communities also enhance soil aggregate stability by releasing sticky extracellular polymers and elongating hyphae, and in the process connect the particles of the earth (Chaachouay et al., 2023). It is now accepted that the capacity of various metabolically active microbial communities to persist is a key determinant of plant productivity, disease suppression, and ecosystem resilience during climatic perturbations (Fig. 4.1).

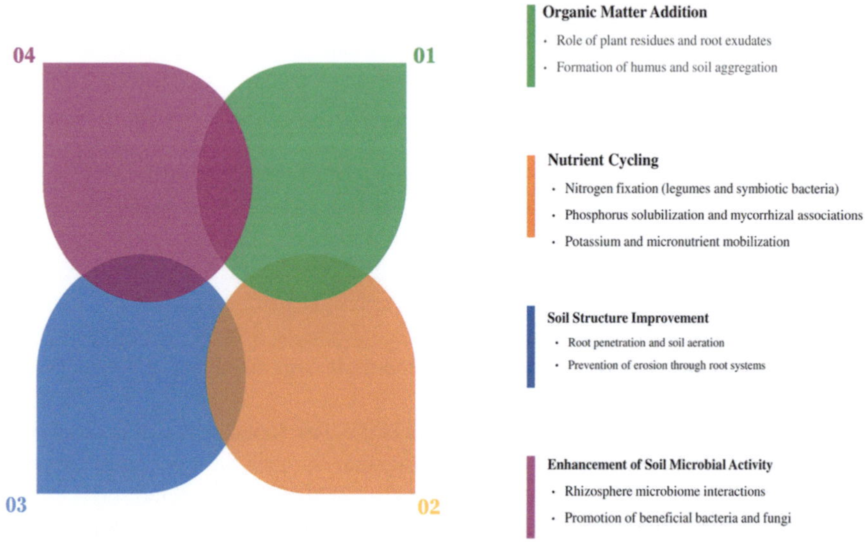

Fig. 4.1 Mechanisms by which plants improve soil health

4.4 Plant-Based Strategies for Soil Health Improvement

Legume cover crops like clover and vetch increase soil fertility by biological nitrogen fixation through a symbiotic relationship with rhizobial bacteria (Zhou et al. 2017). The extensive root systems of perennial grasses, such as rye and oats, help to hold back sediments, reduce erosive forces, and increase the inventory of organic matter in the soil matrix (Magdoff and Weil 2004). Plants such as mustard and radish, also known as brassicas, are unique because they exude natural chemicals that kill soil-borne diseases and pests. In addition to their respective functions, all cover crops reduce soil erosion, enhance rainwater infiltration, and form a living mulch that prevents evaporation and weed growth through competition, rather than relying on herbicides (Waisen 2019). They create a more balanced soil microecosystem as different varieties of microbial communities are fed throughout the year.

Strategic crop rotation disrupts the cycle of soil-borne pests and diseases because the environment that harbors them is constantly changed by alternating host crops. Different crop families support diverse microbial communities in the soil, resulting in a more balanced and resilient ecosystem (Philippot et al. 2021). Crops in rotation systems, e.g., alfalfa or sunflower, are long-rooted and contribute to the recapturing of the nutrients, which have leached deeper into the soil than can be captured by the short-rooted crops (Mohammad 2023). Mixed quality and root architectures across different crops support more predictable soil organic matter dynamics than monocultures (De Long et al. 2019). The shifting structure and other aspects of these diversified systems demonstrate improved nutrient cycling and tolerance to environmental conditions, which underlie the changing inputs of various plant species, reflected in enhanced soil structure and biological activity.

Agroforestry is the association of trees and shrubs with traditional crops and the construction of stratified structures that result in the continuous improvement of soil health. Then the availability of well-established perennial vegetation carries the nutrients found in the subsoil layers to the soil surface through plant litter, thus providing natural fertilization of the system (Tariq et al. 2024). Trees and shrubs have broad root systems that store carbon in the soil over the long term, in addition to providing ground cover year-round (Tariq et al. 2024). These vegetation covers maintain normal soil temperatures, eliminating extremes, and improve the ability to hold water throughout all seasons. Organic input in the form of tree cutting provides ample food for earthworms, which in turn aerate the earth naturally and create channels/pathways through which water can flow by digging through the soil (Chaachouay et al. 2025a, b).

Green manuring is the process of planting plants to eventually add nutrients to the soil, particularly those with rapid growth rates (Yandvinder-Singh et al. 1992). Rapid growers, such as *Crotalaria juncea* L., accumulate large amounts of biomass, with a significant proportion of nitrogen being fixed through natural processes (Barbosa et al. 2020). In addition to being nitrogen-fixing, clover varieties have longer blooming periods, providing valuable pollinator resources. The timing of incorporating green manures is also important; incorporating them at the early flowering stage by cutting and mixing the plants maximizes nutrient availability as the material decomposes (Fageria 2007). The practice yields direct benefits for successive crops while also enhancing soil structure by increasing organic matter (Magdoff and Weil 2004). It also reduces reliance on chemicals, as the canopy growth of green manure crops competes with unwanted weeds, thereby naturally suppressing them (Fig. 4.2).

4.5 Impact of Soil Health on Crop Productivity

Healthy soils act as sponges, boasting strong structural integrity and high organic matter content, which dramatically increases water retention and makes it more available to crops. Pore spaces in well-structured soils, comprising stable aggregates, can facilitate deep water penetration and minimize surface runoff (Liu et al. 2025). Organic matter increases the soil's water-retention capacity, ensuring the soil has moisture when rainfall is scarce (Liu et al. 2025). Such increased water-use efficiency means that plants require less water to survive, and agricultural systems can be more sustainable in water-limited settings, still producing consistent yields under unpredictable rainfall.

A biologically active soil ecosystem restructures the availability of essential nutrients to plants. Healthy soils are composed of communities of microbes that decompose organic materials and mineral particles, thereby releasing nitrogen, phosphorus, and micronutrients to plants (Prasad et al. 2021). Mycorrhizal fungi

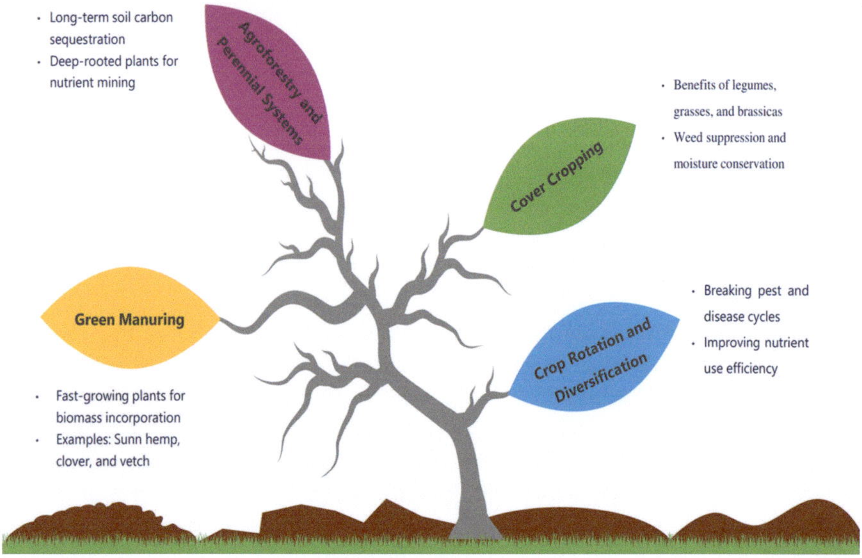

Fig. 4.2 Plant-based strategies for soil health improvement

increase root length, enabling plants to access nutrients beyond the root zone (Hetrick 1991). Natural nutrient cycling minimizes the need for man-made auxiliaries and can produce more balanced plant food at the same time. This results in improved crop and plant growth and quality because plants and crops receive a constant, diverse supply of nutrients tailored to their developmental stage, rather than the periodic increases and decreases of chemical manure.

Healthy soils confer exceptional resilience in crops under environmental stress. Due to the resulting increase in water-absorbing capacity, the plants can tolerate dry conditions, and the diverse microbial population composition enables them to withstand the impacts of salinity stress by eliminating ions through regulation of uptake (Farooqi et al. 2023). The physical properties of healthy soils prevent them from forming a crust that restricts root growth during periods of drought. The symbiotic associations between plants and soil microorganisms trigger plants' natural defence mechanisms, thereby enhancing crops' ability to withstand numerous abiotic stresses (Munir et al. 2022). This hardiness translates into a more consistent yield year after year, despite the volatile environmental factors that would devastate crops planted in degraded soils.

The natural fertility processes are geared to produce a creative response to the demand for crops with better soil health. Phosphorus-solubilizing microbes, nitrogen-fixing bacteria, and other beneficial organisms are increasing the dependency of plants on manufactured fertilizers (Timofeeva et al. 2023). The buffering capacities are high in organic-rich soils, hence enabling a more efficient utilization of any fertilizers applied (Bashir et al. 2021). In the long term, well-managed soils can maintain their high productivity with far fewer synthetic applications, thereby reducing production costs and having minimal environmental impact (Raj et al. 2019). This change reflects the shift from an input-intensive farm production process to a more knowledge-intensive farm management process, where knowledge and soil biological management, rather than chemical fixes, take center stage.

4.6 Case Studies and Examples

It has been proven that plants can significantly improve the quality of degraded soils and increase agricultural output (Ait Elallem et al. 2021; Ayub et al. 2020; Chaachouay et al. 2024). Regenerative methods that incorporate deep-rooted perennials, nitrogen-fixing legumes, and drought-resistant millets have been highly effective in facilitating soil renewal (Finley 2016). Longevity tests with *Medicago sativa* in semi-arid regions have shown that it can fix atmospheric nitrogen and open compacted subsoil layers through its extensive taproot system. Likewise, *Cajanus cajan* intercropping systems in tropical Alfisols have increased phosphorus solubilization (40–60) via root exudate-mediated processes. The use of ancient strains of millet (e.g., finger millet, pearl millet) for restoring drought-resistant soils and as a source of staple food is being rediscovered (Table 4.1).

These studies collectively demonstrate how strategic plant selection can address specific soil constraints, while enhancing crop productivity through ecological intensification. The success stories particularly highlight the untapped potential of (1) deep-rooted perennials for carbon sequestration, (2) legume-rhizobia symbioses for nitrogen economy, and (3) millet-based systems for climate resilience. Studies

Table 4.1 A synthesis of key documented cases

Plant species	Soil improvement mechanism	Productivity impact	Reference
Alfalfa (*Medicago sativa*)	Nitrogen fixation (Rhizobium symbiosis), subsoil biological tillage	Increased subsequent wheat yields by 25–40%	Bagavathiannan and Van Acker (2009), Cangioli et al. (2025)
Pigeon pea (*Cajanus cajan*)	Phosphorus mobilization via root exudates (piscidic acid), drought tolerance	Intercrop maize yields boosted by 1.2–1.8 t/ha	Liu et al. (2022), Vora et al. (2022)
Finger millet (*Eleusine coracana*)	Enhanced mycorrhizal networks, water-use efficiency	30% higher yields than rice under drought	Singh et al. (2019)
Sunn hemp (*Crotalaria juncea*)	Rapid biomass production (8–12 t/ha in 60d), nematode suppression	Soybean yields increased by 15–20% after green manure	Curto et al. (2015), Patel and Dhillon (2017)
Kernza (*Thinopyrum intermedium*)	Perennial roots sequester 3× more soil C than annual wheat	Consistent grain/straw production for 5+ years	Bergquist (2019)

confirm that such plant-driven solutions can regenerate degraded soils within 3–5 growing seasons when properly managed (Drinkwater and Snapp 2022; Negi et al. 2025).

4.7 Challenges and Future Perspectives

Although there are documented benefits of integrating vegetation into land-management regimes to improve soil health, several obstacles have limited the mainstream adoption of these practices. Economic barriers, such as the high cost of planting cover crop seeds or the lack of profitability in perennial systems, deter smallholder farmers from adopting more sustainable systems (Snapp et al. 2005). The knowledge gaps are also a significant problem, as most farmers do not understand the best plant species, optimal planting times, and effective management practices, which vary depending on local conditions. In addition, long-term market and short-run economic forces and demands favor short-run yields and oppose long-run investments in soil, thereby opposing the adoption of regenerative practices (Hoelscher 2022). Unique solutions must be designed to overcome the effects of these barriers by combining financial resources, access to education, and scientifically proven long-term benefits.

Recent advances in the science of plant-soil interactions offer promising solutions to improve soil health and crop yields. Water-holding capacity and microbial activity can be increased by incorporating biomimetic releases into a mixture with plant waste. There are also developments in inoculants to improve plant nutrient and stress-toleration properties, such as mycorrhizal fungi and nitrogen-fixing bacteria (Sammauria et al. 2020). Besides, precision agriculture technologies (e.g., remote sensing, soil microbiome analysis via sequencing) can help monitor soil conditions in real time and inform related changes in management practices. They are found to significantly enhance soil restoration and productivity when applied alongside other traditional vegetative practices such as crop rotation and agroforestry.

Farmers should be educated about supportive policies and programs that help them adopt soil health practices that involve using plants. Regenerative practices should be encouraged through incentive programs that include subsidies, crop insurance, and government- and agricultural-organization carbon credit schemes that reward carbon sequestration in soils (Marks 2019). The knowledge transfer gaps between scientists and farmers can be addressed through direct instructional practices and farmer-to-farmer interactions, with a view to bridging the knowledge and practice gap. In addition, educating schools and online platforms about soil health literacy may make it easier for future generations of farmers to adopt sustainable farming practices. Scientific innovation, policy structures, and community education can be used to transform the agricultural sector to focus on plant utilization as the primary means of generating resilient, productive soils, thereby guaranteeing a future food supply.

4.8 Conclusion

Plants are also critical collaborators in the development and preservation of soil health and, at the same time, increasingly enhance agricultural output. The vegetation mediates the conversion of degraded soils to healthy and productive ecosystems through a complex of synergistic processes, such as the enrichment of organic matter, nutrient cycling, enhancement of soil structure, and biota mobilization. Empirical literature reveals that cover crops in combination with nitrogen-fixing legumes, deep-rooted perennials, and green manure have significant potential to enhance soil water storage capacity, nutrient storage, and carbon fixation. Plant-based methods not only enhance crop yields but also make the system more resilient to weather-stressful conditions like drought and heat. Despite the proven benefits, there are still obstacles, such as economic constraints, lack of knowledge, and the risk of short-run yield decline. These biological mechanisms offer significant opportunities for improvement through updates to microbial inoculants, more accurate agricultural tools, and biochar additions. The promise of advancing the level of these practices is high with context-specific solutions that combine modern scientific knowledge with traditional ecological principles. Climate change and global food security crises are not problems of tomorrow but necessities of today. Transforming depleting agricultural systems to regenerative systems. With a regenerative agrarian system, the shift can be achieved by leveraging plants' soil-engineering capabilities. This paradigm shift must involve all stakeholders, including farmers, researchers, policy-makers, and consumers. It is, further, scientifically valid: by investing in the health of plant-based soils now, future generations will gain numerous benefits in food output, environmental well-being, and climate sustainability. The future of the food system lies in the decisions made today regarding soil management.

References

Abate L, Bachheti A, Bachheti RK, Husen A (2021) Antibacterial properties of medicinal plants: recent trends, progress, and challenges. In: Traditional herbal therapy for the human immune system. CRC Press, pp 13–54

Abate L, Tadesse MG, Bachheti A, Bachheti RK (2022) Traditional and phytochemical bases of herbs, shrubs, climbers, and trees from Ethiopia for their anticancer response. Biomed Res Int 2022. https://doi.org/10.1155/2022/1589877

Abdul Khalil HPS, Hossain MS, Rosamah E, Azli NA, Saddon N, Davoudpoura Y, Islam MN, Dungani R (2015) The role of soil properties and it's interaction towards quality plant fiber: a review. Renew Sust Energ Rev 43:1006–1015. https://doi.org/10.1016/j.rser.2014.11.099

Afridi MS, Javed MA, Ali S, De Medeiros FHV, Ali B, Salam A, Sumaira Marc RA, Alkhalifah DHM, Selim S (2022) New opportunities in plant microbiome engineering for increasing agricultural sustainability under stressful conditions. Front Plant Sci 13:899464

Ait Elallem K, Sobeh M, Boularbah A, Yasri A (2021) Chemically degraded soil rehabilitation process using medicinal and aromatic plants: review. Environ Sci Pollut Res 28(1):73–93. https://doi.org/10.1007/s11356-020-10742-y

Akhila A, Entoori K (2022) Role of earthworms in soil fertility and its impact on agriculture: a review. Int J Fauna Biol Stud 9(3):55–63

Alkorta I, Aizpurua A, Riga P, Albizu I, Amézaga I, Garbisu C (2003) Soil enzyme activities as biological indicators of soil health. Rev Environ Health 18(1). https://doi.org/10.1515/REVEH.2003.18.1.65

Ayub MA, Usman M, Faiz T, Umair M, Ul Haq MA, Rizwan M, Ali S, Zia Ur Rehman M (2020) Restoration of degraded soil for sustainable agriculture. In: Meena RS (ed) Soil health restoration and management. Springer Singapore, pp 31–81. https://doi.org/10.1007/978-981-13-8570-4_2

Bagavathiannan MV, Van Acker RC (2009) The biology and ecology of Feral Alfalfa *(Medicago sativa* L.) and its implications for novel trait confinement in North America. Crit Rev Plant Sci 28(1–2):69–87. https://doi.org/10.1080/07352680902753613

Barbosa IR, Santana RS, Mauad M, Garcia RA (2020) Dry matter production and nitrogen, phosphorus and potassium uptake in *Crotalaria juncea* and *Crotalaria spectabilis*. Pesquisa Agropecuária Trop 50:e61011. https://doi.org/10.1590/1983-40632020v5061011

Bashir O, Ali T, Baba ZA, Rather GH, Bangroo SA, Mukhtar SD, Naik N, Mohiuddin R, Bharati V, Bhat RA (2021) Soil organic matter and its impact on soil properties and nutrient status. In: Dar GH, Bhat RA, Mehmood MA, Hakeem KR (eds) Microbiota and biofertilizers, vol 2: ecofriendly tools for reclamation of degraded soil environs. Springer International Publishing, pp 129–159. https://doi.org/10.1007/978-3-030-61010-4_7

Bergquist GE (2019) Biomass yield and soil microbial response to management of perennial intermediate wheatgrass (Thinopyrum intermedium) as grain crop and carbon sink [Master's Thesis, University of Minnesota]. https://search.proquest.com/openview/d41ecf4d89d5b2efe4143694a2a992e6/1?pq-origsite=gscholar&cbl=18750&diss=y

Bodner G, Mentler A, Keiblinger K (2021) Plant roots for sustainable soil structure management in cropping systems. In: Rengel Z, Djalovic I (eds) The root systems in sustainable agricultural intensification, 1st edn. Wiley, pp 45–90. https://doi.org/10.1002/9781119525417.ch3

Cangioli L, Fagorzi C, Vaccaro F, Varriale S, Amenta ML, Mengoni A, Ingraffia R, Porto AL, Amato G, Giambalvo D, Sangaré JR, Defez R, Bianco C (2025) Soil bacteria and symbiotic rhizobia synergistically promote nitrogen fixation and biomass production of alfalfa (*Medicago sativa* L.) plants even under water shortage conditions without altering the native rhizospheric microbiota. Appl Soil Ecol 213:106283. https://doi.org/10.1016/j.apsoil.2025.106283

Chaachouay N, Azeroual A, Bencharki B, Douira A, & Zidane L (2023) Chapter 3—Hormonal interactions during fruit development and ripening. In A. Husen & W. Zhang (Éds.), Hormonal Cross-Talk, Plant Defense and Development (p. 37-46). Academic Press. https://doi.org/10.1016/B978-0-323-95375-7.00014-8

Chaachouay N, Zidane L, & Husen A (2025b) Enhancement of Seed Germination of Important Crop Plant Species Using Carbon-Based Nanomaterials. In A. Husen (Éd.), Emerging Carbon Nanomaterials for Sustainable Agricultural Practices : Synthesis, Plant Growth, Performance, Production and Protection (p. 137–147). Springer Nature. https://doi.org/10.1007/978-981-97-5104-4_7

Curto G, Dallavalle E, Santi R, Casadei N, D'Avino L, Lazzeri L (2015) The potential of Crotalaria juncea L. as a summer green manure crop in comparison to Brassicaceae catch crops for management of Meloidogyne incognita in the Mediterranean area. Eur J Plant Pathol 142(4):829–841. https://doi.org/10.1007/s10658-015-0655-2

De Long JR, Jackson BG, Wilkinson A, Pritchard WJ, Oakley S, Mason KE, Stephan JG, Ostle NJ, Johnson D, Baggs EM, Bardgett RD (2019) Relationships between plant traits, soil properties and carbon fluxes differ between monocultures and mixed communities in temperate grassland. J Ecol 107(4):1704–1719. https://doi.org/10.1111/1365-2745.13160

Degens BP (1997) Macro-aggregation of soils by biological bonding and binding mechanisms and the factors affecting these: a review. Soil Res 35(3):431–460. https://doi.org/10.1071/s96016

Diacono M, Montemurro F (2011) Long-term effects of organic amendments on soil fertility. In: Lichtfouse E, Hamelin M, Navarrete M, Debaeke P (eds) Sustainable agriculture, vol 2. Springer Netherlands, pp 761–786. https://doi.org/10.1007/978-94-007-0394-0_34

Drinkwater LE, Snapp SS (2022) Advancing the science and practice of ecological nutrient management for smallholder farmers. Front Sustain Food Syst 6. https://doi.org/10.3389/fsufs.2022.921216
Fageria NK (2007) Green manuring in crop production. J Plant Nutr 30(5):691–719. https://doi.org/10.1080/01904160701289529
Farooqi ZUR, Qadir AA, Alserae H, Raza A, Mohy-Ud-Din W (2023) Organic amendment–mediated reclamation and build-up of soil microbial diversity in salt-affected soils: fostering soil biota for shaping rhizosphere to enhance soil health and crop productivity. Environ Sci Pollut Res 30(51):109889–109920. https://doi.org/10.1007/s11356-023-30143-1
Finley S (2016) Sustainable water management in smallholder farming: theory and practice. CABI. https://books.google.fr/books?hl=fr&lr=&id=VwPpDAAAQBAJ&oi=fnd&pg=PR3&dq=The+implementation+of+regenerative+techniques,+which+include+deep-rooted+perennials,+nitrogen-fixing+legumes,+and+drought-resistant+millets,+has+proven+highly+successful+in+promoting+soil+renewal&ots=OO2bDUfiX6&sig=W1lzVSSdf4ilK6Wkf8Jrypi8Bxw
Gerke J (2022) The central role of soil organic matter in soil fertility and carbon storage. Soil Syst 6(2):33. https://doi.org/10.3390/soilsystems6020033
Chaachouay N, Azeroual A, Bencherki B, Douira A, & Zidane L (2024) Chapter 7—Use of melatonin in plants' growth and productions. In A. Husen (Éd.), Biostimulants in Plant Protection and Performance (p. 107-115). Elsevier. https://doi.org/10.1016/B978-0-443-15884-1.00004-X
Gyssels G, Poesen J, Bochet E, Li Y (2005) Impact of plant roots on the resistance of soils to erosion by water: a review. Prog Phys Geograph Earth Environ 29(2):189–217. https://doi.org/10.1191/0309133305pp443ra
Hetrick BAD (1991) Mycorrhizas and root architecture. Experientia 47(4):355–362. https://doi.org/10.1007/BF01972077
Hoelscher HA (2022) Economic evaluation of regenerative agriculture on producer profitability [PhD Thesis]. https://oaktrust.library.tamu.edu/items/cb6a4575-d24c-4aea-8ab8-ce25830fe556
Husen A (2024) Essential minerals in plant-soil system (coordination, signalling, and interaction under adverse situations). Elsevier Inc, Cambridge, MA. https://doi.org/10.1016/C2022-0-02437-5
Husen A (2025) Agricultural crop improvement: plant and soil relationships. Taylor & Francis Group, LLC. https://doi.org/10.1201/9781032630366
Kebede E (2021) Contribution, utilization, and improvement of legumes-driven biological nitrogen fixation in agricultural systems. Front Sustain Food Syst 5. https://doi.org/10.3389/fsufs.2021.767998
Lal R (2012) Climate change and soil degradation mitigation by sustainable management of soils and other natural resources. Agric Res 1(3):199–212. https://doi.org/10.1007/s40003-012-0031-9
Lanfranco L, Bonfante P, Genre A (2016) The mutualistic interaction between plants and Arbuscular Mycorrhizal fungi. Microbiol Spectrum 4(6):4.6.14. https://doi.org/10.1128/microbiolspec.FUNK-0012-2016
Chaachouay N, Zidane L, & Husen A (2025a) Carbon Nanomaterials for Healthier Crop Plants Growth and Enhanced Yield. In A. Husen (Éd.), Emerging Carbon Nanomaterials for Sustainable Agricultural Practices : Synthesis, Plant Growth, Performance, Production and Protection (p. 31-46). Springer Nature. https://doi.org/10.1007/978-981-97-5104-4_2
Lehmann J, Bossio DA, Kögel-Knabner I, Rillig MC (2020) The concept and future prospects of soil health. Nat Rev Earth Environ 1(10):544–553. https://doi.org/10.1038/s43017-020-0080-8
Liu C, Tai Y, Luo J, Wu Y, Zhao X, Dong R, Ding X, Zhao S, Luo L, Liu P, Liu G (2022) Integrated multi-omics analysis provides insights into genome evolution and phosphorus deficiency adaptation in pigeonpea (Cajanus cajan). Hortic Res 9:uhac107. https://doi.org/10.1093/hr/uhac107
Liu Y, Zhu L, Chen C (2025) The mechanisms and optimization strategies of tillage practices affecting soil pore structure. Geograph Res Bullet 4:507–529

Ma W, Tang S, Dengzeng Z, Zhang D, Zhang T, Ma X (2022) Root exudates contribute to below-ground ecosystem hotspots: a review. Front Microbiol 13:937940. https://doi.org/10.3389/fmicb.2022.937940

Magdoff F, Weil RR (2004) Soil organic matter management strategies. In: Soil organic matter in sustainable agriculture, pp 45–65

Marks AB (2019) (Carbon) farming our way out of climate change. Denv L Rev 97:497

Mohammad A (2023) Shelter management of dairy animals for profitable dairy farming. Int Year Millets 2023:60

More SS, Shinde SE, Kasture MC (2020) Root exudates a key factor for soil and plant: an overview. Pharma Innov J 8:449–459

Munir N, Hanif M, Abideen Z, Sohail M, El-Keblawy A, Radicetti E, Mancinelli R, Haider G (2022) Mechanisms and strategies of plant microbiome interactions to mitigate abiotic stresses. Agronomy 12(9):2069

Neemisha, Sharma S (2022) Soil enzymes and their role in nutrient cycling. In: Giri B, Kapoor R, Wu Q-S, Varma A (eds) Structure and functions of pedosphere. Springer Nature Singapore, pp 173–188. https://doi.org/10.1007/978-981-16-8770-9_8c

Negi M, Sharma S, Singh M (2025) Plant-driven soil dynamics: exploring the role of roots, fungi and soil traffic in shaping the soil structure. Commun Soil Sci Plant Anal 56(18):2727–2746. https://doi.org/10.1080/00103624.2025.2524075

Neumann G (2007) Root exudates and nutrient cycling. In: Marschner P, Rengel Z (eds) Nutrient cycling in terrestrial ecosystems, vol 10. Springer Berlin Heidelberg, pp 123–157. https://doi.org/10.1007/978-3-540-68027-7_5

Ogorek LLP, Gao Y, Farrar E, Pandey BK (2025) Soil compaction sensing mechanisms and root responses. Trends Plant Sci 30(5):565–575. https://doi.org/10.1016/j.tplants.2024.10.014

Olanrewaju OS, Ayangbenro AS, Glick BR, Babalola OO (2019) Plant health: feedback effect of root exudates-rhizobiome interactions. Appl Microbiol Biotechnol 103(3):1155–1166. https://doi.org/10.1007/s00253-018-9556-6

Parr JF, Papendick RI, Hornick SB, Meyer RE (1992) Soil quality: attributes and relationship to alternative and sustainable agriculture. Am J Altern Agric 7(1–2):5–11

Patel S, Dhillon NK (2017) Evaluation of sunnhemp (Crotalaria juncea) as green manure /amendment and its biomass content on root knot nematode (Meloidogyne incognita) in successive crop brinjal. J Entomol Zool Stud 5(6):716–720

Philippot L, Griffiths BS, Langenheder S (2021) Microbial community resilience across ecosystems and multiple disturbances. Microbiol Mol Biol Rev 85(2):e00026–e00020. https://doi.org/10.1128/MMBR.00026-20

Prasad S, Malav LC, Choudhary J, Kannojiya S, Kundu M, Kumar S, Yadav AN (2021) Soil microbiomes for healthy nutrient recycling. In: Yadav AN, Singh J, Singh C, Yadav N (eds) Current trends in microbial biotechnology for sustainable agriculture. Springer Singapore, pp 1–21. https://doi.org/10.1007/978-981-15-6949-4_1

Raj A, Jhariya MK, Yadav DK, Banerjee A, Meena RS (2019) Soil for sustainable environment and ecosystems management. In: Jhariya MK, Banerjee A, Meena RS, Yadav DK (eds) Sustainable agriculture, forest and environmental management. Springer Singapore, pp 189–221. https://doi.org/10.1007/978-981-13-6830-1_6

Rajput N, Tiwari A, Chandra MS, Singh S (2022) Soil water content, penetration resistance, crumb ratio and soil carbon restoration through conservation agriculture: an overview. Pharma Innov J 11(6):217–243

Rawal VK, Choudhary R (2023) Managing soil health: new strategies for sustainable agriculture. Novel Approach Agron 50. https://www.researchgate.net/profile/Subhra-Sahoo-2/publication/380998580_Novel_Approaches_in_Agronomy/links/6658854622a7f16b4f62387a/Novel-Approaches-in-Agronomy.pdf#page=61

Raza A, Zahra N, Hafeez MB, Ahmad M, Iqbal S, Shaukat K, Ahmad G (2020) Nitrogen fixation of legumes: biology and physiology. In: Hasanuzzaman IM, Araújo S, Gill SS (eds) The plant family Fabaceae. Springer Singapore, pp 43–74. https://doi.org/10.1007/978-981-15-4752-2_3

Rehman SU, Ijaz SS, Din AMU, Al-Dosary MA, Ansar M, Fatima S, Siddiqa A, Ashraf MN, Haider I, Junaid MB, Raza MA, Yang H (2025) Combined effects of reduced tillage and strip intercropping on soil carbon sequestration in semi-arid environment. J Soil Sci Plant Nutr 25(2):3124–3139. https://doi.org/10.1007/s42729-025-02324-z

Sammauria R, Kumawat S, Kumawat P, Singh J, Jatwa TK (2020) Microbial inoculants: potential tool for sustainability of agricultural production systems. Arch Microbiol 202(4):677–693. https://doi.org/10.1007/s00203-019-01795-w

Shao Y-D, Zhang D-J, Hu X-C, Wu Q-S, Jiang C-J, Xia T-J, Gao X-B, Kuča K (2018) Mycorrhiza-induced changes in root growth and nutrient absorption of tea plants. Plant Soil Environ 64(6):283–289

Shrivastav P, Prasad M, Singh TB, Yadav A, Goyal D, Ali A, Dantu PK (2020) Role of nutrients in plant growth and development. In: Naeem M, Ansari AA, Gill SS (eds) Contaminants in agriculture. Springer International Publishing, pp 43–59. https://doi.org/10.1007/978-3-030-41552-5_2

Singh D, Mathimaran N, Boller T, Kahmen A (2019) Bioirrigation: a common mycorrhizal network facilitates the water transfer from deep-rooted pigeon pea to shallow-rooted finger millet under drought. Plant Soil 440(1–2):277–292. https://doi.org/10.1007/s11104-019-04082-1

Singh NR, Singh A, Devi NP, Kumar YB, Sangma RHC, Philanim WS, Devi MP, Bhutia PL (2024) Agroforestry for soil health. In: Raj A, Jhariya MK, Banerjee A, Jha RK, Singh KP (eds) Agroforestry, 1st edn. Wiley, pp 255–283. https://doi.org/10.1002/9781394231164.ch9

Snapp SS, Swinton SM, Labarta R, Mutch D, Black JR, Leep R, Nyiraneza J, O'Neil K (2005) Evaluating cover crops for benefits, costs and performance within cropping system Niches. Agron J 97(1):322–332. https://doi.org/10.2134/agronj2005.0322a

Srivastav AK, Das P (2025) Sustainable soil health and fertility. In: Srivastav AK, Das P (eds) Biotechnology and IoT in agriculture and food production. Apress, pp 95–104. https://doi.org/10.1007/979-8-8688-1469-3_10

Syers JK (1997) Managing soils for long-term productivity. Philos Trans R Soc Lond B Biol Sci 352(1356):1011–1021. https://doi.org/10.1098/rstb.1997.0079

Sylvain ZA, Wall DH (2011) Linking soil biodiversity and vegetation: implications for a changing planet. Am J Bot 98(3):517–527. https://doi.org/10.3732/ajb.1000305

Tarafdar JC (2022) Role of soil biology on soil health for sustainable agricultural production. In: Giri B, Kapoor R, Wu Q-S, Varma A (eds) Structure and functions of pedosphere. Springer Nature Singapore, pp 67–81. https://doi.org/10.1007/978-981-16-8770-9_3

Tariq A, Graciano C, Sardans J, Zeng F, Hughes AC, Ahmed Z, Ullah A, Ali S, Gao Y, Peñuelas J (2024) Plant root mechanisms and their effects on carbon and nutrient accumulation in desert ecosystems under changes in land use and climate. New Phytol 242(3):916–934. https://doi.org/10.1111/nph.19676

Timofeeva AM, Galyamova MR, Sedykh SE (2023) Plant growth-promoting soil bacteria: nitrogen fixation, phosphate solubilization, siderophore production, and other biological activities. Plants 12(24):4074. https://doi.org/10.3390/plants12244074

Usharani KV, Roopashree KM, Naik D (2019) Role of soil physical, chemical and biological properties for soil health improvement and sustainable agriculture. J Pharmacogn Phytochem 8(5):1256–1267

Vora SM, Ankati S, Patole C, Podile AR, Archana G (2022) Alterations of primary metabolites in root exudates of intercropped Cajanus cajan–Zea mays modulate the adaptation and proteome of Ensifer (Sinorhizobium) fredii NGR234. Microb Ecol 83(4):1008–1025. https://doi.org/10.1007/s00248-021-01818-4

Waisen P (2019) Management of plant-parasitic nematodes and soil health using oil radish (Raphanus sativus) and brown mustard (Brassica juncea) cover crops [PhD Thesis, University of Hawai'i at Manoa]. https://search.proquest.com/openview/dd01403c8dbe080a8f6c3dc242609b54/1?pq-origsite=gscholar&cbl=18750&diss=y

Wang X, Cheng L, Xiong C, Whalley WR, Miller AJ, Rengel Z, Zhang F, Shen J (2024) Understanding plant–soil interactions underpins enhanced sustainability of crop production. Trends Plant Sci 29(11):1181–1190. https://doi.org/10.1016/j.tplants.2024.05.008

Witkamp M (1971) Soils as components of ecosystems. Annu Rev Ecol Syst 2:85–110
Yandvinder-Singh, Bijay-Singh, Khind CS (1992) Nutrient transformations in soils amended with green manures. In: Lal R, Stewart BA (eds) Soil restoration, vol 17. Springer New York, pp 237–309. https://doi.org/10.1007/978-1-4612-2930-8_5
Zhou Y, Zhu H, Yao Q (2017) Improving soil fertility and soil functioning in cover cropped agroecosystems with symbiotic microbes. In: Singh JS, Seneviratne G (eds) Agro-environmental sustainability. Springer International Publishing, pp 149–171. https://doi.org/10.1007/978-3-319-49724-2_8

Chapter 5
Allelopathy: Natural Weed Management Using Plant Interaction

Deepti Singh Vashishth, Shazia Akhtar, Pooja Saklani, and Rinky Yadav

Abstract The term allelopathy was introduced in the early 1990s. It originates from two Greek words: "allelon," meaning "of each other" and "pathos," meaning "to suffer or mutual harm." Lately, the activity has been well known for its adverse effects on neighboring plants, leading to their death. It was considered that it occurs as a defensive response to a threat, but using this mechanism for weed management was not known. In past years, allelopathic activity in plants caught the attention of plant researchers, who used allelopathic plant interactions for weed management and the development of bioherbicides to address the harmful impacts of synthetic chemical herbicides. The chapter explores various types of allelopathic plant interactions, including weed-weed, crop-weed, and medicinal and aromatic plant-weed interactions, which can help in weed management and the development of bioherbicides. Although it is a safer alternative, much more research is needed in the field to develop a sustainable, environmentally friendly, and cost-effective bioherbicide.

Keywords Allelochemicals • Germination • Growth • Interactions • Survival • Weeds

D. S. Vashishth (✉) · S. Akhtar · P. Saklani
Department of Environmental Science, Graphic Era (Deemed to be University), Dehradun, India

R. Yadav
Department of Environmental Science, Alpine Group of Institutions, Dehradun, India

L. A. Worku et al. (eds.), *Plant-Based Solutions for Sustainable Agriculture and Healthcare*, Sustainability Sciences in Asia and Africa,
https://doi.org/10.1007/978-981-95-8910-4_5

5.1 Introduction

Plants are an essential unit of the ecosystem, vital for maintaining its health and survival. They provide oxygen, regulate climate, support biodiversity, conserve soil, and offer various essential resources. Plants perform several physiological processes that govern their growth, survival, and adaptations to the environment. The core processes involved are photosynthesis, respiration, transpiration, mineral nutrition, and hormonal regulation. Primary metabolites produced by plants are directly involved in their growth, development, and reproduction, while secondary metabolites are not directly involved in growth and reproduction but serve in defense, attraction, and stress response through interaction with the environment and other organisms (Taiz et al. 2015). The defense mechanism of plants is directly or indirectly connected with allelopathic activity. The term allelopathy was given by Hans Molisch in 1937. Allelopathy is a natural phenomenon in which a plant produces specific chemicals, known as secondary metabolites or allelochemicals, that influence the germination, growth, and survival of neighboring plants. Earlier, this process was known for its negative impacts, but in recent years, it has been explored for its positive impacts, such as serving as a promising tool for sustainable weed management and providing an eco-friendly alternative to synthetic chemical herbicides (Khamare et al. 2022). These biochemical/allelochemicals are found in different parts of plants, such as roots, stems, leaves, fruits, flowers, and pollen, and can be used for various purposes (Fig. 5.1).

Allelopathy has substantial implications in agriculture, specifically in integrated weed management. Co-opting allelopathic plants, cover crops, mulches, and allelopathic extracts naturally suppresses weeds' growth. Any plant can act

Fig. 5.1 Uses of allelopathy

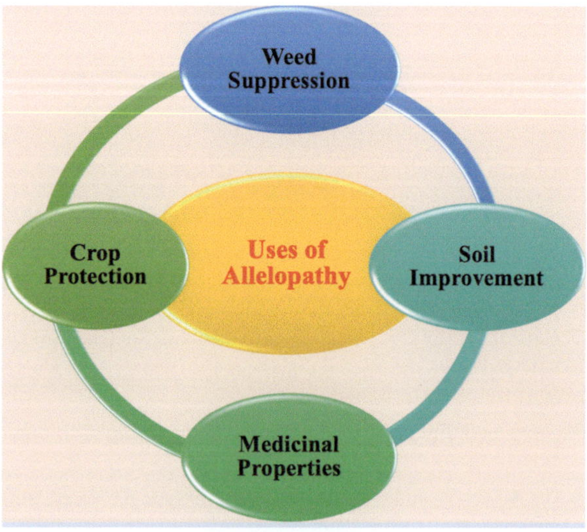

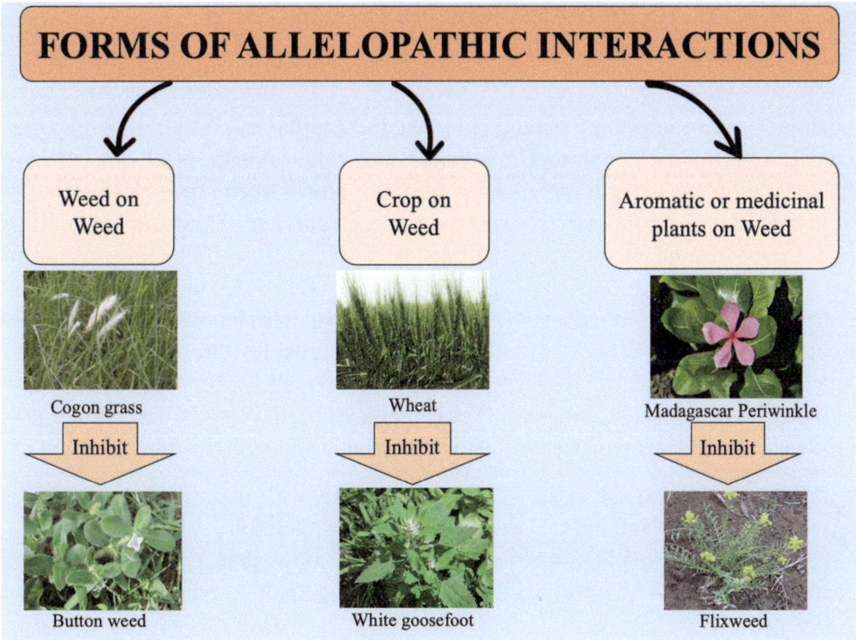

Fig. 5.2 Types of allelopathic plant interactions

as an allelopathic plant for other plants, depending on the presence of allelochemicals in the donor plant and the defence mechanisms of the receiver plants against those allelochemicals. Hence, the types of allelopathic plant interactions for weed management can be classified into three categories: weed-weed interaction, crop-weed interaction, and medicinal and aromatic plant-weed interaction (Fig. 5.2).

India is known for its agriculture because it serves as the economic backbone, provides food security and raw materials for industries, generates employment, which helps rural development, and creates export revenue and environmental stewardship. However, weed infestation is the most important biotic constraint to agricultural production in both developed and developing countries. Weeds are unwanted plants that compete with crops for sunlight, water, nutrients, and space. Also, they nurture insects and pathogens, which attack crop plants. To prevent weed growth, a variety of chemical herbicides are available, which not only affect cash crops and human health but also pollute the environment (Chauhan 2020). In India, weeds account for 25–26% yield loss in kharif crops and 18–25% in rabi crops, inflicting an estimated financial loss of crores annually. Thus, weed management becomes an indispensable part of sustainable agricultural practices (Varshney and Babu 2008).

5.2 Role of Allelopathic Plant Interaction in Weed Management

Allelopathy is an intriguing ecological phenomenon that can fully transform weed management methods through naturally occurring organic plant interactions. Allelopathic plant interactions include several kinds of allelochemicals that influence weed germination, growth, reproduction, and survival. These allelochemicals are released from plants into the environment in various forms, including root exudation, leaf volatilization, leaching from plant residues, decomposition of plant material in soil, and extracts made with organic solvents/water. Understanding allelopathic interactions among various plant species enables the development of a novel bioherbicide using allelochemicals. Common allelochemicals involved in herbicidal activity include phenolic acids, alkaloids, flavonoids, terpenoids, and quinones. The plant interaction utilized in weed management can be through:

- Intercropping: growing allelopathic crops along with main crops to suppress weeds by allelochemical release and resource competition
- Crop rotation: using allelopathic crops in a rotation sequence to reduce specific weed populations by introducing inhibitory compounds and disrupting the weed life cycle
- Cover crops: using allelopathic cover crops, which act as living mulches to suppress weeds
- Mulching: application of plant residue of allelopathic plants to inhibit weeds
- Extract application: using the extract of allelopathic plant parts as a spray to reduce weed density

These plant interactions help in shaping plant community dynamics and in managing invasive weed species (Satyanarayana Murthy et al. 2025; Khamare et al. 2022).

5.3 Allelopathic Plant Interaction Between Weed Plants and Other Weed Plants

Although weeds are a major cause of crop yield losses, some weeds also suppress the growth of other weeds. Certain weed species have the potential to address problems caused by other weed species. In some situations, when other weed control techniques do not work or are not feasible or accessible, the importance of this effect could be clearly illustrated. The allelopathic activity of *Euphorbia heterophylla* was reported against *Cenchrus echinatus*; 100 μL/L of *E. heterophylla* essential oil reduced shoot and root growth and germination of *C. echinatus* (Elshamy et al. 2019). *Parthenium hysterophorus* also exhibits allelopathic activity against *Oscimum americanum*, *Crotalaria mucronata*, *Cassia tora*, and *Oscimum basilicum* because of the presence of parthenin (Araniti et al. 2020). The allelopathic effects of canary grass (*Phalaris minor*) against *Chenopodium album*, *Cirsium arvense*, *Convolvulus arvensis*, *Melilotus indica*, and *Rumex acetosella* were reported by Satyanarayana Murthy et al. (2025).

The cultivation of *Ipomoea tricolor* is recommended before the cultivation of sugarcane because it prevents the growth of several weeds, such as *Amaranthus viridis*, *Euphorbia hirta*, *Phyllanthus niruri*, and *Coccinia indica* (Lotina-Hennsen et al. 2013). Palmitic, stearic, oleic, linoleic, and linolenic acids found in the leaves of *Croton bonplandianum* are responsible for inhibiting the seedling growth of *Medicago hispida*, *Melilotus alba*, and *Vicia sativa* (Singh et al 2015; Sisodia and Siddiqui, 2009). *Acroptilon repens* releases flavonoids that suppress the growth of several weed species, including *Medicago sativa*, *Panicum miliaceum*, *Lactuca sativa*, and *Echinochloa crusgalli* (Sangeetha et al. 2013). *Atriplex bumburyana*, *Maireana georgei*, *Enchylaena tomentosa*, and *Atriplex codonocarpa* are species of the Chenopodiaceae family, which show allelopathic potential against *Lactuca sativa*, *Chenopodium album*, and *Chenopodium murale* (Mushtaq et al. 2020). There are additional studies reported that show the weed-weed allelopathic plant interactions listed in Table 5.1.

5.4 Allelopathic Plant Interaction Between Crop Plants and Weeds

The allelopathic effects of different crop plants can contribute to the development of no-tillage techniques to produce herbicide-free crops. Many crop varieties exhibit allelopathic effects or other traits that can support weed-suppression techniques, such as wheat, barley, rice, sorghum, rye, castor, and mustard. Also, some plants, like wheat, barley, and rice, can suppress others through competitive growth, while others, like sesame, maize, buckwheat, rye, and clover, do so through their dense canopies.

Sorghum/Sunda grass (*Sorghum Bicolor L. Moench*) produces allelopathic chemicals, sorgoleone, that can inhibit the respiration process of weeds. Uddin et al. (2013) found that the synthesized herbicide sorgoleone has weed-removal properties and can inhibit weed growth. Ullah et al. (2022) found that the sorghum water extract and residue can help in suppressing weed density. Dayan et al. (2009) found that sorghum oil extract releases the acid sorgoleone, which can inhibit photosynthesis in young seedlings. Weeds such as *Chenopodium album*, *Phalaris minor*, and *Convolvulus arvensis* can be removed. Sunflower (*Helianthus annuus L.*) contains terpenoids that interfere with seed germination, photosynthesis, and protein synthesis (Rawat et al. 2017). It can effectively remove redroot pigweed (*Amaranthus retroflexus*) and canary grass (*Phalaris minor*). Janusauskaite (2023) concluded that the R extract obtained from unfertilized sunflower donor plants has inhibitory properties on pea seed germination. Flayyih and Almaire (2023) concluded that the phenolic compounds of sunflower residue reduced the weed population.

Table 5.1 Allelopathic weed plants and their targeted weed species

S.No.	Allelopathic weed	Key allelochemicals	Targeted weed	Key references
1.	*Parthenium hysterophorus*	Parthenin, sesquiterpene lactone, phenolics	*Amaranthus viridis, Cassia occidentalis, Echinochloa crusgalli, Phalaris minor, Ageratum conyzoides*	Batish et al. (1997, 2007a, b)
2.	*Lantana Camara*	Lantadene (terpene), flavonoids	*Pontederia crassipes, Microcystis aeruginosa, Phalaris minor, Avena fatua, Chenipodium album, Rumex dentatus*	Kong et al. (2006)
3.	*Artemisia annua*	Arteether, artemisinin	*Lactuca sativa, Amaranthus retroflexus, Ipomoea lacunosa, Portulaca oleracea*	Duke et al. (1987)
4.	*Polygonum sachalinese*	Dehydrovomifoliol, loliolide, emodin, physcion	*Amaranthus viridis, Phleum pratense*	Lun et al. (2023) and Inoue et al. (1992)
5.	*Ranunculus bulbosus*	Phenols, saponins, tannins, ranunculin, protoanemonin	*Echinochloa crusgalli, Ipomoea lacunosa, Sida spinosa*	Serag et al. (2020) and Gander (1998)
6.	*Ruta graveolens*	Valeric acid, methyl salicylate, a-pinene, nonan-2-ol, octanoic acid, 1,8-cineole, alkaloids, coumarins	*Amaranthus retroflexus, Chenopodium album, Cynodon dactylon, Cyperus rotundus, Euphorbia exigua, Portulaca oleracea*	De Feo et al. (2002) and Aliotta and Cafiero (2023)
7.	*Sasa cernua*	p-Coumaric, ferulic, vanillic, and p-hydroxybenzoic acids and p-hydroxybenzaldehyde	*Amaranthus viridis, Echinochloa crusgalli*	Li et al. (1992)
8.	*Senecio jacobaea*	Pyrrolizidine alkaloids	*Lolium perenne*	Kalač and Kaltner (2021) and Ahmed and Wardle (1994)
9.	*Tribulus terrestris*	Flavonoids, flavonol glycosides, steroidal saponins, and alkaloids	*Amaranthus lividus, Brassica tournefortii, Bromus tectorum, Chenopodium murale, Conyza bonariensis, Eruca sativa, Hordeum murinum, Solanum nigrum*	Chhatre et al. (2014) and El-Ghareeb (1991)

Kumar et al. (2025) concluded that allowing cover crops to grow longer in the spring provides more biomass to suppress weeds. Grint et al. (2022) observed that no-till cover crop management reduced weed density by 31% and weed biomass by 61% compared with no-till without crop cover. Rye (*Secale Cereale L.*) releases phenolic and hydroxamic acids from the roots that inhibit the growth of *Avena fatua* and *Chenopodium album* (Pérez, FJ and Ormenoñuñez 1991; Norouzi et al. 2014). It can be used as green manure and a cover crop (Camargo Silva and Bagavathiannan, 2023). Essman et al. (2023) observed that delaying the termination of cereal rye can help in weed removal. They concluded that giant foxtail (*Setaria faberi*) weed density was reduced by the presence of cereal rye. DeSimini et al. (2020) concluded that rye and canola weeds helped reduce horseweed (*Erigeron canadensis*) density and can also suppress giant ragweed (*Ambrosia trifida*). Early termination of the cover crop resulted in higher corn grain yields. Nunes et al. (2024) found that cereal rye mulching was effective in the suppression of weed water hemp (*Amaranthus tuberculatus*). Missaoui et al. (2018) observed that alfalfa (*Medicago sativa*) weed density lowered by 77% in after rye alfalfa plots. Tabaglio et al. (2008) reported high inhibitory activity of benzoxazinoids against redroot pigweed and common purslane (75%).

Castor (*Ricinus communis*) also has strong potential to prevent germination of redroot pigweed (*Amaranthus retroflexus*) and can reduce it by up to 70% (Tarassoli et al. 2021). Islam and Kato-Noguchi (2013) concluded that castor can inhibit the growth of both cress and barnyard grass.

Mustard (*Brassica napus*) releases glucosinolates that decompose into isothiocyanates, which act as a biofumigant upon residue decomposition and are effective against small-seeded annual weeds like *Sonchus asper*, *Matricaria inodora*, *Echinochloa crusgalli*, *Alopecurus myosuroides*, and *Amaranthus hybridus*. These chemicals can disrupt the cell membrane and the enzymatic activities of weeds. Gimsing and Kirkegaard (2006) found in their study that biofumigation of mustard can decrease the weed population. Sencenbaugh et al. (2025) concluded that biofumigation of mustard straw leachate applied as ground meal can help reduce weed density. Bajagain et al. (2024) found that mustard leachate biofumigation with an overwinter cover crop can reduce the density of early-season weeds in Chile pepper. Mustard contains water-soluble substances that inhibit the germination and seed growth of wild oat (Turk and Tawaha 2003; Mahmood et al. 2013).

Rice (*Oryza sativa* L.) plants and their parts release phenolic acid momilactone A & B that can inhibit the growth of roots and shoots of barnyard grass (*Echinochloa crusgalli*) and nutsedge (*Cyperus difformis*). Ho et al. (2020) detected 20 allelochemicals in their study, out of which OM 5930 was the highest in concentration and can be highly effective in weed control. The root exudates of dehulled rice contain vanillin and vanillic acids that have inhibitory effects on seed germination of weeds (Khang et al. 2016). Japonica rice has allelopathic potential against weeds and can inhibit the root growth of barnyard grass (García-Romeral et al. 2024). Momilactone B has inhibitory properties against the root and growth of barnyard grass (Kato-Noguchi et al. 2010, 2011) Wheat (*Triticum aestivum* L.) releases phenolic acids and hydroxamic acids that can suppress weeds like *Agropyron repens*, *Digitaria ciliaris*, *Portulaca oleracea*, *Galinsoga ciliate*, and *Chenopodium album* (González-García et al., 2025).

Barley (*Hordeum vulgare*) releases ferulic and vanillic acids, which have an influential effect on *Echinochloa crusgalli* (Oueslati et al. (2009); Kremer and Ben-Hammouda, 2009). Bundit et al. (2021) concluded that the root and shoot extracts of sun hemp (*Crotalaria juncea*) exhibited different allelopathic effects against weeds at different growth stages and showed maximum effects against weeds around 4 weeks after planting. Pearl millet also contains phenolic acids such as hydroxybenzoic acid, hydroxycinnamic acid, and catechol. It can disrupt N_2-fixation pathways and affect photosynthesis in weeds such as *Bidens pilosa* L., *Euphorbia heterophylla* L., and *Ipomoea grandifolia* (Chandrasekara and Shahidi 2011; Mantovanelli et al. 2025). Khan et al. (2019) concluded that the fresh roots, shoots, and leaves of weeds, including grass, common lambsquarter, field bindweed, broad-leaved dock, amaranth, wild melon, wild jute, and annual sowthistle, were affected by the aqueous extract of millets.

5.5 Allelopathic Plant Interaction Between Medicinal and Aromatic Plants and Weeds

Weeds have long been recognized as a major constraint in agriculture, competing with crops for essential resources and thereby raising production costs. During the early stages of crop domestication, weed-crop interactions played a decisive role, ultimately necessitating the development of weed control strategies (Zimdahl and Basinger 2024; Bachheti et al. 2021; Deepti et al. 2022). Today, herbicides remain the dominant tool in integrated weed management systems (Hesammi 2011; Mohamadi et al. 2012). Over the past few decades, the commercialization of more than 200 chemical herbicides has transformed weed management practices world-wide (Gianessi 2013; Gupta et al. 2021). However, our increasing reliance on synthetic chemicals has raised environmental and health concerns, and there is a high demand for alternatives with eco-friendly features. Medicinal and aromatic plants (MAPs), as well as certain species and mushrooms, contain a wide variety of secondary metabolites, many of which have pesticidal, antifungal, and herbicidal properties. Phenols have long been studied as potential bio-based agrochemicals (Mirmostafaee et al. 2020; Mengstu et al. 2021). Recently, research has uncovered the allelopathic potential of some MAPs. For example, Sadeqifard et al. (2022) demonstrated that aromatic plants used in Iran exhibit strong allelopathic activity and could serve as new bioherbicides for sustainable and eco-friendly agriculture. *Nepeta binaloudensis* (leaf) combined with *Trachyspermum ammi* (fruit) or *Lavandula officinalis* (flower) showed a significant synergistic effect on weed suppression. Moving forward, laboratory and field studies to isolate and optimize selected allelochemicals and to develop bioherbicide formulations are essential.

Using these natural substances in agriculture could reduce reliance on chemical herbicides and promote safer farming practices and public health.

Allelopathic activity of MAPs is mainly due to the presence of secondary metabolites, which are emitted into the environment via volatile emissions, leaching (through rain, dew, or irrigation), root exudates, and/or microbial degradation of plant tissues (Narwal 2006). Essential oils (EOs) containing many volatile compounds are of special interest because they are volatile and are likely to diffuse in the air and influence the neighboring plant. Several studies have confirmed that EOs exhibit strong antimicrobial activity by inhibiting fungal spore germination and bacterial growth (Inderjit 2005). In addition, EOs and their active constituents can inhibit seed germination and seedling establishment of weeds, highlighting their role as natural fumigants in pre- and post-emergent weed control (Oroojalian et al. 2010).

Within the Lamiaceae family, species such as *Lavandula officinalis*, *Salvia ceratophylla*, *S. lavandulifolia*, and *N. binaloudensis* are frequently reported to inhibit neighboring plant germination (Callaway 2003). For example, aqueous extracts of *N. binaloudensis* root and leaves were found to inhibit sunflower seed germination, while the plant-box method revealed the suppressive effect of *L. officinalis* on lettuce seed germination (Fujii and Hiradate 2007). Volatile compounds from MAPs such as N. binaloudensis, L. angustifolia, Pelargonium graveolens, T. ammi, and Salvia species significantly reduced lettuce germination percentage, radicle length, and vigor index (Mirmostafaee et al. 2020). Other studies confirmed that Salvia officinalis completely inhibited lettuce radicle and hypocotyl elongation (Sadeqifard et al. 2022). Beyond allelopathic action, MAPs also display antibacterial, antifungal, antioxidant, and insecticidal properties. For instance, Eryngium platyloba essential oil showed strong antibacterial activity (Hashemi et al. 2013), while aqueous and ethanolic extracts of S. lavandulifolia exhibited high antioxidant potential due to their phenolic content (Bingol and Bursal 2018). Essential oil of Nepeta cataria demonstrated insecticidal activity against Spodoptera littoralis larvae (Pavela 2005), and P. graveolens showed oviposition deterrent and larvicidal effects (Prajapati et al. 2005; Benelli et al. 2017).

In gaseous-phase experiments, essential oils from *P. graveolens*, *P. anisum*, and *Thymus daenensis* were highly effective in inhibiting germination and seedling growth (Mirmostafaee et al. 2020). Interestingly, mixtures of essential oils exhibited variable effects, including synergistic, additive, or antagonistic interactions. For example, *Mentha suaveolens* showed a strong synergistic effect in inhibiting germination and radicle growth, while *P. abrotanoides* was effective in suppressing hypocotyl development. Conversely, the oils of *A. subulatum* and *R. graveolens* demonstrated antagonistic effects when combined with others. These results highlight the complexity of allelopathic interactions and emphasize the importance of testing combinations to optimize bioherbicide development. Significant germination inhibition was reported for *T. daenensis* and *P. anisum* essential oils, attributed to compounds such as thymol, carvacrol, carvone, and trans-anethole, which also possess strong antimicrobial and insecticidal properties (Sharma et al. 2009; Azirak and Karaman 2008; Khalil et al. 2018). Other plant essential oils, such as *Zataria*

Table 5.2 Medicinal plants and their allelopathic effects on radicle growth, hypocotyl growth, and germination percentage

Plant species	Plant part	Effect on radicle growth	Effect on hypocotyl growth	Germination inhibition
Nepeta binaloudensis	Leaf	Strong inhibition	Strong inhibition	Strong inhibition (3.60%)
Achillea nobilis	Flower	Significant inhibition	Not specified	Not specified
Artemisia biebersteinii	Leaf	Significant inhibition	Not specified	Not specified
Phagnalon gnaphalodes	Seed	Significant inhibition	Strong inhibition	Not specified
Eryngium platylobum	Stem	Significant inhibition	Strong inhibition	Not specified
Coriandrum maculatum	Leaf	Significant inhibition	Strong inhibition	Not specified
Foeniculum vulgare	Fruit	Significant inhibition	Strong inhibition	Not specified
Trachyspermum ammi	Fruit	Significant inhibition	Strong inhibition	Not specified
Euphorbia serpens	Leaf	Significant inhibition	Strong inhibition	Not specified
Euphorbia granulata	Leaf	Significant inhibition	Not specified	Not specified
Lycium ruthenicum	Leaf	Significant inhibition	Strong inhibition	Not specified
Limonium depressum	Leaf	Not specified	Strong inhibition	Not specified
Amaranthus hypochondriacus	Leaf	Significant inhibition	Strong inhibition	Not specified
Malva sylvestris	Leaf	Significant inhibition	Strong inhibition	Not specified
Hypericum perforatum	Leaf	Significant inhibition	Strong inhibition	Not specified
Hypericum helianthemoides	Root	Not specified	Strong inhibition	Not specified
Lantana montevidensis	Leaf	Not specified	Strong inhibition	Not specified
Salsola ceratophylla	Leaf	Minimal inhibition (72.76% germination)	Not specified	Least inhibition (72.76%)
Lavandula officinalis	Flower	Not specified	Not specified	Strong inhibition (3.32%)

multiflora and *P. abrotanoides*, showed allelopathic effects against weeds with broad-spectrum activity, demonstrating their potential in allelopathy-based weed management strategies (Saharkhiz et al. 2010; Amini et al. 2014, 2016).

Medicinal and aromatic plants offer a good, eco-friendly alternative to synthetic herbicides. Their essential oils and secondary metabolites are essential for their synergistic interactions with herbicides. Their essential oils and secondary metabolites, especially when used together, may be a sustainable tool for weed management and reduced pesticide use (Table 5.2).

Allelopathic activity of MAPs is primarily attributed to secondary metabolites, which are released into the environment through volatilization, leaching (via rain, dew, or irrigation), root exudation, or microbial degradation of plant tissues (Narwal

2006). Essential oils (EOs) rich in volatile compounds are of particular importance because they can diffuse into the air and affect neighboring plants. Several studies confirm that EOs exhibit strong antimicrobial activity by inhibiting fungal spore germination and bacterial growth (Inderjit 2005). Additionally, EOs and their active constituents can suppress seed germination and seedling growth in weeds, highlighting their potential role as natural fumigants in pre- and post-emergence weed management (Oroojalian et al. 2010). Within the Lamiaceae family, species such as *Lavandula officinalis*, *Salvia ceratophylla*, *S. lavandulifolia*, and *N. binaloudensis* are frequently reported to inhibit neighboring plant germination (Callaway 2003). For example, aqueous extracts of *N. binaloudensis* root and leaves were found to inhibit sunflower seed germination, while the plant-box method revealed the suppressive effect of *L. officinalis* on lettuce seed germination. The results revealed that sunflowers can be an allelopathic species, posing a risk to crop rotation. The suppressive effect of sunflowers on other plants including weeds can be applied in sustainable agriculture (Fujii and Hiradate 2007). Volatile compounds from MAPs *such as N. binaloudensis*, *L. angustifolia*, *Pelargonium graveolens*, *T. ammi*, and *Salvia* species significantly reduced lettuce germination percentage, radicle length, and vigor index (Mirmostafaee et al. 2020). Other studies confirmed that *Salvia officinalis* completely inhibited lettuce radicle and hypocotyl elongation (Sadeqifard et al. 2022). Beyond allelopathic action, MAPs also display antibacterial, antifungal, antioxidant, and insecticidal properties. For instance, *Eryngium platyloba* essential oil showed strong antibacterial activity (Hashemi et al. 2013), while aqueous and ethanolic extracts of *S. lavandulifolia* exhibited high antioxidant potential due to their phenolic content (Bingol and Bursal 2018). Essential oil of *Nepta cataria* demonstrated insecticidal activity against *Spodoptera littoralis* larvae (Pavela 2005), and *P. graveolens* showed oviposition-deterrent and larvicidal effects (Prajapati et al. 2005; Benelli et al. 2017). In gaseous-phase experiments, essential oils of *P. graveolens*, *P. anisum*, and *Thymus daenensis* were highly effective in inhibiting germination and seedling growth (Mirmostafaee et al. 2020). Interestingly, combinations of essential oils displayed variable synergistic, additive, or antagonistic interactions. For example, *Mentha suaveolens* exhibited strong synergistic inhibition of germination and radicle growth, while *P. abrotanoides* was effective in hypocotyl suppression. Conversely, oils of *A. subulatum* and *R. graveolens* showed antagonistic effects when combined with others. Such results underscore the complexity of allelopathic interactions and highlight the importance of optimizing combinations for bioherbicide development. Strong germination inhibition was reported for *T. daenensis* and *P. anisum* essential oils, attributed to compounds such as thymol, carvacrol, carvone, and trans-anethole, which also possess potent antimicrobial and insecticidal properties (Sharma et al. 2009; Azirak and Karaman 2008; Khalil et al. 2018). Other MAPs, including *Zataria multiflora* and P. *abrotanoides*, exhibited broad-spectrum phytotoxic and antifungal effects, confirming their importance in allelopathy-based weed management strategies (Saharkhiz et al. 2010; Amini et al. 2014, 2016). Medicinal and aromatic plants present a promising eco-friendly alternative to synthetic herbicides. Their essential oils and secondary metabolites are particularly useful when used in synergistic herbicides. Their essential oils and secondary metabolites, particularly when used in synergistic combinations, could serve as sustainable tools in weed suppression while reducing chemical pesticide use (Table 5.2).

5.6 Weed and Medicinal Plants Competition

Proper spacing of plants allows for efficient absorption of nutrients and water from the soil and, in some cases, enables proper penetration of sunlight and soil coverage (Spitters 1989). The most important principle is to achieve a closed canopy to restrict weed growth while ensuring proper crop development (Dabney et al. 2001). The plant density depends on whether the target product is green biomass, seeds, or bioactive compounds (Einhellig 1995). Importantly, higher plant productivity per unit area does not necessarily mean more plants lead to higher yields or better quality; in fact, overcrowding may decrease seed or fruit yield (Onat et al. 2017).

Plant density and weed pressure also influence the accumulation of secondary metabolites in medicinal plants (Borges et al. 2017). Therefore, defining the production objective—whether to maximize biomass or enhance metabolite content—while in others, metabolite production increases alongside yield (Fig. 5.3). For example, thyme shows reduced metabolite levels yield rises, whereas basil demonstrates a positive correlation between yield and metabolite content (Sifola and Barbieri 2006). In fennel, which is valued for its seed oils, higher essential oil concentrations were obtained at lower planting densities (Khorshidi et al. 2009). Similarly, mixed cultivation systems may alter competition both within and between species, affecting yield and chemical composition (Craine and Dybzinski 2013). Hence, proper inter-row and intra-row spacing in medicinal plant production is vital for minimizing weed competition and maximizing quality outcomes (Sedara and Sedara 2020).

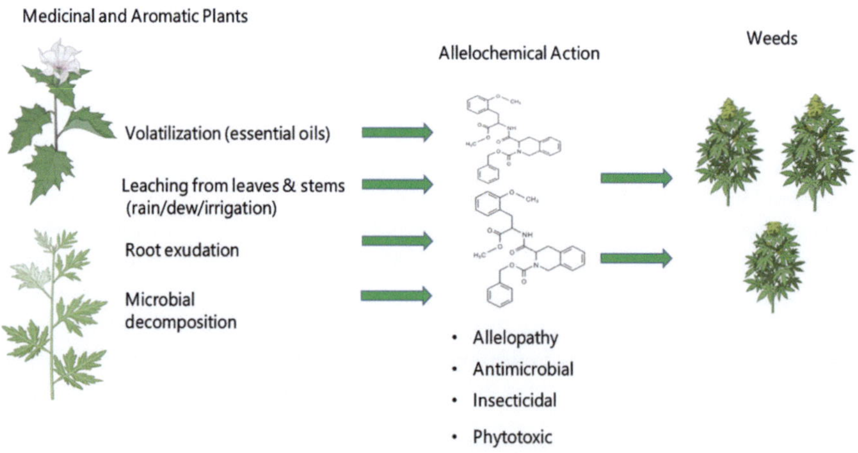

Fig. 5.3 The interaction between medicinal plants and aromatic plants and weeds

Table 5.3 Allelopathic plants, their main allelochemicals, and target weeds/crops

Allelopathic plant	Main allelochemical(s)	Sensitive/target plants	Reference
Black walnut (*Juglans nigra* L.)	Juglone	Horseweed (*Conyza canadensis*), hairy fleabane (*Conyza bonariensis*), purslane (*Portulaca oleracea*), tall annual morning glory (*Ipomoea purpurea*)	Shrestha (2009)
Eucalyptus (*Eucalyptus* spp.)	1,8-cineole	Common amaranth (*Amaranthus retroflexus*), purslane (*Portulaca oleracea*)	Azizi and Fujii (2006)
Ailanthus altissima	Ailanthone	Common amaranth (*A. retroflexus*), garden cress (*Lepidium sativum*), foxtail (*Setaria glauca*), barnyard grass (*Echinochloa crusgalli*), corn (*Zea mays*)	Heisey (1996)
Fine fescue grasses (*Festuca* spp.)	M-tyrosine	Rape (*Brassica nigra*), birdsfoot trefoil (*Lotus corniculatus*), large crabgrass (*Digitaria sanguinalis*), white clover (*Trifolium repens*), common dandelion (*Taraxacum officinale*)	Peters (1968), Stephenson and Posler (1988), and Bertin et al. (2007)
Sorghum bicolor L.	Sorgoleone	False cleavers (*Galium spurium*), Japanese dock (*Rumex japonicus*), Indian jointvetch (*Aeschynomene indica*), common amaranth (*A. retroflexus*)	Uddin et al. (2013)
Artemisia annua L.	Arteether, Artemisinin	Lettuce (*Lactuca sativa*), common amaranth (*A. retroflexus*), pitted morning glory (*Ipomoea lacunosa*), purslane (*P. oleracea*)	Duke et al. (1987)
Callistemon citrinus	Leptospermone	Hairy crabgrass (*Digitaria sanguinalis*), yellow foxtail (*Setaria glauca*), amaranth (*A. retroflexus*), red oat (*Avena sativa*), Indian mustard (*Brassica juncea*), curly dock (*Rumex crispus*)	Cornes (2005)
Callistemon viminalis		*Angallis arvensis*, *Lepidium didymium*, and *Vicia sativa*	Vashishth et al. (2023)
Piper longum L.	Sarmentine	Field bindweed (*Convolvulus arvensis*), wild mustard (*Sinapis arvensis*), horseweed (*Conyza canadensis*), sprangletop (*Leptochloa chinensis*)	Dayan et al. (2015)
Azadirachta indica A. Juss.	Nimbolide B, Nimbic acid B	Garden cress (*Lepidium sativum*), lettuce (*Lactuca sativa*), alfalfa (*Medicago sativa*)	Salam and Kato-Noguchi et al. (2010)
Sunflower (*Helianthus annuus* L.)	Helianthus	Hairy crabgrass (*D. sanguinalis*), prickly sida (*Sida spinosa*), tumbleweed (*Amaranthus album*)	Frans and Semidey (1992)
Murrya koenigii		Angallis arvensis, Lepidium didymium, and Vicia sativa	Vashishth et al. (2023)

Allelopathy research is important for understanding how secondary metabolites released by medicinal and aromatic plants suppress weed germination and growth. Woody species such as *Juglans nigra*, *Ailanthus altissima*, *Azadirachta indica*, and *Callistemon citrinus* inhibit weeds such as *Conyza*, *Amaranthus*, and *Digitaria* with compounds including juglone, ailanthone, and leptospermone. Several allelochemicals, including sorgoleone, momilactones, and M-tyrosine, are produced by *Sorghum*, *Festuca*, and crabgrass that inhibit the growth of *Echinochloa*, *Galium*, and *Brassica. Piper longum* and Neem contain sarmentine, tricolorin, and nimbolides, which have significant bioherbicidal activity. In a triple action (Table 5.3), *Eucalyptus* and *Artemisia annua* utilize volatile cineole and artemisinin to inhibit *Lactuca*, *Portulaca*, and *Amaranthus*.

The allelopathic potential of Eucalyptus species has been the focus of extensive scientific research (Sasikumar et al. 2002; Bajwa and Nazi 2005; El-Khawas and Shehata 2005). These species release various allelochemicals, mainly allelopathic PDE phenolic acids and volatile terpenes, which are found in different parts of the plant, such as leaves, bark, and roots (May and Ash 1990). The foliage also contains essential oils and resinous compounds that can have a direct or indirect inhibitory effect on nearby plants and seeds, as well as on soil microorganisms (Ruwanza et al. 2014). Volatile compounds like 1,8-cineole, limonene, and α- and β-pinene have been identified as major contributors to the allelopathic activity of *Eucalyptus* species (Muller et al. 1964).

Similarly, allelopathic interactions have been described between Ailanthus altissima (tree of heaven). First, Voigt and Mergen (1962) documented the inhibitory effects of its foliage and stem extracts on nearby vegetation growth. Phytochemical studies reveal that different parts of the plant contain alkaloids, terpenoids, steroids, flavonoids, phenolics, and quassinoids (Albouchi et al. 2013; Kim et al. 2015; Ni et al. 2018). Comparative research shows that root bark is the most phytotoxic, while wood tissues have minimal inhibitory effects (Heisey 1990a). Soil incorporation trials with oven-dried root bark demonstrated strong suppression of seedling emergence, biomass, and survival of *Lepidium sativum* (garden cress) (Heisey 1990b). The compound responsible for these effects is ailanthone, an indole-3-acetate antagonist classified as a quassinoid, which has been reported to have herbicidal activity before and after emergence (Heisey 1996; Sladonja et al. 2015). Ailanthone has been shown to effectively inhibit germination and growth across various species, including both monocots and dicots. Greenhouse experiments demonstrated that treatment rates as low as 0.5 kg/ha could control species such as *Amaranthus retroflexus*, *Lepidium sativum*, *Setaria glauca*, *Echinochloa crusgalli*, and *Zea mays* (Heisey 1996). Interestingly, Ailanthus altissima seedlings remained unaffected, indicating an innate mechanism of self-protection against autotoxicity (Heisey 1996). Due to challenges in isolation, rapid soil prevalence, and nonselective action, ailanthone has not been commercialized despite its potent herbicidal properties (Bhowmik and Inderjit 2003; Heisey 1999). Nonetheless, it remains a promising natural herbicide for organic farming and small-scale weed control if formulation issues can be resolved.

The allelopathic ability of fine *fescue* grasses (*Festuca* spp.) is well documented. These grasses release root exudates containing phytotoxic substances that inhibit the growth of competing plants (Bertin et al. 2009; Peters and Mohammed Zam 1981). Extracts from dry shoots and roots of Festuca arundinacea have shown significant suppressive effects on the germination and growth of *Brassica nigra* and *Lotus corniculatus* in both laboratory and field conditions (Peters 1968; Stephenson and Posler 1988). Conversely, *Festuca rubra* has proven to be an effective weed suppressant when used as living mulch or killed sod in no-tillage systems (Weston 1990). The main allelochemical in fine fescue roots is M-tyrosine. This non-protein amino acid inhibits germination and growth of *Digitaria sanguinalis*, *Trifolium repens*, and *Taraxacum officinale* (Bertin et al. 2007). In sorghum (*Sorghum* bicolor), weed suppression is attributed to both hydrophilic and hydrophobic allelochemicals, such as phenolic acids, aldehydes, and sorgoleone (Czarnota et al. 2003). Sorgoleone was initially isolated from root exudates by Netzly and Butler (1986) and accounts for approximately 90% of the hydrophobic exudates from sorghum roots, synthesized in root hair cells (Yang et al. 2004). Sorgoleone exhibits strong bioherbicidal activity, surpassing that of classic allelochemicals like juglone and other phenolics (Uddin et al. 2013). Incorporating sorghum residues—roots, stems, and leaves—into soil or using them as surface mulch has been shown to reduce weed biomass by 25–50% (Cheema 1988). These findings highlight sorghum's potential as a natural, sustainable bioherbicide in integrated weed management systems. *Callistemon citrinus* secretes the allelochemical leptospermone, which suppresses the germination and development of weeds such as *Digitaria sanguinalis*, *Setaria glauca*, *Amaranthus retroflexus*, *Avena sativa*, *Brassica juncea*, and *Rumex crispus* (Cornes 2005).

Callistemon viminalis exhibits strong phytotoxicity against *Anagallis arvensis*, *Lepidium didymum*, and *Vicia sativa*, attributed to phenolic and terpenoid compounds (Vashishth et al. 2023), making it a promising natural bioherbicide for sustainable weed control in the future.

5.7 Weed Autopathic Interactions (Autotoxicity)

Some weeds exhibit self-inhibition. Autotoxicity is little different from allelopathy, as it is self-toxic and cannot be stimulated. The weeds have such self-toxic biochemicals in their seeds and tissues. The nature, cause, and purpose of this type of self-toxicity remain questionable and are not fully understood (Foy 2024). Whittaker and Feeny (1971) stated that it is an evolutionary paradox. It plays a major role in the survival strategies of some plant species in the wild by controlling their natural populations, particularly in harsh environments with limited resources.

5.8 Conclusion

Weeds in medicinal plant fields create stress through direct competition, and the severity depends on the medicinal plant species, the weed flora present, and the level of infestation. This competition can result in significant yield and quality losses, as weeds may negatively influence the color, taste, aroma, and medicinal efficacy of these plants. Critical weed-free periods, especially during crop establishment, are critical; if weeds dominate during this stage, yield losses may reach up to 60%. To address this challenge, preventive and cultural practices such as sowing at the optimal time, using mulch, and soil solarization should be prioritized. Mechanical, physical, allelopathic, and biological control methods are other sustainable strategies. While herbicides can be effective, their use should be cautious because they may adversely affect the secondary metabolites that determine the medicinal value. With the increasing demand for organically grown medicinal plants, agroecological weed management practices have become more important. Future research should focus on understanding the role of biological control, particularly allelopathic interactions, in weed management strategies while maintaining crop yield and metabolite profiles to achieve sustainable, high-quality production.

References

Ahmed M, Wardle DA (1994) Allelopathic potential of vegetative and flowering ragwort (*Senecio jacobaea* L.) plants against associated pasture species. Plant Soil 164(1):61–68

Albouchi F, Hassen IE, Casabianca H & Hosni K (2013) Phytochemicals, antioxidant, antimicrobial and phytotoxic activities of Ailanthus altissima (Mill.) Swingle leaves. S Afr J Bot 87:164–174. https://doi.org/10.1016/j.sajb.2013.04.003

Aliotta G, Cafiero G (2023) Biological properties of rue (*Ruta graveolens* L.): potential use in sustainable agricultural systems. In: Principles and practices in plant ecology. CRC Press, pp 551–563

Amini S, Azizi M, Joharchi MR, Shafei MN, Moradinezhad F, Fujii Y (2014) Determination of allelopathic potential in some medicinal and wild plant species of Iran by dish pack method. Theor Exp Plant Physiol 26:189–199. https://doi.org/10.1007/s40626-014-0019-z

Amini S, Azizi M, Joharchi MR, Moradinezhad F (2016) Evaluation of allelopathic activity of 68 medicinal and wild plant species of Iran by sandwich method. Int J Hortic Sci Technol *3*(2):243–253. https://doi.org/10.22059/ijhst.2016.62923

Araniti F, Lupini A, Sorgonà A, Conforti F, Marrelli M, Statti GA et al (2020) Allelopathic activity and identification of phytotoxic compounds from residues and aqueous extracts of Calendula arvensis. Agronomy 10(2):227

Azirak S, & Karaman S (2008) Allelopathic effect of some essential oils and components on germination of weed species. Acta Agriculturae Scandinavica, Section B – Soil and Plant Science, 58(1):88–92

Azizi M, Fujii Y (2006) Allelopathic effect of some medicinal plant substances on seed germination of *Amaranthus retroflexus* and *Portulaca oleracea*. Acta Hortic *699*:61–68. https://doi.org/10.17660/actahortic.2006.699.5

Bachheti, Archana, Deepti, Bachheti RK, Husen A (2021) Medicinal plants and their pharmaceutical properties under adverse environmental conditions. In: Harsh environment and plant resilience. Springer, Cham, pp 457–502

Bajagain A, Lehnhoff EA, Creamer R, Steiner R, Schutte BJ (2024) Timing termination of a biofumigant cover crop for weed suppression in chile pepper. HortTechnology 34(2):142–152

Bajwa R, Nazi I (2005) Allelopathic effects of *Eucalyptus citriodora* on growth, nodulation and AM colonization of *Vigna radiata* (L.) Wilczek. Allelopath J 15(2):237–246

Batish DR, Kohli RK, Singh HP, Saxena DB (1997) Studies on herbicidal activity of parthenin, a constituent of *Parthenium hysterophorus*, towards billgoat weed (*Ageratum conyzoides*). Curr Sci:369–371

Batish DR, Arora K, Singh HP, Kohli R (2007a) Potential utilization of dried powder of *Tagetes minuta* as a natural herbicide for managing rice weeds. Crop Prot 26:566–571. https://doi.org/10.1016/j.cropro.2007.01.008. [apastyle.apa.org], [paperpile.com]

Batish DR, Pal Singh H, Kohli RK, Kaur S, Saxena DB, Yadav S (2007b) Assessment of phytotoxicity of parthenin. Z Naturforsch C 62(5–6):367–372. https://doi.org/10.1515/znc-2007-5-609

Benelli G, Pavela R, Canale A, Cianfaglione K, Ciaschetti G, Conti F, Nicoletti M, Senthil-Nathan S, Mehlhorn H, Maggi F (2017) Acute larvicidal toxicity of five essential oils (…) against the filariasis vector *Culex quinquefasciatus*: synergistic and antagonistic effects. Parasitol Int 66:166–171. https://doi.org/10.1016/j.parint.2017.01.005

Bertin C, Weston LA, Huang T, Jander G, Owens T, Meinwald J et al (2007) Grass roots chemistry: *meta*-tyrosine, an herbicidal nonprotein amino acid. Proc Natl Acad Sci 104(43):16964–16969. https://doi.org/10.1073/pnas.0707198104

Bertin C, Senesac A, Rossi F, DiTommaso A, Weston L (2009) Evaluation of selected fine-leaf fescue cultivars for their turfgrass quality and weed suppressive ability in field settings. HortTechnology 19(3):660–668. https://doi.org/10.21273/hortsci.19.3.660

Bhowmik PC, Inderjit (2003) Challenges and opportunities in implementing allelopathy for natural weed management. Crop Prot 22(4):661–671. https://doi.org/10.1016/S0261-2194(02)00242-9

Bingol MN, Bursal E (2018) LC-MS/MS analysis of phenolic compounds and in vitro antioxidant potential of *Stachys lavandulifolia* var. *brachydon* Boiss. Int Lett Nat Sci 72:28–36. https://doi.org/10.18052/www.scipress.com/ILNS.72.28

Borges CV, Minatel IO, Gomez-Gomez HA, & Lima GPP (2017) Medicinal plants: Influence of environmental factors on the content of secondary metabolites. In Medicinal plants and environmental challenges (pp. 259–277). Cham: Springer International Publishing

Bundit A, Ostlie M, Prom-U-Thai C (2021) Sunn hemp (Crotalaria juncea) weed suppression and allelopathy at different timings. Biocontrol Sci Technol 31(7):694–704

Camargo Silva G, & Bagavathiannan M (2023) Mechanisms of weed suppression by cereal rye cover crop: A review. Agronomy Journal, 115(4):1571–1585

Chandrasekara A, Shahidi F (2011) Determination of antioxidant activity in free and hydrolyzed fractions of millet grains and characterization of their phenolic profiles by HPLC-DAD-ESI-MSn. J Funct Foods 3(3):144–158

Chauhan BS (2020) Grand challenges in weed management. Front Agron 1:3

Cheema ZA (1988) Weed control in wheat through sorghum allelochemicals. Ph.D. thesis. Agronomy Department, University of Agriculture, Faisalabad, Pakistan

Chhatre S, Nesari T, Somani G, Kanchan D, Sathaye S (2014) Phytopharmacological overview of *Tribulus terrestris*. Pharmacogn Rev 8(15):45

Cornes D (2005) “Callisto: a very successful maize herbicide inspired by allelochemistry,” in Proceedings of the Fourth World Congress on Allelopathy (Wagga Wagga, Australia: NSW: Australia. Centre for Rural Social Research, Charles Sturt University;), 2:2636

Craine JM, & Dybzinski R (2013) Mechanisms of plant competition for nutrients, water and light. Functional Ecology, 27(4):833–840

Czarnota MA, Paul RN, Weston LA, Duke SO (2003) Anatomy of sorgoleone-secreting root hairs of sorghum species. Int J Plant Sci 164(6):861–866. https://doi.org/10.1086/378661

Dabney SM, Delgado JA, & Reeves DW (2001) Using winter cover crops to improve soil and water quality. Communications in Soil Science and Plant Analysis, 32(7–8):1221–1250

Dayan FE, Howell JL, & Weidenhamer JD (2009) Dynamic root exudation of sorgoleone and its in planta mechanism of action. J Exp Bot 60(7):2107–2117

Dayan FE, Owens DK, Watson SB, Asolkar RN, Boddy LG (2015) Sarmentine, a natural herbicide from *Piper* species with multiple herbicide mechanisms of action. Front Plant Sci 6:222. https://doi.org/10.3389/fpls.2015.00222

De Feo V, De Simone F, Senatore F (2002) Potential allelochemicals from the essential oil of *Ruta graveolens*. Phytochemistry 61(5):573–578

Deepti, Bachheti AJ, Chauhan K, Bachheti RK, Husen A (2022) Impact of UV radiation on the growth and pharmaceutical properties of medicinal plants. In: Environmental pollution and medicinal plants. CRC Press, pp 47–59

DeSimini SA, Gibson KD, Armstrong SD, Zimmer M, Maia LO, Johnson WG (2020) Effect of cereal rye and canola on winter and summer annual weed emergence in corn. Weed Technol 34(6):787–793

Duke SO, Vaughn KC, Croom EM Jr, Elsohly HN (1987) Artemisinin, a constituent of annual wormwood (Artemisia annua), is a selective phytotoxin. Weed Sci 35(4):499–505

Einhellig FA (1995) Allelopathy: Current status and future goals. In Inderjit, K. M. M. Dakshini, & FA Einhellig (Eds.), Allelopathy: Organisms, Processes, and Applications (ACS Symposium Series 582, pp. 1–24). American Chemical Society. https://doi.org/10.1021/bk-1995-0582.ch001

El-Ghareeb RM (1991) Suppression of annuals by *Tribulus terrestris* in an abandoned field in the sandy desert of Kuwait. J Veg Sci 2(2):147–154

El-Khawas SA, Shehata MM (2005) The allelopathic potentialities of *Acacia nilotica* and *Eucalyptus rostrata* on monocot (*Zea mays* L.) and dicot (*Phaseolus vulgaris* L.) plants. Biotechnology 4(1):23–34. https://doi.org/10.3923/biotech.2005.23.34

Elshamy AI, Abd-ElGawad AM, El Gendy AENG, Assaeed AM (2019) Chemical characterization of *Euphorbia heterophylla* L. essential oils and their antioxidant activity and allelopathic potential on *Cenchrus echinatus* L. Chem Biodivers 16(5):e1900051

Essman AI, Loux MM, Lindsey AJ, Dobbels AF (2023) The effects of cereal rye cover crop seeding rate, termination timing, and herbicide inputs on weed control and soybean yield. Weed Sci 71(4):387–394

Flayyih TM, Almarie AA (2023) Total phenolic exudation and allelopathic potential of sunflower residues as sustainable weed management. Iraqi J Sci:2215–2222

Foy CL (2024) Weed allelopathy, its ecological impacts. In: Allelopathy in agroecosystems, vol 43. CRC Press

Frans RE, Semidey N (1992) The role of allelopathic sunflower in cotton production [weed suppression]. In: 1st international weed control congress, vol 2. Weeds Science Society of Victoria Inc., Melbourne, Australia, February 17, pp 171–174

Fujii Y, Hiradate S (2007) Allelopathy: new concepts & methodology. CRC Press

Gander JR (1998) Potential of allelopathic weeds for weed control in soybean. University of Arkansas

García-Romeral J, Castanera R, Casacuberta J, Domingo C (2024) Deciphering the genetic basis of allelopathy in japonica rice cultivated in temperate regions using a genome-wide association study. Rice 17(1):22

Gianessi LP (2013) The increasing importance of herbicides in worldwide crop production. Pest Manag Sci 69:1099–1105. https://doi.org/10.1002/ps.3484

Gimsing AL, Kirkegaard JA (2006) Glucosinolate and isothiocyanate concentration in soil following incorporation of Brassica biofumigants. Soil Biol Biochem 38(8):2255–2264

Grint KR, Arneson NJ, Oliveira MC, Smith DH, Werle R (2022) Cereal rye cover crop terminated at crop planting reduces early-season weed density and biomass in Wisconsin corn–soybean production. Agrosyst Geosci Environ 5(1):e20245

González-García E, Sánchez-Moreiras AM, & Vieites-Álvarez Y (2025) Allelopathic potential and chemical profile of wheat, rice and barley against the herbicide-resistant weeds Portulaca oleracea L. and Lolium rigidum Gaud. BMC Plant Biology, 25(1):624

Gupta S, Verma D, Tufchi N, Kamboj A, Bachheti A, Bachheti RK, Husen A (2021) Food, fodder and fuelwoods from forest. In: Husen A, Bachheti RK, Bachheti A (eds) Non-timber forest products: food, healthcare and industrial applications. Springer, Cham, pp 383–425

Hashemi M, Ehsani A, Jazani NH, Aliakbarlu J, Mahmoudi R (2013) Chemical composition and in vitro antibacterial activity of essential oil and methanol extract of *Echinophora platyloba* DC against some food-borne pathogenic bacteria. Vet Res Forum Int Q J 4:123–127

Heisey R (1990a) Allelopathic and herbicidal effects of extracts from tree of heaven (*Ailanthus altissima*). Am J Bot 77(5):662–670

Heisey RM (1990b) Evidence for allelopathy by tree-of-heaven (Ailanthus altissima). Journal of Chemical Ecology, 16(6):2039–2055

Heisey R (1996) Identification of an allelopathic compound from *Ailanthus altissima* (Simaroubaceae) and characterization of its herbicidal activity. Am J Bot 83(2):192–200. https://doi.org/10.1002/j.1537-2197.1996.tb12697.x

Heisey RM (1999) Development of an allelopathic compound from tree-of-heaven (Ailanthus altissima) as a natural product herbicide. In JT, Romeo (Ed.), Biologically Active Natural Products: Agrochemicals (pp. 57–68). CRC Press, Boca Raton, FL

Hesammi E (2011) The allelopathic effects of *Sorghum halepense* and *Amaranthus retroflexus* extract on the germination of corn grain. Aust J Basic Appl Sci 5:2249–2253

Ho TL, Nguyen TT, Vu DC, Nguyen NY, Nguyen TT, Phong TN et al (2020) Allelopathic potential of rice and identification of published allelochemicals by cloud-based metabolomics platform. Metabolites 10(6):244

Inderjit (2005) Soil microorganisms: an important determinant of allelopathic activity. Plant Soil 274:227–236. https://doi.org/10.1007/s11104-005-0748-7

Inoue M, Nishimura H, Li HH, Mizutani J (1992) Allelochemicals from *Polygonum sachalinense* Fr. Schm. (Polygonaceae). J Chem Ecol 18(10):1833–1840

Islam AM, Kato-Noguchi H (2013) Allelopathic prospective of Ricinus communis and Jatropha curcas for bio-control of weeds. Acta Agric Scand Sect B Soil Plant Sci 63(8):731–739

Janusauskaite D (2023) The allelopathic activity of aqueous extracts of Helianthus annuus L., grown in boreal conditions, on germination, development, and physiological indices of Pisum sativum L. Plants 12(9):1920

Kalač P, Kaltner F (2021) Pyrrolizidine alkaloids of European Senecio/Jacobaea species in forage and their carry-over to milk: a review. Anim Feed Sci Technol 280:115062

Kato-Noguchi H (2011) The chemical cross talk between rice and barnyardgrass. Plant Signal Behav 6(8):1207–1209

Kato-Noguchi H, Hasegawa M, Ino T, Ota K, Kujime H (2010) Contribution of momilactone A and B to rice allelopathy. J Plant Physiol 167(10):787–791

Khalil N, Ashour M, Fikry S, Singab AN, Salama O (2018) Chemical composition and antimicrobial activity of the essential oils of selected Apiaceous fruits. Future J Pharm Sci 4:88–92

Khamare Y, Chen J, Marble SC (2022) Allelopathy and its application as a weed management tool: a review. Front Plant Sci 13:1034649

Khan A, Imam Malik MW, Hussain I, Malik MH, Nadim MA, Baloch MS (2019) Allelopathic potential of aqueous leaf extracts of pearl millet (Pennisetum typhoides S. & H.) on germination and growth of some selected weeds. Pak J Weed Sci Res 25(3)

Khang DT, Anh LH, Thu Ha PT, Tuyen PT, Quan NV, Minh LT et al (2016) Allelopathic activity of dehulled rice and its allelochemicals on weed germination. Int Lett Nat Sci 58:1

Khorshidi J, Tabatabaei MF, Omidbaigi R & Sefidkon F (2009) Effect of densities of planting on yield and essential oil components of fennel (Foeniculum vulgare Mill var. Soroksary). J Agric Sci 1(1):152

Kim HM, Kim SJ, Kim HY, Ryu B, Kwak H, Hur J et al (2015) Constituents of the stem barks of *Ailanthus altissima* and their potential to inhibit LPS-induced nitric oxide production. Bioorg Med Chem Lett 25(5):1017–1020. https://doi.org/10.1002/chin.201527253

Kong CH, Wang P, Zhang CX, Zhang MX, Hu F (2006) Herbicidal potential of allelochemicals from Lantana camara against Eichhornia crassipes and the alga Microcystis aeruginosa. Weed Res 46(4):290–295

Kremer RJ, & Ben-Hammouda M (2009) Allelopathic Plants. 19. Barley (Hordeum vulgare L). Allelopathy Journal, 24(2)

Kumar V, Singh V, Flessner ML, Reiter MS, Kumari A, Price A et al (2025) Impact of cover crops and termination timing on weed management in corn. Weed Technol 39:e69

Li HH, Nishimura H, Hasegawa K, Mizutani J (1992) Allelopathy of *Sasa cernua*. J Chem Ecol 18(10):1785–1796

Lotina-Hennsen B, King-Díaz B, Pereda-Miranda R (2013) Tricolorin A as a natural herbicide. Molecules 18(1):778–788

Lun TL, Tojo S, Teruya T, Kato-Noguchi H (2023) Allelopathic activity of the invasive plant Polygonum chinense Linn. and its allelopathic substances. Plants 12(16):2968

Mahmood K, Khaliq A, Cheema ZA, Arshad M (2013) Allelopathic activity of Pakistani wheat genotypes against wild oat (Avena fatua L.). Pak J Agri Sci 50(2):169–176

Mantovanelli GC, Silva AA, Ricardo LL, Kagami FL, de Almeida JD, Barbosa MC et al (2025) Pearl millet cover crop extract inhibits the development of the weed Ipomoea grandifolia by inducing oxidative stress in primary roots and affecting photosynthesis efficiency. Plants 14(2):222

May FE, Ash JE (1990) An assessment of the allelopathic potential of *Eucalyptus*. Aust J Bot 38(3):245–254. https://doi.org/10.1071/BT9900245

Mengstu A, Bachheti A, Abate L, Bachheti RK, Husen A (2021) Health-promoting benefits, value-added products, and other uses of banana. In: Husen A, Bachheti RK, Bachheti A (eds) Non-timber forest products: food, healthcare and industrial applications. Springer, Cham, pp 339–364

Mirmostafaee S, Azizi M, Fujii Y (2020) Study of allelopathic interaction of essential oils from medicinal and aromatic plants on seed germination and seedling growth of lettuce. Agronomy 10(2):163. https://doi.org/10.3390/agronomy10020163

Missaoui A (2018) Allelopathic effects of cereal rye on weed suppression and forage yield in Alfalfa. Am J Plant Sci

Mohamadi N, Rajaie P, Fahimi H (2012) The allelopathic assay of *Eucalyptus camaldulensis* Labill on morphological and physiological parameters on monocot and dicot plants. ACECR Sci Inf Database 25:456–464

Muller C, Muller W, Haines B (1964) Volatile growth inhibitors produced by aromatic shrubs. Science 143(3605):471–473. https://doi.org/10.1126/science.143.3605.471

Mushtaq W, Adnan M, Hameed R, Ali HHA, Rehman TU, Hussain J et al (2020) Weed allelopathy: mechanisms, methodologies, and significance in weed management. Plants 9(3):357

Narwal SS (2006) Allelopathy in ecological sustainable agriculture. In Allelopathy: a physiological process with ecological implications (pp. 537–564). Dordrecht: Springer Netherlands

Netzly DH, Butler LG (1986) Roots of sorghum exude hydrophobic droplets containing biologically active components. Crop Sci 26(4):775–778. https://doi.org/10.2135/cropsci1986.0011183x002600040031x

Ni J, Shi J, Tan Q, Chen Q (2018) Two new compounds from the fruit of *Ailanthus altissima*. Nat Prod Res 33(1):101–107. https://doi.org/10.1080/14786419.2018.1437434

Norouzi N, Mohammadi GR, & Nosratti I (2014) Allelopathic activity of rye (Secale cereale) and common vetch (Vicia villosa) on corn and some weed species. Int J Plant Anim Environ Sci Vol. 4, No. 3:409–412 ref. 29

Nunes JJ, Arneson NJ, Smith D, Ruark M, Conley S, Werle R (2024) Elucidating waterhemp (Amaranthus tuberculatus) suppression from cereal rye cover crop biomass. Weed Sci 72(3):284–295

Onat B, Bakal H, Gulluoglu L, & Arioglu H (2017) The effects of row spacing and plant density on yield and yield components of peanut grown as a double crop in Mediterranean environment in Turkey. Turk J Field Crops 22(1):71–80

Oroojalian F, Kasra-Kermanshahi R, Azizi M, Bassami MR (2010) Phytochemical composition of the essential oils from three Apiaceae species and their antibacterial effects on food-borne pathogens. Food Chem 120:765–770. https://doi.org/10.1016/j.foodchem.2009.10.024

Oueslati O, Ben-Hammoudam H, Ghorbel M, El Gazzeh M, & Kremer RJ (2009) Role of phenolic acids in expression of barley (Hordeum vulgare) autotoxicity. Allelopathy Journal, 23(1)

Pavela R (2005) Insecticidal activity of some essential oils against larvae of *Spodoptera littoralis*. Fitoterapia 76:691–696. https://doi.org/10.1016/j.fitote.2005.02.004

Pérez FJ, & Ormenoñuñez J (1991) Difference in hydroxamic acid content in roots and root exudates of wheat (Triticum aestivum L.) and rye (Secale cereale L.): Possible role in allelopathy. J Chem Ecol 17(6):1037–1043

Peters EJ (1968) Toxicity of tall fescue to rape and birdsfoot trefoil seeds and seedlings. Crop Sci 8(6):650–653. https://doi.org/10.2135/cropsci1968.0011183x000800060003x

Peters EJ, Mohammed Zam A (1981) Allelopathic effects of tall fescue genotypes. Agron J 73(1):56–58. https://doi.org/10.2134/agronj1981.00021962007300010013x

Prajapati V, Tripathi AK, Aggarwal KK, Khanuja SPS (2005) Insecticidal, repellent and oviposition-deterrent activity of selected essential oils against *Anopheles stephensi*, *Aedes aegypti* and *Culex quinquefasciatus*. Bioresour Technol 96:1749–1757. https://doi.org/10.1016/j.biortech.2004.11.008

Rawat LS, Maikhuri RK, Bahuguna YM, Jha NK, & Phondani PC (2017) Sunflower allelopathy for weed control in agriculture systems. J Crop Sci Biotechno 20(1):45–60

Ruwanza S, Gaertner M, Esler KJ, Richardson DM (2014) Allelopathic effects of invasive *Eucalyptus camaldulensis* on germination and early growth of four native species in the Western Cape, South Africa. South For J For Sci 77(2):91–105. https://doi.org/10.2989/20702620.2014.965985

Sadeqifard S, Mirmostafaee S, Joharchi MR, Zandavifard J, Azizi M, Fujii Y (2022) Evaluation of allelopathic activity interactions of some medicinal plants using fractional inhibitory concentration and isobologram. Agronomy 12(12):3001. https://doi.org/10.3390/agronomy12123001

Saharkhiz MJ, Smaeili S, Merikhi M (2010) Essential oil analysis and phytotoxic activity of two ecotypes of *Zataria multiflora* Boiss. Growing in Iran. Nat Prod Res 24(17):1598–1609. https://doi.org/10.1080/14786411003754280

Sangeetha C, & Baskar P (2015) Allelopathy in weed management: A critical review. African Journal of Agricultural Research, 10(9):1004–1015

Salam MA, Kato-Noguchi H (2010) Evaluation of allelopathic potential of neem (*Azadirachta indica* A. Juss) against seed germination and seedling growth of different test plant species. Int J Sustain Agric 2(2):20–25

Sasikumar K, Vijayalakshmi C, Parthiban KT (2002) Allelopathic effects of *Eucalyptus* on blackgram (*Phaseolus mungo* L.). Allelopath J 9(2):205–214

Satyanarayana Murthy I, Jolly GE, John AP (2025) Allelopathy in weed management: a comprehensive review. Int J Plant Soil Sci 37(5):96–104

Sedara AM, & Sedara OS (2020) A review on inter-row crops mechanical weeder. Turk J Agr Eng Res (TURKAGER), 1(1):200–221

Sencenbaugh L, Mangold JM, Ulrich DE, Rew LJ (2025) Efficacy of multiple Brassica biofumigation techniques in the suppression of non-native and native grass seedling emergence and productivity. Weed Res 65(1):e12670

Serag MS, Khedr AEHA, El-Amier YA, El-Afify SM (2020) Bioactive constituents and allelopathic activities of the invasive weed *Ranunculus sceleratus* L. Nile Delta, Egypt. J Exp Sci 11:1–4

Sharma PK, Raina AP, Dureja P (2009) Evaluation of the antifungal and phytotoxic effects of various essential oils against *Sclerotium rolfsii* (Sacc) and *Rhizoctonia bataticola* (Taub). Arch Phytopathol Plant Protect 42(1):65–72. https://doi.org/10.1080/03235400600940897

Shrestha A (2009) Potential of a black walnut (*Juglans nigra*) extract product (Naturecur®) as a pre- and post-emergence bioherbicide. J Sustain Agric 33(8):810–822. https://doi.org/10.1080/10440040903303397

Singh NK, Seth A, & Maurya SK (2015) Croton bonplandianum Baill.: A rich source of essential fatty acids, linoleic and linolenic acid. Der Pharma Chemica, 7(3):85–88

Sisodia S, & Siddiqui MB (2009) Allelopathic potential of rhizosphere soil of Croton bonplandianum on growth and establishment of some crop and weed plants. African Journal of Agricultural Research, 4(5):461–467

Sladonja B, Sušek M, Guillermic J (2015) Review on invasive tree of heaven (*Ailanthus altissima* (Mill.) Swingle) conflicting values: assessment of its ecosystem services and potential biological threat. Environ Manag 56(4):1009–1034. https://doi.org/10.1007/s00267-015-0546-5

Soltys D, Kwasniewski M, Sliwinska E, Ciereszko I (2013) Allelopathic potential of wheat varieties against great burnet (Sanguisorba officinalisL.). Allelopath J 31(1):117–128

Spitters CJT (1989) Weeds: Population dynamics, germination and competition. In Weed Research (Vol. 29, pp. 303–314). Blackwell Scientific Publications

Stephenson R, Posler G (1988) The influence of tall fescue on the germination, seedling growth and yield of birdsfoot trefoil. Grass Forage Sci 43(3):273–278. https://doi.org/10.1111/j.1365-2494.1988.tb02152.x

Tabaglio V, Gavazzi C, Schulz M, Marocco A (2008) Alternative weed control using the allelopathic effect of natural benzoxazinoids from rye mulch. Agron Sustain Dev 28(3):397–401

Taiz L, Zeiger E, Møller IM, Murphy A (2015) Plant physiology and development, 6th edn. Sinauer Associates, Sunderland

Tarassoli Z, Labbafi M, Jokar Shoorijeh F (2021) Allelopathic effect of herbal formulation containing Ferula assa-foetida L. essential oil and castor oil (Ricinus communis L.) as an herbicide on Amaranthus retroflexus L. seed germination. J Med Plants 20(80):69–82

Turk MA, Tawaha AM (2003) Allelopathic effect of black mustard (Brassica nigra L.) on germination and growth of wild oat (Avena fatua L.). Crop Prot 22(4):673–677

Uddin MR, Park SU, Dayan FE, Pyon JY (2013) Herbicidal activity of formulated sorgoleone, a natural product of sorghum root exudate. Pest Manag Sci 70(2):252–257. https://doi.org/10.1002/ps.3550

Ullah R, Aslam Z, Attia H, Sultan K, Alamer KH, Mansha MZ,... & Zaman QU (2022) Sorghum allelopathy: alternative weed management strategy and its impact on mung bean productivity and soil rhizosphere properties. Life, 12(9):1359

Varshney JG, Babu MBBP (2008) Future scenario of weed management in India. Indian J Weed Sci 40(1):1–9

Vashishth DS, Bachheti A, Bachheti RK, Husen A (2023) Allelopathic effect of *Callistemon viminalis* leaf extract on weeds, soil features, and growth performance of wheat and chickpea plants. J Plant Interact 18(1):2248172. https://doi.org/10.1080/17429145.2023.2248172

Voigt G, Mergen F (1962) Seasonal variation in toxicity of *Ailanthus* leaves to pine seedlings. Bot Gaz 123(4):262–265. https://doi.org/10.1086/336160

Weston LA (1990) Cover crop and herbicide influence on row crop seedling establishment in no-tillage culture. Weed Sci 38(2):166–171. https://doi.org/10.1017/s0043174500056320

Whittaker RH, Feeny PP (1971) Allelochemicals: chemical interactions between species. Science 171:757–770

Yang X, Scheffler BE, Weston LA (2004) *SOR1*, a gene associated with bioherbicide production in sorghum root hairs. J Exp Bot 55(406):2251–2259. https://doi.org/10.1093/jxb/erh252

Zimdahl RL, Basinger NT (2024) Fundamentals of weed science, 6th rev edn. Academic Press

Chapter 6
Phytoremediation: Using Plants to Address Environmental Challenges

Marie Yayinie, Limenew Abate Worku, Teshale Adane, and Tadesse Bassie Gelaw

Abstract Heavy metals, radionuclides, pesticides, medications, fertilisers, and other organic substances are examples of persistent contaminants that frequently become stuck in sediments, soils, and mineral deposits in natural environments. This renders the pollutants biologically unavailable and contaminates the lithosphere and hydrosphere. Although there are conventional ways to prevent those pollutants, such as extraction, phytoremediation is a contemporary strategy. It successfully tackles these problems and has been used for decades as a new green technology. Naturally, plants are used in phytoremediation, an economical and environmentally friendly technique, to immobilise, eliminate, degrade, stabilise, or reduce the toxicity of hazardous materials discharged into the environment from various sources. Phytostabilisation, phytoextraction, phytodegradation, phytovolatilisation, rhizodegradation, and rhizofiltration are the principal mechanisms at play in this process. The limitations of this technique can be mitigated with various supportive measures, such as genetic engineering, natural and chemical adjustments, and enhancement of natural microbial activity.

Keywords Phytoremediation · Pollutants · Contaminants · Toxic metals

M. Yayinie · L. A. Worku (✉) · T. B. Gelaw
Department of Chemistry, College of Natural and Computational Science, Debre Tabor University, Debre Tabor, Ethiopia

T. Adane
Department of Chemical Engineering, Gafat Institute of Technology, Debre Tabor University, Debre Tabor, Ethiopia

L. A. Worku et al. (eds.), *Plant-Based Solutions for Sustainable Agriculture and Healthcare*, Sustainability Sciences in Asia and Africa,
https://doi.org/10.1007/978-981-95-8910-4_6

6.1 Introduction

6.1.1 Definition and Overview of Phytoremediation

Both organic and inorganic pollutants are released into the environment by human activity and by natural processes (Bayabil et al. 2022; Hou 2023; Puri et al. 2023). Potentially toxic substances (PTEs) like various toxic metals like chromium (Cr), arsenic (As), cobalt, cadmium (Cd), lead (Pb), mercury (Hg), nickel (Ni), and zinc (Zn) along with other pollutants like nitrates, nitrites, chlorides, bicarbonates, sulphates, and phosphates, mix with groundwater and enter into the soil from various sources such as municipal wastes and geogenic sources, including mining and industrial effluents (Li et al. 2019; Martínez-Alcalá et al. 2010; Vardhan et al. 2019). In contrast, most persistent organic pollutants, including polycyclic aromatic hydrocarbons (PAHs) and per- and polyfluoroalkyl substances (PFAS), primarily originate from waste generated by public activities (Guillotin and Delcourt 2022; Kumar et al. 2021).

Speedy industrialisation, municipal waste, and natural processes release large amounts of highly toxic inorganic and organic pollutants, such as toxic metals, radioactive materials, PAHs, petroleum hydrocarbons, polychlorinated biphenyls (PCBs), halogenated and nitroaromatic compounds, polychlorinated dibenzo-p--dioxins and dibenzofurans, solvents, explosives, and pesticides, into aquatic and soil ecosystems (Kamran et al. 2021; Saleem et al. 2020b; Hu et al. 2010; Prakash et al. 2013; Malik et al. 2017). These pollutants have gained global attention for their toxic effects on living organisms. They are extremely persistent in the environment. Traditional biochemical and physical treatment methods are expensive and often generate noxious by-products. These contaminants were removed using an alternate method known as bioremediation, which uses particular microbial species like fungi and algae, and phytoremediation, which uses hyperaccumulator plants (Wu et al. 2016; Anyakora et al. 2013).

Phytoremediation is an environmentally friendly method that uses plants and their natural processes to degrade, remove, or transform pollutants into harmless by-products (Chigbo and Batty 2014; Samardjieva et al. 2015; Iqbal et al. 2015; Shmaefsky and Husen 2023; Yayinie et al. 2025). The term *phytoremediation* combines the Greek *phyto* ('plant') and the Latin *remedium* ('cure or restoring balance') and was first introduced in 1994 by Ilya Raskin (Vamerali et al. 2010). This method employs different plant species to reclaim contaminated soils and waters (Ali et al. 2012), building on evidence that plants can extract, metabolise, and accumulate pollutants (Bollag et al. 1994). Plants respond to soil contaminants by absorbing them, signalling their presence, or releasing them at the surface. Restoring, cleansing, and rescuing places contaminated with varying levels of environmental toxins can be accomplished through phytoremediation, a controlled, cost-effective approach that uses green plants to remediate the impacted areas (Georgin et al. 2025) . Plants should be able to endure in hazardous and demanding environments. Restoring, cleansing, and rescuing places contaminated with varying levels of environmental toxins can be accomplished through phytoremediation, a controlled, cost-effective approach using green plants (Babu et al. 2021). Plants must be able to survive in hazardous, stressful environments to restore impacted areas. As autotrophs, they store pollutants in their tissues and contribute to environmental detoxification (Philp et al. 2005)

Phytoremediation is the process of containing, eliminating, or neutralising harmful toxins using green plants and the microorganisms associated with them, along with agronomic methods and soil reclamation or amendments. As long as plant species are properly selected based on the type of contamination, land features, and climate, these special qualities make phytoremediation an affordable and sustainable alternative to physical and chemical approaches (Sahoo et al. 2025). Advances in biotechnology have further enabled the use of genetically engineered or specifically selected plants to remove or reclaim both organic and inorganic contaminants in soil and water (Azad et al. 2014). Pollutant uptake, translocation from root to shoot, mineralisation, transformation, and, occasionally, compartmentalisation are the steps usually involved in the process. Although its affordability and sustainability make it appealing, its limited efficacy in completely cleaning contaminated environments remains limited. Exposure of plants to organic and inorganic pollutants often results in reduced growth due to oxidative stress.

Overproduction of reactive oxygen species (ROS) causes stress by disrupting the balance between free radicals and antioxidants, with cytotoxic consequences. This stress arises from the overproduction of ROS, which disrupts the equilibrium between free radicals and antioxidants, leading to cytotoxic effects. Hyperaccumulator plants and microbes used in bioremediation can counteract ROS by stimulating antioxidant defence mechanisms. These systems comprise non-enzymatic antioxidants, including proline, carotenoids (β-carotene), ascorbic acid, α-tocopherol, and reduced glutathione, as well as enzymatic antioxidants such as peroxidase, glutathione, catalase, reductase, polyphenol oxidase, and ascorbate peroxidase. This chapter explores in detail how these antioxidant systems are activated in plants and microbes during the bioremediation of organic contaminants (Tanwir et al. 2021).

6.1.2 Importance of Phytoremediation in Modern Environmental Management

First, plants are typically thought of as a source of fibre, food, and wood. Second, in addition to their 'altruistic' provision of habitat for other species, humans can also value their presence for aesthetic reasons. However, their importance as an ecological counterbalance to industrialised processes is increasingly recognised. Fossil fuel combustion, waste production (inorganic and organic waste, sewage, and effluents), and general water flow and processing are some of the activities considered to be beneficial to plants. Plants have long been recognised for their consumption of CO, and, more recently, of other gaseous industrial by-products. Both the scientific and general press have recently acknowledged their contributions to reducing global warming. Their use as a final water treatment and for the disposal of sludge resulting from wastewater treatment is centuries-old. Another new emerging technology of using plants to clean up soil contamination caused by organic or inorganic contaminants is termed phytoremediation (Simonich and Hites 1994).

As an environmentally benign option for cleaning up contaminated areas, phytoremediation has become well known. As a sustainable, eco-friendly method, phytoremediation has attracted significant interest as a potential way to reduce the negative effects of organic chemicals and heavy metals on the environment. This new and advanced method uses plants to naturally immobilise, detoxify, or remove contaminants, assisting in the restoration of polluted environments. Common inorganic pollutants include phosphates, nitrates, radionuclides, heavy metals, and metalloids (Nedjimi 2021). Excessive amounts of Hg, Pb, As, and Cd can cause cancer, developmental delays, neurological damage, chronic anaemia, renal damage, and cognitive impairment in both humans and wildlife. Because they are persistent and detrimental, organic pollutants from municipal, industrial, and farming processes also pose serious risks (González-Mille et al. 2018). Because of their hydrophobic properties, substances such as insecticides, phenols, PAHs, and PCBs can bioaccumulate in the environment, cause toxicity to species, and disturb ecosystems (Aryal 2024).

Using plants as in situ purifying agents to remediate soil and water systems is called phytoremediation. Hydrocarbons, pesticides, heavy metals, radionuclides, and, more recently, pharmaceuticals have all been the subject of extensive research on phytoremediation. It is possible to reach the key point that ensures the method's success and efficiency by choosing plants that are suited to the type and quantity of pollutants, as well as the attributes of the contaminated site. Plants used in phytoremediation processes should be highly adaptable and inherently tolerant of pollutant exposure, according to previous studies. Additionally, a large biomass yield, ease of distribution and harvesting, and a strong ability to take up and mount up pollutants are required (Ishikawa et al. 2006; Gahlawat and Gauba 2016; Singh et al. 2008).

The plant's capacity to extract pollutants from the substrate and shift them from subterranean to aboveground plant parts is frequently influenced by two key factors. The first metric, the bioconcentration factor (BCF), is explained as follows:

$$BCF_{\text{roots}} = C_{\text{roots}} / C_{\text{medium}}$$
$$BCF_{\text{shoots}} = C_{\text{shoots}} / C_{\text{medium}},$$

where C_{roots} is the concentration of pollutant in roots, C_{shoots} is shoot contaminant quantity, and C_{medium} is pollutant's amount in growing area (i.e. soil or aqueous solution).

Another method for determining the second parameter is the translocation factor (TF).

$$\text{TF} = C_{\text{shoots}} / C_{\text{roots}},$$

where C_{roots} is the concentration of a contaminant in roots, and C_{shoots} is the concentration of a contaminant in shoots (Grzegórska et al. 2020).

6.1.3 Scope and Objectives of the Chapter

The most current advancements in phytoremediation are thoroughly reviewed in this chapter. The various environmental contaminants, mainly those of the lithosphere and hydrosphere, along with their sources, were presented in detail. Phytoremediation principles and mechanisms used as a remediation technology, the various plant species used in phytoremediation, its application approaches, and the environmental, economic, and social benefits were elaborated. Various case study scenarios will be discussed to clarify and exemplify phytoremediation as a remediation technique. The main elements that affect phytoremediation's effectiveness, as well as the most recent, promising remedies, will also be thoroughly covered. Lastly, we address the difficulties and constraints associated with phytoremediation and highlight potential guidelines in this area, such as the prospective technique for processing post-harvested plants to produce bioenergy.

Hence, the principal objective of this book chapter is to:

- Define and present a comprehensive outline of phytoremediation
- Highlight the usefulness of phytoremediation in environmental management
- Describe the kinds of environmental pollutants that are suitable for phytoremediation
- Describe the basic workings and scientific underpinnings of phytoremediation
- Distinguish the selection criteria for plant species that are employed in phytoremediation
- Describe how phytoremediation projects are practically implemented and investigate methods for increasing the effectiveness of phytoremediation
- Discuss the unique benefits of phytoremediation for the environment, economy, and society, as well as assess the findings of actual case studies
- Analyse current issues and consider potential avenues for phytoremediation application and future study

6.2 Background and Context

6.2.1 Overview of Environmental Contaminants Affecting Soils and Waters

Around the globe, surface waters and groundwater are increasingly contaminated by agricultural, industrial, military, and anthropogenic activities. Toxic chemicals, including metals, radionuclides, and organic contaminants, accumulate rapidly due to contamination, which is often caused by ignorance, negligence, a lack of forethought, or the high cost of effective waste disposal. In addition to depleting natural resources, this pervasive contamination severely stresses ecosystems (Paz-Alberto and Sigua 2013; Kafle et al. 2022). Human activities, including fast urbanisation, intensive agriculture, and industrialisation, mainly cause environmental contamination. Organic compounds, heavy metals, radionuclides, metalloids, oil spills, and agrochemicals are all pollutants. The leading causes of soil pollution are high fertiliser and pesticide usage, irrigation with contaminated water, mining, and the discharge of industrial and household wastewater (Wu et al. 2016; Saleem et al. 2020a).

Mining activities significantly affect soil electrical conductivity, pH, and cation exchange capacity. The concentrations of various heavy metals and organic contaminants in soil and sediments increase as a result of these modifications. Toxic metals such as Pb, Cd, Cr, copper (Cu), Ni, iron (Fe), Zn, and manganese (Mn) are frequently found at high concentrations. Additionally, it is common to find organic contaminants, including phenols, hydrocarbons, aromatic chemicals, and organochlorinated pesticides (Liu et al. 2017; Song et al. 2017).

The microbial population, their vital functions, and their overall activity are all severely disturbed by the accumulation of heavy metals in soil. The chemical and toxicological properties of heavy metals determine the specific impact on microbial populations and enzymatic activities. For instance, Pb inhibits invertase, ureases, catalases, and acid phosphatase; As targets sulphatase and phosphatase; and Cd influences enzymes such as ureases, proteases, and basic phosphatases. High levels of Cr (VI) are clearly harmful and have an adverse influence on the metabolism of soil microbial cells (Huang et al. 2009; Singh and Kalamdhad 2011; Karaca et al. 2010).

Toxic heavy metals bioaccumulate in organisms as a result of consuming contaminated food crops. The health of both animals and humans may be severely threatened by eventual biomagnification at higher levels of the food chain. The bioaccumulation of such metals in the human body can result in several severe health issues, including:

- Chronic exposure to Cd can cause Itai-itai disease.
- Zn may cause anaemia and renal problems.
- Copper causes harm to the central nervous system, hepatic system failure, and mucosal corrosion.
- Ni is commonly linked to nervous system issues and skin discomfort.
- Pb can result in neurological system damage and cardiovascular issues.
- Hg can be associated with skin rashes, psychological breakdown, and abortion.
- Cr, a known carcinogen, can cause cancer (Zohdi et al. 2012; Singh and Kalamdhad 2011; Pan et al. 2010).

Major organic contaminants polluting the soil include petroleum hydrocarbons, chlorinated solvents, PCBs, explosives, PAHs (such as pyrene), and trinitrotoluene (TNT). Various pesticides commonly detected in contaminated soils include the following:

- Atrazine
- Dieldrin
- Hexachlorobenzene
- Dichlorodiphenyltrichloroethane (DDT)
- 1,2,4-Trichlorobenzene

These chemicals can have serious environmental impacts, harming soil-dwelling organisms, reducing access to essential nutrients, and diminishing overall soil productivity (Alkorta and Garbisu 2001). Similar to soil, several contaminants have polluted water resources. Toxic heavy metals are introduced by urbanisation and industrialisation expansions through sewage, overflows, and seepages, which account for the majority of water contamination sources. Approximately 58% of health problems, including diarrhoea, a foremost cause of early childhood mortality, are linked to the untreated discharge of almost 80% of the world's wastewater into the environment (Parveen et al. 2020; Saleem et al. 2020a; Connor et al. 2017).

6.3 Principles of Phytoremediation

Air, soil, and water pollution have increased dramatically due to the rapid growth of global businesses, including energy, agriculture, mining, metallurgy, and chemical production and transportation (Grzegórska et al. 2020). Concurrently, pro-ecological initiatives to enhance the quality and purity of the environment have been bolstered by rising public awareness of environmental issues. Promoting cleaner production methods and implementing efficient treatment techniques that lower contamination levels are crucial to addressing the issue. Conventional techniques for cleaning up contaminated soils and scums include stabilisation, incineration, vitrification, isolation, thermal treatment, chemical oxidation, and solvent extraction. These techniques, however, are overpriced and usually result in secondary waste. On the other hand, phytoremediation provides a more ecologically friendly alternative by using plants and the microorganisms that live within them to eliminate, degrade, or immobilise hazardous materials in the environment (Kang 2014; Grzegórska et al. 2020).

6.3.1 Mechanisms of Phytoremediation

Using plants as an in situ cleansing agent to refine aquatic and soil systems is known as phytoremediation. When plants capture the useful nutrients and water from the soil through their roots, they may actively convey or passively diffuse toxic substances or pollutants. Numerous methods, such as phytodegradation, phytovolatilisation, phytoextraction, phytostabilisation, rhizofiltration, or rhizodegradation, may be applied in the purification mechanism (Hou 2023; Jin et al. 2021). Figure 6.1 illustrates the primary mechanism of phytoremediation, using examples of contaminants that might be treated using comparable techniques.

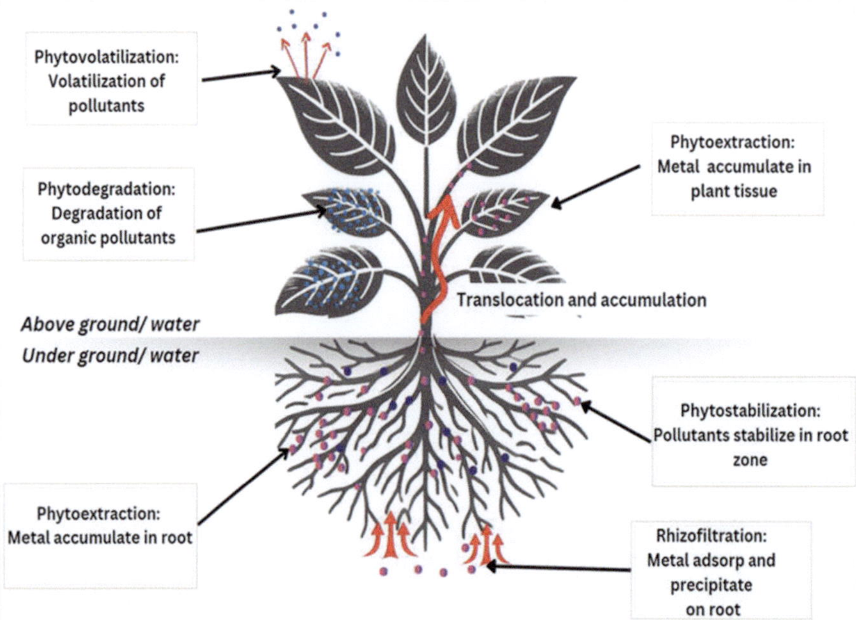

Fig. 6.1 Plant interaction (phytoremediation) of various contaminants, for each mechanism. (Adapted from Ahmad (2024))

6.3.2 Phytoextraction

Phytoextraction, as one method of phytoremediation, by which plants extract hazardous substances from water or soil, is primarily responsible for heavy metals that have a high density and may be poisonous to living things and radionuclides, even at relatively minimal amounts. The plants used for phytoextraction are referred to as hyperaccumulators because their tissues contain incredibly high levels of heavy metals. The heavy metals that plants extract are also toxic to them. Plants that absorb fewer pollutants can also carry out phytoextraction. However, because of their swift growth and biomass production, they may be able to remove a significant quantity of toxins from the soil (Guidi Nissim et al. 2018).

The process of phytoextraction involves several phases, beginning with the uptake of pollutants from soil or water. Afterwards, these toxins move from the roots to the plant tissues above ground, where they gradually accumulate. Plants that have accumulated pollutants are then harvested. In the phytoextraction process, plants with both TF and BCF >1 may be used. Because heavy metals can bind to key proteins required for basic functions, they pose a serious hazard to living things. They may also trigger the breakdown of other molecules, producing highly reactive species, such as ROS, that further disrupt biological processes. In addition to depleting important chemicals, these interactions produce highly reactive radicals such as O and OH (Brennan and Shelley 1999; Yoon et al. 2006a).

A plant needs to undergo several processes to avoid toxic metals in water or soil. The metal must first exist in a form that plant roots can absorb. The roots then absorb the dissolved metal. The plant then chelates the metal, binding it to organic substances either before or after absorption to protect itself and improve mobility. After that, the chelated metal is delivered to particular tissues for secure storage. Last but not least, the plant needs to build defences against and ways to mitigate any harm caused by the movement and storage of these metals. The key mechanisms in a hyperaccumulator plant are those that transport and accumulate heavy metals (Seth et al. 2012; Rascio and Navari-Izzo 2011).

6.3.3 Phytostabilisation

One method frequently used to remediate soils, wetlands, and mining wastes that contain metals such as Zn, Pb, Cu, Cd, Mn, Ni, As, Fe, and Cr is phytostabilisation (Farahat and Galal 2018). By immobilising metals, this method lowers the bioavailability of contaminants and stops leaching, erosion, airborne movement, and the spread of harmful compounds to nearby areas. The procedure depends on rhizosphere-driven processes such as complexation, sorption, precipitation, and adsorption. Excluder plants are important because they limit metal uptake by keeping metals in their roots (BCF >1) and by showing minimal translocation to shoots (TF <1) (Alkorta et al. 2010; Nouri et al. 2011).

By encouraging root accumulation, assisted phytostabilisation, which involves adding organic or inorganic amendments such as manure, charcoal, or industrial by-products, improves plant growth, increases metal immobilisation, and reduces metal transfer to shoots. Mutual amendments include composted manure, biochar, limestone, furnace slag, biosolids, bone meal, diatomite, sand, activated carbon, red mud, chalcedonite, and dolomite (Radziemska et al. 2017; Phusantisampan et al. 2016). *Miscanthus sinensis* and *Pteridium aquilinum* were shown to be more effective at phytostabilising Pb/Zn mine tailings when furnace slag, bottom ash, bone meal, and red mud were added. Compared with *P. aquilinum*, M. sinensis exhibited lower TFs and predominantly accumulated metals in its roots (Pavel et al. 2014). Furthermore, the quantities of soluble and extractable toxic metals were considerably reduced by Fe-rich additions, such as furnace slag and red mud. Another way to enhance phytostabilisation is to inoculate roots with mycorrhizal fungi, which live in symbiotic relationships with plants. Studies show that this type of inoculation increases plant growth and resilience to metal stress, enhances root accumulation, and reduces metal transfer to shoots (Gu et al. 2017).

6.3.4 Phytodegradation

Phytodegradation is the scheme by which poisonous substances are broken down and removed by plants within their tissues. This method can clean up organic contaminants, such as pesticides, by engaging various plant parts to break them down or transform them. In the direct process of phytodegradation, pollutants are captured by the plant and then metabolised or broken down into smaller, less hazardous molecules, which are distributed throughout the plant tissue. Because plants lack active transporters for organic contaminants, they are absorbed passively. Insecticides, herbicides, petroleum hydrocarbons, medications, and surfactants in soil and water can all be broken down by either phytodegradation or phytotransformation. It involves contaminant uptake and breakdown into intermediates with low molecular weight through plant metabolic methods, catalysed by enzymes such as dehalogenases, oxygenases, and reductases (Grzegórska et al. 2020; Kafle et al. 2022).

Susarla et al. (1999) found that *Myriophyllum aquaticum* degraded perchlorate five times faster in aqueous media than in sand, likely due to chloride ion release, with degradation occurring stepwise to chloride. About 70% of the TNT in the water was eliminated by the same plant (Vanderford et al. 1997; Yoon et al. 2006a, b). *Arabidopsis thaliana* degraded 95% of 2,4-dinitrotoluene (DNT) in 15 days, incorporating part of the transformed compounds into tissues (e.g. lignin, cellulose), while generating monoamino-nitrotoluene and other intermediates.

6.3.5 Rhizofiltration

Rhizofiltration is a phytoremediation method that concentrates and precipitates pollutants within root tissues using plants to adsorb or absorb them from the rhizosphere. For successful restoration, plants with large surface areas and long, fibrous roots are favoured. Sunflower, spinach, tobacco, rye, and Indian mustard are a few examples of these plants. This method is usually utilised for wastewater or groundwater contaminated with radionuclides, including U, Ra, and Cs, as well as poisonous heavy metals such as Cd, Cr, Ni, Pb, and Zn. Rhizofiltration is a phytoremediation method that concentrates and precipitates pollutants within root tissues by using plant roots to adsorb or absorb them from the rhizosphere. For successful restoration, plants with large surface areas and long, fibrous roots are favoured. Sunflower, tobacco, spinach, rye, and Indian mustard are a few examples of these plants. This technique is mainly practicable to groundwater tainted with the heavy toxic metals and radionuclides listed above (Lee and Yang 2010; Mikheev et al. 2017).

Studies show Helianthus annuus can remove up to 80% of uranium within 24 h, while Phaseolus vulgaris achieves 60–80% under acidic conditions (pH 3–5), where adsorption, precipitation, and sorption dominate the mechanism. *Brassica juncea* and *Chenopodium amaranticolor* showed lower uranium uptake (20–23% and 13%, respectively). Helianthus annuus also removed ~50% of U and ~70% of Ra in

2 days, mainly retained in roots with limited translocation. Phaseolus vulgaris removed up to 98.9% U at pH 5 within 72 h, with uranium forming insoluble precipitates at neutral pH (Lee and Yang 2010; Yang et al. 2015). For heavy metals, Pistia stratiotes showed strong tolerance and accumulation, removing up to 95% of Cd and Pb in 7 days, with metal levels in roots about 10 times higher than in leaves (Vesely et al. 2011).

6.4 Plant Species Used in Phytoremediation

6.4.1 Plant Characteristics Required for Effective Phytoremediation

Several variables affect the process of cleaning up contaminated soils, including the application of plants. The most significant of these are the characteristics, concentration, and phytoavailability of contaminants in the soil, as well as plant species (Grzegórska et al. 2020). The unsurpassed plant species for phytoremediation should be hardy, harvest very large biomasses, be resistant to the injurious effects of metals and pollutants, be simple to grow, have an elevated captivation capacity, and be unappealing to herbivores (Shabani and Sayadi 2012). The toxin, the plant type, and the soil all affect how well tidying works. The effectiveness of restoring and cleansing depends heavily on soil carbon concentrations, pH, electrical conductivity, microbial processes, and other soil factors. It is influenced by plant biomass and metabolism. Plant species' capacity for their phytoremediation can be assessed by (i) the potential to accumulate several toxic metals, (ii) large root systems for the breakdown of organic contaminants, (iii) suitability for growth in Nordic climates, and (iv) potential for post-harvest biomass reuses (Grzegórska et al. 2020; Paladino et al. 2025).

6.4.2 Selection Criteria for Phytoremediator Plants in Various Ecosystems

The process of phytoremediation varies depending on a plant's tendency to take up pollutants through its roots, which are subsequently transported to the leaves for eventual disposal. The two ways nutrients are absorbed are through passive processes, such as transpiration, and through active transport via membrane proteins that use apoplastic and symplastic pathways in the xylem. Moreover, because plants absorb unnecessary inorganic materials, it is crucial to normalise the intake of vital minerals like copper, Zn, and magnesium, which can become hazardous in excess. Environmental conditions have a decisive effect on the success of phytoremediation. The extent to which plants can absorb and handle pollutants depends on numerous factors, including soil type, pH, moisture content, and climate (Aryal 2024).

An assortment of plant species is commonly used in phytoremediation, including grasses, vegetables, willows, hybrid poplars, and hyperaccumulators. Specific plants can be chosen based on the contamination type and its location. For example, mustard and alfalfa (horse feed) are effective for treating contaminants in surface soil, whereas crossbreed poplars are essential for remediating groundwater. Moreover, therapeutic and fragrant plants are vital, as they help reduce contamination in the food chain. Fragrant or aromatic species from families such as *Lamiaceae*, *Poaceae*, *Asteraceae*, and *Geraniaceae* have shown promise in phytoremediation of heavy metals. Plants like *Echinacea purpurea* and *Callistephus chinensis* are also important for phytoremediation due to their capacity to release compounds that aid in the breakdown of PAHs. Edible crops are generally unsuitable due to potential health risks associated with contamination or defilement; however, other plant families offer innovative, cost-effective solutions for phytoremediation (Saxena et al. 2019; Pandey et al. 2019).

A phytoremediation project's success depends on the choice of plant species. To ensure the plant can flourish in contaminated areas, effectively absorb toxic substances, and rehabilitate the site, several criteria must be considered. The parameters listed next are regularly used when selecting plants for phytoremediation mechanisms.

Contaminant Tolerance

Plants used for phytoremediation must be able to thrive in contaminated areas and exhibit strong resistance to toxins. Because of their proven resistance to micropollutants and their outstanding ability to absorb and store such contaminants, species like *Typha angustifolia* and *Juncus effusus* are frequently used. Furthermore, plants such as *Vetiveria zizanioides*, which have rapid growth and high biomass, are preferred because they can absorb more heavy metals while still thriving in contaminated soils. To achieve more focused and efficient remediation, the selection of appropriate plant species should be based on their resistance to specific contaminants, such as organic compounds or toxic heavy metals. In general, choosing plants entails assessing their capacity for increasing contamination, biomass throughput, and pollution tolerance. Both ecological stability and effective repair results are promoted by this all-encompassing decision method. For example, sweet potato cultivars have demonstrated the capacity to tolerate elevated Pb concentrations and to accumulate the metal without appreciably impairing growth (Tyagi et al. 2023; Daurov et al. 2024).

Root Structure and Depth

Because it influences how well plants can absorb, transmit, and store pollutants, the composition of root structures is important when choosing plants for phytoremediation. Plants with extensive, fibrous root systems, such as *Phragmites australis* and *Vetiveria zizanioides*, are generally favoured because they can efficiently penetrate the soil and improve pollutant absorption. In addition to their physical attributes, the root systems must be able to form symbiotic interactions with soil microbes and exhibit significant resistance to contaminants. These microbes enhance the mechanisms that break down and stabilise pollutants. In urban and industrial settings, Balsam poplar (*Populus balsamifera*) is notable for its robust root system, which helps stabilise soil and absorb pollutants. Because pollution has no detrimental effects on the microbial populations associated with *Balsam Poplar* roots, this plant is beneficial in areas with extremely polluted environments (Mei et al. 2022; Amabogha et al. 2023).

Biomass Production

Biomass output is a crucial consideration for selecting plant species, particularly those employed in extensive phytoremediation initiatives. Plants with high biomass, like sunflower (*Helianthus annuus*) and giant reed (*Arundo donax*), are valued for their capacity to both create a substantial amount of organic matter and absorb vast amounts of pollutants. Water hyacinth (*Eichhornia crassipes*) and willow (*Salix viminalis*) are examples of high-output biomass plants that offer two benefits: they enhance soil or water quality by removing pollutants, and the collected biomass can be used for bioenergy or other renewable energy sources. This dual functionality, fabricating biomass and removing contaminants, significantly increases the remediation effort's overall sustainability, making it a crucial factor in plant selection (Nahar and Sunny 2024).

Adaptability to Local Conditions

Environmental adaptation is an important consideration when choosing plant species for phytoremediation. If plants are well suited to the confined temperature, soil types, and pollution levels, they are more likely to flourish and effectively absorb or break down pollutants. For instance, *Pinus halepensis* shows significant resilience in soils contaminated with various heavy metals. In contrast, *Dittrichia viscosa* is an excellent choice for very dry areas because it can withstand both arid conditions and pollutants such as As and Cd. Using plants that are suitable for the specific site conditions improves the restoration process sustainability by reducing the need for additional inputs such as fertiliser and water. Furthermore, localised species often form closer relationships with native microbial communities, thereby increasing the efficacy of phytoremediation (Kharazian et al. 2024).

Contaminant-Specific Uptake Mechanisms

In phytoremediation, contaminant-specific absorption systems play a major role in a plant's capacity to absorb and neutralise various poisons. The processes of organic pollutants and heavy metals differ greatly. To assist in the absorption and transfer of heavy metals such as Cd, Pb, and Zn into plant tissues, plants use specific transport proteins and enzymes. Chelation, in which metals bind to organic molecules and are then compartmentalised within vacuoles, is often used to reduce cellular toxicity. In contrast, enzymatic pathways—particularly those involving cytochrome P450 enzymes—absorb organic pollutants and convert them into less hazardous molecules. These breakdown products may then be added to the plant's biomass or released into the atmosphere (Liu et al. 2022).

Non-Invasiveness and Ecological Impact
When choosing plants for phytoremediation, it is important to confirm that they won't become invasive and jeopardise the local ecology. Invasive species can outperform native plants, disrupt biodiversity, and cause long-term ecological problems. To prevent non-native species from spreading outside the restoration area, monitoring and control measures must be put in place if they are chosen. Phytoremediation is an environmentally friendly, non-invasive method. It uses plants to remove pollutants without seriously upsetting the ecosystem. Assisting in habitat restoration, improving soil health, and preserving biodiversity provide an environmentally friendly alternative to conventional remediation methods (Sharma et al. 2023).

6.4.3 Hyperaccumulator Plants and Their Traits

Certain plant species are known as hyperaccumulators because they can absorb and store extremely high levels of metals or other contaminants in their tissues. They are essential to the phytoextraction process, which aims to confiscate metals from polluted water or soil. Hyperaccumulators differ from most plant species in that they can tolerate and concentrate toxic substances, often accumulating them in their leaves and stems. Specialised metabolic pathways that deposit metals in vacuoles or cell walls, thereby reducing their toxicity, enable this characteristic. For example, they produce metal-binding proteins that help detoxify metals, such as phytochelatins and metallothioneins. These plants generally flourish in environments that are low in nutrients but high in metal content, where other plants cannot thrive (Grzegórska et al. 2020; Rai et al. 2021) (Fig. 6.2).

These plants efficiently take up large quantities of one or more toxic metals from the soil. Unlike most species, they do not retain these metals in their roots; instead, the metals are efficiently transported to the shoots and collected in aboveground tissues, particularly the leaves, at concentrations 100–1000 times larger than those found in non-hyperaccumulating plants. Remarkably, they show no cyphers of phytotoxicity. Hyperaccumulation is characterised by three main characteristics that distinguish it from associated non-hyperaccumulating groups: (i) significantly higher capacity for metal uptake from the ground; (ii) a quicker and more efficient root-to-shoot mobility of the toxic metals; and (iii) an improved capacity to detoxify and sequester significant quantities of the toxic heavy metals in their leaves. Thus far, about 450 species of angiosperms are hyperaccumulators of noxious metals such as Pb, Cd, Co, As, Ti, and Se (Eslava-Silva et al. 2023; Rascio and Navari-Izzo 2011).

Table 6.1: Examples of hyperaccumulator species that are extensively utilised in phytoremediation.

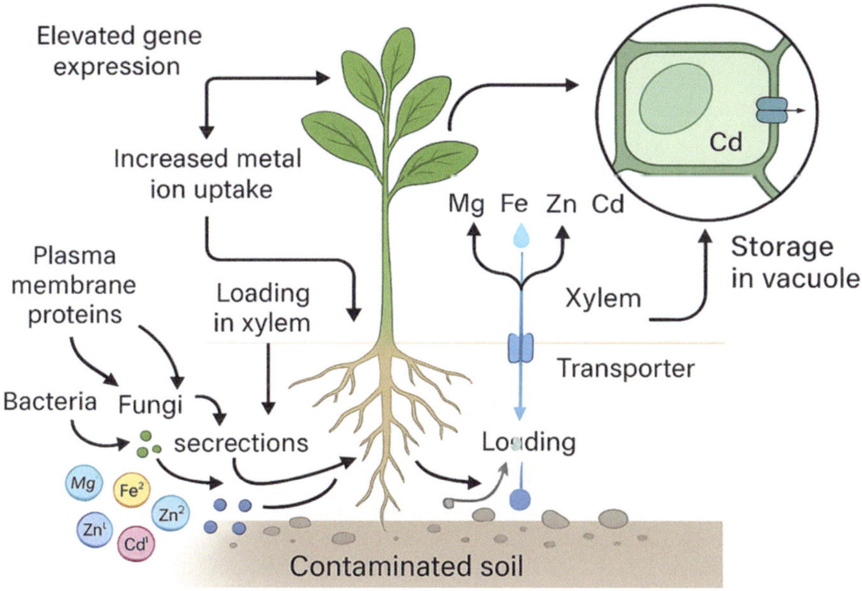

Fig. 6.2 Hyperaccumulators as a prelude to solving the heavy metal toxicity. (Adapted from Bhat et al. (2025))

6.4.4 Site Assessment and Preparation

Phytoremediation is a cost-effective and ecologically beneficial approach that eliminates toxic contaminants from the hydrosphere, lithosphere, and atmosphere. It concerns plants' innate capacity to absorb, detoxify, and accumulate pollutants. Before selecting a specific phytoremediation method, the characteristics of the metal contaminants must be assessed. Consequently, it is essential to understand the sources and the geographical distribution of metal contaminants before implementing phytoremediation at a site. The effectiveness of this approach depends on proper site preparation and comprehensive site assessment. The site assessment is an essential initial stage for collecting data on site features and contaminants, which affects the overall remediation plan (Mukherjee et al. 2025). Stereotypically, the site assessment, as reported by various researchers (Zhuang 2023; Chen et al. 2010; Oh et al. 2014), comprises the following:

- *Characterisation of Contaminants*: Having information about the type, amount, and spatial distribution of toxic impurities, for instance, pesticides, toxic metallic elements, and noxious organic molecules, is vital because it enables one to select the best phytoremediation strategy.
- *Site Characteristics*: To examine the site's nature, the parameters include soil organic matter content, texture, pH, nutrient levels, and water table depth. These elements can affect their survival and the accessibility of contaminants to organisms. Moreover, it is helpful to assess the current native vegetation and the local

Table 6.1 Plants that can act as hyperaccumulators of various heavy metals

Heavy metals	Hyperaccumulator plants	Bioaccumulation (mg/kg)	Reference
Ni	Alyssoides utriculata,	>1000	Roccotiello et al. (2015)
	Bromus hordeaceus	1467	D'Souza and Paul (2014)
	Linaria spartea	492	D'Souza and Paul (2014)
Cr (VI) Cr^{+6}	Pteridium aquilinum	11,854	Eslava-Silva et al. (2023)
	Phragmites australis	4825	Solomou et al. (2022)
Pb	Lonicera periclymenum	318	Solomou et al. (2022)
	Cistus salvifolius	548	Solomou et al. (2022)
As	Populus nigra	22×10^6	Tangahu et al. (2011)
	Pteris vittata	23,000	Solomou et al. (2022)
Zn	Sedum alfredii	13,799	Solomou et al. (2022)
	Noccaea caerulescens	19,410	Solomou et al. (2022)
	Verbascum phrygium	17044.54	Akin (2021)

environment, including temperature, rainfall, and the length of the growing season. Therefore, plants should be suitable for their environment.

- *Risk Assessment*: Determining appropriate repair and cleaning objectives requires assessing the potential hazards posed by the pollution to human health and the ecosystem.
- *Plant Selection*: Selecting the ideal plant species while considering the numerous venoms and site characteristics is also an important part of the evaluation. The ideal plants have properties such as pollution tolerance, a deep and broad root system, significant biomass production, and a non-invasive growth style.

After selecting the appropriate procedure and plant species, site preparation is essential to provide optimal conditions for plant survival and effective contaminant removal. The primary step in site preparation is physical preparation, including clearing land, weed removal, tilling or ploughing, and fencing. The other part is soil amendments that require fertilisation (putting essential nutrients), pH adjustment (with chemicals like lime or sulphur to attain an appropriate medium), and the application of chelating agents. The third action is planting, which involves sowing seeds, transplanting seedlings or mature trees, adjusting irrigation systems, and monitoring tools such as wells (Grzegórska et al. 2020; Aryal 2024; Oh et al. 2014).

6.4.5 Monitoring and Evaluating Remediation Progress

All environmental renovation plans need to monitor and assess the cleanup paths to ensure that the selected approach is effective and compliant with project and regulatory objectives. Here, the transition from implementation to validation occurs. Monitoring involves sampling, regular observation, and testing of environmental features such as soil, water, and plants to detect variations in pollutant levels over time. Evaluation targets on quantifying the effectiveness of the remediation process by comparing the obtained data with the baseline original values and preset objectives. Some of the vital factors include the rate of contaminant removal, the degree of ecosystem restoration, and any undesirable side effects (Ali et al. 2013; Oh et al. 2014).

6.5 Enhancing the Proficiency of Phytoremediation

To maximise the overall effectiveness of phytoremediation, it is essential to improve the absorption of contaminants by plants and their subsequent transport throughout plant tissues. Upgraded phytoremediation typically encompasses several methods: Chemical-assisted phytoextraction is used to enhance the mobility of heavy metals in soil by using surfactants or chelating agents such as citric acid and EDTA. Another innovative method for enhancing phytoremediation performance is the application of genetically modified organisms (GMOs). This technique uses molecular cloning and targeted gene expression to boost the ability of high-biomass plants to absorb additional metals from the soil. It also boosts the biomass production of naturally occurring hyperaccumulator species. The introduction of genes that produce xenobiotic degradation and heavy metal accumulation can significantly speed up environmental remediation processes (Oh et al. 2014).

Moreover, agricultural management practices can enhance phytoremediation performance by promoting better plant development and increasing microbial activity. This can be accomplished by optimising cultivation systems, adding carbon sources, and applying the right fertilisers. Particularly, in the rhizosphere, the area of soil around plant roots, plant-microbe interactions are essential for enhancing phytoremediation. Immunising this zone with a specific bacterium that degrades contaminants more effectively than indigenous microbes can improve plant growth and microbial efficiency. Mycorrhizal fungi have also been shown to augment a plant's ability to absorb, transport, and store heavy metals from polluted soils (Oh et al. 2014).

An environmentally acceptable method that increases a plant's capacity to purify contaminated soils is phytoremediation with soil amendments. The method augments overall plant production by loading and dispersing beneficial substances that significantly modify the contaminant's bioavailability, absorption, translocation, and bioaccumulation (Aryal 2024; Mukherjee et al. 2025). Adjusting the soil environment for plant growth and reducing tension with water-retaining materials (WRMs) and controlled-release fertilisers (CRFs) is another effective approach to improve phytoremediation. The goal of this approach is to increase the plant's biomass and health, both of which are essential for efficient pollutant absorption. Because contaminated soils are often nutrient-deficient, plants are stressed, and less biomass is required for restoration. CRFs improve phytoremediation by providing a consistent, optimal supply of vital nutrients over an extended period. CRFs are granules covered in a sulphur or polymer shell that gradually decomposes and releases nutrients only when the plant needs them. This coating offers the benefits of sustained growth, decreased stress, and lower leaching (Cartmill et al. 2014).

6.6 Case Studies and Real-World Applications

Restoring Pb-Contaminated Soils Through Phytoremediation

Universal environmental contamination from human activity is a growing concern, particularly from chemical by-products such as Pb. Pb released into the environment pollutes air, soil, and water, contributing to a weakening of an individual's mental, cognitive, and physical health. An environmentally friendly way to reduce Pb levels in soil is phytoremediation, a technique that uses plants to clean up contaminated areas. To see this, research was conducted in the Philippines by Paz-Alberto, A.M., and his associate (Paz-Alberto and Sigua 2013). The target of the research was as follows: (i) to assess the three grass species' vegetative traits and survival rates of various grasses such as carabao (*Paspalum conjugatum* L.), vetiver (*Vetiveria zizanioides* L.), and cogon (*Imperata cylindrica* L.); to assess those fully-fledged in soils containing diverse concentrations of Pb; and (ii) to assess and compare their potential as phytoremediators based on Pb accumulation in plant tissues.

The three test species, namely mentioned above, were cultivated separately in plastic bags filled with soils containing dissimilar amounts of Pb as a pollutant: 75 $mg{\cdot}kg^{-1}$ (37.5 $kg{\cdot}ha^{-1}$) and 150 $mg{\cdot}kg^{-1}$ (75 $kg{\cdot}ha^{-1}$) of Pb. The Pb concentrations in both the plants and soils were determined before and after treatment using an atomic absorption spectrophotometer. The experiment used a 3 × 2 factorial design in a completely randomised design.

The outcome or findings of the study showed that vetiver grass produced the driest matter overall (33.85–39.39 $Mg{\cdot}ha^{-1}$) and had the best vegetative performance of all the test plants. Conversely, carabao grass produced the least herbage, yielding 14.12 and 5.72 $Mg{\cdot}ha^{-1}$ in soils amended with 75 and 150 $mg{\cdot}Pb{\cdot}kg^{-1}$, respectively. Furthermore, the grass species named 'vetiver' had the main plant persistence rate, suggesting that it is more resistant to Pb contamination in the soil. Furthermore, it recorded the highest *Pb absorption rate* (10.16 ± 2.81 $mg{\cdot}kg^{-1}$), followed by *cogon grass* (2.34 ± 0.52 $mg{\cdot}kg^{-1}$) and *carabao grass* (0.49 ± 0.56 $mg{\cdot}kg^{-1}$).

The combined shoots and roots of the three grasses showed no appreciable variation in Pb accumulation between the two Pb treatments (75 and 150 $mg{\cdot}kg^{-1}$). On the other hand, large amounts of Pb were captured by the grass vetiver in both its roots and shoots, and it produced the greatest biomass. Its comprehensive root system, with many root hairs, which improve fertiliser uptake, increase soil contact, and encourage Pb absorption, accounts for its exceptional performance. The plants' overall biomass and phytoremediation performance appear to be correlated, suggesting that greater biomass production is associated with increased Pb uptake.

6.7 Environmental, Economic, and Social Benefits of Phytoremediation

As we have seen so far, 'phytoremediation' refers to a range of inexpensive, ecologically safe green technologies that use various biological processes to remove contaminants, primarily from the lithosphere and hydrosphere. It is a valuable substitute for traditional therapy approaches. Surfactants, radionuclides, noxious heavy metals, pharmaceuticals, agricultural-related chemicals like herbicides, and PAHs are just a few of the contaminants this strategy can target. Phytoremediation uses a wide variety of plants, including legumes, grass, deciduous and coniferous trees, and aquatic and marsh plants. Compared with traditional remediation approaches such as solidification, extraction, or vitrification, phytoremediation is much more cost-effective and environmentally friendly (Grzegórska et al. 2020).

Based on the findings done by Wan et al. (2016), the overall cost of phytoremediation for the lithosphere contaminated with toxic heavy metals like Cd, As, and Pb was \$37.7 per m^3, compared to the prices for solidification (\$87–\$190 per m^3), extraction (\$240–\$290 per m^3), and vitrification (\$75–\$425 per ton). In situ phytoremediation also has the benefit of reducing the spread of pollutants through the air or water and preventing secondary contamination. Although it has not yet been utilised extensively globally, phytoremediation offers several benefits over traditional cleanup methods, such as reduced labour and equipment costs, effective soil erosion control, and enhanced air quality (Grzegórska et al. 2020).

6.8 Challenges and Future Perspectives of Phytoremediation

Even though phytoremediation is a viable, affordable, and eco-friendly approach to cleaning up the hydrosphere and lithosphere, various barriers prevent its broad application and effectiveness. One of the foremost drawbacks of this approach is the time required for therapy, which may take months or years to yield obvious improvements. Another disadvantage of phytoremediation is its dependence on seasonal conditions; for example, in winter, when plant growth is limited, it is less effective. Agronomic factors, such as soil conditions, can also significantly affect remediation efficacy. Low nutrient levels and poor soil structure can hinder plant growth, so field preparation, including fertiliser and irrigation, is essential. Thus, the overall cost of phytoremediation may increase due to these additional requirements (Robinson et al. 2015). The limited rooting depth of most plant species is another major obstacle to phytoremediation. Contaminants are usually found at deeper levels, out of reach of plant roots, since they only reach a depth of 1–3 m. This limitation makes phytoremediation less effective at eliminating contaminants deeper in the soil profile (Aryal 2024).

Much of the previous research on phytoremediation was conducted in tightly controlled greenhouse environments, with variables including temperature, humidity, and photoperiod.

However, the success of phytoremediation in real-world applications may depend on various constraints, including coexisting pollutants, fertilisation, soil pH, soil texture, and meteorological conditions. Achieving high removal efficiency may be difficult, as it is hard to replicate the controlled conditions of greenhouse experiments in field settings. Another major problem is disposing of harvested plants that have accumulated toxins, as these materials are often categorised as hazardous waste and require proper management and disposal (Grzegórska et al. 2020).

In view of the ongoing research and innovation aimed at advancing the application proficiency of the phytoremediation approach, it has a bright future. Recent advancements and trends have made the field a more significant component of sustainable environmental management. A significant area of progress is genetic engineering, which enables the modification of specific plant traits to enhance their capacity to absorb, tolerate, and detoxify pollutants. It is possible to modify plants to achieve specific corrective objectives genetically. Another fascinating direction for future research is the combination of phytoremediation with other remediation technologies, such as chemical and microbiological methods. Merging these complementary techniques might yield more efficient solutions for complex contamination setups that are difficult to manage with a single approach. Moreover, by using nanotechnology to improve the speed and thoroughness of pollutant removal, nanophytoremediation is emerging as a cutting-edge technique that enhances the overall efficacy and efficiency of the process.

6.9 Conclusion

The rapid expansion of international industries, such as energy, agriculture, mining, metallurgy, chemical production, and transportation, has resulted in a sharp rise in air, soil, and water pollution. When plants suck water and nutrients from the soil with their roots, they may actively transport or passively diffuse toxic substances or pollutants, and can be used for environmental rehabilitation. This method is named phytoremediation. The approach refers to a range of low-cost, environmentally safe green solutions that employ various biological processes to eliminate toxic chemicals, mostly from water and soil environments. The purification process can employ a variety of techniques, including phytodegradation, phytovolatilisation, phytoextraction, phytostabilisation, rhizofiltration, or rhizodegradation. Several variables affect the process of cleaning up contaminated soils by green plants. The most significant of these are the characteristics of contaminants, their concentrations, the phytoavailability of these pollutants in the soil, and the plant species.

Specific plants can be chosen based on the type of contamination and its location. Parameters such as contaminant tolerance, root structure and depth, biomass production, and adaptability to local conditions can be used to choose the right plant species for the applied phytoremediation technique. Typically, phytoremediation techniques can be enhanced by using multiple approaches, for instance, chemical-assisted phytoextraction, surfactants or chelating agents, and agricultural management techniques. Among the main drawbacks of this approach, the time required for therapy is long, and the difficulty of field application of those developed in controlled greenhouse-like laboratory environments is high.

References

Ahmad H (2024) Phytoremediation: the green solution. In: Bioremediation for environmental sustainability, p 1205602. https://doi.org/10.5772/intechopen.1007671

Akin B (2021) In vitro germination and phytoremediation potential of endemic plant species Verbascum phrygium Bornm. Growing under zinc stress. Pol J Environ Stud 30(2):1513

Ali I, Asim M, Khan TA (2012) Low-cost adsorbents for the removal of organic pollutants from wastewater. J Environ Manag 113:170–183

Ali H, Khan E, Sajad MA (2013) Phytoremediation of heavy metals—concepts and applications. Chemosphere 91(7):869–881

Alkorta I, Garbisu C (2001) Phytoremediation of organic contaminants in soils. Bioresour Technol 79(3):273–276

Alkorta I, Becerril JM, Garbisu C (2010) Phytostabilization of metal contaminated soils. Rev Environ Health 25(2):135–146

Amabogha ON, Garelick H, Jones H, Purchase D (2023) Combining phytoremediation with bioenergy production: developing a multi-criteria decision matrix for plant species selection. Environ Sci Pollut Res 30(14):40698–40711

Anyakora C, Ehianeta T, Umukoro O (2013) Heavy metal levels in soil samples from highly industrialized Lagos environment. Afr J Environ Sci Technol 7:917–924

Aryal M (2024) Phytoremediation strategies for mitigating environmental toxicants. Heliyon 10(19):e38683

Azad MAK, Amin L and Sidik NM (2014) Genetically engineered organisms for bioremediation of pollutants in contaminated sites. Chin Sci Bull 59(8):703–714

Bayabil HK, Teshome FT, Li YC (2022) Emerging contaminants in soil and water. Front Environ Sci. https://doi.org/10.3389/fenvs.2022.873499

Bhat BA, Rather MA, Bilal T, Nazir R, Qadir RU, Mir RA (2025) Plant hyperaccumulators: a state-of-the-art review on mechanism of heavy metal transport and sequestration. Front Plant Sci 16:1631378

Brennan MA, Shelley ML (1999) A model of the uptake, translocation, and accumulation of lead (Pb) by maize for the purpose of phytoextraction. Ecol Eng 12(3–4):271–297

Babu SOF, Hossain MB, Rahman MS, Rahman M, Ahmed AS, Hasan MM, Rakib A, Emran TB, Xiao J and Simal-Gandara J (2021) Phytoremediation of toxic metals: a sustainable green solution for clean environment. Appl Sci 11(21):10348

Cartmill AD, Cartmill DL, Alarcón A (2014) Controlled release fertilizer increased phytoremediation of petroleum-contaminated sandy soil. Int J Phytoremediation 16(3):285–301

Chen J, Shi Z, Hu LB, Su Y, Han FX, Monts DL (2010) Site assessment, long-term monitoring and regulatory concerns for application of phytoremediation technology for remediation of heavy metal/metalloid-contaminated soils. In: Faerber T, Herzog J (eds) Solid waste management and environmental remediation. Nova Science Publishers Inc.

Chigbo C, Batty L (2014) Phytoremediation for co-contaminated soils of chromium and benzo [a] pyrene using Zea mays L. Environ Sci Pollut Res 21(4):3051–3059

Connor R, Renata A, Ortigara C, Koncagül E, Uhlenbrook S, Lamizana-Diallo BM, Zadeh SM, Qadir M, Kjellén M, Sjödin J (2017) The United Nations world water development report 2017 wastewater: the untapped resource. United Nations World Water Dev Rep 5:184

D'Souza R, Paul MS (2014) Phytoremediation of soils contaminated with metals and metalloids at mining areas: potential of native flora. In: Environmental risk assessment of soil contamination. Intech Open, p 485

Daurov D, Lim YH, Park SU, Kim YH, Daurova A, Sapakhova Z, Zhapar K, Abilda Z, Toishimanov M, Shamekova M, Zhambakin K (2024) Selection and characterization of lead-tolerant sweet-potato cultivars for phytoremediation. Plant Biotechnol Rep 18(3):327–339

Eslava-Silva FDJ, Muñíz-Díaz de León ME, Jiménez-Estrada M (2023) Pteridium aquilinum (Dennstaedtiaceae), a novel hyperaccumulator species of hexavalent chromium. Appl Sci 13(9):5621

Farahat EA, Galal TM (2018) Trace metal accumulation by Ranunculus sceleratus: implications for phytostabilization. Environ Sci Pollut Res 25(5):4214–4222

Gahlawat S, Gauba P (2016) Phytoremediation of aspirin and tetracycline by Brassica juncea. Int J Phytoremediation 18(9):929–935

González-Mille DJ, Espinosa-Reyes G, del Carmen Cuevas-Díaz M, Martínez-Toledo Á, Carrizalez-Yáñez L, García-Arreola ME, Ilizaliturri-Hernández CA (2018) Evaluation of the biological activity of soil in a gradient concentration of arsenic and lead in Villa de la Paz, San Luis Potosi, Mexico. In: Soil contamination and alternatives for sustainable development. Intech Open

Grzegórska A, Rybarczyk P, Rogala A, Zabrocki D (2020) Phytoremediation—from environment cleaning to energy generation—current status and future perspectives. Energies 13(11):2905. https://doi.org/10.3390/en13112905

Gu HH, Zhou Z, Gao YQ, Yuan XT, Ai YJ, Zhang JY, Zuo WZ, Taylor AA, Nan SQ, Li FP (2017) The influences of *Arbuscular mycorrhizal* fungus on phytostabilization of lead/zinc tailings using four plant species. Int J Phytoremediation 19(8):739–745

Guidi Nissim W, Palm E, Mancuso S, Azzarello E (2018) Trace element phytoextraction from contaminated soil: a case study under Mediterranean climate. Environ Sci Pollut Res 25(9):9114–9131

Guillotin S, Delcourt N (2022) Studying the impact of persistent organic pollutants exposure on human health by proteomic analysis: a systematic review. Int J Mol Sci 23(22):14271. https://doi.org/10.3390/ijms232214271

Georgin J, SP Franco D, De Oliveira JS, Dehmani Y, El Messaoudi N, Miyah Y, Alruwaili A, Knani S and Hwang Y, (2025) Decontamination of pollutants present in water, air, and soil through phytoremediation: a critical review. Int J Phytoremediation 1–27

Hou D (2023) Sustainable soil management for food security. Soil Use Manag 39(1):1. https://doi.org/10.1111/SUM.12883

Hu Q-H, Weng J-Q, Wang J-S (2010) Sources of anthropogenic radionuclides in the environment: a review. J Environ Radioact 101:426–437

Huang S-H, Bing P, Yang Z-H, Chai L-y, Zhou L-c (2009) Chromium accumulation, microorganism population and enzyme activities in soils around chromium-containing slag heap of steel alloy factory. Transact Nonferr Metal Soc China 19:241–248

Iqbal M, Ahmad A, Ansari MKA, Qureshi MI, Aref IM, Khan PR, Hegazy SS, El-Atta H, Husen A, Hakeem KR (2015) Improving the phytoextraction capacity of plants to scavenge metal(loid)-contaminated sites. Environ Rev 23:44–65. https://doi.org/10.1139/er-2014-0043

Ishikawa S, Noriharu AE, Murakami M, Wagatsuma T (2006) Is Brassica juncea a suitable plant for phytoremediation of cadmium in soils with moderately low cadmium contamination? Possibility of using other plant species for Cd-phytoextraction. Soil Sci Plant Nutr 52(1):32–42

Jin Y, Wang L, Song Y, Zhu J, Qin M, Wu L, Hu P, Li F, Fang L, Chen C, Hou D (2021) Integrated life cycle assessment for sustainable remediation of contaminated agricultural soil in China. Environ Sci Technol 55(17):12032–12042. https://doi.org/10.1021/acs.est.1c02535

Kafle A, Timilsina A, Gautam A, Adhikari K, Bhattarai A, Aryal N (2022) Phytoremediation: mechanisms, plant selection and enhancement by natural and synthetic agents. Environ Adv 8:100203

Kamran M, Danish M, Saleem MH, Malik Z, Parveen A, Abbasi GH, Jamil M, Ali S, Afzal S, Riaz M (2021) Application of abscisic acid and 6-benzylamino-purine modulated morpho-physiological and antioxidative defense responses of tomato (Solanum lycopersicum L.) by minimizing cobalt uptake. Chemosphere 263:128169

Kang JW (2014) Removing environmental organic pollutants with bioremediation and phytoremediation. Biotechnol Lett 36(6):1129–1139

Karaca A, Cetin SC, Turgay OC, Kizilkaya R (2010) Effects of heavy metals on soil enzyme activities. In: Soil heavy metals. Springer, pp 237–262

Kharazian P, Cappai G, Boi ME, Porceddu M, Piredda M, De Giudici G, Bacchetta G (2024) Greenhouse investigation on the phytoremediation potential of pioneer tree Pinus halepensis Mill. in abandoned mine site. Int J Phytoremediation 26(5):773–783

Kumar M, Bolan NS, Hoang SA, Sawarkar AD, Jasemizad T, Gao B, Keerthanan S, Padhye LP, Singh L, Kumar S, Vithanage M, Li Y, Zhang M, Kirkham MB, Vinu A, Rinklebe J (2021) Remediation of soils and sediments polluted with polycyclic aromatic hydrocarbons: to immobilize, mobilize, or degrade? J Hazard Mater 420:126534. https://doi.org/10.1016/j.jhazmat.2021.126534

Lee M, Yang M (2010) Rhizofiltration using sunflower (Helianthus annuus L.) and bean (Phaseolus vulgaris L. var. vulgaris) to remediate uranium contaminated groundwater. J Hazard Mater 173(1–3):589–596

Li C, Zhou K, Qin W, Tian C, Qi M, Yan X, Han W (2019) A review on heavy metals contamination in soil: effects, sources, and remediation techniques. Soil Sed Cont: Int J 28(4):380–394. https://doi.org/10.1080/15320383.2019.1592108

Liu Y, Zeng G, Zhong H, Wang Z, Liu Z, Cheng M, Liu G, Yang X, Liu S (2017) Effect of rhamnolipid solubilization on hexadecane bioavailability: enhancement or reduction? J Hazard Mater 322:394–401

Liu X, Yang A, Kümmel S, Richnow HH (2022) Uptake and metabolization of HCH isomers in trees examined over an annual growth period by compound-specific isotope analysis and enantiomer fractionation. Environ Sci Technol 56(14):10120–10130

Malik Z, Ahmad M, Abassi GH, Dawood M, Hussain A, Jamil M (2017) Agrochemicals and soil microbes: interaction for soil health. In: Xenobiotics in the soil environment. Springer, pp 139–152

Martínez-Alcalá I, Walker DJ, Bernal MP (2010) Chemical and biological properties in the rhizosphere of Lupinus albus alter soil heavy metal fractionation. Ecotoxicol Environ Saf 73(4):595–602. https://doi.org/10.1016/j.ecoenv.2009.12.009

Mei X, Wang Y, Li Z, Larousse M, Pere A, da Rocha M, Zhan F, He Y, Pu L, Panabières F, Zu Y (2022) Root-associated microbiota drive phytoremediation strategies to lead of Sonchus Asper (L.) Hill as revealed by intercropping-induced modifications of the rhizosphere microbiome. Environ Sci Pollut Res 29(16):23026–23040

Mikheev AN, Lapan OV, Madzhd SM (2017) Experimental foundations of a new method for rhizofiltration treatment of aqueous ecosystems from 137Cs. J Water Chem Technol 39(4):245–249

Mukherjee S, Leri AC, Bandaranayaka C, Vázquez-Núñez E, Barros R, Khan AHA, Zhou P, Zhang T, Bernal MP, Clemente R, Bolan N (2025) Sustainable management of post-phytoremediation biomass. Energy Ecol Environ:1–35

Nahar K, Sunny SA (2024) Co-benefits of Eichhornia Crassipes (water hyacinth) as sustainable biomass for biofuel production and aquatic ecosystem phytoremediation. Fuels 5(3):317–333

Nedjimi B (2021) Phytoremediation: a sustainable environmental technology for heavy metals decontamination. SN Appl Sci 3(3):286

Nouri J, Lorestani B, Yousefi N, Khorasani N, Hasani AH, Seif F, Cheraghi M (2011) Phytoremediation potential of native plants grown in the vicinity of Ahangaran lead–zinc mine (Hamedan, Iran). Environ Earth Sci 62(3):639–644

Oh K, Cao T, Li T, Cheng H (2014) Study on application of phytoremediation technology in management and remediation of contaminated soils. J Clean Energy Technol 2(3):216–220

Paladino GL, Dupaul G, Jonsson A, Haller H, Eivazi A, Hedenström E (2025) Selecting effective plant species for the phytoremediation of persistent organic pollutants and multielement contaminated fibrous sediments. Environ Sci Eur 37(1):1–20

Pan J, Plant JA, Voulvoulis N, Oates CJ, Ihlenfeld C (2010) Cadmium levels in Europe: implications for human health. Environ Geochem Health 32:1–12

Pandey J, Verma RK, Singh S (2019) Suitability of aromatic plants for phytoremediation of heavy metal contaminated areas: a review. Int J Phytoremediation 21(5):405–418

Parveen A, Saleem MH, Kamran M, Haider MZ, Chen J-T, Malik Z, Rana MS, Hassan A, Hur G, Javed MT (2020) Effect of citric acid on growth, ecophysiology, chloroplast ultrastructure, and phytoremediation potential of jute (Corchorus capsularis L.) seedlings exposed to copper stress. Biomolecules 10:592

Pavel PB, Puschenreiter M, Wenzel WW, Diacu E, Barbu CH (2014) Aided phytostabilization using Miscanthus sinensis× giganteus on heavy metal-contaminated soils. Sci Total Environ 479:125–131

Paz-Alberto AM, Sigua GC (2013) Phytoremediation: a green technology to remove environmental pollutants. Am J Clim Chang 2(1):71–86. https://doi.org/10.4236/ajcc.2013.21008

Phusantisampan T, Meeinkuirt W, Saengwilai P, Pichtel J, Chaiyarat R (2016) Phytostabilization potential of two ecotypes of Vetiveria zizanioides in cadmium-contaminated soils: greenhouse and field experiments. Environ Sci Pollut Res 23(19):20027–20038

Prakash D, Gabani P, Chandel AK, Ronen Z, Singh OV (2013) Bioremediation: a genuine technology to remediate radionuclides from the environment. Microb Biotechnol 6:349–360

Puri M, Gandhi K, Kumar MS (2023) Emerging environmental contaminants: a global perspective on policies and regulations. J Environ Manag 332:117344. https://doi.org/10.1016/j.jenvman.2023.117344

Philp JC, Bamforth SM, Singleton I and Atlas RM (2005) Environmental pollution and restoration: a role for bioremediation. Bioremediation: Appl Microb Solut R-World Environ Cleanup 1–48

Radziemska M, Gusiatin ZM, Bilgin A (2017) Potential of using immobilizing agents in aided phytostabilization on simulated contamination of soil with lead. Ecol Eng 102:490–500

Rai GK, Bhat BA, Mushtaq M, Tariq L, Rai PK, Basu U, Dar AA, Islam ST, Dar TU, Bhat JA (2021) Insights into decontamination of soils by phytoremediation: a detailed account on heavy metal toxicity and mitigation strategies. Physiol Plant 173(1):287–304

Rascio N, Navari-Izzo F (2011) Heavy metal hyperaccumulating plants: how and why do they do it? And what makes them so interesting? Plant Sci 180(2):169–181

Robinson BH, Anderson CWN, Dickinson NM (2015) Phytoextraction: where's the action? J Geochem Explor 151:34–40

Roccotiello E, Serrano HC, Mariotti MG, Branquinho C (2015) Nickel phytoremediation potential of the Mediterranean Alyssoides utriculata (L.) Medik. Chemosphere 119:1372–1378

Saleem MH, Ali S, Kamran M, Iqbal N, Azeem M, Tariq Javed M, Ali Q, Zulqurnain Haider M, Irshad S, Rizwan M (2020a) Ethylenediaminetetraacetic acid (EDTA) mitigates the toxic effect of excessive copper concentrations on growth, gaseous exchange and chloroplast ultrastructure of Corchorus capsularis L. and improves copper accumulation capabilities. Plants 9:756

Saleem MH, Ali S, Rehman M, Rana MS, Rizwan M, Kamran M, Imran M, Riaz M, Soliman MH, Elkelish A (2020b) Influence of phosphorus on copper phytoextraction via modulating cellular organelles in two jute (Corchorus capsularis L.) varieties grown in a copper mining soil of Hubei Province, China. Chemosphere 248:126032

Samardjieva KA, Gonçalves RF, Valentão P, Andrade PB, Pissarra J, Pereira S, Tavares F (2015) Zinc accumulation and tolerance in Solanum nigrum are plant growth dependent. Int J Phytoremediation 17(3):272–279

Saxena G, Purchase D, Mulla SI, Saratale GD, Bharagava RN (2019) Phytoremediation of heavy metal-contaminated sites: eco-environmental concerns, field studies, sustainability issues, and future prospects. Rev Environ Contam Toxicol 249:71–131

Seth CS, Remans T, Keunen E, Jozefczak M, Gielen H, Opdenakker K, Weyens N, Vangronsveld J, Cuypers A (2012) Phytoextraction of toxic metals: a central role for glutathione. Plant Cell Environ 35(2):334–346

Shabani N, Sayadi MH (2012) Evaluation of heavy metals accumulation by two emergent macrophytes from the polluted soil: an experimental study. Environmentalist 32(1):91–98

Sharma JK, Kumar N, Singh NP, Santal AR (2023) Phytoremediation technologies and their mechanism for removal of heavy metal from contaminated soil: an approach for a sustainable environment. Front Plant Sci 14:1076876

Shmaefsky BR, Husen A (2023) Phytoremediation potential of medicinal plants. In: Husen A, Iqbal M (eds) Medicinal plants – their response to abiotic stress. Springer Nature Singapore Pte Ltd, Singapore, pp 443–469. https://doi.org/10.1007/978-981-19-5611-9_17

Simonich SL, Hites RA (1994) Importance of vegetation in removing polycyclic aromatic hydrocarbons from the atmosphere. Nature 370(6484):49–51

Singh J, Kalamdhad AS (2011) Effects of heavy metals on soil, plants, human health and aquatic life. Int J Res Chem Environ 1(2):15–21

Singh S, Eapen S, Thorat V, Kaushik CP, Raj K, D'souza SF (2008) Phytoremediation of 137cesium and 90strontium from solutions and low-level nuclear waste by Vetiveria zizanoides. Ecotoxicol Environ Saf 69(2):306–311

Solomou AD, Germani R, Proutsos N, Petropoulou M, Koutroumpilas P, Galanis C, Maroulis G, Kolimenakis A (2022) Utilizing mediterranean plants to remove contaminants from the soil environment: a short review. Agriculture 12(2):238

Song B, Zeng G, Gong J, Zhang P, Deng J, Deng C, Yan J, Xu P, Lai C, Zhang C (2017) Effect of multi-walled carbon nanotubes on phytotoxicity of sediments contaminated by phenanthrene and cadmium. Chemosphere 172:449–458

Susarla S, Bacchus ST, Wolfe NL, McCutcheon SC (1999) Phytotransformation of perchlorate and identification of metabolic products in Myriophyllum aquaticum. Int J Phytoremediation 1(1):97–107

Sahoo A, Chhotaray SP, Meher I, Behera SP, Pal A, Meena M, Swapnil P, Yadav A and Bhardwaj R (2025) Phytoremediation for a sustainable future: Integrating plant based strategies in soil and wastewater remediation. Bioresour Technol Rep 102266

Tangahu BV, Sheikh Abdullah SR, Basri H, Idris M, Anuar N, Mukhlisin M (2011) A review on heavy metals (As, Pb, and Hg) uptake by plants through phytoremediation. Int J Chem Eng 2011(1):939161

Tyagi SK, Kumar P, Tanveer N, Pandey N, Rajput Y (2023) Recent advances in phytoremediation of hazardous substances using plants: a tool for soil reclamation and sustainability. AgroEnviron Sustain 1(2):180–191

Vamerali T, Bandiera M, Mosca G (2010) Field crops for phytoremediation of metal-contaminated land. A review. Environ Chem Lett 8(1):1–17

Vanderford M, Shanks JV, Hughes JB (1997) Phytotransformation of trinitrotoluene (TNT) and distribution of metabolic products in *Myriophyllum aquaticum*. Biotechnol Lett 19(3):277–280

Vardhan KH, Kumar PS, Panda RC (2019) A review on heavy metal pollution, toxicity and remedial measures: current trends and future perspectives. J Mol Liq 290:111197. https://doi.org/10.1016/j.molliq.2019.111197

Veselý T, Tlustoš P, Száková J (2011) The use of water lettuce (Pistia stratiotes L.) for rhizofiltration of a highly polluted solution by cadmium and lead. Int J Phytoremediation 13(9):859–872

Wan X, Lei M, Chen T (2016) Cost–benefit calculation of phytoremediation technology for heavy-metal-contaminated soil. Sci Total Environ 563:796–802

Wu Q, Zhou H, Tam NF, Tian Y, Tan Y, Zhou S, Li Q, Chen Y, Leung JY (2016) Contamination, toxicity and speciation of heavy metals in an industrialized urban river: implications for the dispersal of heavy metals. Mar Pollut Bull 104:153–161

Yang M, Jawitz JW, Lee M (2015) Uranium and cesium accumulation in bean (Phaseolus vulgaris L. var. vulgaris) and its potential for uranium rhizofiltration. J Environ Radioact 140:42–49

Yayinie M, Worku LA, Bachheti RK, Bachheti A (2025) Chitin-based nanoparticles for soil remediation. In: Chitin-based nanoparticles for the agriculture sectors. Springer Nature Singapore, Singapore, pp 85–102

Yoon J, Cao X, Zhou Q, Ma LQ (2006a) Accumulation of Pb, Cu, and Zn in native plants growing on a contaminated Florida site. Sci Total Environ 368(2–3):456–464

Yoon JM, Oliver DJ, Shanks JV (2006b) Phytotransformation of 2, 4-dinitrotoluene in Arabidopsis thaliana: toxicity, fate, and gene expression studies in vitro. Biotechnol Prog 22(6):1524–1531

Zhuang G (2023) Current phytoremediation technologies and applications. J Phys Conf Ser 2608(1):012054. IOP Publishing

Zohdi H, Emami M, Shahverdi HR (2012) Galvanic corrosion behavior of dental alloys. In: Environmental and industrial corrosion-practical and theoretical aspects. Intech Open

Chapter 7
Medicinal Plants: Bridging Traditional Medicine and Modern Healthcare

Aneela Bashir, Ayman M. Al-Qaaneh, Asfandyar, Tahira Younis, Ali Raza Ishaq, Abdel Nasser B. Singab, and Heba A. S. El-Nashar

Abstract Nature offered several benefits in the form of herbal plants to treat a variety of disorders. This chapter provides an overview of the secondary metabolites, which are unique organic compounds produced by plants, microorganisms, and fungi that are not directly involved in primary growth but play important roles in their ecological adaptation. Their important characteristics, such as structural diversity and biosynthetic processes, are discussed, and their functions in defense, ecological interactions, and evolutionary adaptations are emphasized. The chapter also focused on the important role of secondary metabolites in diabetes control and antioxidant, anti-inflammatory, and enzyme-inhibiting properties. Important classes such as phenolic compounds, flavonoids, alkaloids, terpenoids, saponins, polysaccharides, steroids, and coumarins are also explained for their role in blood glucose

A. Bashir
Key Laboratory of Biorheological Science and Technology of Ministry of Education, Bioengineering College, Chongqing University, Chongqing, China

A. M. Al-Qaaneh
Faculty of Allied Medical Sciences, Al-Balqa Applied University (BAU), Al-Salt, Jordan

Asfandyar · A. R. Ishaq (✉)
State Key Laboratory of Biocatalysis and Enzyme Engineering, Engineering Research Center for Microbial Cell Factory of Hubei Province, Environmental Microbial Technology Center of Hubei Province, College of Life Sciences, Hubei University, Wuhan, China

T. Younis
Department of Biochemistry and Biotechnology, The Women University of Multan, Punjab, Pakistan

A. N. B. Singab
Department of Pharmacognosy, Faculty of Pharmacy, Ain Shams University, Cairo, Egypt

Center for Drug Discovery Research and Development, Faculty of Pharmacy, Ain Shams University, Cairo, Egypt

H. A. S. El-Nashar (✉)
Department of Pharmacognosy, Faculty of Pharmacy, Ain Shams University, Cairo, Egypt
e-mail: heba_pharma@pharema.asu.edu.eg

L. A. Worku et al. (eds.), *Plant-Based Solutions for Sustainable Agriculture and Healthcare*, Sustainability Sciences in Asia and Africa,
https://doi.org/10.1007/978-981-95-8910-4_7

control, improving insulin sensitivity, and preventing diabetes-related processes. The review covers metabolites of plant, microbial, and fungal origin, including specific cases, therapeutic possibilities, and mechanisms of action. By focusing on biochemical, pharmacological, and ecological sides, this chapter will explain the secondary metabolites as a potential source for creating new antidiabetic drugs.

Keywords Secondary metabolites • Phenolic compounds • Antidiabetic activity • Plant-derived compounds • Microbial metabolites

7.1 Introduction

Nature provides us with many potent solutions to treat a variety of diseases, from chronic to acute, by offering natural compounds. Recent studies proved that human beings have been using natural compounds to treat various kinds of diseases for 60,000 years. Microbes, fungi, and plants produce secondary metabolites (organic chemicals) that are not involved in their growth, development, or reproduction (El-Nashar et al. 2024c; Eldahshan et al. 2025), known as traditional medicine (TM). But unlike primary metabolites, such as proteins, carbohydrates, and nucleic acids, which are essential for basic metabolic processes and cellular functioning, secondary metabolites are not required for an organism's basic life (Pettit 2011). They fill essential ecological and adaptive roles. These substances often support defense mechanisms that protect the organism from infections, predators, and environmental stressors. (Demain and Fang 2000). By inhibiting the growth of competing organisms, they also contribute to interspecies competition through allelopathic activities. Chemical signaling, which facilitates communication between and within species, also depends on secondary metabolites (El-Nashar et al. 2022), such as symbiotic interactions, pollinator attraction, and herbivore deterrence. Secondary metabolites are crucial not only for other living things but also for their own species (Erb and Kliebenstein 2020).

Recent research revealed that secondary metabolites are specialized chemicals that differ greatly from primary metabolites in terms of physiological functions (El-Nashar et al. 2021, 2024a). Furthermore, these chemicals serve important functions in organisms' immunity, allowing them to survive significant changes in the habitat (Bagal et al. 2023). Furthermore, these chemicals are found in trace levels and are restricted to the tissues of specific plant organs during their embryonic stages (El-Nashar et al. 2024b). The level of secondary metabolites is totally linked with abiotic factors like light, temperature, and biotic stress, which are among the environmental variables (El-Nashar et al. 2025a, b). Likewise, those compounds are not important to maintain the cellular system of an organism (Wawrosch and

Zotchev 2021). Secondary metabolites have a wide range of applications in various industries, including pharmaceuticals, agriculture, and biotechnology. Medicinal plants are a rich source of secondary metabolites that are purely organic and useful for human health (Bennett and Wallsgrove 1994). This chapter covers the different types of secondary metabolites, how they are made, and their potential use in diabetes treatment. We will focus on plant-based compounds that have been shown to lower blood glucose levels and explore their future applications.

7.2 Medicinal Plants: A House of Secondary Metabolites

Medicinal plants include a high concentration of secondary metabolite chemicals that perform important roles as antioxidants, antidiabetics, and anticancer agents. These chemicals are also required for human development, reproduction, survival, and adaptation. These include polyketides, phenols, terpenoids, and alkaloids. For instance, secondary metabolites, including flavonoids and other phenolics, act as skin protective agents to protect skin against UV damage and participate in signaling. Polyketides are important in microbial competition and medicine. The diversity of these compounds makes them valuable both ecologically and for practical applications in agriculture, pharmacology, and biotechnology (Pichersky and Gang 2000). Secondary metabolites are made through specialized biosynthetic pathways originating from primary metabolites. Polyketides are produced via the polyketide pathway, terpenoids through the mevalonate pathway, and phenolics via the shikimate pathway. These pathways are linked to secondary metabolism to primary metabolism because they are about the ecological and economic benefits of these compounds (Ferrer et al. 2008). Secondary metabolites are in specific species or families, like nicotine, which is found in *Nicotiana tabacum*, and capsaicin in chili peppers (Bennett and Wallsgrove 1994). These compounds are produced in small quantities but still defend plants against herbivores and against other pathogens that are harmful to them (Wink 2003). In a study by Fraenkel (1959), alpine plants produced phenolic compounds that absorb UV radiation, which play the role of secondary metabolites in environmental adaptation beneficial for their survival (Fraenkel 1959). Secondary metabolites have an ecological role and are known for their medicinal properties. Fleming (2001) reported that penicillin, a fungal polyketide, is used for the treatment of bacterial infections, and Taxol from yew trees became a major cancer therapy. Flavonoids present in fruits and vegetables have antioxidant activity and reduce chronic diseases. According to these studies, secondary metabolites continue to provide a bridge between traditional remedies and healthcare benefits (Fleming 2001). This section focuses on the diversity, biosynthesis, and biological roles of secondary metabolites, with emphasis on plant-derived compounds that have potential therapeutic effects, including blood glucose-lowering properties, as reported in several recent studies.

7.3 Role of Secondary Metabolites in Diabetes Mellitus (DM)

Secondary metabolites control diabetes mellitus (DM) due to their biological effects, as they have antidiabetic, antioxidant, and anti-inflammatory properties. Its different categories, like alkaloids, flavonoids, terpenoids, saponins, and phenolic compounds, have been researched in depth for their functions in managing blood glucose levels and inhibiting complications associated with diabetes (Sukhikh et al. 2023). Alkaloids have been used for treating hyperglycemia to increase insulin production, and they also improve glucose absorption. Flavonoids aid in reducing oxidative stress, retinopathy, and neuropathy because of their antioxidant qualities. Another family of terpenoids is known to inhibit enzymes involved in glucose absorption, especially α-amylase and α-glucosidase, and also alter the metabolism of carbohydrates. Saponins are beneficial in managing type 2 diabetes because they enhance lipid metabolism and insulin sensitivity. Last but not the least, phenolic chemicals work in many ways, such as increasing insulin receptor activation, decreasing inflammation, and protecting pancreatic β-cells from oxidative stress/damage (Shehadeh et al. 2021). These secondary metabolites provide therapy for diabetes mellitus by reducing long-term consequences and promoting glycemic control. Their efficiency proves them as essential elements in creating innovative therapeutic approaches for the treatment of DM (Hashim et al. 2024; Parveen et al. 2021) (Fig. 7.1).

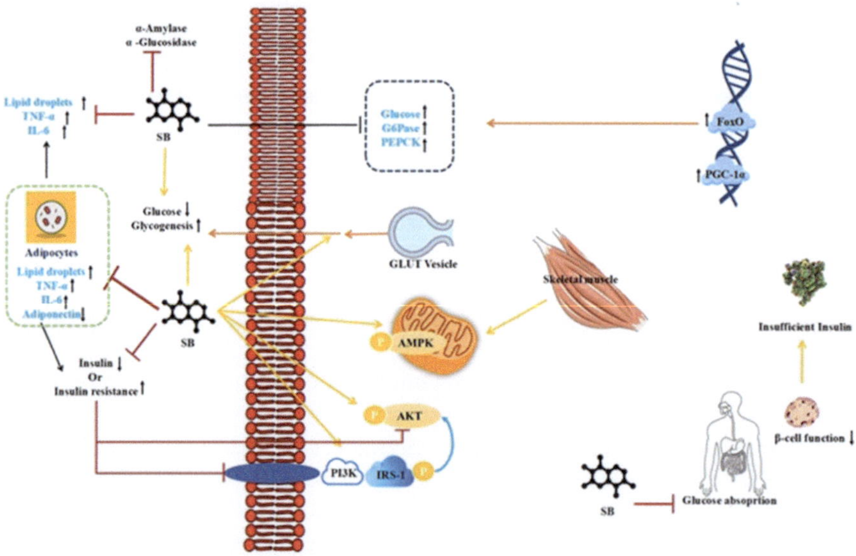

Fig. 7.1 Mechanism of action of secondary metabolites in hyperglycemia. ↑; upregulation, ↓; downregulation, ⊥; Inhibition

7.3.1 Antioxidant Activity

Oxidative stress plays an essential role in diabetes mellitus (DM) and its complications. It is produced when there are more reactive oxygen species (ROS) in our body than it can neutralize. This imbalance damages cells and worsens diabetic symptoms. Natural compounds, including phenolic compounds, are strong antioxidant agents to reduce the concentration of intracellular superoxides for maintaining the cellular stress (Brownlee 2001). By protecting pancreatic β-cells from ROS, phenolic compounds support insulin secretion and slow diabetes progression. Its antioxidant property also lowers diabetic neuropathy and retinopathy, which are the effects of long-term oxidative damage. This two-sided role is protecting β-cells and reducing complications and shows their potential for improving diabetes care and patient outcomes (Krawczyk et al. 2023).

7.3.2 Insulin Sensitization

A lot of secondary metabolites are involved to improve insulin sensitivity and reduce insulin resistance. The most active ones are flavonoids and terpenoids. They influence some pathways like AMP-activated protein kinase (AMPK) and peroxisome proliferator-activated receptors (PPARs) because both these are important for glucose metabolism and energy control. For example, quercetin, which is a flavonoid, promotes glucose uptake by increasing glucose transporters in tissues. Curcumin, which is a terpenoid from turmeric, lowers insulin resistance by regulating PPAR activity and reducing inflammation, which is also linked to metabolic disorders. From the above-mentioned conclusion, we can say that secondary metabolites own a strong ability to reduce the glucose level in blood, proving them as strong antidiabetic agents (Dhanya et al. 2017).

7.3.3 Inhibition of Carbohydrate-Hydrolyzing Enzymes

Secondary metabolites, such as alkaloids and flavonoids, have antidiabetic effects by inhibiting the activity of glucose-digesting enzymes such as α-amylase and α-glucosidase. Furthermore, these enzymes transform complicated polysaccharides into simple monosaccharides, which help to increase blood glucose levels. Thus, inhibition of those enzymes is required to halt the conversion in order to combat hyperglycemia. Furthermore, this type of regulation reduces the warnings for other metabolic syndromes. Thus, medicinal plants offer the highest chance of treating diabetes by altering the action of carbohydrate-utilizing enzymes (Kumar et al. 2011b).

7.3.4 Anti-Inflammatory Effects

Medicinal plants include a high concentration of phytochemicals, which function as anti-inflammatory agents to treat chronic inflammation. Diabetes mellitus leads to increased insulin resistance and impaired β-cell activity. Natural substances, including curcumin, resveratrol, and catechins, can improve insulin resistance by regulating the NF-κB signaling system, which modulates pro-inflammatory cytokines. Furthermore, by inhibiting NF-κB activation, they reduce the release of inflammatory chemicals such as interleukins and TNF-α. Reducing inflammation improves glucose metabolism, protects β-cells, and lowers insulin resistance in diabetic circumstances. This anti-inflammatory effect of secondary metabolites demonstrates their promise as strong treatments for diabetes and prevention (Khan et al. 2024; Yu et al. 2024).

7.3.5 Protection Against Diabetic Complications

Nephropathy, neuropathy, and cardiovascular disorders (CVD) are due to diabetes mellitus (DM) that can be treated by secondary metabolites. The underlying processes are targeted by these natural substances and have strong biological activity. Resveratrol, which is a polyphenol found in grapes and red wine, has been shown to improve kidney health in diabetic nephropathy by reducing oxidative damage and improving renal function (Ferrer et al. 2008; Wen et al. 2013). Epigallocatechin gallate (EGCG), the main catechin in green tea, improves vascular function and reduces inflammation, lowering cardiovascular risk. Such results show the therapeutic potential of secondary metabolites in controlling problems associated with diabetes and enhancing patient outcomes (Potenza et al. 2020).

7.3.6 Pancreatic β-Cell Protection and Regeneration

Maintaining healthy pancreatic β-cells is essential for blood glucose control since these cells produce insulin. Many secondary metabolites protect and even regenerate β-cells damaged by OS. Compounds such as gymnemic acids (from Gymnema sylvestre) and berberine (an alkaloid) have shown strong protective effects. They reduce harmful impacts of reactive oxygen species (ROS) and inflammatory cytokines that often cause β-cell dysfunction (Lv et al. 2021). In addition to protection, these substances also promote β-cell growth and renewal, which restore insulin--producing capacity. Their ability to support β-cell survival and insulin balance provides an important direction to researchers for developing future diabetes therapies and medication (Choudhury et al. 2018; Egbuna et al. 2021).

7.4 Plant-Based Phenolic Compounds

Phenolic compounds are a large group of secondary metabolites that are easily/commonly found in plants. They have one or more hydroxyl groups attached to their aromatic rings. These compounds play many roles in plant life, for example, protection from UV radiation, defense against infections, and participation in ecological processes, for example, pollination and seed dispersal. Phenolic compounds are mainly produced through the phenylpropanoid and shikimate pathways. Because these connect their formation to basic plant metabolism. This link shows their importance for plant growth as well as their survival (Dai and Mumper 2010). Based on their structure and biological role, phenolic compounds can be divided into several main types. Flavonoids, for example, give bright colors to fruits and flowers, attract pollinators, and act as strong antioxidants. Similarly, these chemicals are beneficial to plant health; tannins protect plants from insects and pathogens. Stilbenes, such as resveratrol, help plants defend themselves and have a positive impact on human health. Lignans reinforce plant cell walls and allow plants to withstand environmental stress. (Dai and Mumper 2010) (Fig. 7.2).

7.4.1 Functions of Phenolic Compounds

Phenolic compounds have the ability to maintain the plant's health from abiotic changes (Abdelghffar et al. 2022). These roles can be divided into several main functions.

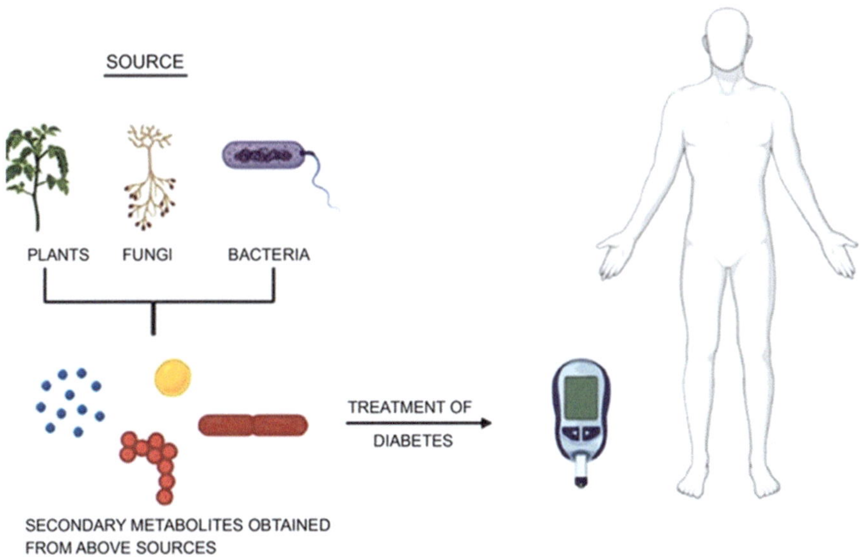

Fig. 7.2 Different sources of secondary metabolites useful for management of diabetes mellitus

7.4.1.1 Defense Mechanisms

Phenolic compounds act as strong protectors against biological and environmental stress. They help plants deal with oxidative stress caused by UV radiation, pollution, and harsh conditions. These compounds work as antioxidants, neutralizing reactive oxygen species (ROS) that can damage DNA, proteins, and lipids inside cells. By reducing ROS, phenolics keep plant cells healthy and maintain normal functions. Besides antioxidant action, phenolic molecules such as flavonoids and tannins defend plants from herbivores and microbes. They can make plant tissues taste bitter or hard to digest, discouraging insects and animals from eating them (Sabir et al. 2025; Yan et al. 2023). Many phenolics also show antibacterial and antifungal properties, blocking the growth of harmful bacteria and fungi. This double protective role against both biotic and abiotic stress shows how important phenolic compounds are for plant defense and survival. Their protective potential also makes them valuable in crop improvement and biotechnology (Daglia 2012).

7.4.1.2 Structural Integrity

Phenolic compounds like lignin are key to plant strength and structure. Lignin, formed from phenolic building blocks, is found in secondary cell walls of vascular plants. It gives stiffness and mechanical support, helping plants withstand wind, water flow, and physical injury. In woody plants, lignin is abundant in xylem tissues, improving water transport and overall stability. Lignin also acts as a protective barrier, blocking microbial invasion and disease. By strengthening cell walls, it helps plants resist infection and damage. This dual role is providing strength and defense; it also shows how essential phenolic compounds are for growth and adaptation (Neuhaus et al. 2021).

7.4.1.3 Ecological Interactions

Phenolic pigments such as flavonoids and anthocyanins help plants communicate with animals and other organisms. These compounds give flowers and fruits bright colors, from red and purple to blue, which attract pollinators like bees, butterflies, and birds. Such interactions increase pollination success and fruit development. Colorful fruits also draw animals that eat and disperse seeds, ensuring plant reproduction and maintaining biodiversity. Through these interactions, phenolic pigments support both plant and animal life, playing an important role in ecosystem balance (Tanaka et al. 2008).

7.4.1.4 Allelopathy

Allelopathy is a natural process where plants release chemicals in soil to affect nearby plant growth. Phenolic acids like ferulic acid and caffeic acid are active in this process. These compounds can enter the soil during the season of leaves falling, root secretions, and decaying plant material. Once released, they may slow down germination, root elongation, and growth of nearby plants, giving the source plant a better chance to access nutrients, water, and sunlight. Allelopathic effects depend on soil microbes, plant species, and environmental factors. This interaction affects the plant community and offers a natural method for weed control. In sustainable agriculture, phenolic-based allelopathy is used to manage weeds without chemicals (Ninkuu et al. 2021).

7.4.2 Plant-Derived Secondary Metabolites with Antidiabetic Activity

Natural compounds (from plants, fungi, or other sources) show good potential in diabetes control. As they act on different parts of insulin regulation and glucose metabolism. These are bioactive compounds that belong to many chemical groups and help reduce blood sugar in many ways. Some improve insulin sensitivity, some are involved in increasing glucose uptake, and others block enzymes that digest carbohydrates. And these above-mentioned ways help to maintain normal blood sugar levels and reduce diabetes problems (Table 7.1) (Cappelli et al. 2022). Plants contain many secondary metabolites with strong antidiabetic activity. These active molecules are often present in roots, leaves, or seeds and can influence enzymes linked to glucose metabolism and insulin response. Their strength is usually measured by IC_{50} values, which indicate how effectively they block enzyme activity. Research on various medicinal plants has identified key bioactive compounds, their action mechanisms, and possible uses in diabetes therapy. This information, summarized in Table 7.2, highlights the great potential of plant-based compounds in developing safe and effective treatments for diabetes (Ardalani et al. 2021).

7.5 Microorganisms and Secondary Metabolites

Secondary metabolites are considered specialized metabolites, and these microbial metabolites carry peculiar structures and have a very significant impact on the health, food, and economy of our society (Monciardini et al. 2014). These compounds are usually low molecular mass products of secondary metabolism, which take place out of step with the main microbial growth phase. They include antibiotics, pigments, toxins, enzyme inhibitors, and immunomodulators; effectors of ecological competition or symbiosis; or compounds with hormonal activity or particular

Table 7.1 Natural compound classes owing to the antidiabetic activity

No	Compound class	Name	Mechanism of action	References
1	Phenolic compounds	Quercetin, chlorogenic acid	Act as antioxidants, reducing oxidative stress that impairs insulin signaling Inhibit enzymes like α-amylase and α-glucosidase, which delay carbohydrate digestion and reduce postprandial glucose spikes Improve pancreatic β-cell function by reducing apoptosis	Hunyadi et al. (2012)
2	Flavonoids	Kaempferol, catechin	Enhance glucose uptake in peripheral tissues by activating AMP-activated protein kinase (AMPK) Improve insulin sensitivity by modulating the PI3K/Akt signaling pathway Reduce inflammation in insulin-resistant states	Franco et al. (2020), Hossain and Shahidi (2023)
3	Alkaloids	Berberine, metformin	Activate AMPK, improving glucose utilization and reducing hepatic gluconeogenesis Inhibit glucose absorption in the intestine, lowering blood glucose levels Regulate gut microbiota, contributing to improved metabolic health	Ahmad and Haque (2024), Al-Kuraishy et al. (2022)
4	Terpenoids	Ginsenosides (from ginseng), gymnemic acid	Stimulate insulin secretion by pancreatic β-cells Inhibit glucose absorption in the intestine Modulate PPARγ (peroxisome proliferator-activated receptor gamma), enhancing lipid and glucose metabolism	Guo et al. (2025)
5	Saponins	Diosgenin, momordicosides (from bitter melon)	Improve insulin secretion and sensitivity Lower blood glucose levels by inhibiting gluconeogenesis and promoting glycogen synthesis Exhibit antioxidant and anti-inflammatory effects, reducing diabetes-related complications	Elekofehinti et al. (2017), Jeepipalli et al. (2020)

(continued)

Table 7.1 (continued)

No	Compound class	Name	Mechanism of action	References
6	Polysaccharides	Polysaccharides from aloe vera, ganoderma lucidum	Enhance insulin sensitivity and glucose uptake Modulate immune responses, reducing inflammation Delay gastric emptying, reducing the glycemic response to meals	Kumar et al. (2011a)
7	Steroids	β-sitosterol, withaferin A	Improve glucose homeostasis by enhancing insulin signaling pathways Inhibit inflammation and oxidative stress, which impair metabolic function Enhance lipid metabolism, reducing diabetic dyslipidemia	Jayaraman et al. (2025)
8	Coumarins	Esculin, aesculetin	Improve insulin secretion and sensitivity. Reduce oxidative stress and inflammation in diabetic conditions Regulate glucose uptake in tissues	Kadakol et al. (2016), Yeram and Kulkarni (2022)

↑; upregulation, ↓; downregulation, ⁱ; Inhibition

effects on lipid or carbohydrate metabolism. Some have already established themselves as antimicrobials, antivirals, antioxidants, antitumoral, vaso-relaxants or constrictives, diuretics, or laxatives. Others are used as colorants, pesticides, or growth promoters for animals or plants (Fouillaud et al. 2016). Approximately 53% of the FDA-approved natural product-based drugs originate from microorganisms (Patridge et al. 2016).

Substances such as pigments, alkaloids, antibiotics, and terpenes, which only occur in certain organisms, are the products of secondary metabolism. They are thus distinct from primary metabolites (products of primary metabolism) involved in energy metabolism and growth. Primary metabolites consist of a large group of compounds such as glycolysis and tricarboxylic acid (TCA) cycle intermediates, amino acids, proteins, purines, pyrimidine bases, polysaccharides, and fatty acids. In certain microorganisms, secondary metabolites are usually synthesized at the end of the exponential growth (trophophase) or at the beginning of the stationary phase (idiophase), and their formation is usually repressed during rapid growth. The risk of metabolic suicide diminishes when cell division is complete before toxic secondary products (e.g., antibiotics) are produced. The operation of secondary metabolism is advantageous by keeping metabolism running at a low rate, rather than closing it down completely after growth ceases. The unbalanced growth hypothesis suggests that mechanisms controlling primary metabolism are not adequate to prevent overproduction of some compounds when balanced growth ceases. Since these compounds may be toxic to the cell, secondary metabolism diverts synthesis to the

Table 7.2 Plant-derived secondary metabolites with antidiabetic activity

Plant name	Part	Active phytochemicals	IC_{50}	Mechanism of action	References
Gymnema sylvestre (gymnema)	Leaves	Gymnemic acids	0.18 mg/mL for α-amylase inhibition	Inhibits glucose absorption in the intestine, delays carbohydrate digestion, and enhances insulin secretion	Ardalani et al. (2021)
Momordica charantia (bitter melon)	Fruit	Charantin, polypeptide-p, momordicosides	2.4 μg/mL for α-glucosidase inhibition	Reduces blood glucose by enhancing insulin sensitivity, promoting glucose uptake, and suppressing gluconeogenesis	Grover and Yadav (2004)
Trigonella foenum-graecum (fenugreek)	Seeds	4-Hydroxyisoleucine, trigonelline	0.45 mg/mL for α-amylase inhibition	Enhances insulin secretion and sensitivity, reduces intestinal glucose absorption, and lowers postprandial glucose levels	Sur et al. (2001)
Syzygium cumini (jamun)	Seeds	Ellagic acid, anthocyanins	1.9 μg/mL for DPP-IV inhibition	Reduces oxidative stress, improves pancreatic β-cell function, and inhibits enzymes involved in carbohydrate digestion	Sharma et al. (2006)
Aloe vera	Leaf gel	Polysaccharides, aloesin	3.2 μg/mL for α-glucosidase inhibition	Increases glucose uptake, improves insulin sensitivity, and reduces inflammation	Elkomy et al. (2023)
Cinnamomum verum (cinnamon)	Bark	Cinnamaldehyde, polyphenols	1.1 μg/mL for PTP1B inhibition	Enhances insulin receptor signaling, improves glucose uptake, and reduces oxidative stress	Deshmukh et al. (2023), Maltese and Sinclair (2024)
Ocimum sanctum (holy basil)	Leaves	Eugenol, rosmarinic acid	2.8 μg/mL for α-glucosidase inhibition	Improves insulin secretion, inhibits enzymes involved in carbohydrate digestion, and reduces lipid peroxidation	Choudhury et al. (2014)

production of harmless products, which are excreted. According to this theory, secondary metabolism should increase long-term viability (Singh et al. 2023). Microbial secondary metabolites, like antibiotics, pigments, growth hormones, antitumor agents, and others, are not essential for the growth and development of microorganisms, but they have shown great potential for human and animal health (Ruiz et al. 2010). Among the microorganisms producing the above-mentioned compounds, bacteria, including actinobacteria, and fungi produce a diverse array of bioactive small molecules with significant potential to be used in medicine (Sacks et al. 2018). These bioactive compounds are mainly produced by the activation of cryptic gene clusters, which are not active under normal conditions, and thus, the expression of these clusters would be helpful in the exploitation of the chemical diversity of microorganisms (Pettit 2011).

Microbes produce specialized chemicals called secondary metabolites. Proteins involved in crucial cellular processes like energy production and growth are known as primary metabolites; these chemicals are separate from them. These secondary metabolites are synthesized by microbes during specific stages of their life cycles, such as the stationary phase or post-exponential growth (Monciardini et al. 2014). Their molecular weight is often quite low. Antibiotics, colors, toxins, enzyme inhibitors, immune modulators, and compounds like lipids and carbohydrates that impact ecological interactions or metabolic processes are all part of this broad category. Metabolites are compounds that considerably affect many facets of human civilization, such as the food production process, individual health, and the economy. They have many different uses, including antibacterial, antiviral, antioxidant, anticancer, vaso-relaxant, diuretic, laxative, colorant, pesticide, and growth promoter for animals and plants (Fouillaud et al. 2016). Microbes are the source of around 53% of the FDA-approved drugs produced from natural sources, which highlights the importance of microorganisms in the pharmaceutical business (Patridge et al. 2016).

When microbial growth is inhibited, secondary metabolites are typically produced. Because of this, there is less chance of metabolic interference, often known as "metabolic suicide," in which harmful compounds, like antibiotics, could harm the microbe that produces them. The precise moment that the cellular resources are sufficient to maintain the synthesis of these specialized metabolites is determined by this timing. This is achieved without interfering with the essential cellular processes required for development and maintenance. Secondary metabolism is in charge of maintaining a reduced rate of metabolic activity instead of totally turning it off when primary growth stops. It is crucial that this function be performed. It is supported by the uneven growth hypothesis. According to this view, even when balanced growth stops happening, the systems that control primary metabolism may not be able to stop the overproduction of harmful compounds completely. To lessen the danger posed by the carcinogenic molecules, secondary metabolism guides synthesis toward the production of harmless chemicals that may be excreted from the cell. According to this theory, microorganisms' long-term survival is enhanced when resources are redirected toward secondary metabolism (Singh et al. 2023).

Antibiotics, pigments, growth hormones, and anticancer medicines are all examples of metabolites that have demonstrated great promise for enhancing human and animal health, even though they are not essential for the basic creation and growth of bacteria (Ruiz et al. 2010).

Among the microorganisms producing the above-mentioned compounds, bacteria, including actinobacteria, and fungi produce a diverse array of small bioactive molecules with significant potential to be used in medicine. Actinobacteria and other fungi and bacteria produce these bioactive chemicals at an alarming rate. Most of the time, these compounds are made by synthesizing cryptic gene clusters, which normally don't do anything. In order to tap into the vast chemical diversity of microbes and harness their potential in fields such as pharmaceuticals and others, it is crucial to activate these gene clusters.

A rising global health concern, diabetes mellitus (DM) is afflicting both developed and emerging nations at alarming rates. In 2019, 463 million individuals were living with diabetes, according to data collected by the International Diabetes Federation. Research by Saeedi et al. (2019) indicates that the worldwide prevalence of diabetes is projected to increase to 579 million by 2030 and hit 700 million by 2045 (Saeedi et al. 2019). A significant complication that can occur due to diabetes is diabetic nephropathy (DN), which has emerged as a primary cause of chronic kidney disease (Raval et al. 2020). Short-chain fatty acids (SCFAs) are crucial in biological regulation by activating G protein-coupled receptors and suppressing histone deacetylases (HDACs). Short-chain fatty acids (SCFAs) are crucial in biological regulation by activating G protein-coupled receptors and suppressing histone deacetylases (HDACs). These mechanisms assist in controlling inflammatory reactions and lowering mean arterial pressure (Magliocca et al. 2022). Acetate, propionate, and butyrate are types of SCFAs generated when gut bacteria ferment carbohydrates that the human digestive system is unable to break down (Cong et al. 2022). Bacteroidetes mainly produce acetate and propionate, whereas Firmicutes are primarily responsible for butyrate production (Levy et al. 2016).

SCFAs like acetic acid, propionic acid, and butyric acid are crucial metabolites generated by gut microbiota. Propionate is essential in metabolic processes as it decreases gluconeogenesis. It stimulates AMP-activated protein kinase (AMPK), which suppresses enzymes such as glucose-6-phosphatase and phosphoenolpyruvate carboxykinase, both linked to glucose synthesis. Research targeting the propionate receptor GPR43 using small interfering RNA (siRNA) revealed that this blockage of gluconeogenesis was counteracted and the activation of AMPK induced by propionate was hindered (Yoshida et al. 2019). By regulating cellular and physiological pathways, SCFAs, especially propionate, affect metabolic processes and could have potential in treating diabetes and its complications, including diabetic nephropathy. Type 2 diabetes (T2D) is a metabolic disorder with elevated blood glucose and decreased insulin production. It is a major global health issue, driven by poor diet and inactivity, and leads to significant healthcare costs (Huang et al. 2020).

7.5.1 *Role of Intestinal Flora Metabolites in Treating Diabetic Nephropathy (DN)*

Diabetes mellitus (DM) has become one of the fastest-growing public health problems in both developed and developing countries. According to the 2019 International Diabetes Federation statistics, 463 million people had diabetes in 2019, and 579 million are expected to have diabetes by 2030, and 700 million (51%) by 2045 (Saeedi et al. 2019). Diabetic nephropathy (DN) is one of the most serious complications of diabetes, and it is also emerging as a major cause of chronic kidney disease (Raval et al. 2020). Short-chain fatty acids play a role in bioregulation by inhibiting histone deacetylases (HDACs) and activating G protein receptors, thereby reducing the inflammatory response and lowering mean arterial pressure (doi:10.3390/ijms23105354). SCFAs include acetate, propionate, and butyrate, which are produced by microbial communities through the fermentation of indigestible carbohydrates (Cong et al. 2022). Acetate and propionate are mainly produced by Bacteroidetes, whereas butyrate is mainly produced by Firmicutes (Levy et al. 2016). Acetic acid, propionic acid, and butyric acid are the main SCFA metabolites produced by gut microbes. Propionate can reduce gluconeogenesis by activating AMP-activated protein kinase (AMPK) to downregulate glucose-6-phosphatase and phosphoenolpyruvate carboxykinase. In addition, siRNA-mediated knockdown of the propionate receptor GPR43 prevented propionate-induced AMPK activation and eliminated gluconeogenesis inhibition (Yoshida et al. 2019).

7.6 Conclusion and Future Remarks

Medicinal plants are factories of phytochemicals that are really helpful to treat different kinds of diseases. Phytochemicals exhibit a wide range of biological activities, including anticancer, antitumor, antimicrobial, antidiabetic, and anti-inflammatory effects. Consequently, edible plants have attracted considerable research interest for the discovery of novel compounds that underscore their therapeutic potential. In the future, there will be chances to improve the identification, synthesis, and distribution of secondary metabolites with antidiabetic potential thanks to the integration of contemporary technologies like metabolomics, genomics, synthetic biology, and nanotechnology. In addition, ethical bioprospecting, medicinal plant biodiversity protection, and sustainable sourcing will be essential to guaranteeing these bioactive resources' long-term availability.

To sum up, secondary metabolites have a lot of potential as a natural tool in the fight against diabetes, and further interdisciplinary research, along with developments in pharmaceutical formulation and clinical evaluation, will be essential to converting their potential into workable, affordable, and efficient treatment options.

References

Abdelghffar EAR, Mostafa NM, El-Nashar HAS, Eldahshan OA, Singab ANB (2022) Chilean pepper (Schinus polygamus) ameliorates the adverse effects of hyperglycaemia/dyslipidaemia in high fat diet/streptozotocin-induced type 2 diabetic rat model. Ind Crop Prod 183:114953

Ahmad R, Haque M (2024) Metformin: beyond type 2 diabetes mellitus. Cureus 16(10):e71730

Al-Kuraishy HM, Al-Gareeb AI, Alexiou A, Papadakis M, Nadwa EH, Albogami SM, Alorabi M, Saad HM, Batiha GE (2022) Metformin and growth differentiation factor 15 (GDF15) in type 2 diabetes mellitus: a hidden treasure. J Diabetes 14(12):806–814

Ardalani H, Hejazi Amiri F, Hadipanah A, Kongstad KT (2021) Potential antidiabetic phytochemicals in plant roots: a review of in vivo studies. J Diabetes Metab Disord 20(2):1837–1854

Bagal D, Chowdhary AA, Mehrotra S, Mishra S, Rathore S, Srivastava V (2023) Metabolic engineering in hairy roots: an outlook on production of plant secondary metabolites. Plant Physiol Biochem 201:107847

Bennett RN, Wallsgrove RM (1994) Secondary metabolites in plant defence mechanisms. New Phytol 127(4):617–633

Brownlee M (2001) Biochemistry and molecular cell biology of diabetic complications. Nature 414(6865):813–820

Cappelli SL, Domeignoz-Horta LA, Loaiza V, Laine AL (2022) Plant biodiversity promotes sustainable agriculture directly and via belowground effects. Trends Plant Sci 27(7):674–687

Choudhury SS, Bashyam L, Manthapuram N, Bitla P, Kollipara P, Tetali SD (2014) Ocimum sanctum leaf extracts attenuate human monocytic (THP-1) cell activation. J Ethnopharmacol 154(1):148–155

Choudhury H, Pandey M, Hua CK, Mun CS, Jing JK, Kong L, Ern LY, Ashraf NA, Kit SW, Yee TS, Pichika MR, Gorain B, Kesharwani P (2018) An update on natural compounds in the remedy of diabetes mellitus: a systematic review. J Tradit Complement Med 8(3):361–376

Cong J, Zhou P, Zhang R (2022) Intestinal microbiota-derived short chain fatty acids in host health and disease. Nutrients 14(9):1977

Daglia M (2012) Polyphenols as antimicrobial agents. Curr Opin Biotechnol 23(2):174–181

Dai J, Mumper RJ (2010) Plant phenolics: extraction, analysis and their antioxidant and anticancer properties. Molecules 15(10):7313–7352

Demain AL, Fang A (2000) The natural functions of secondary metabolites. Adv Biochem Eng Biotechnol 69:1–39

Deshmukh SK, Agrawal S, Gupta MK (2023) Fungal metabolites: a potential source of antidiabetic agents with particular reference to PTP1B inhibitors. Curr Pharm Biotechnol 24(8):927–945

Dhanya R, Arya AD, Nisha P, Jayamurthy P (2017) Quercetin, a Lead compound against type 2 diabetes ameliorates glucose uptake via AMPK pathway in skeletal muscle cell line. Front Pharmacol 8:336

Egbuna C, Awuchi CG, Kushwaha G, Rudrapal M, Patrick-Iwuanyanwu KC, Singh O, Odoh UE, Khan J, Jeevanandam J, Kumarasamy S, Chukwube VO, Narayanan M, Palai S, Găman MA, Uche CZ, Ogaji DS, Ezeofor NJ, Mtewa AG, Patrick-Iwuanyanwu CC, Kesh SS, Shivamallu C, Saravanan K, Tijjani H, Akram M, Ifemeje JC, Olisah MC, Chikwendu CJ (2021) Bioactive compounds effective against type 2 diabetes mellitus: a systematic review. Curr Top Med Chem 21(12):1067–1095

Eldahshan OA, Nilofar N, Zengin G, El-Nashar HAS (2025) Chemical profile of essential oil from Pimenta racemosa leaves, antioxidant potential, and its enzyme inhibitory properties. Sci Rep 15(1):30705

Elekofehinti OO, Ejelonu OC, Kamdem JP, Akinlosotu OB, Adanlawo IG (2017) Saponins as adipokines modulator: a possible therapeutic intervention for type 2 diabetes. World J Diabetes 8(7):337–345

Elkomy N, El-Shaibany A, Elnagar GM, Abdelkhalek AS, Al-Mahbashi H, Elaasser MM, Raweh SM, Aldiyarbi MA, Raslan AE (2023) Evaluation of acute oral toxicity, anti-diabetic and antioxidant effects of Aloe vera flowers extract. J Ethnopharmacol 309:116310

El-Nashar HAS, Mostafa NM, El-Shazly M, Eldahshan OA (2021) The role of plant-derived compounds in managing diabetes mellitus: a review of literature from 2014 to 2019. Curr Med Chem 28(23):4694–4730
El-Nashar HAS, Mostafa NM, Eldahshan OA, Singab ANB (2022) A new antidiabetic and anti-inflammatory biflavonoid from Schinus polygama (Cav.) Cabrera leaves. Nat Prod Res 36(5):1182–1190
El-Nashar HAS, Al-Qaaneh AM, Bhuia MS, Chowdhury R, Abdel-Maksoud MA, Ebaid H, Malik A, Torequl Islam M, Aufy M, Elhawary EA (2024a) UPLC-ESI/MS(n) metabolic profiling of Cedrela odorata L. and Toona ciliata M. Roem and in vitro investigation of their anti-diabetic activity supported with molecular docking studies. Front Chem 12:1462309
El-Nashar HAS, Eldahshan OA, Nilofar N, Zengin G (2024b) Chemical profiling and enzyme inhibitory properties of essential oil isolated from Artemisia judaica grown in Egypt: GC-MS analysis and in-vitro studies. Nat Prod Res 40:1–5
El-Nashar HAS, Taleb M, El-Shazly M, Zhao C, Farag MA (2024c) Polysaccharides (pectin, mucilage, and fructan inulin) and their fermented products: a critical analysis of their biochemical, gut interactions, and biological functions as antidiabetic agents. Phytother Res 38(2):662–693
El-Nashar HAS, Eldahshan OA, El Hassab MA, Zengin G, Elhawary EA (2025a) UPLC/MS(n) analysis of Bougainvillea glabra leaves and investigation of antioxidant activities and enzyme inhibitory properties. Sci Rep 15(1):28272
El-Nashar HAS, Zengin G, Tarek A, Eldahshan OA (2025b) Comparative metabolic profiling of Cordia myxa leaves and fruits based on UPLC-ESI/MS-MS analysis and investigation of antioxidant activities and enzyme inhibitory properties. Sci Rep 15(1):19296
Erb M, Kliebenstein DJ (2020) Plant secondary metabolites as defenses, regulators, and primary metabolites: the blurred functional trichotomy. Plant Physiol 184(1):39–52
Ferrer JL, Austin MB, Stewart C Jr, Noel JP (2008) Structure and function of enzymes involved in the biosynthesis of phenylpropanoids. Plant Physiol Biochem 46(3):356–370
Fleming A (2001) On the antibacterial action of cultures of a penicillium, with special reference to their use in the isolation of B. influenzae. 1929. Bull World Health Organ 79(8):780–790
Fouillaud M, Venkatachalam M, Girard-Valenciennes E, Caro Y, Dufossé L (2016) Anthraquinones and derivatives from marine-derived fungi: structural diversity and selected biological activities. Mar Drugs 14(4):64
Fraenkel GS (1959) The raison d'ĕtre of secondary plant substances; these odd chemicals arose as a means of protecting plants from insects and now guide insects to food. Science 129(3361):1466–1470
Franco RR, Ribeiro Zabisky LF, de Lima P, Júnior J, Mota Alves VH, Justino AB, Saraiva AL, Goulart LR, Espindola FS (2020) Antidiabetic effects of Syzygium cumini leaves: a non hemolytic plant with potential against process of oxidation, glycation, inflammation and digestive enzymes catalysis. J Ethnopharmacol 261:113132
Grover JK, Yadav SP (2004) Pharmacological actions and potential uses of Momordica charantia: a review. J Ethnopharmacol 93(1):123–132
Guo R, Zhang S, Li A, Zhang P, Peng X, Lu X, Fan X (2025) Ginsenoside Rb1 and berberine synergistically protect against type 2 diabetes mellitus via GDF15/HAMP pathway throughout the liver lobules: insights from spatial transcriptomics analysis. Pharmacol Res 215:107711
Hashim M, Badruddeen AJ, Khan MI, Ahmad M, Islam A, Ahmad A (2024) Diabetic neuropathy: an overview of molecular pathways and protective mechanisms of phytobioactives. Endocr Metab Immune Disord Drug Targets 24(7):758–776
Hossain A, Shahidi F (2023) Persimmon leaves: nutritional, pharmaceutical, and industrial potential-a review. Plants (Basel) 12(4):937
Huang W, Man Y, Gao C, Zhou L, Gu J, Xu H, Wan Q, Long Y, Chai L, Xu Y, Xu Y (2020) Short-chain fatty acids ameliorate diabetic nephropathy via GPR43-mediated inhibition of oxidative stress and NF-κB signaling. Oxidative Med Cell Longev 2020:4074832

Hunyadi A, Martins A, Hsieh TJ, Seres A, Zupkó I (2012) Chlorogenic acid and rutin play a major role in the in vivo anti-diabetic activity of Morus alba leaf extract on type II diabetic rats. PLoS One 7(11):e50619

Jayaraman S, Prasad M, Natarajan SR, Krishnamoorthy R, Alshuniaber MA, Gatasheh MK, Veeraraghavan VP, Rajagopal P, Palanisamy CP (2025) Molecular mechanisms underlying the effects of beta-sitosterol on TGF-β1/Nrf2/SIRT1/p53-mediated signaling in the kidney of a high-fat diet and sucrose-induced type-2 diabetic rat. Chem Biol Interact 411:111443

Jeepipalli SPK, Du B, Sabitaliyevich UY, Xu B (2020) New insights into potential nutritional effects of dietary saponins in protecting against the development of obesity. Food Chem 318:126474

Kadakol A, Sharma N, Kulkarni YA, Gaikwad AB (2016) Esculetin: a phytochemical endeavor fortifying effect against non-communicable diseases. Biomed Pharmacother 84:1442–1448

Khan MZ, Chen W, Liu X, Kou X, Khan A, Khan RU, Zahoor M, Wang C (2024) An overview of bioactive compounds' role in modulating the Nrf2/Keap1/NF-κB pathway to alleviate lipopolysaccharide-induced endometritis. Int J Mol Sci 25(19):10319

Krawczyk M, Burzynska-Pedziwiatr I, Wozniak LA, Bukowiecka-Matusiak M (2023) Impact of polyphenols on inflammatory and oxidative stress factors in diabetes mellitus: nutritional antioxidants and their application in improving antidiabetic therapy. Biomolecules 13(9):1402

Kumar R, Sharma B, Tomar NR, Roy P, Gupta AK, Kumar A (2011a) In vivo evaluation of hypoglycemic activity of Aloe spp. and identification of its mode of action on GLUT-4 gene expression in vitro. Appl Biochem Biotechnol 164(8):1246–1256

Kumar S, Narwal S, Kumar V, Prakash O (2011b) α-glucosidase inhibitors from plants: a natural approach to treat diabetes. Pharmacogn Rev 5(9):19–29

Levy M, Thaiss CA, Elinav E (2016) Metabolites: messengers between the microbiota and the immune system. Genes Dev 30(14):1589–1597

Lv X, Zhao Y, Yang X, Han H, Ge Y, Zhang M, Zhang H, Zhang M, Chen L (2021) Berberine potentiates insulin secretion and prevents β-cell dysfunction through the miR-204/SIRT1 signaling pathway. Front Pharmacol 12:720866

Magliocca G, Mone P, Di Iorio BR, Heidland A, Marzocco S (2022) Short-chain fatty acids in chronic kidney disease: focus on inflammation and oxidative stress regulation. Int J Mol Sci 23(10):5354

Maltese G, Sinclair AJ (2024) Comment on Kutz et al. Comparative cardiovascular effectiveness and safety of SGLT-2 inhibitors, GLP-1 receptor agonists, and DPP-4 inhibitors according to frailty in type 2 diabetes. Diabetes Care 2023;46:2004-2014. Diabetes Care 47(4):e32–e33

Monciardini P, Iorio M, Maffioli S, Sosio M, Donadio S (2014) Discovering new bioactive molecules from microbial sources. Microb Biotechnol 7(3):209–220

Neuhaus WC, Jemison AL, Kozlowski MC (2021) Oxidative dehydrogenative couplings of alkenyl phenols. Org Biomol Chem 19(38):8205–8226

Ninkuu V, Zhang L, Yan J, Fu Z, Yang T, Zeng H (2021) Biochemistry of terpenes and recent advances in plant protection. Int J Mol Sci 22(11):5710

Parveen A, Sultana R, Lee SM, Kim TH, Kim SY (2021) Phytochemicals against anti-diabetic complications: targeting the advanced glycation end product signaling pathway. Arch Pharm Res 44(4):378–401

Patridge E, Gareiss P, Kinch MS, Hoyer D (2016) An analysis of FDA-approved drugs: natural products and their derivatives. Drug Discov Today 21(2):204–207

Pettit RK (2011) Small-molecule elicitation of microbial secondary metabolites. Microb Biotechnol 4(4):471–478

Pichersky E, Gang DR (2000) Genetics and biochemistry of secondary metabolites in plants: an evolutionary perspective. Trends Plant Sci 5(10):439–445

Potenza MA, Iacobazzi D, Sgarra L, Montagnani M (2020) The intrinsic virtues of EGCG, an extremely good cell guardian, on prevention and treatment of diabesity complications. Molecules 25(13):3061

Raval N, Kumawat A, Kalyane D, Kalia K, Tekade RK (2020) Understanding molecular upsets in diabetic nephropathy to identify novel targets and treatment opportunities. Drug Discov Today 25(5):862–878

Ruiz B, Chávez A, Forero A, García-Huante Y, Romero A, Sánchez M, Rocha D, Sánchez B, Rodríguez-Sanoja R, Sánchez S, Langley E (2010) Production of microbial secondary metabolites: regulation by the carbon source. Crit Rev Microbiol 36(2):146–167

Sabir IA, Manzoor MA, Khan I, Hu X, Chen J, Qin Y (2025) Emerging trends in secondary metabolite research in Caryophyllales: Betalains and their roles in plant adaptation and defense mechanisms. J Agric Food Chem 73(4):2249–2265

Sacks D, Baxter B, Campbell BCV, Carpenter JS, Cognard C, Dippel D, Eesa M, Fischer U, Hausegger K, Hirsch JA, Shazam Hussain M, Jansen O, Jayaraman MV, Khalessi AA, Kluck BW, Lavine S, Meyers PM, Ramee S, Rüfenacht DA, Schirmer CM, Vorwerk D (2018) Multisociety consensus quality improvement revised consensus statement for endovascular therapy of acute ischemic stroke. Int J Stroke 13(6):612–632

Saeedi P, Petersohn I, Salpea P, Malanda B, Karuranga S, Unwin N, Colagiuri S, Guariguata L, Motala AA, Ogurtsova K, Shaw JE, Bright D, Williams R (2019) Global and regional diabetes prevalence estimates for 2019 and projections for 2030 and 2045: results from the International Diabetes Federation Diabetes Atlas, 9(th) edition. Diabetes Res Clin Pract 157:107843

Sharma SB, Nasir A, Prabhu KM, Murthy PS (2006) Antihyperglycemic effect of the fruit-pulp of Eugenia jambolana in experimental diabetes mellitus. J Ethnopharmacol 104(3):367–373

Shehadeh MB, Suaifan G, Abu-Odeh AM (2021) Plants secondary metabolites as blood glucose-lowering molecules. Molecules 26(14):4333

Singh NK, Baranwal J, Pati S, Barse B, Khan RH, Kumar A (2023) Application of plant products in the synthesis and functionalisation of biopolymers. Int J Biol Macromol 237:124174

Sukhikh S, Babich O, Prosekov A, Kalashnikova O, Noskova S, Bakhtiyarova A, Krol O, Tsvetkova E, Ivanova S (2023) Antidiabetic properties of plant secondary metabolites. Meta 13(4):513

Sur P, Das M, Gomes A, Vedasiromoni JR, Sahu NP, Banerjee S, Sharma RM, Ganguly DK (2001) Trigonella foenum graecum (fenugreek) seed extract as an antineoplastic agent. Phytother Res 15(3):257–259

Tanaka Y, Sasaki N, Ohmiya A (2008) Biosynthesis of plant pigments: anthocyanins, betalains and carotenoids. Plant J 54(4):733–749

Wawrosch C, Zotchev SB (2021) Production of bioactive plant secondary metabolites through in vitro technologies-status and outlook. Appl Microbiol Biotechnol 105(18):6649–6668

Wen D, Huang X, Zhang M, Zhang L, Chen J, Gu Y, Hao CM (2013) Resveratrol attenuates diabetic nephropathy via modulating angiogenesis. PLoS One 8(12):e82336

Wink M (2003) Evolution of secondary metabolites from an ecological and molecular phylogenetic perspective. Phytochemistry 64(1):3–19

Yan XM, Zhou SS, Liu H, Zhao SW, Tian XC, Shi TL, Bao YT, Li ZC, Jia KH, Nie S, Guo JF, Kong L, Porth IM, Mao JF (2023) Unraveling the evolutionary dynamics of the TPS gene family in land plants. Front Plant Sci 14:1273648

Yeram PB, Kulkarni YA (2022) Glycosides and vascular complications of diabetes. Chem Biodivers 19(10):e202200067

Yoshida H, Ishii M, Akagawa M (2019) Propionate suppresses hepatic gluconeogenesis via GPR43/AMPK signaling pathway. Arch Biochem Biophys 672:108057

Yu X, Pu H, Voss M (2024) Overview of anti-inflammatory diets and their promising effects on non-communicable diseases. Br J Nutr 132(7):898–918

Chapter 8
Bioactive Compounds from Plants for Disease Prevention and Management

Iriny M. Ayoub, Nehal Ibrahim, Yosra A. El Khodary, and Abdel Nasser B. Singab

Abstract Plant-derived bioactive compounds represent a wide variety of naturally occurring chemical substances that have shown significant potential in preventing and curing human illnesses. This chapter presents a comprehensive overview of the structural classes and pharmacological actions of major phytochemicals, particularly phenolics, flavonoids, alkaloids, terpenoids, and organosulfur compounds. Examples such as curcumin, resveratrol, berberine, and allicin are discussed in terms of their mechanisms of action in combating oxidative stress, inflammation, microbial infections, neurodegenerative conditions, and cancer. In addition, this chapter highlights their methods of extraction and purification, recent advancements in analytical characterization techniques, and how they function on both cellular and molecular levels. The challenges of poor bioavailability, metabolic instability, and lack of standardization are critically examined, along with emerging strategies to enhance delivery and therapeutic efficacy, including nanoformulations. This integrative approach highlights the relevance of plant-derived bioactives as potential candidates for disease prevention and therapeutic intervention, joining the gap between traditional knowledge and recent drug discovery.

Iriny M. Ayoub and Nehal Ibrahim contributed equally to this chapter.

I. M. Ayoub (✉) · N. Ibrahim · Y. A. El Khodary
Department of Pharmacognosy, Faculty of Pharmacy, Ain Shams University, Cairo, Egypt
e-mail: irinyayoub@pharma.asu.edu.eg

A. N. B. Singab
Department of Pharmacognosy, Faculty of Pharmacy, Ain Shams University, Cairo, Egypt

Center for Drug Discovery Research and Development, Faculty of Pharmacy, Ain Shams University, Cairo, Egypt

L. A. Worku et al. (eds.), *Plant-Based Solutions for Sustainable Agriculture and Healthcare*, Sustainability Sciences in Asia and Africa,
https://doi.org/10.1007/978-981-95-8910-4_8

Keywords Bioactive compounds · Antioxidants · Medicinal plants · Chronic diseases · Inflammation

8.1 Introduction

Bioactive constituents are naturally derived compounds with demonstrated biological activity, produced by plants, animals, microorganisms, or even marine organisms as secondary metabolites in small quantities. They exert significant roles in maintaining health and preventing diseases. Depending on their chemical structure and biological function, bioactive compounds can be categorized into several classes. They can also be grouped into essential or nonessential bioactive compounds according to their influence on vital functions in human body. Essential compounds play vital biological roles in the human body, and their deficiency can contribute to disease progression. Whereas nonessential compounds are not involved in essential biological functions, their deficiency generally does not have a negative impact on health (Martín Ortega and Segura Campos 2019). For instance, vitamins A, B, C, D, and E are among the essential bioactive compounds. Vitamin C has been investigated for its therapeutic efficacy in diverse functions, including allergy management, conversion of cholesterol to bile acids, synthesis of neurotransmitters, bone development, wound healing, and connective tissue preservation (Chambial et al. 2013). In addition, fatty acids have been shown to be involved in numerous physiological and pathological processes, besides their nutritional aspect. Polyunsaturated fatty acids are considered the most beneficial class of fatty acids. One of the most significant examples belonging to this class is omega-3 that possesses potent anti-inflammatory, neuroprotective, and cardioprotective actions (Simonetto et al. 2019). In contrast, phytochemicals are plant-derived bioactive compounds that are considered nonessential. They are biosynthesized via the secondary metabolism inside plant cells of medicinal plants, fruits, vegetables, cereal grains, and algae. Secondary metabolism denotes biosynthetic pathways that yield metabolites not essential for the organism's fundamental processes of growth or reproduction.

8.2 Classification of Bioactive Compounds

Phytochemicals possess diverse classes that are categorized according to their molecular structures into: polyphenolics, phytosterols, alkaloids, terpenoids, and organosulfur compounds (Somani et al. 2015; Abate et al. 2022). They can also be categorized according to different characteristics including the number of carbons or number of rings. Medicinal plants such as *Papaver somniferum*, *Cinchona officinalis*, and *Eugenia aromatica*, along with a huge variety of plants and flowers, are considered a rich mine of phytochemicals with potential health benefits (Zhang et al. 2015; Husen 2021, 2022, 2023a, b, 2024a, b; Husen and Pant 2024a, b; Pant et al. 2025; Pant and Husen 2025). Leading sources of phytochemicals are found across several food groups: these include fruits such as strawberries, wild berries, blackberries, blueberries, and pomegranates; vegetables like red peppers, spinach, broccoli, and asparagus, which exhibit high polyphenol content and significant antioxidant capacity; and cereals, including red, black, and purple rice, which are noted for possessing high concentrations of bioactive compounds (Liu 2013).

Sources of bioactive compounds are very diverse and include plants, animals, and marine organisms. They are mainly classified into three categories: terpenoids, comprising approximately 25,000 types; phenolic compounds, around 8000 types; and alkaloids, with about 12,000 types (Carbonell-Capella et al. 2014). In this work, we are concerned with plant-based bioactive compounds due to the great significance of medicinal plants as basis of several traditional medicine systems which use medicinal plants for treating specific health conditions from ancient times and offer significant resources of novel drug-lead molecules at the present time. Traditional medicine starts with Sumerian civilization with more than 5000 years of antiquity, followed by Chinese traditional medicine (2500 BC) which describes the role of herbs and medicinal roots in treating various health conditions. Many plant preparations are still in use until today including cinnamon bark, ginseng, and ephedra. Indian and Egyptian civilizations as well created books dating back more than a 1000 years describing the therapeutic use of plants, spices, and foods (Petrovska 2012). Furthermore, thousands of phytochemicals have been proven as pharmacologically active entities, and some of them have unlocked a new prospect for the management of several ailments (Dereli 2023).

Plants yield a wide variety of organic compounds, generally categorized as primary or secondary metabolites. Primary metabolites, including carbohydrates, proteins, lipids, and nucleic acids, play a key role in maintaining fundamental physiological processes and are found in all plants. On the other hand, secondary metabolites arise from alternative metabolic pathways to perform specialized functions. They are typically restricted to certain taxonomic groups and play key roles in helping plants adapt to environmental changes and withstand biotic and abiotic stresses. Additionally, secondary metabolites are largely responsible for the medicinal properties of plants (Koche et al. 2016). Herein, we provide a detailed overview of diverse categories of secondary metabolites, their subclasses, and pharmacological activities.

8.2.1 *Terpenes or Terpenoids*

Terpenoids constitute the most abundant group of naturally occurring bioactive compounds present in nearly all living organisms. They possess a unique structural diversity, being derived from multiple isoprene units, C5 (2-methylbuta-1,3-diene), that undergo structural rearrangement yielding linear and cyclic compounds (Fig. 8.1). Based on the number of isoprene units in their structure, terpenes can be

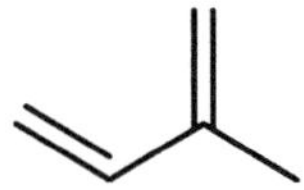

Fig. 8.1 Methylbuta-1,3--diene structure (isoprene unit)

classified as monoterpenes (two isoprene units), sesquiterpenes (three units), diterpenes (four units), triterpenes (six units), tetraterpenes (eight units), and so on.

Moreover, terpenoids can be classified into six major classes: sterols, tocopherols, taxanes, artemisinins, cannabinoids, and ingenanes (Adefegha et al. 2022). Terpenes contribute to the essence of plants and can function as insect repellents or attractants. Also, they play a key role in plant defense against pathogens and herbivores as they become toxic at high concentrations. Besides, evidence from in vivo and clinical studies has demonstrated that terpenes possess both preventive and therapeutic potential in managing numerous disorders exhibiting antibacterial, antifungal, antiparasitic, anticancer, antiviral, anti-inflammatory, antispasmodic, antiallergic, antihyperglycemic, and immunomodulatory activities (Cho et al. 2017). In addition, they possess great environmental and economic significance and are extensively utilized in various industries, including pharmaceuticals, cosmeceuticals, food, and chemicals.

8.2.1.1 Biosynthetic Pathways

The terpenoid biosynthetic pathway involves two major mechanisms: the mevalonate (MVA) pathway and the methylerythritol phosphate (MEP) pathway, in addition to glycolysis, which converts glucose into dimethylallyl diphosphate (DMAPP) and isopentenyl diphosphate (IPP), the five-carbon compounds and universal precursors of terpenoids. Both the MVA and MEP pathways comprise an 18-enzyme cascade (Chatzivasileiou et al. 2019). Concerning the mevalonate pathway, it starts with the key metabolite, the acetyl-CoA, which is converted to acetyl-SCoA that initiates a series of Claisen reactions resulting in the condensation of three molecules of acetyl-SCoA to mevalonic acid. Mevalonic acid goes through phosphorylation followed by decarboxylation with elimination of a phosphate group, which in turn produces IPP and its isomer DMAPP. Methylerythritol phosphate (MEP) pathway can also produce IPP and DMAPP within plant plastids and chloroplasts. It utilizes pyruvate and glyceraldehyde-3-phosphate derived from glycolysis to yield deoxyxylulose-5-phosphate, which is subsequently reduced to MEP. A series of reactions then converts MEP into IPP and DMAPP. This pathway is particularly prominent in bacteria (Volke et al. 2019; Perreca et al. 2020). It is important to note that C5 units produced via the MVA pathway are primarily utilized in the biosynthesis of sesquiterpenes, triterpenes, and sterols, whereas C5 units generated through the MEP pathway contribute to the formation of isoprene, monoterpenes, diterpenes, carotenoids, and chlorophyll.

The head-to-tail condensation of DMAPP and IPP, initiated by the nucleophilic attack of the terminal double bond of IPP on the allylic carbon of DMAPP, yields geranyl diphosphate (GPP), a 10-carbon intermediate that serves as the precursor of all monoterpenes. Another IPP compound is added to GPP's allylic carbon and leads to the formation of farnesyl pyrophosphate (FPP), the precursor of sesquiterpenes (15 carbons). Further addition of an IPP unit in the same head-to-tail manner generates geranylgeranyl diphosphate (GGPP), the 20-carbon precursor of diterpenes. All these nucleophilic condensations follow a "head-to-tail" pattern, in which the C1 carbon of one isoprene unit reacts with the C4 carbon of the next. In contrast, triterpenes (30 carbons) and tetraterpenes (40 carbons) are formed through the "tail-to-tail" dimerization of FPP and GGPP, respectively. The structural diversity of monoterpenes and sesquiterpenes arises largely from the high reactivity of the carbocations produced after diphosphate elimination from GPP and FPP. Additionally, cyclase enzymes catalyze the formation of cyclic structures. For instance, GPP may undergo isomerization to form neryl diphosphate (NPP) or rearrangement to produce linalyl diphosphate (LPP), both of which serve as intermediates to produce numerous structures (Fig. 8.2) (Câmara et al. 2024).

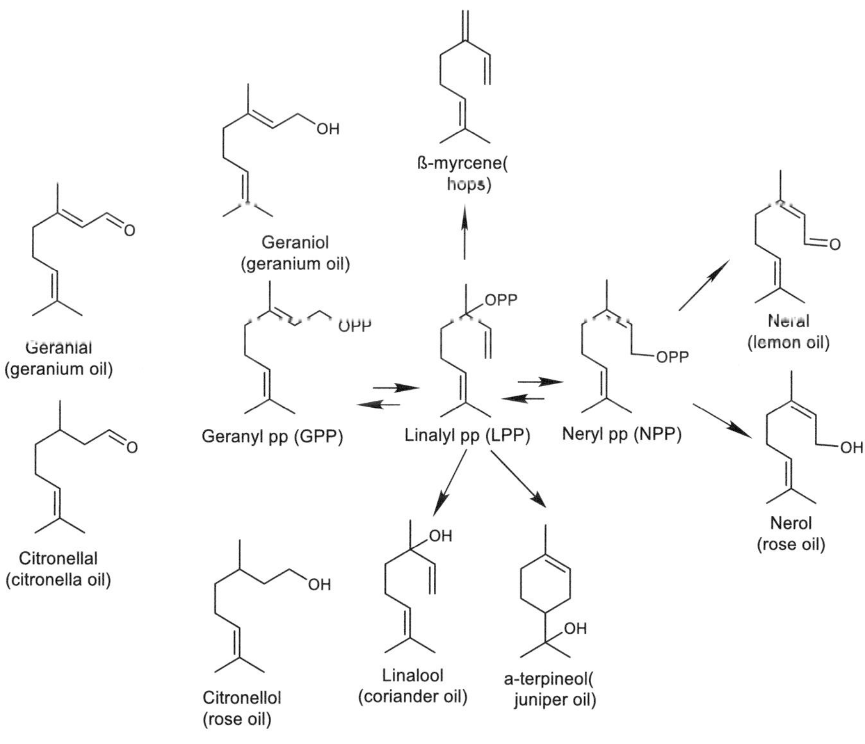

Fig. 8.2 Isomerization of GPP and its different derivatives

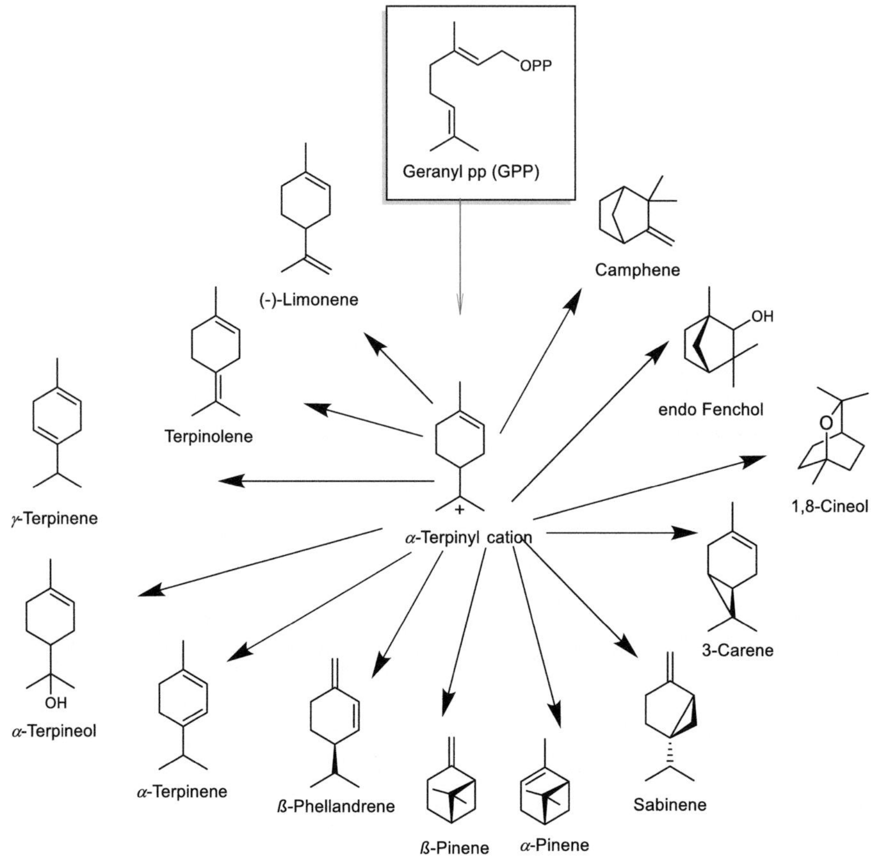

Fig. 8.3 Examples of monoterpenes structural diversity

Moreover, myriad array of monoterpenes (10-carbon) can result from GPP, as it serves as a substrate for numerous enzyme-catalyzed reactions, including cyclization, dehydrogenation, hydroxylation, isomerization, oxidation and/or reduction, and conjugation (Fig. 8.3) (Câmara et al. 2024).

Similarly, diverse group of sesquiterpene derivatives (15-carbon) can arise from the rearrangement of farnesyl pyrophosphate (FPP), the primary precursor of sesquiterpenes, via sesquiterpene synthase. More than 200 different sesquiterpene compounds can result, either acyclic structures (farnesene, farnesol, and nerolidol) or a wide range of cyclic structures, including monocyclic (humulene, germacrene, longifolene, and bisabolene), bicyclic and tricyclic structures, as well as polycyclic spiro structures (Fig. 8.4) (Xu and Dickschat 2020).

The biological significance of terpenoid compounds toward human health or the parent plant is discussed herein.

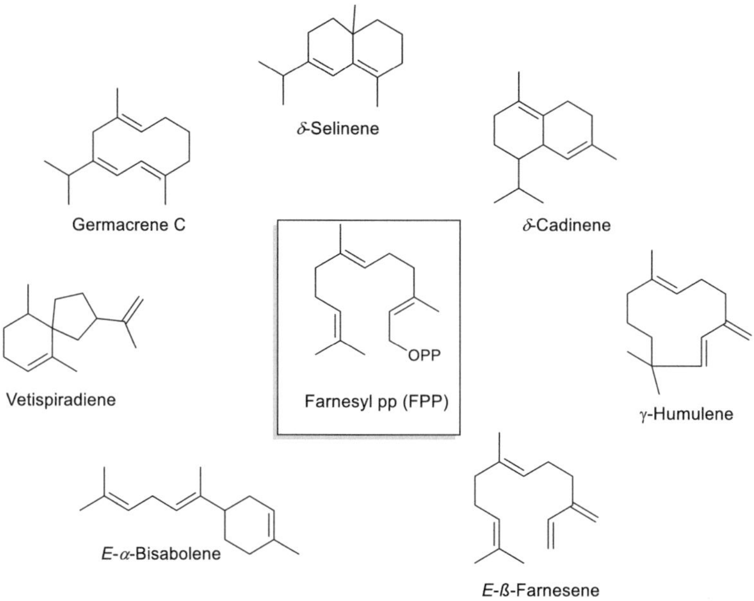

Fig. 8.4 Examples of sesquiterpenes structural diversity

8.2.1.2 Terpenoids Activity in Plants

These compounds contribute to important environmental functions, such as signaling or defense. They release terpenoid volatile compounds that affect pollinators and seed dispersers, neighboring plants, besides herbivores and predators. Terpenoids can be promising candidates for the development of novel bioinsecticides via understanding their spectrum and mode of action on insect targets (Jankowska et al. 2018).

8.2.1.3 Terpenoids Repellent Properties

Terpenoids serve as a chemical defense system in plants by synthesizing and releasing terpenoid volatile compounds that can influence insect physiology and behavior through their toxic or repellent properties. Thymol, eugenol, α-pinene, β-caryophyllene, and citronellol are extensively exploited as insect repellents, particularly against ticks, termites, and mosquitoes. Furthermore, numerous studies show that terpenoid compounds exert neurotoxic effect which causes hyperactivity in the early stage then death later on. One of the most prominent mechanisms of terpenoids neurotoxic action is the acetylcholinesterase (AChE) enzyme's inhibition which is implicated in neuro–neuronal and neuromuscular junctions in both insects and mammals. Carvacrol, limonene, and α- and β-pinene are among the most well-known AChE competitive inhibitors. Another mechanism is through the augmentation of GABA's neurotransmitter impact on mammalian receptors exerted by monoterpenoids such as thymol and menthol. Furthermore, Terpenoids are able to work via the insect octopaminergic system, which is a neurotransmission, neuromodulator, and neurohormone system that regulates insect activity, social behavior, and stress response. Thymol, eugenol, carvacrol, and α-terpineol are among the most common monoterpenes mediating this mechanism of action (Jankowska et al. 2018).

8.2.1.4 Terpenoids Activity in Human Health

Terpenoids exhibit a vast array of biological activities including neuroprotective activities and antiaging effects (Masyita et al. 2022). Besides, several examples including menthol, carvone, ascaridole, volkensin, and methyl eugenol can exert significant antimicrobial effects against both antibiotic-susceptible and antibiotic-resistant bacterial strains. Myrcene, a well-known monoterpene, can exert anti-inflammatory activity in acute and chronic inflammatory disorders (Bai et al. 2019). Several terpenoids also possess anticancer and antimalarial activities. On the other hand, caution should be taken into consideration if terpenoids are ingested in large amounts due to the risk of toxicity. Symptoms of toxicity include GIT irritation, altered mental status, seizures, and even coma (Kłos and Chlubek 2022). It is worth mentioning that terpenoids can exert synergistic effect when used in combinations, which enhance efficacy, improve bioavailability, and reduce side effects to be effectively applied in traditional medicine practices and modern therapeutic approaches. These interactions may involve pharmacodynamic mechanisms or pharmacokinetic mechanisms. The combination of menthol and eucalyptol presents an example of synergy. Menthol serves as a freshening agent and promotes eucalyptol diffusion hence, enhancing its anti-inflammatory activity and contributing to more effective pain management (Câmara et al. 2024). Herein, we provide a detailed overview of the most potent biological activities produced by terpenoids.

Anticancer Activity of Terpenoids

Terpenoids are proven to be potent antitumorigenic agents against various kinds of cancer cells, both at the in vitro and in vivo levels. They are proven to exert opposing effects against the following processes: unrestrained cell growth, malfunction of apoptosis, stimulation of invasion, induction of angiogenesis, and metastasis, which are all involved in the complex process of tumorigenesis (Abu-Izneid et al. 2020). Numerous terpenoids are reported to have anticancer activity against different kinds of cancerous cells (Table 8.1).

Table 8.1 Examples of terpenoid compounds possessing anticancer activity and their mechanisms of action

Terpenoid compound	Class	Cancer cells and mechanism of action	References
Betulinic acid 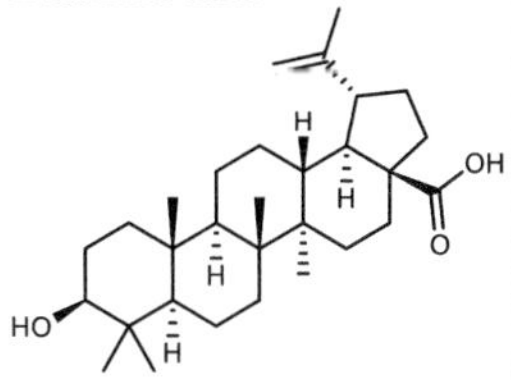	Triterpene	A375 melanoma cells/dose--dependent inhibition of mitochondrial respiration and glycolysis Induces mitochondrial dysfunction (10μM)	Coricovac et al. (2021)
Borneol 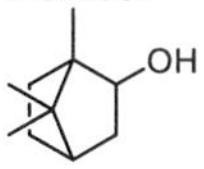	Monoterpene	Glioma cells/promotes apoptosis by regulation of HIF-1α expression via the mTORC1/eIF4E pathway Esophageal squamous cell carcinoma cells/enhances the apoptotic response to paclitaxel by inhibiting the PI3K/AKT signaling pathway	Meng et al. (2018)
Carvone 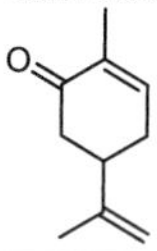	Monoterpene	MCF-7 breast ductal carcinoma/tumor-protective activity (IC_{50} 14.22μM)	Abbas et al. (2020)
Citral (nanostructured lipid carrier) 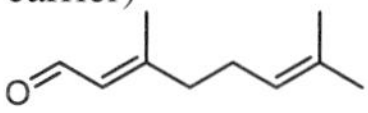	Monoterpene	MBA-231 breast cancer cells/displays in vitro cytotoxic and antitumor effects	Nordin et al. (2019)
Limonene (chitosan nanoparticle formulation comprising limonene and limonene-rich essential oils) 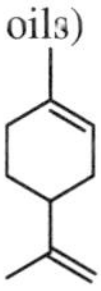	Monoterpene	Melanoma and breast cancers/potential phytotherapeutic anticancer agents	Estévez--Sarmiento et al. (2020)
Myrcene 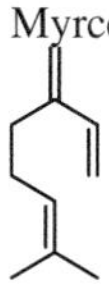	Monoterpene	A549 lung adenocarcinoma/increased apoptosis via caspase induction (IC50 0.5μg/mL), MTT assay	Bai and Tang (2020)

(continued)

Table 8.1 (continued)

Terpenoid compound	Class	Cancer cells and mechanism of action	References
Ursolic acid 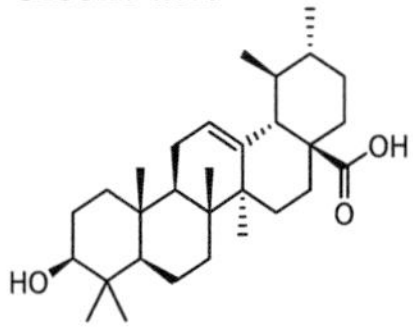	Triterpene	NCI-60 cancer cells/demonstrate significant antiproliferative activity and cytotoxicity	Khusnutdinova et al. (2021)
Oleanolic acid 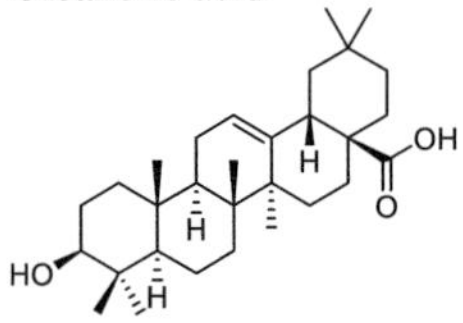			
α-Phellandrene 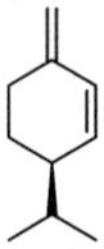	Monoterpene	Murine (S-180) and melanoma (B-16/F-10) cells/antinociceptive effect and tumor-reducing activity (IC_{50} 217.9 and 436.0μg/mL); MTT assay	Pinheiro-Neto et al. (2021)
α-Terpineol 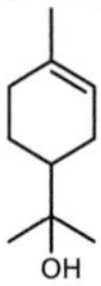	Monoterpene	Murine sarcoma 180 cancer cell line/displays broad antitumor activity (colorectal, breast, lung, leukemia) via NF-κB pathway suppression.	Negreiros et al. (2021)

Cardioprotective Activity of Terpenoids

Terpenoids have been shown to exhibit potent therapeutic effects in the management of cardiovascular diseases (CVDs) by alleviating complications such as hypertension, atherosclerosis, and impaired cardiac function. These effects arise from their ability to control inflammation and apoptosis, primarily through modulation of the mitogen-activated protein kinase (MAPK) signaling pathway. Two key molecular mechanisms contribute to CVD development: the NF-κB pathway and the NLRP3 inflammasome. The NF-κB pathway, in particular, regulates the expression of genes involved in immune responses, inflammation, and cell survival. Meanwhile, the NLRP3 inflammasome is a multiprotein complex that stimulates caspase-1, resulting in the maturation and secretion of inflammatory cytokines IL-1β and IL-18. Herein, we provide some examples of terpenoids possessing CVD-alleviating activity; oridonin can work as natural inhibitor of NF-κB, which is a critical factor in the regulation of CVD-associated inflammation (Haines et al. 2024).

On the other side, β-caryophyllene and artemisinin suppress NLRP3 inflammasome activation, thereby reducing excessive inflammation associated with myocardial infarction, heart failure, atrial fibrillation, myocardial ischemia-reperfusion injury, and hypertension. This highlights the NLRP3 inflammasome as a critical therapeutic target in the management of CVDs. However, no clinically approved NLRP3 inflammasome inhibitors are currently available. Consequently, terpenoids, owing to their modulatory effects on the NLRP3 inflammatory pathway, represent promising candidates for future cardiovascular therapies (Câmara et al. 2024).

Neuroprotective Effects of Terpenoids

Terpenoids are frequently associated with neuroprotective activity, although the precise mechanisms involved remain incompletely understood. Inflammation is a major contributor to cellular damage, including those within the nervous system. The phosphoinositide 3-kinase (PI3K)/Akt signaling pathway plays a key role in promoting cell survival and maturation in response to external stimuli. Therefore, its dysregulation results in the incidence of several neurological disorders. Terpenoids have been extensively investigated for their neuroprotective activity (Xu et al. 2022). Terpenoids possess strong antioxidant and anti-inflammatory activities, supporting their potential role in protecting neural cells. Recent research illustrates that about 33 monoterpenoid compounds such as carvacrol, eucalyptol (1,8-cineole), citronellol, menthol, myrtenal, myrtenol, (+)-linalool, (−)-linalool, sobrerol, α-limonene, thymol, rotundifolone (piperitenone oxide), α-terpinen-4-ol, α-terpineol, perillyl alcohol, and *p*-cymene exhibit promising potential in preventing and managing neurological disorders. Moreover, compounds such as celastrol, erythrodiol, oleanolic acid, ginsenosides, asiatic acid, ursolic acid, and various triterpenoid saponins have been shown to mitigate neuroinflammation and oxidative stress. Terpenoids are also known to act on multiple signaling pathways simultaneously, contributing to their broad neuroprotective effects. Therefore, they are considered promising multitarget candidates for the protection and management of neurodegenerative disorders (González-Cofrade et al. 2019).

8.2.2 *Alkaloids*

Alkaloids are nitrogen-containing organic compounds of natural origin, predominantly present in flowering plants with strong or weak basic character. Alkaloids are extremely abundant comprising a huge class of natural products with about 12,000 compounds (Tabakam and Makhafola 2024). Alkaloids derive their name from being basic in nature in the same way as inorganic alkalis. Therefore, they can react with acids to yield salts.

8.2.2.1 Classification of Alkaloids

Alkaloids are one of the most multifarious and broadly investigated classes of natural compounds. Therefore, their classification seems to be challenging due to a huge variety in their chemical structures, biological origins, and biosynthetic pathways. Nevertheless, they are commonly classified based on a few key criteria.

Classification According to Biosynthetic Precursor

This is a widely used method that classifies alkaloids based on the amino acid from which they are biosynthesized as follows:

- *Ornithine-derived alkaloids*: Examples include pyrrolizidine and tropane alkaloids (atropine and cocaine).
- *Lysine-derived alkaloids*: Examples include piperidine and quinolizidine alkaloids. Piperine, a major alkaloid of black pepper, is a piperidine alkaloid.
- *Tyrosine-derived alkaloids*: This group includes isoquinoline alkaloids, for example, opium alkaloids (codeine, morphine, and papaverine).
- *Tryptophan-derived alkaloids*: This is a large and important group that includes indole, quinoline, and ergoline alkaloids. Examples are strychnine, quinine, and reserpine, respectively.
- *Phenylalanine-derived alkaloids*: Examples include ephedrine and mescaline.
- *Histidine-derived alkaloids*: Pilocarpine is an example.
- *Nicotinic acid-derived alkaloids*: Nicotine is a well-known example.

Classification According to Chemical Structure

This classification depends on the nature of the nitrogen-containing heterocyclic ring in the alkaloid structure.

- *Pyrrolidine alkaloids*: Characterized by a five-membered heterocyclic ring with a single nitrogen atom (e.g., nicotine).
- *Piperidine alkaloids*: Characterized by a six-membered ring containing one nitrogen atom (e.g., coniine, lobeline, pelleteirine, and piperine).
- *Tropane alkaloids*: They are derived from L-ornithine or arginine amino acids. This class is characterized by a bicyclic structure containing one nitrogen atom (e.g., atropine, scopolamine, and cocaine).
- *Indole alkaloids*: They are biosynthesized from tryptophan amino acid. Alkaloids belonging to this class are characterized by containing an indole ring system, which is a fusion of a benzene and a pyrrole ring. This class represents one of the major classes of alkaloids, which is further divided into different subclasses. The most common alkaloids belonging to this class include reserpine, ergotamine, and strychnine.

- *Quinoline alkaloids*: They are biosynthesized from tryptophan amino acid and characterized by a quinoline ring system (e.g., quinine and camptothecin).
- *Isoquinoline alkaloids*: They are biosynthesized from tyrosine amino acid. This group is characterized by an isoquinoline ring system. Morphine, codeine, papaverine, emetine, and tubocurarine are typical examples.
- *Steroidal alkaloids*: These alkaloids have a basic steroid nucleus. Examples include solanidine and solasidine from potato.
- *Terpenoidal alkaloids*: These alkaloids possess a terpenoid basic nucleus. Taxol is the most important example of this class.
- *Purine alkaloids*: These are not derived from amino acids but from purine nucleotides. They are characterized by a purine ring system and include caffeine, theobromine, and theophylline.

Classification Based on Type

Hegnauer proposed this classification, which divides alkaloids into three major groups depending on their chemical and biosynthetic characteristics:

- *True Alkaloids*: These compounds include a heterocyclic ring containing a nitrogen atom and are derived from amino acids. They are typically physiologically active and often have a bitter taste. Examples include morphine, nicotine, and atropine.
- *Protoalkaloids*: These compounds are also derived from amino acids, but the nitrogen atom is incorporated in a straight side chain rather than inserted in a heterocycle. They are often called biological amines. Examples include mescaline, capsaicin, and ephedrine.
- *Pseudoalkaloids*: These compounds contain a nitrogen atom within a heterocyclic ring, yet they are not biosynthesized from amino acids. Instead, their carbon skeleton is derived from other pathways, such as terpenoids or purines. Examples include caffeine and theobromine (purine based) and taxol (terpenoid based) (Aniszewski 2015). *Examples of different alkaloids belonging to different classes are illustrated in* Table 8.2.

8.2.2.2 Alkaloids Biological Activities

Alkaloids are considered potent bioactive compounds due to their vast array of biological activities, which have been extensively applied in both traditional medicine and modern treatment systems. They have the ability to exert potent biological action at low doses. They are widely used in pain relief, CNS stimulation, treatment of malaria, treatment of cancer, antimicrobial, and several other categories of treatment. Herein, a detailed discussion of the main pharmacological effects exerted by major alkaloids is provided.

Table 8.2 Examples of alkaloids belonging to different classes and their reported pharmacological activities

Alkaloid	Structure	Source	Biological activity	References
1. True alkaloids				
A. Pyridine alkaloids				
Nicotine	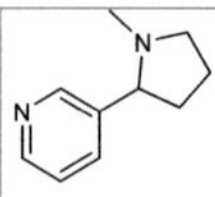 Nicotine	*Nicotiana tobacum* (Fam. Solanaceae)	CNS stimulant Behavior-modifying action	Benowitz (2009)
Arecoline	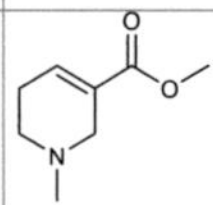 Arecoline	*Areca catechu* (Fam. Arecaceae)	M1 muscarinic receptor agonist, increases acetylcholine level in brain, so, brain function is activated Promote peristalsis and glandular secretions	Liu et al. (2016)
B. Piperidine alkaloids				
Piperine	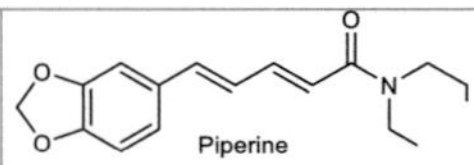 Piperine	*Piper nigrum* (Fam. Piperaceae)		
Lobeline	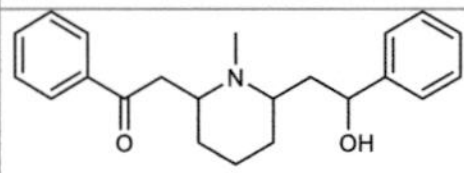 Lobeline	*Lobelia inflata* (Fam. Lobiliaceae)	Treatment of chronic bronchitis and asthmatic bronchitis	Stansbury (2013)
Pelletierine	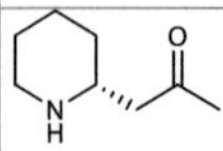 Pelletierine	*Punica granatum* (Fam. Lythraceae).	Antiparasitic Anthelmintic	Al-Shahwany et al. (2013)
C. Tropane alkaloids				
Atropine	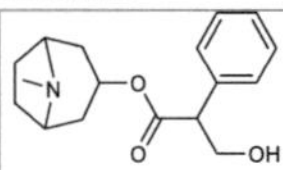 Atropine	*Atropa belladonna* (Fam. Solanaceae)	Atropine is anticholinergic It reduces the secretion such as sweat, saliva, and gastric juice It exerts tachycardia by blocking M2 receptors on the SA node, thereby reducing vagal tone	Aniszewski (2015)
Scopolamine (Hyoscine)	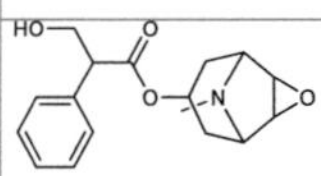 Scopolamine	*Hyocyamus niger* (Fam. Solanaceae)	It is considered nonselective muscarinic antagonist It produces sedative, antiemetic, and amnestic effects	Aniszewski (2015)

(continued)

Table 8.2 (continued)

Alkaloid	Structure	Source	Biological activity	References
Cocaine	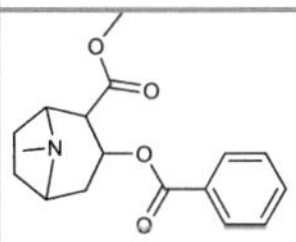 Cocaine	*Erythroxylon coca* (Fam. Erythroxylaceae)	It exerts sympathetic effect which cause CNS stimulation It is used as local anesthetic	Debnath et al. (2018)
D. Indole alkaloids				
Brucine	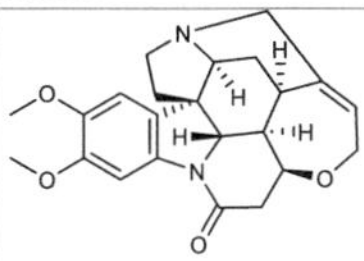 Brucine	*Strychnos nux-vomica* (Fam. Loganiaceae)	Analgesic and anti-inflammatory effects as it reduces 5-hydroxytryptamine (5-HT) level while increases 5-hydroxytryindole-3-acetic acid level	Chen et al. (2014)
Ergotamine	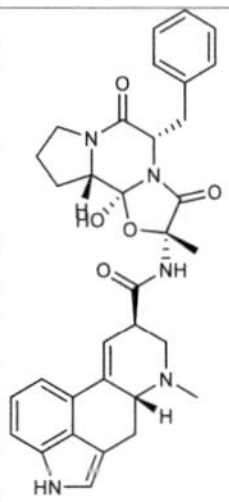 Ergotamine	*Claviceps purpurea* (Fam. Clavicipitaceae)	Ergotamine is a vasoconstrictor of peripheral and cranial blood vessels α-Adrenoreceptor antagonist with additional effects on tryptaminergic and dopaminergic receptors, leading to depression of central vasomotor centers	De Costa (2002), Duringer et al. (2005)
Reserpine	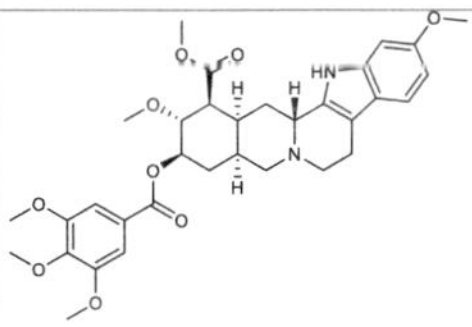 Reserpine	*Rauwolfia serpentine* (Fam. Apocyanaceae)	Antihypertensive	Strawbridge et al. (2023)
E. Quinoline alkaloids				
Quinine	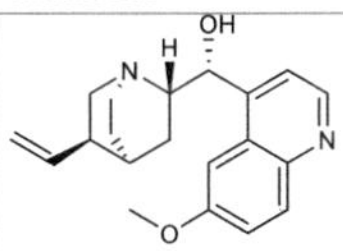 Quinine	*Cinchona succirubra*, *C. ledgeriana*, *C. officinalis* (Fam. Rubiaceae)	Effective antimalarial drug against *Plasmodium vivax* and *Plasmodium malariae*	Achan et al. (2011)
Camptothecin	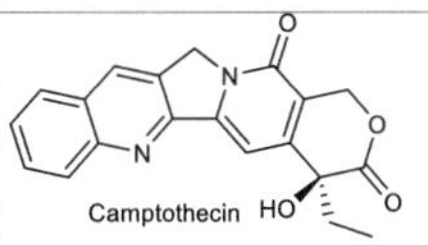 Camptothecin	*Camptotheca acuminata*	It is used as antitumor as it inhibits DNA topoisomerase 1	Martino et al. (2017)

(continued)

Table 8.2 (continued)

Alkaloid	Structure	Source	Biological activity	References
F. Isoquinoline alkaloids				
Morphine	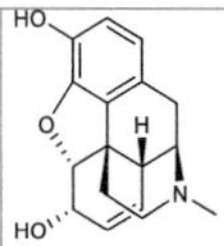 Morphine	*Papaver somniferum* (Fam. Papaveraceae)	Potent analgesic	Debnath et al. (2018)
Papaverine	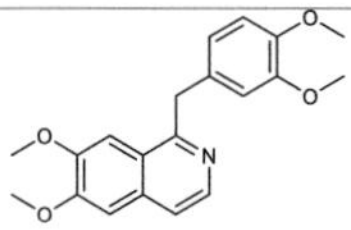 Papaverine		Smooth muscle relaxant Vasodilator	Shimizu et al. (2000)
Codeine	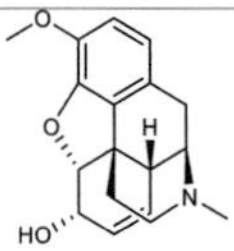 Codeine		Antitussive Mild analgesic	Debnath et al. (2018)
Berberine	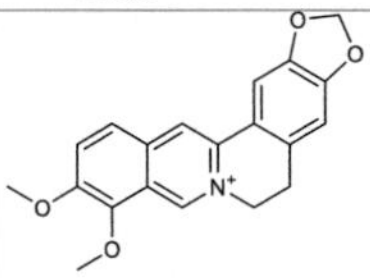 Berberine	*Berberis aristata*, *B. lyceum*, *B. petiolaris* (Fam. Berberidaceae).	Antidiabetic Antioxidant Anti-inflammatory Anti-hypertensive Hepatoprotective Antidepressant	Turner et al. (2008), Shanbhag et al. (1970)
Emetine	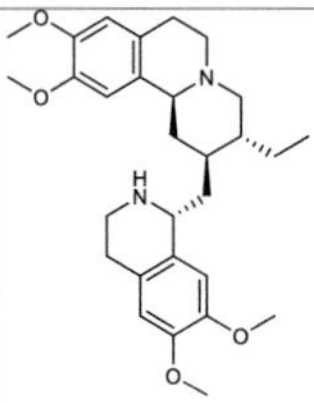 Emetine	*Psychotria ipecacuanha* (Fam. Rubiaceae).	It disrupts ribosomal and mitochondrial protein synthesis and impairs DNA and RNA synthesis and function. Therefore, it possesses a vital role in many biological activities such as antiviral, anticancer, antiparasitic, and contraceptive	Akinboye and Bakare (2011)
G. Imidazole alkaloid				
Pilocarpine	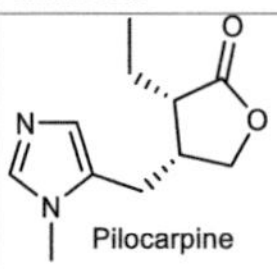 Pilocarpine	*Pilocarpus microphyllus* (Fam. Rutaceae)	Parasympathomimetic Glaucoma treatment (decrease intraocular pressure) Sialagogue (treatment of xerostomia throat cancer caused by chemotherapy) Cardiovascular effects: small doses decrease blood pressure, while higher doses elevate blood pressure	Santos and Moreno (2004)

(continued)

Table 8.2 (continued)

Alkaloid	Structure	Source	Biological activity	References
H. Steroidal alkaloids				
Solanidine	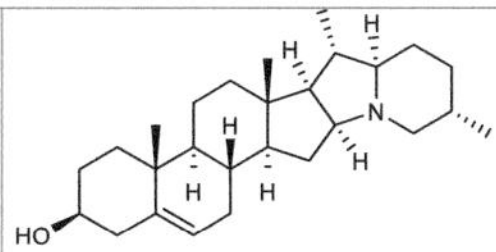 Solanidine	*Solanum tuberosum* (Fam. Solanaceae)	Cytotoxic against different tumor cell lines including human hepatocarcinoma (HepG2), human large intestine cancer cell line (LS-174), and human gastric carcinoma cell line (SGC-7901) Starting material for the synthesis of steroidal drugs	Hameed et al. (2017)
I. Terpenoidal alkaloids				
Taxol	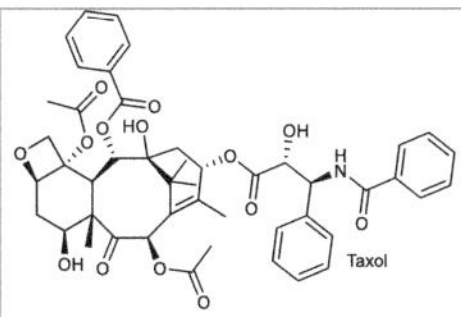	*Taxus bravifolia* (Fam. Taxaceae)	Treatment of ovarian and breast cancer	Malonga et al. (2005)
2. Proto alkaloids				
Ephedrine	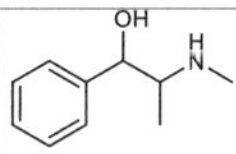 Ephedrine	*Ephadra sinica*, *E. vulgaris*, *E. major* (Fam. Ephadraceae)	Sympathetic stimulant directly acts on α and β receptors It is used to prevent hypotension during spinal anesthesia and also serves as a bronchodilator in asthma	Ma et al. (2007)
Capsaicine	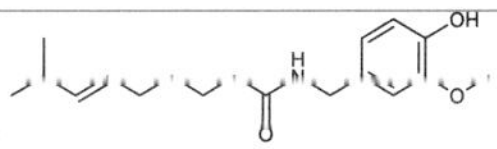 Capsaicine	*Capsicum annum* (Fam. Solanaceae)	It can be used as analgesic as it bonds to transient receptor potential vanilloid 1 (TRPV1), which is expressed in the sensory neurons Improve digestion, used for weight loss	Reyes--Escogido et al. (2011)
Colchicine	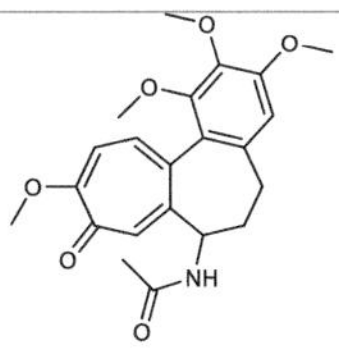 Colchicine	*Colchicum autumnale* (Fam. Liliaceae)	Potent anti--inflammatory for the treatment of pericarditis and postsurgical atrial fibrillation Management and prophylaxis of acute gout and Hereditary Mediterranean Fever	Slobodnick et al. (2015)

(continued)

Table 8.2 (continued)

Alkaloid	Structure	Source	Biological activity	References
3. Pseudo alkaloids				
Caffeine	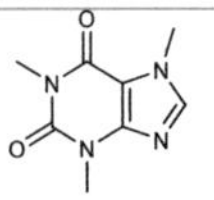 Caffeine	*Theobroma cocoa* (Fam. Malvaceae) *Thea sinensis* (Fam. Theaceae)	Analgesic, anorectant and CNS stimulant It acts on the peripheral adenosine receptor (A1), inhibits adenylate cyclase activity, and suppresses lipolysis	Cappelletti et al. (2015)
Theobromine	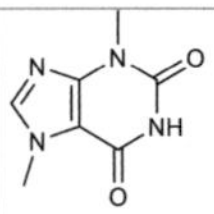 Theobromine	*Theobroma cocoa* (Fam. Malvaceae)	CNS stimulant Diuretic Mood enhancement	Salihović et al. (2014)
Theophylline	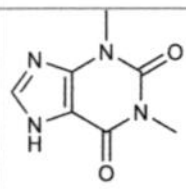 Theophylline	*Thea sinensis* (Fam. Theaceae)	Respiratory tract smooth muscles relaxant Diuretic	Salihović et al. (2014)

Analgesic and Anesthetic Activities

- *Morphine*: It belongs to the opium analgesics, which are extracted from *Papaver somniferum* (opium poppy). It works on opioid receptors in the CNS and hence it interferes with the relay of pain signals and the experience of pain.
- *Codeine*: It is another opium alkaloid, which is derived from the opium poppy, and is reported to exert antitussive activity in addition to mild-to-moderate analgesia (Debnath et al. 2018).
- *Mitragynine*: It is abundant in *Mitragyna speciosa*, fam. Rubiaceae. Mitragynine exerts antinociceptive activity as it acts on CB1cannabinoid receptor and modulates opioid receptors (Shamima et al. 2012).
- *Aconitine*: It occurs in all *Aconitum* species, fam. Ranunculaceae. Aconitine, the C_{19} diterpenoid alkaloid, exerts potent antinociceptive activity in different analgesic assays as it targets site II of Na^+ channels (Anwar et al. 2003).

Antimalarial Activity

- *Quinine*: It is derived from cinchona bark. It was the first antimalarial drug discovered in early 1600s, with rapid schizonticidal activity against intra-erythrocytic parasites and gametocytocidal effects against both *Plasmodium vivax* and *Plasmodium malariae* (Achan et al. 2011).
- *Artemisinin*: Artemisinin and its derivatives are isolated from *Artemisia annua*, the sweet wormwood plant. It exerts a powerful antimalarial activity mediated via reactive oxygen species generation that kill the malaria parasite (Tu 2011).

Anticancer Activity

- *Vincristine* and *Vinblastine*: These alkaloids, which are obtained from Madagascar periwinkle (*Catharanthus roseus*), are used in the treatment of numerous cancer types including breast cancer, Kaposi's sarcoma, Hodgkin's disease, and testicular cancer. Vincristine can specifically cure acute childhood lymphoblastic leukemia. The antitumor activity of vincristine and vinblastine is similar as they both hinder the ability of tubulin to form microtubules, leading to microtubular crystallization, mitotic arrest, and cancer cells' apoptosis (Kumar 2016).
- *Taxol (Paclitaxel)*: It is isolated from the bark of *Taxus brevifolia*, the Pacific yew tree. Taxol has been reported to stabilize microtubules and hence prevent depolymerization, as part of cell division, leading to cancer cells' apoptosis (Malonga et al. 2005).
- *Sanguinarine*: It is isolated from the root of *Argemone mexicana* and *Sanguinaria canadensis*, family Papaveraceae. It is used as chemotherapeutic agent that may cause cell apoptosis through its effect on the Na+-K+-ATPase pump. Therefore, it is considered an essential candidate for the development of successful immunotherapeutic approaches for selected carcinomas (Malikova et al. 2006).
- *Pancratistatin*: It is naturally found in all Amaryllidaceae family plants that are usually called "Spider lily." Pancratistatin exhibits strong apoptotic activity against a wide range of cancer cell lines. Remarkably, it is noncytotoxic to normal cells while selectively targeting cancer cells (Griffin et al. 2011).
- *Cynarin*: It is isolated from leaves and flower buds of *Cynara cardunculus*, fam. Asteraceae. Cynarin exerts significant antiproliferative and antioxidant effects (Velez et al. 2012).

Antibacterial and Antiviral Activities

- *Berberine*: It is isolated from goldenseal plant (*Hydrastis canadensis*) and barberries (*Berberis vulgaris*). It exerts its antimicrobial activity against several types of bacteria by interfering with bacterial cell membrane and inhibition of DNA synthesis. Moreover, berberine has shown promising antiviral activity via inhibition of viral replication of SARS-CoV-2, the causative agent of COVID-19 infection (Chu et al. 2014).
- *Sanguinarine*: It is isolated from bloodroot (*Sanguinaria canadensis*) and reported to possess good antibacterial activity. It is used in some oral preparations to treat gingivitis (Malikova et al. 2006).
- *Stachydrine*: It was isolated from the flower of *Chrysanthemum morifolium* family Asteraceae. It possesses potent antibacterial activity against *Staphylococcus aureus*, *Bacillus proteus*, *Sonne bacillus*, *B. paratyphosus*, *B. typhosus*,

β-hemolytic streptococci, *Pseudomonas aeruginosa*, and *Vibrio cholerae* (Taiwo et al. 2013).

- *Ricinine*: It is one of the main toxic alkaloids found in the seeds of *Ricinus communis* (castor bean) from the family Euphorbiaceae. At low doses, it exhibits significant antibacterial activity against *Staphylococcus aureus*, *Escherichia coli*, *Salmonella newport*, *Bacillus subtilis*, *Streptococcus pyogenes*, *Pseudomonas aeruginosa*, *Serratia marcescens*, *Shigella flexneri*, *Klebsiella pneumonia*, and *Proteus vulgaris* (Nath et al. 2011).
- *Bufotenine*: It is isolated from *Amanita muscaria* (fly agaric mushroom) which belongs to the Amanitaceae family. It exerts a dose-dependent antiviral activity against rabies virus infection on mammalian cultured cells. It shows synergistic antiviral activity with a synthetic tetrapeptide based on ocellatin-F1, a natural antimicrobial peptide, mimicking the rabies virus glycoprotein region, and prevents rabies symptoms in 40% of infected mice (Vigerelli et al. 2018).

Cardiovascular Activity

- *Reserpine*: It is isolated from the Indian snakeroot *Rauwolfia serpentina*, fam. Apocynaceae. It is utilized in the management of hypertension. It acts by removing catecholamines and serotonin from nerve endings, suppressing sympathetic response, and consequently reducing blood pressure (Strawbridge et al. 2023).
- *Quinidine*: It is a stereoisomer of quinine. It is mainly used as an antiarrhythmic as it inhibits sodium influx, increases the stability of the cardiac membrane, as well as stabilizes heart rhythm.
- *Sparteine*: It is isolated from *Lupinus luteus* and *Cystisus scoparium* belonging to family Fabaceae. It exhibits marked antiarrhythmic activity (Tasso et al. 2023).
- *Veratramine*: It is isolated from *Veratrum album* and *Veratrum viride* rhizomes (family Liliaceae) and serves as an effective antihypertensive agent. Its activity is mediated through reflex suppression of the heart and carotid sinus pressor receptors (Pelletier and Pelletier 1983).

Respiratory Activity

- *Ephedrine*: It is present in *Ephedra* species. Ephedrine has been widely employed to treat asthma and bronchitis. It acts on α and β adrenergic receptors, which cause vasodilatation and relaxation of bronchial smooth muscle that result in undisturbed airway (Ma et al. 2007).
- *Codeine*: It is principally used as powerful antitussive as it depresses respiratory center in a dose-dependent manner besides its pain relieving property (Debnath et al. 2018).

- *Lobeline*: It is identified in herbs and seeds of *Lobelia inflate*, family Lobeliaceae. It exerts significant activity in the management of asthmatic and chronic bronchitis by targeting pulmonary sensory endings, stimulating serotonin and phenyldiguanide receptors, inhibiting dopamine uptake, and promoting dopamine release from presynaptic storage vesicles (Stansbury 2013).
- *Noscapine and Narceine*: They are opium alkaloids isolated from *Papaver somniferum*, family Papaveraceae. Noscapine suppresses cough through a central action on the medullary cough center (Kumar et al. 2022). Narceine exerts both analgesic and antitussive activities (Viesca and Gómez 2023).
- *Theophylline*: It is isolated from *Camellia sinensis* leaves (fam. Theaceae). It acts as a phosphodiesterase inhibitor that exerts respiratory tract smooth muscles relaxant effect and alleviates the contraction of diaphragm (Salihović et al. 2014).

Central Nervous System Activity

- *Galantamine*: It is produced from bulbs of specific types of *Galanthus* species. It is chiefly employed in treating Alzheimer's disease (AD). Moreover, it is an acetylcholinesterase inhibitor that enhances acetylcholine level in brain, consequently enhancing cognition (Janssen and Schäfer 2017).
- *Physostigmine*: It is extracted from the seeds of Calabar bean or *Physostigma venenosum*. It is used to treat Alzheimer's disease via competitive antagonism with acetylcholinesterase, which in turn potentiates cholinergic transmission and improves cognition. Additionally, it can be used for the management of glaucoma (Arens et al. 2018).
- *Huperzine A*: It is an alkaloid obtained from *Huperzia serrata*. It is explored for its potential to improve cognitive function and suppress the progression of Alzheimer's disease (Debnath et al. 2018).
- *Catuabine*: It is obtained from *Trichilia catigua* bark. It has four derivatives: catuabine A, B, C, and D. It exerts antidepressant-like effects (Aniszewski 2015).
- *Cuscohygrine*: It is isolated from *Atropa belladonna*, *Datura innoxia*, or *Mandragora officinarum*. It exerts GABA-like activity resulting in a decrease in neuronal activity, inhibition of nerve cells from hyperstimulation, and creation of antistress activity (Okhunov et al. 2016).
- *Conessine*: It is a steroid alkaloid present in the bark and dried seeds of *Holarrhena antidysenterica*, fam. Apocynaceae. It acts as an H3-receptor antagonist, blocking histamine action and helping to alleviate depression (Santora et al. 2008).
- *Caffeine*: Present in coffee beans, tea, cocoa, and other sources. Caffeine acts on the central nervous system via targeting adenosine receptors to overstimulate neurons to release dopamine and norepinephrine, neurotransmitters that are associated with wakefulness and alertness (Cappelletti et al. 2015).

- *Theobromine*: It is isolated from *Theobroma cacao*, fam. Malvaceae. Theobromine acts by inhibiting phosphodiesterases and blocking adenosine receptors, leading to CNS stimulation (Salihović et al. 2014).
- *Nicotine*: It is found in tobacco plants. Nicotine elicits the release of dopamine and acetylcholine neurotransmitters by interacting with nicotinic acetylcholine receptors. This leads to enhanced concentration, alertness, and overall feeling of wellbeing (Benowitz 2009).
- *Mescaline*: It is reported in various types of cacti, for example, *Lophophora williamsii*, family Cactaceae. Mescaline acts on serotonin receptors as a nonselective agonist. Serotonin receptors in the CNS include both G-protein–coupled receptors and ligand-gated ion channels, mediating neurotransmission through the release of neurotransmitters and hormones. They play key roles in regulating mood, memory, learning, cognition, anxiety, sleep, and nausea. Therefore, they are considered targets of therapeutic agents and illicit drugs, including hallucinogens (Kovacic and Somanathan 2009).
- *Psilocybin*: It is the major psychoactive alkaloid isolated from some species of mushrooms, for example, *Psilocybe mexicana*, family Strophariaceae. It also induces sympathomimetic effect (Passie et al. 2002).
- *Lysergic acid*: It is isolated from rye ergot fungus, *Claviceps purpurea*, fam. Clavicipitaceae. It is used to synthesize lysergic acid diethylamide (LSD); a semisynthetic drug derived from lysergic acid. LSD can significantly exert acute sympathomimetic and neurocognitive effects and alter the state of consciousness (Passie et al. 2008).

Gastrointestinal Activity

- *Papaverine*: It is a naturally occurring opium alkaloid, which is well known for its use as gastrointestinal smooth muscles relaxant. This action is mediated by the inhibition of phosphodiesterase and subsequent relaxation of smooth muscles (Shimizu et al. 2000).
- *Atropine*: It is isolated from *Atropa belladonna*. It exerts its gastrointestinal activity by the prevention of smooth muscles spasms and secretion (Aniszewski 2015).
- *Conessine*: It is a steroidal alkaloid found in the dried seeds and bark of *Holarrhena antidysenterica*, fam. Apocynaceae. It exhibits antidysenteric, antidiarrheal, and antiamoebic activities against *Salmonella enteritidis*, *Salmonella typhimurium*, *Escherichia coli*, *Vibrio cholerae*, and *Shigella flexneri*. (Ballal et al. 2000).
- *Boldine*: It is an aporphine alkaloid that is isolated from Boldo leaves, *Peumus boldus*, fam. Monimiaceae. It exerts hepatoprotective activity by stimulating the transcriptional regulator of the bile salt export pump through its interaction with the farnesoid X receptor (Cermanova et al. 2015).

Antidiabetic Activity

- *Berberine*: It is isolated from *Berberis aristata*, *B. lyceum*, and *B. petiolaris*. It activates adenosine monophosphate (AMP)-activated protein kinase (AMPK) and enhances insulin sensitivity (Turner et al. 2008).
- *Hygrine*: It strongly inhibits α/β-glucosidase, α/β-mannosidase, and β-galactosidase enzymes (Gutiérrez-Grijalva et al. 2021).

8.2.3 Phenolic Compounds

Plant polyphenols are considered among the most ubiquitous and broadly distributed chemical components. They have attracted significant attention because they are naturally found in various sources, including vegetables, fruits, cereals, tea, coffee, and many other plants, and have important medicinal applications. Plant polyphenols are usually categorized into phenolic acids, flavonoids, anthocyanins, tannins, and phenylpropanoids. These compounds share phenolic rings as the basic nucleus from which a vast array of variable structures can arise. Phenolic compounds have numerous functions in the plant, especially providing a defense mechanism against pathogens and predators. The biological activities of polyphenols have been extensively studied through various in vitro and in vivo experiments, demonstrating their potential to promote health and prevent, delay, or mitigate many chronic disorders. Polyphenols are particularly noted for their antioxidant, anti-inflammatory, and anticancer properties (Ghiringhelli et al. 2012). They also have an ameliorative effect on severe health conditions such as diabetes, osteoporosis, and cardiovascular and neurodegenerative disorders (García-Conesa and Larrosa 2020). Other studies have established that polyphenols exert beneficial actions on intestinal health, and hence, they are considered significant prebiotics that can reduce inflammation and prevent cancer by modulating the intestinal flora (Thilakarathna et al. 2018). Moreover, their natural ubiquity and biocompatibility make them excellent candidates for addition to functional foods and for exerting their positive impacts on human health. Functional foods rich in polyphenols play a crucial role in preventing various chronic illnesses, including diabetes, hypertension, and cancer. Rising public awareness of health and wellness is driving increased demand for polyphenols (Zhang et al. 2022b). Herein, a detailed overview of the most significant reported biological activities is provided.

8.2.3.1 Biological Activities of Polyphenols

Antioxidant Activity

Plant polyphenolics generally possess remarkable antioxidant activity owing to their specific structures that contain many hydroxyl groups. Polyphenols as antioxidants act by capturing free radicals, hence protecting DNA from oxidative damage and preventing several diseases. Much research has investigated the antioxidant activity of different phenolic compounds. A randomized clinical trial involving 308 participants treated with curcumin was conducted to evaluate its antioxidant potential. The results showed that curcumin has significant antioxidant activity and can reduce malondialdehyde (MDA) levels. Additionally, the antioxidant capacities of glutathione (GSH), ascorbic acid, and catechins were compared. Catechins demonstrated the strongest scavenging activity against 2,2′-azino-bis (3-ethylbenzothiazol ine-6-sulfonic acid) (ABTS) radicals, the highest reducing equivalents for trivalent iron ions, and effectively prevented dihydrorhodamine oxidation. The excellent antioxidant properties of catechins support them as ultimate candidates as antioxidants for oxidative stress prophylaxis and therapy (Grzesik et al. 2018). Furthermore, many other polyphenolic compounds are well-known as powerful antioxidants, such as resveratrol, gallic acid, and daidzein (Canedo-Santos et al. 2022).

Anti-inflammatory Activity

Plant polyphenolic compounds exhibit significant inhibitory effects on various inflammatory cells, either by modulating cytokine signaling or controlling cytokine release. Rutin, in particular, has demonstrated notable anti-inflammatory activity, comparable to standard drugs (Himesh Soni et al. 2013). Hesperidin was also evaluated for its anti-inflammatory activity in vitro using RAW264.7 cells and in vivo in a CCl_4-induced acute liver injury model. It was found to significantly suppress tumor necrosis factor (TNF-α), interleukin-6 (IL-6), and nitric oxide (NO) in both in vitro and in vivo studies (Ding et al. 2017). Similarly, numerous polyphenolic compounds have been shown to possess strong anti-inflammatory activity, including curcumin, resveratrol, baicalin, and silybin (Stanca et al. 2013; Xu et al. 2021).

Anticancer Activity

Polyphenols possess good anticancer activity against many kinds of cancer. They act by inhibiting tumor propagation and exerting lethal effects on cells, inducing apoptosis. Resveratrol has been reported to inhibit cell proliferation, induce apoptosis, and exhibit antioxidant properties, contributing to the prevention of cancer and related diseases (Lee and Lee 2021; Fayez et al. 2022). Quercetin has been reported to help prevent and treat esophageal cancer (Davoodvandi et al. 2020). Additionally, silymarin has been shown to provide both prophylactic and curative effects against liver diseases, and it can induce apoptosis in liver cancer cells. Curcumin as well as epigallocatechin gallate (EGCG) exhibit significant anticancer activity against breast cancer (Liu and Ho 2018; Bimonte et al. 2020).

Antimicrobial Activity

Polyphenols also exert potent antibacterial activity on a wide range of microorganisms. Flavonoids exclusively showed greater antibacterial activity compared to other polyphenolic compounds among other biological activities (Zhang et al. 2022b). Curcumin-loaded chitosan film showed significant antibacterial activity against *Rhizopus solani* and *Staphylococcus aureus*. Tea polyphenols such as rutin and silymarin have been extensively reported and confirmed to possess antibacterial activities (Yang and Zhang 2019).

Pro-Oxidant Effect

Although polyphenols exhibit strong antioxidant activity, some may result in DNA damage or induce apoptosis at high doses. Resveratrol, for instance, has been reported to have pro-oxidant effects, accelerating aging and reducing brewer's yeast chronological lifespan (Canedo-Santos et al. 2022). Similarly, gallic acid induces oxidative loss of amine and thiol groups, affecting protein structure and function (Cao et al. 2016). High doses of soybean in pig feed induced oxidative changes in the liver, fat, and plasma and after 64 days, increased superoxide dismutase (SOD) activity and total antioxidant capacity, indicating a potent pro-oxidant effect (Chen et al. 2016). These findings suggest that proper dosage is crucial when using polyphenolic compounds for medicinal purposes.

Antidiabetic Activity

A diet rich in polyphenols may help reduce the risk of developing diabetes. Research studies have shown that polyphenols modulate the insulin pathway and enhance insulin sensitivity in peripheral tissues (Dominguez Avila et al. 2017). Other polyphenolic compounds have been found to inhibit the activity of the enzymes α-amylase and α-glucosidase, which in turn control the intestinal absorption of glucose and preserve normal blood sugar level (Vayalil 2012; Yehia et al. 2023). Tea, which is rich in catechins, showed excellent antioxidant and antidiabetic activities. Quercetin has been recognized as an antidiabetic agent, demonstrating both preventive and therapeutic effects against diabetes in in vitro and in vivo models (Eid and Haddad 2017). Thus, polyphenolic compounds hold significant promise in preventing and managing the development of diabetes.

Antihypertensive Activity

Cocoa, a well-known source of polyphenols, is abundant in catechins and proanthocyanidins, which can significantly reduce oxidation of low-density lipoproteins, improve endothelial function, and promote vasodilation (Williamson 2017). Phenolic acids, flavanols, resveratrol, anthocyanins, tannins, and other polyphenolics exert vasodilation effect and other beneficial properties of blood pressure control (Huang et al. 2013). Curcumin has been identified as an antihypertensive agent that may be useful for both preventing and treating hypertension. Vasodilatory effects were investigated for curcumin, amlodipine, and curcumin-amlodipine combination on isolated rat aortic rings and results showed that hypertensive patients treated with amlodipine could consume curcumin or turmeric without diminishing the antihypertensive effects of amlodipine (Lee et al. 2021).

Antiobesity Activity

Polyphenols can combat obesity through different mechanisms such as promoting lipolysis and fat oxidation, inhibiting adipocyte differentiation, and enhancing adipocyte apoptosis. Green tea is considered a rich source of catechins and gallocatechin gallate, besides other valuable polyphenolic compounds, which suggests its significant role in obesity control (Sung et al. 2010). Through different mechanisms, soybean also exerts an antiobesity effect via suppressing the action of pancreatic lipase and pancreatic protein lipase, which results in preventing adipocyte proliferation and stimulating hormone-sensitive lipase to initiate lipolysis (Guo et al. 2009). Additionally, quercetin has been reported to strongly inhibit adipose tissue accumulation in obese rats, indicating promising antiobesity effects (Ting et al. 2018).

Antiatherosclerotic Activity

Plant polyphenols can exert protective activity of the cardiovascular system via different mechanisms including decreasing low-density lipoprotein, increasing high-density lipoprotein, and preventing the oxidation of low-density lipoprotein. Notably, resveratrol and ellagic acid have been reported to exert significant antiatherosclerotic activity by enhancing endothelial barrier function (Bahramsoltani et al. 2019). In addition, EGCG can display certain antiatherosclerotic activities via specific mechanism of action (Tanaka et al. 2012).

8.2.3.2 Classification of Plant Polyphenols

Polyphenols are compounds characterized by a benzene ring bearing at least two phenolic hydroxyl groups. They are commonly categorized into simple phenols, which include phenolic acids and coumarins, and complex polyphenols, which

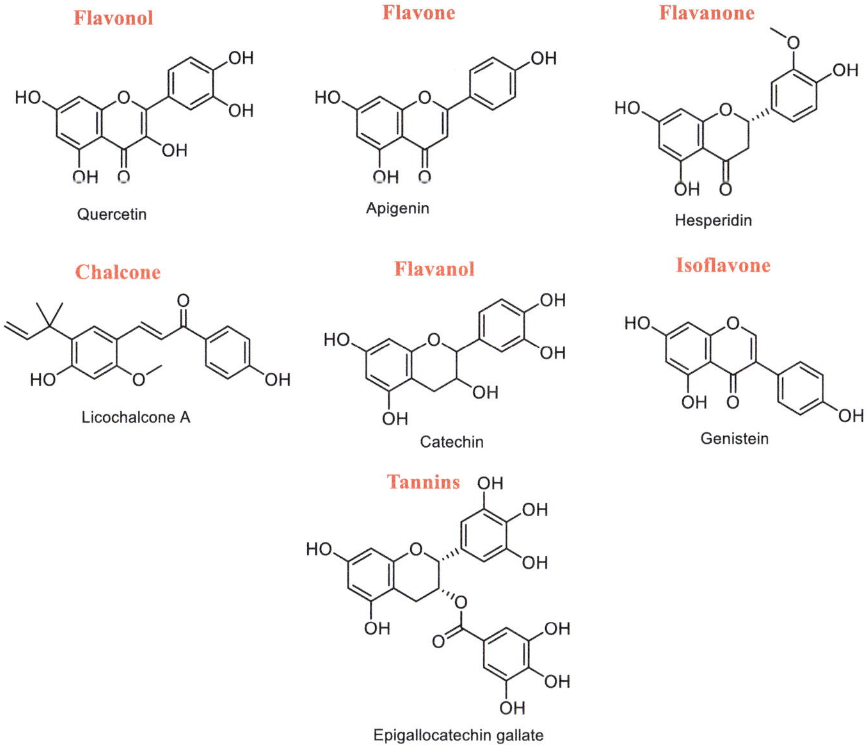

Fig. 8.5 Phenolic compounds belonging to different classes

encompass flavonoids, hydrolyzable and condensed tannins, lignans, and anthocyanins according to their structural characteristics (Goleniowski et al. 2013).

Flavonoids

Flavonoids represent the major class of phenolic compounds and are found predominantly as glycosides in plant cells. Structurally, they share a basic C6–C3–C6 skeleton composed of three cyclic rings (El Khodary et al. 2024). Based on differences in their chemical structures, flavonoids are further classified into flavanones, flavanols, flavones, flavonols, isoflavones, and anthocyanins (Fig. 8.5) (George et al. 2025). Flavanols are widely recognized for their strong antioxidant activity. They mainly include catechins and proanthocyanidins. Flavonols, such as quercetin, are widely reported to possess significant antioxidant and anti-inflammatory properties. Flavones, such as apigenin and luteolin, are best known for their antioxidant and anti-inflammatory activities. Flavanones, predominantly occurring in citrus fruits, possess notable antioxidant properties. Finally, isoflavones, known for their estrogenic activity, are mainly found in soy products (Li and Beta 2013). Examples of compounds belonging to various classes of flavonoids are listed in Table 8.3.

Table 8.3 Examples of different flavonoids and their biological activities

Flavonoids	Structure	Pharmacological activity	References
Flavonols			
Quercetin	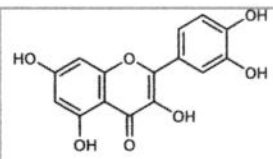	Antioxidant Antidiabetic Anti-inflammatory Anticancer Immunoprotective	Di Petrillo et al. (2022)
Rutin	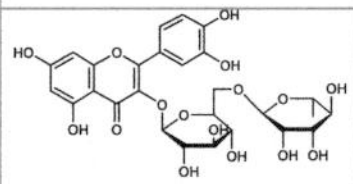	Anti-inflammatory Antiallergic Antiviral Antioxidant	Semwal et al. (2021)
Myricetin	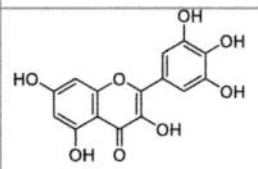	Anti-inflammatory Antitumor Antimutagenic Anti-dental caries Antioxidant	Imran et al. (2021)
Fisetin	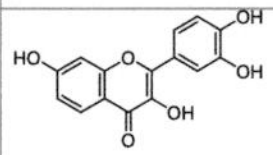	Antioxidant Anti-inflammatory Antispasmodic Treatment of diabetic kidney injury Anticoagulant and antithrombotic	Naeimi and Alizadeh (2017)
Baicalin	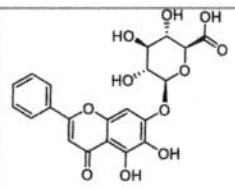	Anti-inflammatory Antibacterial Antithrombotic Anti-hyperlipidemic Diuretic Asthma relief Hemostasis Detoxification	Ibrahim et al. (2022)
Baicalein 5,6-dimethyl ether	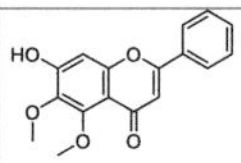	Neuroprotective activity Antiaging activity	Brinza et al. (2024), Ayoub et al. (2024)
Morin	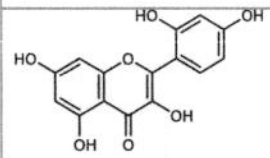	Anti-inflammatory Immunomodulatory Antitumor Antioxidant	Thakur et al. (2020)
Flavones			
Apigenin	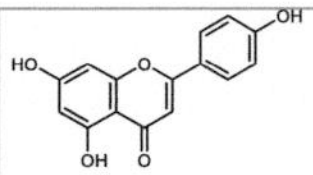	Anticancer Antiviral Anti-inflammatory Antioxidant Antihypertensive Sedative	Zhou et al. (2017)

(continued)

Table 8.3 (continued)

Flavonoids	Structure	Pharmacological activity	References
Luteolin	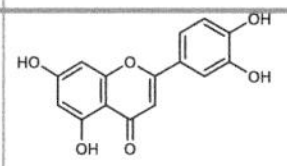	Antioxidant Anti-inflammatory Antiallergy Antitumor Uric acid lowering Treatment of demyelinating disease	Manzoor et al. (2019)
Tangeretin	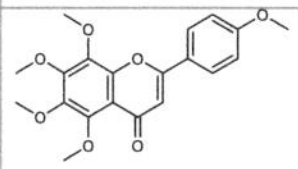	Antifungal effect Antitumor Smooth muscles relaxation	Singh et al. (2020)
Flavanones			
Liquiritigenin	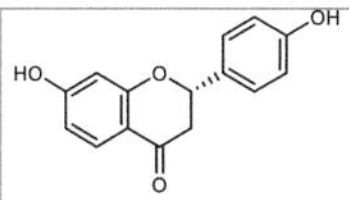	Antibacterial Antiulcer Antispasmodic Hepatocyte monoamine oxidase inhibitor	Ramalingam et al. (2018)
Hesperidin	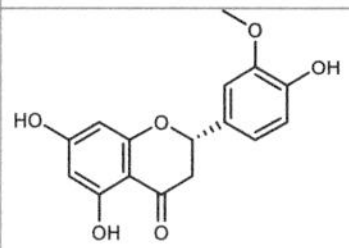	Antiallergy Anti-inflammatory Lower blood pressure Antibacterial Treatment of hepatitis Treatment of lower bone density Antitumor	Li and Schluesener (2017)
Isoflavones			
Daidzein	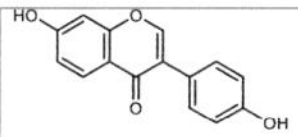	Antioxidant Estrogenic effects	Mayo et al. (2019)
Genistein	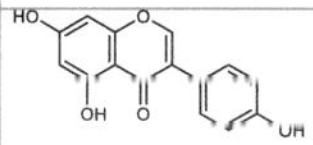	Antioxidant Estrogenic properties Anticancer activity	Leis et al. (2021)
Biochanin A	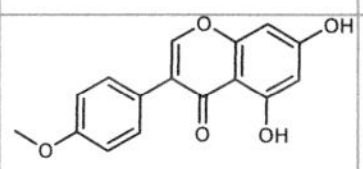	Estrogen-like effect Antifungal Antitumor Anti-hyperlipidemic	Raheja et al. (2018)

Flavonoids in plants play an essential role in growth, progress, flowering, and fruiting. In addition, they exert valuable effects on the human body owing to their diverse biological activities, which include antioxidant, anti-inflammatory, antibacterial, and anticancer effects, besides disease prevention potential (Zhang et al. 2022b).

Phenolic Acids

Phenolic acids are an important subclass of phenolic compounds and are commonly distributed throughout the plant kingdom. Phenolic acids occur in plants as conjugates with other natural compounds such as terpenes, larger polyphenols (flavonoids), structural constituents of the plant (cellulose, proteins, and lignin), or different organic molecules (glucose, quinic, maleic, and tartaric acids). They are considered potent antioxidants, which protect the human body against oxidative damage disorders including stroke, CVDs, and cancers when consumed in vegetables and fruits. Phenolic acids, or phenol carboxylic acids, possess a phenolic ring and at least one carboxylic acid group. They are classified according to the carbon atoms in the side chain as C6–C3, C6–C2, and C6–C1 compounds, with the most prominent being C6–C3 (hydroxycinnamic acid [HCA] derivatives) and C6–C1 (hydroxybenzoic acid derivatives). Both categories are biosynthesized from the shikimate pathway. Phenolic acids share a common basic skeleton, differing only in the number and position of hydroxyl groups attached to the aromatic ring. Most phenolic acids in plants are found as derivatives of cinnamic and benzoic acids, while only a minor portion occurs in the free acid state. Owing to their widespread presence in plants, phenolic acids have been extensively studied in chemical, biological, agricultural, and medical research (Goleniowski et al. 2013).

Hydroxybenzoic Acids

Hydroxybenzoic acids, characterized by a C6–C1 skeleton, are derivatives from benzoic acid, such as *p*-hydroxybenzoic, salicylic, 2,3-dihydroxybenzoic, 2,5-dihydroxybenzoic, gallic, and vanillic acids.

Hydroxycinnamic Acids

Hydroxycinnamic acids comprise *p*-coumaric, caffeic, sinapic, and ferulic acids, as well as their esterified conjugates, such as chlorogenic acid (5-O-caffeoylquinic acid) and ferulic acid esters, which represent the most common naturally occurring HCA. Hydroxycinnamic acids are biosynthesized from *p*-coumaric acid as starting material that undergoes hydroxylation and methylation reactions to produce different derivatives. Hydroxycinnamates occur in nature in various forms, including monomers, dimers, esters, monosaccharides, disaccharides, and polymers. They can also exist as amides, formed by esterification with amino acids and amines, primarily found in coffee and cocoa (Goleniowski et al. 2013). Different phenolic acids' structures are illustrated in Fig. 8.6.

Functions of Phenolic Acids in Plants
Phenolic acids are associated with diverse plant functions, such as photosynthesis, enzyme activity, protein biosynthesis, nutrient uptake, structural support, and allelopathy. Moreover, simple phenolic acids serve as important signaling molecules in plant-microbe interactions. For example, methyl salicylate is synthesized at the

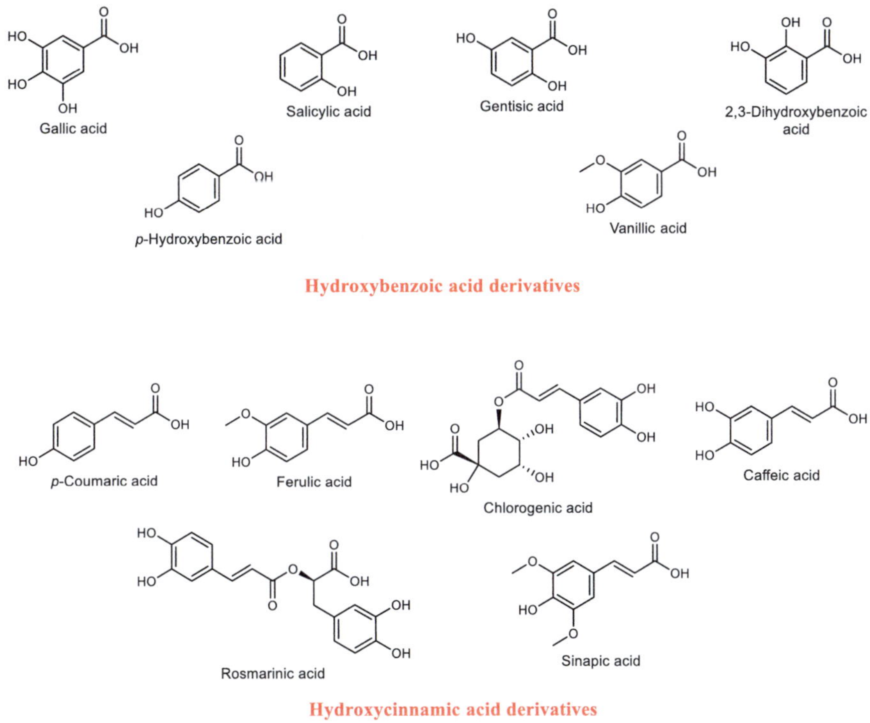

Fig. 8.6 Representative structures of phenolic acids

primary infection site, helping to protect the plant from further spread and enhance future resistance (Mandal et al. 2010).

Functions of Phenolic Acids in Human Health

Phenolic acids have been stated to possess significant biological and pharmacological activities. Moreover, it was assumed that diets rich in phenolic acids help in health maintenance and disease prevention. The high antioxidant content in fruits and vegetables helps prevent diseases associated with oxidative stress, such as stroke, coronary heart disease, and cancers. Herein, we provide the detailed physiological activities related to each phenolic acid solely.

Caffeic Acid

- It selectively blocks leukotrienes biosynthesis, hence prevents immunoregulation diseases, allergic reactions, and asthma.
- Caffeic acid and its esters display anticancer activity against colon carcinogenesis.
- It suppresses aflatoxin production by *Aspergillus flavus* without interfering with fungal proliferation.
- It exhibits antibacterial activity against *Pseudomonas aeruginosa* and *Staphylococcus aureus*.

- Caffeic acid derivatives, such as dicaffeoylquinic acid and dicaffeoyltartaric acid, have been demonstrated to selectively inhibit human immunodeficiency virus type 1 (HIV-1) integrase, the enzyme responsible for incorporating viral DNA into host chromatin (Goleniowski et al. 2013).

Gallic Acid

- It exhibits strong anti-inflammatory activity by inhibiting the AP-1 transcription factor, which regulates inflammation, cell proliferation, and differentiation (Falsaperla et al. 2005).
- It displays a significant anticancer activity against human prostate cancer as well as exerting antimelanogenic activity through ROS-mediated mechanism.
- A phenolics-rich fraction from evening primrose, containing 55% gallic acid, has demonstrated significant antitumor effect.
- It has been shown to be effective against cerebral tumors by reducing glioma cell viability and inhibiting their propagation, invasion, and angiogenic activity.
- It stimulates apoptosis and/or necrosis of HeLa cervical cancer.
- It has been reported to possess considerable antiviral properties against *Herpes simplex* virus (Goleniowski et al. 2013).

Ferulic Acid

- Feruic acid possesses several physiological activities including antioxidant, anti-inflammatory, anticancer, antimicrobial, and antithrombotic activities (Fujisawa et al. 2002).
- It also has protective effect against coronary disease in addition to antihyperlipidemic effect.
- It has been shown to protect against cancer, osteoporosis, and hot flashes in postmenopausal women.
- It protects the skin from ultraviolet damage.
- It possesses antidiabetic activity by reducing blood glucose levels. It is widely used in food and cosmetic industries owing to its distinctive properties and low toxicity.
- It is utilized as the starting material in the production of vanillin and preservatives and as an ingredient in functional foods and skin protection products (Goleniowski et al. 2013).

Salicylic Acid

- It is incorporated principally in aspirin and other pharmaceutical products, which exert anti-inflammatory, antipyretic, and analgesic effects.
- It exhibits keratolytic, antiseptic, and antifungal properties, making it useful for managing various skin conditions, including acne, psoriasis, dandruff, and seborrheic dermatitis (Stuart et al. 2000).
- It helps plants to resist stressful environmental conditions and pathogen attacks via acting as hormonal mediator that results in the formation of long-term disease resistance against different fungi, bacteria, and viruses (Heil and Bostock 2002).

Cinnamic Acid

- Cinnamic acid exerts greater antioxidant activity compared to its benzoic acid counterparts, including *p*-hydroxy, *p*-hydroxymethoxy, *p*-hydroxydimethoxy, and dihydroxy benzoic acid (Goleniowski et al. 2013).

3-Hydroxybenzoic Acid

- It exhibits antifungal, antimutagenic, and estrogenic activities (Pugazhendhi et al. 2005).

Gentisic Acid

- It exerts an analgesic, anti-inflammatory, antirheumatic, and antiarthritic activities.
- It inhibits the oxidation of low-density lipoprotein in human plasma (Ashidate et al. 2005).
- It exhibits powerful anticarcinogenic activity, which is mediated by Fibroblast Growth Factor inhibition (Fernández et al. 2010).

Vanillic Acid

- It possesses antisickling and anthelmintic activities.
- It suppresses hepatic fibrosis associated with chronic liver injury.
- It has also been reported to inhibit the 5′-nucleotidase enzyme in snake venom (Goleniowski et al. 2013).

Sinapic Acid

- It exhibits antioxidant ability via metal chelation.
- It possesses the ability to reverse arsenic-induced hepatotoxicity.
- It exerts good antidiabetic activity (Goleniowski et al. 2013).

Rosmarinic Acid

Rosmarinic acid and its derivatives represent promising chemical entities that exert a vast array of biological activities including:

- In vitro antioxidant activity against superoxide and hydroxyl radicals
- Antiviral, antithrombotic, antiplatelet, and antiallergic actions
- Treatment of rheumatoid arthritis, peptic ulcers, cancer, cataract, and bronchial asthma
- Enhancement of cognitive functions and prevention against Alzheimer's disease progression
- Cardioprotective effects, cancer chemoprevention, and amelioration of kidney diseases.
- Topical anti-inflammatory drug
- Significant antibacterial properties against *Bacillus subtilis*, *Escherichia coli*, and *Micrococcus luteus* (Goleniowski et al. 2013)

Tannins

Condensed Tannins

Condensed tannins form a broad class of polyphenols, including flavan-3-ols, such as catechin and proanthocyanidins, flavan-4-ols, and flavan-3,4-diols. They encompass a C15 (C6–C3–C6) general structure composed of a benzopyran nucleus (A and C rings) attached to an aromatic ring (B ring) at carbon C-2 of the C ring. Flavanols are found in the form of monomers, oligomers, and polymers, which differ primarily in the hydroxylation or methoxylation patterns of their rings as well as the degree of galloylation. They arise from the shikimate pathway through the condensation of phenylalanine with malonyl-CoA, forming naringenin chalcone, the key precursor in the biosynthesis of flavonoids. They have been reported to possess various biological effects, including antioxidant and anticancer activities (Oliveira et al. 2013).

Classification of Condensed Tannins

1. *Flavanol Monomers*

   Hydroxyl groups are present at C-3 in flavan-3-ols, at C-4 in flavan-4-ols, and at both C-3 and C-4 in flavan-3,4-diols, whereas flavans lack hydroxylation on the pyran ring. Flavan-3-ols are abundant in nature while other types are rarely found in nature. Flavan-3-ols, such as catechin, epicatechin, afzelechin, epiafzelechin, gallocatechin, and epigallocatechin exist in free form, as gallic acid esters, or in glycosylated forms.
2. *Proanthocyanidins*

   Proanthocyanidins are present as flavan-3-ols oligomers or polymers that are highly diverse according to monomer units features, the type of interflavanic bond, esterification with gallic acid, and the degree of polymerization. Based on ring A, proanthocyanidins can be divided into phloroglucinol and resorcinol derivatives. Phloroglucinol type is much more abundant in nature, which can be subdivided into procyanidins and prodelphinidins with catechins (3,4-dihydroxylated ring B) and gallocatechins (3,4,5-trihydroxylated ring B), respectively.

   Proanthocyanidin dimers and trimers are classified based on the type of interflavanic bond linking their monomeric units. Extensive polymerization leads to structures with a high number of isomers, arising from differences in bonding patterns and monomer stereochemistry. Oligomeric proanthocyanidins have six monomers at maximum, while polymers have more than six condensed units (Oliveira et al. 2013). Representative structures of different types of condensed tannins are illustrated in Fig. 8.7.

Condensed Tannins Physicochemical Properties

- Proanthocyanidins can interact with and precipitate alkaloids and proteins. Therefore, this property can exert precipitation of salivary proteins in the oral cavity, which describes the astringent character associated with tannin-rich foods.

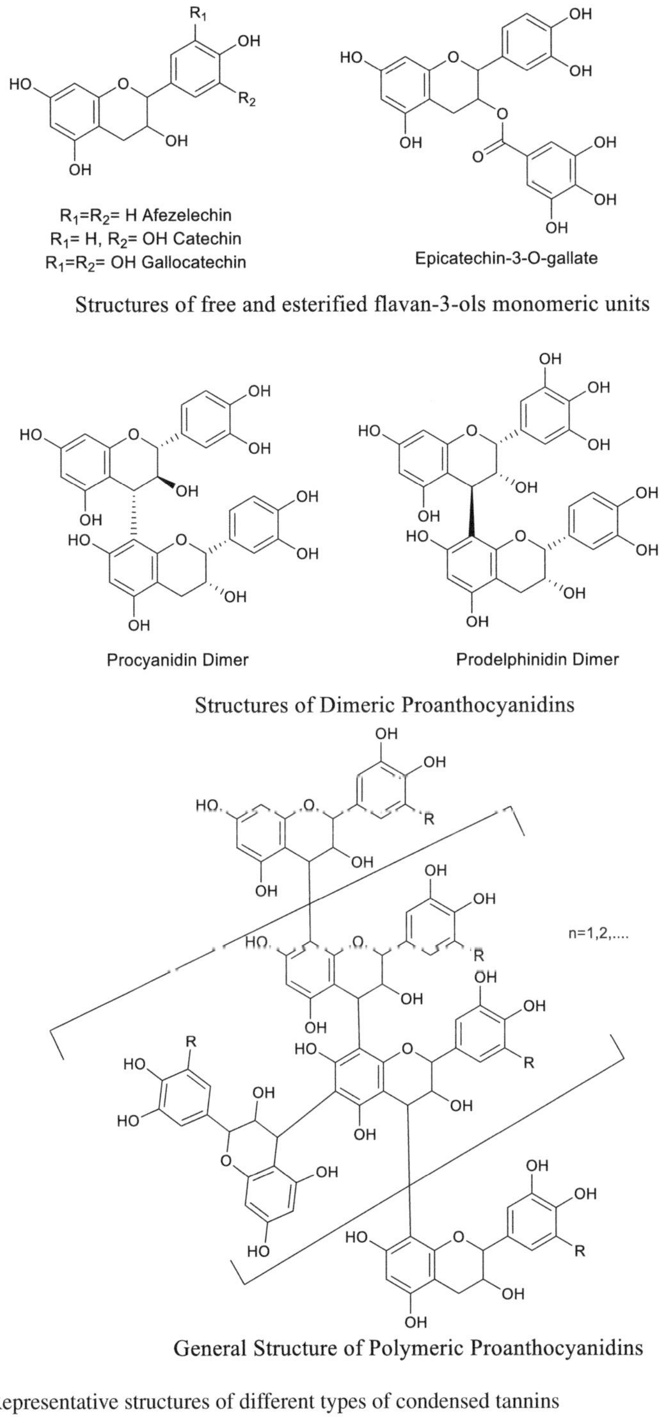

Fig. 8.7 Representative structures of different types of condensed tannins

- Flavanols contribute to the color of certain foods, such as red wines, either through copigmentation with anthocyanins, which enhances color brightness, or by reacting with anthocyanins to yield new colored compounds.
- Flavanols can form stable complexes with metal ions, thereby affecting the bioavailability of various minerals (Oliveira et al. 2013).

Activity of Condensed Tannins in Plants
Flavanols exert a protection activity against both abiotic (e.g., sunlight) and biotic stress such as predators and pathogens in plants (Oliveira et al. 2013).

Activity of Condensed Tannins in Human Health
- Flavanols are considered reducing agents, exhibiting significant antioxidant and free radicals scavenging activities.
- Proanthocyanidins prevent lipid peroxidation, inhibit platelet aggregation, and lower low-density lipoprotein levels, which are major contributors to atherosclerosis and CVDs.
- Proanthocyanidins exert a preventive activity against different types of cancers.
- Proanthocyanidins have been reported to inhibit proteases, lipases, and glycosidases, demonstrating antinutritional activity (Oliveira et al. 2013).

Hydrolyzable Tannins
Hydrolyzable tannins generally comprise a central sugar unit, usually glucose, esterified with gallic acid or its derivatives. *β*-Glucogallin represents the simplest form of glucosyl gallate and serves as the primary precursor of this class. It was first isolated from Chinese rhubarb (*Rheum officinale*) in 1903. Diverse hydrolyzable tannin structures can result from different patterns of bond formation between glucose and gallic acid. Hydrolyzable tannins are further divided into gallotannins, which possess depsidically linked galloyl units, and ellagitannins, distinguished by the presence of hexahydroxydiphenoyl moieties. The *C*-glucosidic ellagitannins are a unique subclass of hydrolyzable ellagitannins that includes C–C linkage between C-1 of glucose core and C-2 of the 2-*O*-galloyl unit (Jourdes et al. 2013). Representative structures of different types of hydrolyzable tannins are illustrated in Fig. 8.8.

Activity of Hydrolyzable Tannins in Human Health
- Like other tannins, hydrolyzable tannins effectively scavenge reactive oxygen species, conferring strong antioxidant properties.
- They can provide protective effects against senescence-associated conditions such as neurodegeneration, cancer, and cardiovascular disorders.
- They have also been shown to mitigate hemorrhages, internal inflammation, and gastrointestinal disorders owing to their ability to interact with macromolecules including proteins and polysaccharides.
- Ellagitannins are also capable of exerting significant antibacterial and antiviral activities.
- Antitumor action via suppression of carcinogens and tumor progression.

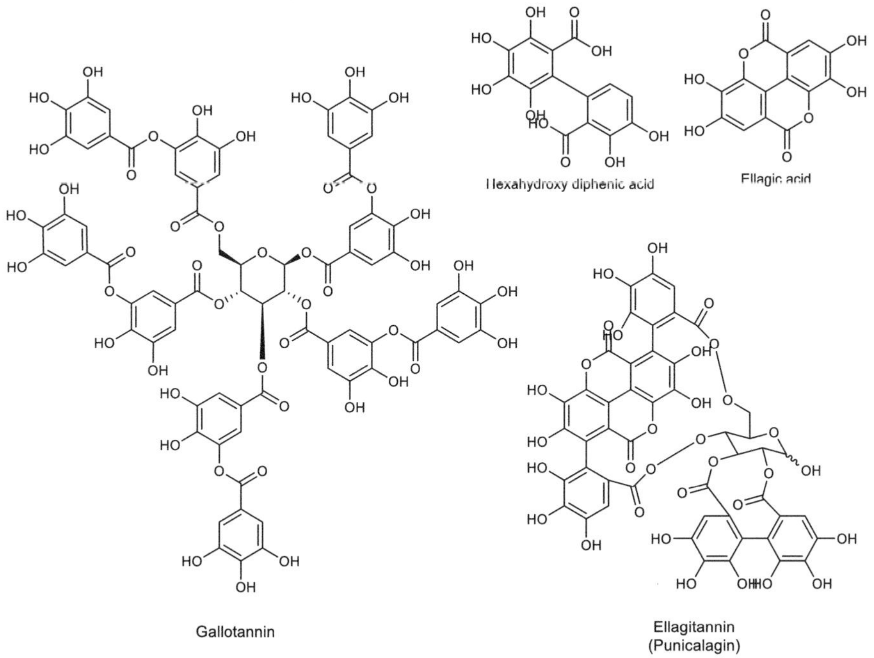

Fig. 8.8 Representative structures of different types of hydrolyzable tannins

Hydrolyzable Tannins in Folk Medicine

Different plant species containing ellagitannins and gallotannins have been used in folk medicine, especially in Asia. Representative examples are to be discussed in more detailed form.

- Geraniin, which is present in a considerable amount in the aerial parts of *Geranium thumbergii*, is used as an antidiarrheal and preventive agent against constipation. Moreover, it exerts significant antioxidant activity.
- Agrimoniin, the first dimeric ellagitannin to be isolated, is primarily employed for its anticancer activity, besides it exhibits antidiarrheal and hemostatic properties.
- Punicalagin is an ellagitannin which is one of the main constituents of pomegranate juice. Pomegranate juice owes its substantially high antioxidant activity compared to red wine and green tea to its punicalagin content.
- Acutissimins contained in *Quercus* and *Castanea* species have the potential to treat gastritis, gastric ulcer, diarrhea, and several inflammation conditions (oral, genital, and skin). They are also used as antitussive and tonic (Jourdes et al. 2013).

8.3 Isolation Techniques of Bioactive Compounds

The chemical diversity of different classes of bioactive compounds makes it challenging for both their isolation and characterization. Efficient separation techniques emerged to overcome this issue, which are mainly based on chromatography. In this context, a comprehensive overview of the most significant and recent separation approaches of different categories of common bioactive compounds will be presented. Subsequent steps of extraction of bioactive compounds from raw material are illustrated in (Fig. 8.9).

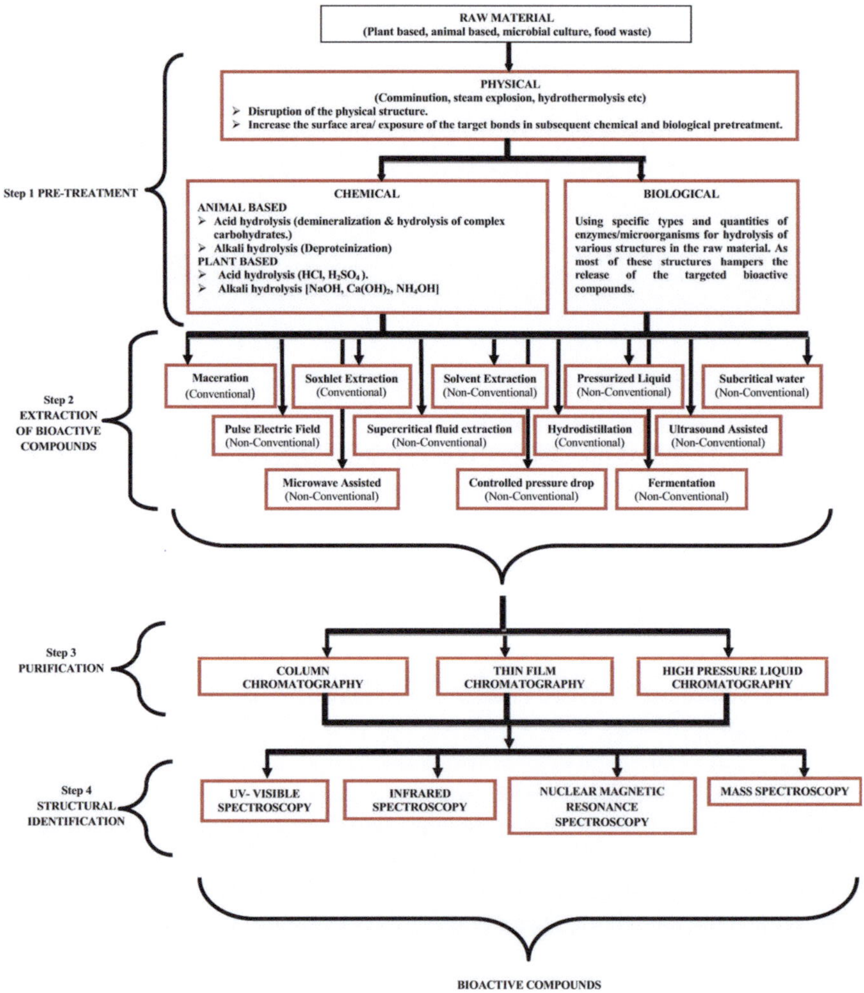

Fig. 8.9 Detailed steps of bioactive compounds' extraction from raw materials (Azeez et al. 2017; Altemimi et al. 2017)

Extraction Methods

Extraction can be defined simply as the process of combining the raw material containing the bioactive compound of interest with appropriate suitable solvent to ensure its whole dissolution while leaving the other components undissolved. After that, solvent can be evaporated, and the bioactive compound is obtained. Extraction process is influenced by different factors, including solvent nature, temperature, pressure, sample type (plant, animal, microbes, etc.), and amount of the raw material. Extraction methods are generally classified into two principal groups: conventional group that includes maceration, Soxhlet extraction, and hydrodistillation and nonconventional group that includes microwave-assisted, pulse electric field, solvent extraction, supercritical fluid extraction, subcritical water, fermentation, controlled pressure drop, ultrasound-assisted, and pressurized liquid techniques. In addition, each type of bioactive compounds has a specific technique for its extraction (Azeez et al. 2017).

1. *Maceration*: Ground plant material is percolated in a suitable solvent within a closed vessel for a defined period to extract bioactive compounds. The resulting solution is then separated, and the remaining plant residue (marc) is pressed to recover any residual extract.
2. *Soxhlet Extraction*: Using a distillation unit, dried plant material and a suitable solvent are subjected to repeated heating and condensation until the bioactive compounds are fully extracted, after which the solvent containing the compounds is collected (Azeez et al. 2017).
3. *Hydrodistillation*: This method involves using water as a solvent to extract bioactive compounds, either by immersing the plant material in boiling water or by injecting it with steam. The process includes hydrolysis, diffusion, and heat-induced decomposition to facilitate extraction.
4. *Pulse Electric Field*: Raw material is exposed to an electric field of 500–1000 V/cm for 10–20 s to increase cell permeability by disrupting cell membranes. This technique significantly enhances extraction efficiency (Martín Ortega and Segura Campos 2019).
5. *Solvent Extraction*: Different solvents such as hexane, ether, acetone, methanol, and ethanol are employed to extract bioactive compounds based on their polarity (Martín Ortega and Segura Campos 2019).
6. *Supercritical Fluid Extraction*: Bioactive compounds are extracted by placing the target material in solvents like nitrogen, methane, or ethylene under controlled temperature and pressure. In their supercritical state, these solvents exhibit improved transport and extraction properties owing to gas-like viscosity and liquid-like density (Martín Ortega and Segura Campos 2019).
7. *Pressurized Liquid Extraction*: High-temperature (50–200 °C) and high-pressure (1450–2175 psi) conditions are employed in this extraction to reduce solvent polarity. Adjusting the temperature ensures polarity compatibility between the solvent and target compounds, while pressure enhances solvent penetration into the plant matrix (Martín Ortega and Segura Campos 2019).

8. *Subcritical Water*: The method operates by lowering the dielectric constant of water through high-pressure (10–60 bar) and high-temperature (100–374 °C) conditions, thereby enabling water to efficiently extract bioactive compounds in place of traditional organic solvents (Martín Ortega and Segura Campos 2019).
9. *Fermentation*: This technique can be classified into submerged fermentation (performed in nutrients-containing liquid medium) or solid-state fermentation (on solid matrix with or without using direct water). Several bioactive compounds are produced as secondary metabolites resulting from full growth of certain microorganisms on the raw material or when one of the main nutrients is consumed from the media as a result of their growth (Martín Ortega and Segura Campos 2019).
10. *Controlled Pressure Drop*: Material is treated with saturated steam briefly, then subjected to a rapid pressure drop that vaporizes volatiles and cools the material, preventing heat-induced degradation. This process enhances mass transfer and compound recovery (Martín Ortega and Segura Campos 2019).
11. *Microwave Assisted*: Microwave treatment evaporates intracellular water, generating pressure that ruptures cell structures, and increases material porosity. This facilitates solvent access to target compounds, thereby improving extraction yield (Martín Ortega and Segura Campos 2019).
12. *Ultrasound Assisted*: Through repeated ultrasonic cycles, bubbles form in the solvent via acoustic cavitation. Their collapse generates shock waves that rupture cell walls, significantly enhancing the extraction of bioactive compound (Martín Ortega and Segura Campos 2019).

8.4 Mechanisms of Action of Bioactive Compounds from Plants for Disease Prevention and Management

8.4.1 Antioxidant Activity

Polyphenols are the most abundant secondary metabolites found in the plants. The primary reported activity of polyphenolic compounds is their scavenging ability of free radicals and reactive oxygen species (ROS) and their antioxidant power (Leopoldini et al. 2011). Free radicals can damage vital biomolecules, namely, proteins, nucleic acids, and lipids, leading to membrane peroxidation and initiation of inflammatory cascades. These events are linked to aging and the pathogenesis of several disorders such as diabetes, cardiovascular and neurodegenerative diseases, and cancer (Chandimali et al. 2025).

Phenolic antioxidants include phenolic acids (e.g., caffeic, ferulic, gallic, and rosmarinic acids), flavonoids (e.g., quercetin and myricetin), phenolic diterpenes (e.g., carnosic acid and carnosol), tocopherols (e.g., α- and γ-tocopherol), and volatile oils (e.g., eugenol and thymol). The antioxidant characteristics exhibited by phenolic substances are mediated by three mechanisms; two of them are related to the direct neutralization of free radicals either by single electron transfer (SET) or hydrogen atom transfer (HAT) in addition to metal chelation. Direct neutralization of free radicals by different phenolic compounds such as flavonoids yields less reactive aromatic radical species that are more stabilized by electron delocalization over the whole molecule (Leopoldini et al. 2011). The position and number of hydroxyl groups on the flavonoid basic structure determine its radical scavenging potential with the substitution on B-ring is the most crucial (Brewer 2011).

Most polyphenolic compounds tend to scavenge free radicals via hydrogen atom transfer, as revealed by computational and experimental studies, and the systems displaying ortho-dihydroxy functionality are the most efficient for radical scavenging and metal chelation as in cyanidin, quercetin, epicatechin, protocatechuic acid, caffeic acid, gallic acid, and hydroxytyrosol (Brewer 2011; Leopoldini et al. 2011). In these systems, the generated radical is stabilized by intramolecular H-bonds in addition to resonance. On the other hand, SET mechanism is more encountered with structures showing planar conformation and wide electron delocalization as in resveratrol and tocopherols (Leopoldini et al. 2011).

By chelating transition metals like Fe^{2+}, phenolics inhibit Fenton reactions responsible for producing highly reactive hydroxyl radicals (Leopoldini et al. 2011). Moreover, metal ions of iron, copper, and zinc tend to bind to amyloid β (Aβ) in brain tissues leading to its aggregation, plaques formation, and eventually neurotoxicity (Olufunmilayo et al. 2023). Indeed, significantly high iron content and accumulation of free radicals in certain brain tissues were correlated with neurodegenerative disorders, for example, Parkinson's disease (PD) and Alzheimer's disease (AD) (Olufunmilayo et al. 2023). The existence of several hydroxyl groups in flavonoids structure together with the carbonyl group in ring C offers numerous sites for metal chelation which alter metal ions redox potential.

Among various phenolic compounds, catechins proved the most active to inhibit microsomal lipid peroxidation (Brewer 2011). Meanwhile, the ability of hydroxycinnamic acids to delay rancidity of fish muscle was found to be mostly dependent on their capacity to donate electrons (Medina et al. 2007).

Rapid in vitro assays are frequently employed to evaluate the antioxidant potential of natural products such as the 2,2-diphenyl-1-picrylhydrazyl (DPPH) radical scavenging, 2,2′-azino-bis (3-ethylbenzothiazoline-6-sulfonic acid) (ABTS) radical scavenging, oxygen radical absorbance capacity (ORAC), and Folin Ciocalteu (FC) assays. While the ABTS, DPPH, and FC assays are based on SET mechanism, ORAC assay relies on HAT. The ORAC assay, in contrast to ABTS, DPPH, and FC, measures a decrease in fluorescence rather than decolorization and is preferred in food industries (Platzer et al. 2022). Several studies reviewed the structural requirements of phenolic compounds in relation to their antioxidant efficacy (Lang et al. 2024; Munteanu and Apetrei 2021; Platzer et al. 2022). The number of hydroxyl groups in phenolic compounds was found to display the most pronounced effect on antioxidant activity. Additionally, glycosylation had no effect or slightly decreased antioxidant activity. No clear correlation could be established between different assays suggesting that they are influenced by different structural criteria (Platzer et al. 2021; Platzer et al. 2022). However, multivariate analysis revealed that DPPH and FC assays showed some degree of correlation than ABTS and ORAC assays which in turn are somewhat correlated (Platzer et al. 2022).

Phenolic compounds can interfere with the production of free radicals through the inhibition of several enzymes. Flavonoids inhibit nitric oxide synthase (NOS) responsible for the biosynthesis of nitric oxide (NO). Besides its inherent radical nature, NO reacts with the superoxide radical to yield peroxynitrite, a powerful oxidant and cytotoxic species (Prolo et al. 2024). Moreover, flavonoids such as quercetin, genistein, daidzein, chrysin, luteolin, baicalein, and puerarin are reported as potential inhibitors of xanthine oxidase. This enzyme plays a key role in inflammatory signaling and oxidative injury. Xanthine oxidase catalyzes the oxidation of hypoxanthine to xanthine and then to uric acid, producing superoxide. Xanthine oxidase-derived oxidative stress is implicated in endothelial dysfunction and atherosclerosis (Battelli et al. 2014; Xue et al. 2023). Flavonoids, for example, quercetin can protect the myocardium from oxidative damage by inhibiting the NF-κB pathway (Lv et al. 2021). Furthermore, this flavonol was reported to modulate mitochondrial membrane permeability and prevent its oxidative damage (Daniel et al. 2018). Another flavonol, kaempferol, was found to inhibit mitochondrial ROS-dependent ferroptosis by decreasing mitochondrial iron overload, increasing mitochondrial membrane potential, and suppressing lipid peroxidation, thereby protecting against doxorubicin-induced myocardial damage (Zhang et al. 2025).

Various findings confirmed the antioxidant potential of proanthocyanidins in culture, animal, and clinical studies through the suppression of oxidative stress-induced DNA damage and induction of DNA repair mechanisms. Proanthocyanidins can directly scavenge free radicals, induce DNA repair enzymes, and enhance nucleotide excision repair mechanisms (Yang et al. 2018). Additionally, proanthocyanidins prevent lipid peroxidation, downregulate various signaling pathways implicated in oxidative stress, for example, nuclear factor-κB (NF-B) and mitogen-activated protein kinase (MAPK) pathways, and activate nuclear factor erythroid 2-related factor 2 (Nrf2) pathway, thereby promoting the expression of cytoprotective genes (Yang et al. 2018). Proanthocyanidins were reported to protect against hydrogen peroxide-induced oxidative damage in retinal cells, increase glutathione (GSH) content and superoxide dismutase (SOD) activity, and lower malondialdehyde (MDA) production (Ma et al. 2024).

Resveratrol, a natural stilbene, is recognized for its potent antioxidant and anti-inflammatory properties. Resveratrol modulates the expression of critical antiaging and antioxidant genes. Additionally, this stilbene provided strong neuroprotective effect against memory loss in Alzheimer's disease (Cosín-Tomàs et al. 2019; Corpas et al. 2019).

8.4.2 Anti-inflammatory Activity

Inflammation is the first line of defense against damaging biological, chemical, or physical stimuli such as pathogens, chemicals agents, or physical injury with the aim of eliminating pathogens, resolving infections, and repairing tissue damage. However, the overexpression of inflammatory mediators, for example, interleukin (IL)-1, IL-6, tumor necrosis factor-α (TNF-α), NO, and prostaglandin E_2 (PGE_2), leads to excessive tissue damage and the development of inflammatory disorders, for example, asthma, arthritis, cardiovascular complications, diabetes, inflammatory bowel diseases, and cancer. Since the discovery of salicylic acid, a vast array of anti-inflammatory compounds has been identified in the plant kingdom including terpenoids which hold great promise in treating inflammatory diseases.

Terpenoids are among the most plentiful and chemically diverse natural products including monoterpenes, sesquiterpenes, diterpenes, sesterterpenes, and triterpenes (Del Prado-Audelo et al. 2021). Several in vitro studies documented the anti-inflammatory activity of terpenes in various cell lines (Viveiros et al. 2022; Blevins et al. 2022; Mazzantini et al. 2024). Monoterpenes, such as limonene, terpinolene, α-phellandrene, linalool, and borneol, as well as triterpene glycosides, were reported to inhibit the expression of the pro-inflammatory cytokines IL-1, IL-6, and TNF-α in RAW 264.7 macrophage cell culture and upregulate the expression of the anti-inflammatory cytokine IL-11 (Del Prado-Audelo et al. 2021; Kim et al. 2020).

The findings of in vitro studies cannot be extrapolated to living organisms due to lack of efficacy and toxicity data in biological systems. Different in vivo inflammatory models induced by lipopolysaccharides (LPS), carrageenan, dextran sulfate sodium (DSS), or ovalbumin confirmed the downregulation of NF-κB, MUC5AC gene expression, and other signaling pathways implicated in inflammation (Chen et al. 2019; Zhang et al. 2022a; Wang et al. 2024; De Morais Oliveira-Tintino et al. 2018). Meanwhile, upregulation of Nrf2 activity was associated with α-pinene, linalool, borneol, 1,8-cineole, and β-caryophyllene (Kim et al. 2020).

β-Caryophyllene is a known agonist of cannabinoid-2 receptors, which are mainly expressed in immune cells. The anti-inflammatory behavior of β-caryophyllene in cisplatin-induced nephropathy and DSS-induced colitis models was mediated by cannabinoid-2 receptors. Other in vivo studies explored the inhibitory potential of terpenes on the generation of pro-inflammatory mediators in neuroinflammation. Linalool and D-limonene decreased the levels of COX-2, iNOS, p38 MAPK, IL-1β, and NO in AD transgenic mouse model and amyloid β (Aβ)-expressing fly heads (Kim et al. 2020; Sabogal-Guáqueta et al. 2016).

Additionally, the monoterpenes limonene, α-pinene, and myrcene effectively reduced the expression of various kinases of the MAPK family in mouse macrophages, human chondrocytes, and mice models. These kinases play a crucial role in regulating cellular responses to stimuli and the production of pro-inflammatory mediators (Del Prado-Audelo et al. 2021).

Another mechanism by which terpenes exert their anti-inflammatory activity is by modulating autophagy. Dysregulation of this self-digestion process plays an essential role in neurodegenerative diseases such as AD and PD and in cancer. A plethora of monoterpenes, for example, limonene, linalyl acetate, geraniol, citral, terpinene-4-ol, α-phellandrene, borneol, camphor, *p*-cymene, and carvacrol, sesquiterpenes, for example, β-elemene, and triterpenes, for example, cucurbitacin E, were reported to activate autophagy by various mechanisms with beneficial neuroprotective and anticancer applications (Ashrafizadeh et al. 2020; Stacchiotti and Corsetti 2020).

Terpenes also display anti-inflammatory activities indirectly through antioxidant behavior. Oxidative stress causes tissue injury which may result in inflammation. The in vitro anti-inflammatory properties of terpinolene and α-phellandrene were mediated by the significant suppression of superoxide and NO generation in macrophages. In an inflammatory rat model, camphene increased GSH content, inhibited the generation of NO and ROS, restored mitochondrial membrane potential, and improved cell viability (Kim et al. 2020). β-Caryophyllene decreased the generation of ROS and the related respiratory activities in arthritic rat model (Ames-Sibin et al. 2018).

Triterpenes exhibit notable anti-inflammatory activity. Betulinic acid, a lupane-type pentacyclic triterpene, has been shown to modulate the activity of various mediators and cell types involved in inflammation, both in vitro and in vivo, across multiple inflammatory disease models. Several pro-inflammatory cytokines production, for example, Il-1ß, IL-6, IL-8, IL-12, and TNF, can be suppressed by betulinic acid chiefly by blocking MAPK and NF-κB signaling pathways and downregulating critical pro-inflammatory genes. Moreover, betulinic acid enhances the production of the essential anti-inflammatory cytokine IL-10. Additionally, betulinic acid inhibits myeloperoxidase and COX-2 activity and controls the production of PGE_2 (Oliveira-Costa et al. 2022).

The labdane diterpene andrographolide isolated from *Andrographis paniculate* was extensively studied for its anti-inflammatory properties and is being clinically investigated for the remedy of osteoarthritis (Hancke et al. 2019). Triptolide is another potent anti-inflammatory diterpene triepoxide isolated from the traditional Chinese medicinal plant *Tripterygium wilfordii.* Triptolide inhibits NF-κB-dependent transcription and effectively suppresses the production of IL-6 and TNF-α at $IC_{50} < 30$ nM in LPS-induced RAW 264.7 cells. This compound holds promise for the treatment of rheumatoid arthritis and is undergoing clinical trials (Fan et al. 2018); likewise, its derivative, (5R)-5-hydroxytriptolide, has also exhibited similar potential (Ge et al. 2022; Zhou et al. 2021).

Curcumin, the primary bioactive phenolic ingredient in *Curcuma longa* (turmeric), is a strong antioxidant, anti-inflammatory, and anticancer molecule with significant potential for treating multiple diseases. The anti-inflammatory properties of curcumin are the origin of its multifaceted biological activities. These include the inhibition of pro-inflammatory T helper (Th)17 cells and upregulation of immunosuppressive T regulatory (Treg) cells (Moudgil and Venkatesha 2022). Curcumin regulates diverse inflammatory signaling pathways, for example, NF-κB, MAPK, activator protein 1 (AP-1), and peroxisome proliferator-activated receptor γ (PPARγ), and decreases the generation of pro-inflammatory mediators, for example, IL-1, IL-1β, IL-6, IL-8, IL-17, IL-27, TNF-α, iNOS, NO, granulocyte colony-stimulating factor, regulated upon activation of normal T cell expressed and secreted factor, and monocyte chemotactic protein-1. In clinical trials, curcumin effectively reduces plasma C-reactive protein levels. Additionally, curcumin controls inflammation by its antioxidant attributes. It enhances antioxidant enzymes activity and decreases ROS production by regulating nicotinamide adenine dinucleotide phosphate oxidase. The preclinical and clinical applications of curcumin extend to arthritis, atherosclerosis, inflammatory bowel disease, psoriasis, and COVID-19 (Peng et al. 2021).

Resveratrol is a strong anti-inflammatory stilbene, which modulates immune cells, T cell differentiation, and autoantibody production and suppresses the secretion of biochemical inflammatory mediators and pro-inflammatory cytokines. The complex anti-inflammatory activities of resveratrol are mediated by the activation of sirtuin-1 (a protein deacetylase implicated in immune functions), suppression of NF-κB, and targeting of adenosine monophosphate-activated protein kinase (AMPK). The latter is also a target of other natural anti-inflammatory agents, for example, soy isoflavones, epigallocatechin-3-gallate (EGCG), triptolide, and curcumin. Additionally, resveratrol suppresses the activity and differentiation of Th17 and Th1, thus altering the balance of Th1/Th2 in favor of anti-inflammatory and immunoregulatory events. By targeting NF-κB, COX-2, and MAPK pathways, resveratrol interferes with the production of pro-inflammatory cytokines triggered by the buildup of advanced glycation end products, which takes place in course of senescence and cellular aging (Moudgil and Venkatesha 2022).

In preclinical and clinical trials on rheumatoid arthritis, resveratrol reduces pain and blocks the propagation of joint inflammation and bone loss by interfering with the interplay of chondrocytes and macrophages (Mittal et al. 2021; Karimi et al. 2022). Resveratrol's beneficial effects extend to prophylactic arthritis models (Moudgil and Venkatesha 2022).

Resveratrol is also being actively investigated for its potential in other autoimmune disorders including multiple sclerosis, type 1 diabetes, psoriasis, and colitis. In multiple sclerosis, several pathways are invoked by resveratrol including the activation of sirtuin-1 and estrogen receptor (ER) with downregulation of NF-κB and reduction of pro-inflammatory cytokines. Moreover, resveratrol improves the production of Treg cells. The identification of a new mechanism including micro-RNA added novel dimensions to the mechanisms of resveratrol protective action in multiple sclerosis (Moudgil and Venkatesha 2022; Gandy et al. 2019).

8.4.3 Antimicrobial Activity

Most of the antimicrobial plant secondary metabolites are phenols or their derivatives including phenolic acids, coumarins, quinones, tannins, flavonoids, lignans, and stilbenes. Besides, other plant-derived compounds elicit antimicrobial properties such as organosulfur compounds, terpenoids, alkaloids, and saponins.

The number and position of phenolic hydroxyl groups in polyphenolic compounds are considered as the key structural feature determining their antimicrobial activity. Hydroxyl groups at C-5 and C-7 of ring A in the flavonoid backbone were found to improve the antibacterial activities of flavones and flavanes (Linden et al. 2020). Similarly, hydroxylation or *O*-glycosylation at position 3 of ring C is crucial for better antibacterial activity, which could explain why flavanols interact with microbial membranes better than the corresponding flavones. Meanwhile, the interplay of hydroxyl groups and electron delocalization is critical for the antimicrobial activity and justifies the strong antimicrobial properties of thymol and carvacrol compared to menthol (Chen et al. 2024). Polyphenols with pyrogallol ring in their structure are better antimicrobials than their counterparts with catechol and resorcinol moieties (Phan et al. 2014).

These hydroxyl groups play the role of proton exchangers leading to rapid depolarization of microbial membranes, dysregulation of the proton motive force, collapse of microbial membrane integrity, and ultimately leakage of cellular components. Phenolic acids such as gallic and ferulic acids can cross microbial cell membranes resulting in localized hyperacidification and irreversible cell membrane damage (Gyawali and Ibrahim 2014; Chen et al. 2024). Gallic acid and phenolic berry extracts were shown to disintegrate the outer membrane of Gram-negative bacteria, for example, *Salmonella*, by chelating its divalent cations. Additionally, catechins such as epigallocatechin gallate interact with porin protein channels in the outer membrane of the Gram-negative bacteria *Escherichia coli* disrupting the transportation of nutrients across the membrane through these channels (Chen et al. 2024).

Polyphenolic compounds could also inhibit the synthesis of vital microbial macromolecules and disrupt energy metabolism. Proanthocyanins from cranberry were reported to downregulate proteins linked to the synthesis of ATP, DNA, RNA, and citric acid cycle proteins in *Pseudomonas aeruginosa* (Chen et al. 2024). Cyanin from pomegranate was found to suppress the activity of ATP synthase in *E. coli.* Polyphenols can affect DNA replication through inhibition of DNA gyrase and dihydrofolate reductase (Lin et al. 2022). In an in silico study, DNA gyrase and topoisomerase IV were postulated as potential pharmacological targets for hydroxyanthraquinones from *Rubia* species: purpurin, pseudopurpurin, and xanthopurpurin. These compounds could strongly bind to DNA-enzyme complexes leading to potential inhibition of bacterial nucleic acid synthesis (Alov et al. 2022). Similarly, pinocembrin and other manuka honey constituents were reported as potential inhibitors of *E. coli* DNA gyrase, based on in silico docking study, while not affecting *B.*

subtilis DNA gyrase which confirm the probiotic effect of manuka honey (Ibrahim et al. 2026).

Eugenol is a phenylpropene found in the volatile oils of many plants of the Myrtaceae, Lamiaceae, and Lauraceae families such as *Syzygium aromaticum* (clove) and the genus *Cinnamomum*. Eugenol shows broad antimicrobial actions against Gram positive and Gram negative bacteria and fungi. In bacterial cells, eugenol targets the cytoplasmic membrane leading to disruption of membrane architecture, increase in nonspecific permeability, and disturbance in ATP and ions transport. The bacterial ultrastructural modification by eugenol is mediated by changes in fatty acid profile of cell membrane as evidenced in various bacterial strains, for example, *S. aureus*, *E. coli*, *Salmonella typhimurium*, and *Pseudomonas fluorescens* (Marchese et al. 2017). Furthermore, eugenol induced cell membrane and DNA damage through the production of ROS leading to cell death in *S. aureus* clinical isolates from pus including vancomycin-resistant *S. aureus* (Das et al. 2016). Additionally, eugenol affects vital bacterial enzymes, for example, ATPase, protease, and amylase (Marchese et al. 2017).

A range of fungal species are sensitive to eugenol such as *Aspergillus niger*, *Candida albicans*, *Saccharomyces cerevisiae*, *Penicillum italicum*, *Penicillum glabrum*, and *Fusarium oxysporum* (Ulanowska and Olas 2021). In the fungal cell lipid bilayer, eugenol can be positioned between the fatty acyl chains, facilitated by its lipophilic character, leading to disruption of membrane permeability and fluidity, fungal adhesion to host cells, and hyphal morphological transition. Additionally, eugenol creates oxidative stress state leading to fungal membrane lipid peroxidation (Marchese et al. 2017).

Eugenol elicits antiviral activity against a variety of viruses including influenza A virus, severe acute respiratory syndrome coronavirus 2 (SARS-CoV-2), hepatitis viruses, transmissible gastroenteritis virus, Ebola virus, and West Nile virus (Chen et al. 2025). The antiviral effect of eugenol against influenza A virus does not include neuraminidase direct inhibition. Alternatively, this effect is mediated by suppression of p38 MAPK, ERK1/2, and IKK/NF-κB pathways leading to antiautophagic events. In silico studies evidenced that eugenol has a strong affinity for several SARS-CoV-2 components including the main protease, papain-like protease, and RNA polymerase (Chen et al. 2025). Another in silico docking study illustrated the affinity of *trans*-anethole, estragole, and other fennel constituents to hepatitis A virus protease and hepatitis C virus RNA polymerase (Ibrahim et al. 2021). Still, these results need to be confirmed clinically. The antiviral potential of eugenol against herpes simplex viruses 1 and 2 was established in vitro by inhibiting the viral replication and in synergism with acyclovir.

The use of eugenol in oral disinfection and as adjunctive therapy for periodontitis is well known. Eugenol is also used in gastrointestinal and respiratory tract infections where it manages upper respiratory mucosa inflammation. Additionally, the use of eugenol-based antiseptic products can diminish the risk of airborne viral transmission, for example, SARS-CoV-2 (Chen et al. 2025).

Due to its antimicrobial action against many food spoilage and foodborne pathogens, clove oil and eugenol are effective natural substitutes to artificial food preservatives. The antibacterial effect of clove volatile oil and eugenol was extensively studied against numerous foodborne pathogens, for example, *S. aureus*, *E. coli*, *B. cereus*, *S. typhimurium*, and *L. monocytogenes*. Interestingly, eugenol inhibited the growth of *Salmonella typhi* at Minimum Inhibitory Concentration (MIC) as low as 0.0125%. Clove oil showed very promising inhibitory effects against the growth of *Pseudomonas putida* in meat with MIC 0.1% (w/v). Clove oil also inhibited the growth of *Yersinia enterocolitica* and *Salmonella enterica* in soft cheese with MIC of 2.5μg/ml (Marchese et al. 2017). The antimicrobial effect of clove oil was shown against seven strains of pathogenic and spoilage bacteria and fungi with MIC reaching 0.001% (w/v) (Ahmed et al. 2021). Eugenol has a significant antifungal effect against food decaying fungi, for example, *Aspergillus ochraceus*, *Rhizopus oryzae*, *Penicillium citrinum*, and *Cladosporium* species. The antifungal applications of eugenol extend to inhibit the growth of foodborne pathogenic fungi, for example, *P. roqueforti* and *A. carbonarius*, in watermelon slices (Marchese et al. 2017).

Many alkaloids display antimicrobial effects with established role in the management of diverse infectious diseases. Piperine is a piperidine alkaloid isolated from *Piper nigrum* (black pepper) and *Piper longum* (long pepper) fruits. Piperine increases cell membrane permeability of *Candida albicans* and disrupts its mitochondrial membrane potential while increasing the intracellular accumulation of ROS. Moreover, piperine can attenuate *C. albicans* virulence factors by regulating its morphological transitions and inhibiting biofilm formation. The antifungal activity of piperine is mediated by regulating genes related to adhesion, biofilm, and hyphae. The proven synergistic antifungal activity of piperine with fluconazole against *C. albicans* underscores its potential use as natural adjuvant antifungal therapy (Priya and Pandian 2022; Bravo-Chaucanés et al. 2023). Additionally, piperine displayed antibacterial effect against *Mycobacterium tuberculosis* in vitro and in vivo and inhibited efflux pumps in *Staphylococcus aureus* and methicillin-resistant *S. aureus* (MRSA) (Haq et al. 2021; Khameneh et al. 2019). Other studies on *P. aeruginosa* showed that piperine treatment reduced cell surface hydrophobicity and microbial motility, increased the accumulation of ROS, decreased the expression of the quorum sensing gene (lasI), inhibited biofilm formation, and disintegrated already existing biofilm (Das et al. 2023).

Berberine is the major bioactive compound in plants of genus *Berberis* and in *Coptis chinensis* (*Rhizoma coptidis*), a cornerstone in traditional Chinese medicine well-known for its diverse pharmacological activities. Berberine is an isoquinoline alkaloid extensively studied for its antimicrobial, anticancer, and health promoting benefits in metabolic disorders (Och et al. 2020; Singh et al. 2024). The broad anti-infectious action of this alkaloid extends against bacteria, fungi, viruses, and protozoa. The antimicrobial action of berberine is due to its inhibition of the temperature-sensitive mutant Z protein, which is vital for bacterial filamenting and cell division (Khameneh et al. 2019). Other mechanisms of berberine include targeting cell division through the inhibition of RNA polymerase, gyrase, and topoisomerase IV as well as DNA intercalation (Khameneh et al. 2019).

The broad-spectrum antimicrobial capabilities of organosulfur compounds were extensively studied. These sulfur-containing compounds include isothiocyanates, sulfides, and polysulfides. Plants of genera like *Allium* (such as garlic and onion) and *Brassica* (such as cabbage and mustard) are rich sources of organosulfur compounds. Indeed, the use of *Allium sativum* L (garlic) as antimicrobial and antiseptic goes back to Louis Pasteur (Borlinghaus et al. 2014).

The antibacterial efficacy of garlic is exclusively ascribed to allicin, a diallyl thiosulfinate. Allicin antibacterial activity was established against various pathogens, for example, *Staphylococcus epidermidis*, *P. aeruginosa*, *Streptococcus agalactiae*, *S. typhimurium*, *Vibrio cholerae* and antibiotic-resistant strains, for example, MRSA. The efficacy of allicin is comparable to conventional antibiotics, for example, penicillin and kanamycin. However, it offers broader spectrum of activity. Different thiosulfinate derivatives of allicin were tested for their antibacterial actions against 20 bacterial isolates revealing that branching of the alkyl chain attached to the thiosulfinate core reduced the activity. Interestingly, alkyl chain elongation improved the bacteriostatic action against Gram-positive bacteria but decreased the activity against Gram-negative (Borlinghaus et al. 2014).

Being a reactive sulfur species (Itani et al. 2008), allicin is thought to exert most of its effects via redox-dependent mechanisms. The mechanism of allicin antibacterial action is due to its redox reaction with thiol groups in enzymes and glutathione. Allicin can inhibit the synthesis of biological macromolecules by the inhibition of sulfhydryl-dependent enzymes such as RNA polymerase, alcohol dehydrogenase, and thioredoxin reductase (Lanzotti et al. 2014; Khameneh et al. 2019).

Besides, allicin displays strong antifungal activity by inhibiting spore germination and hyphal growth both in vitro and in vivo with promising applications in the management of lung aspergillosis and *Candida* infections (efficacy comparable to fluconazole), as well as in green agriculture practices (Borlinghaus et al. 2014; Khodavandi et al. 2011).

Isothiocyanates (ITCs) are another class of volatile organosulfur compounds. ITCs are formed by the action of myrosinase enzyme on plant glucosinolates particularly from the Brassicaceae family. The potent antibacterial action of ITCs against *Helicobacter pylori* may be attributed to urease inhibition and suppression of inflammation (Fahey et al. 2013). The ITCs' antimicrobial action derives from their highly electrophilic nature. The carbon atom of –N=C=S moiety tends to attack amines, hydroxyls, and most importantly sulfhydryl groups of amino acid residues, disrupting protein structure, stabilization, and regulatory functions. Sulforaphane, one of the ITCs, showed potent antibacterial properties against *S. aureus*, *Listeria monocytogenes*, and *H. pylori*. Allyl ITC was reported to impair cell wall integrity resulting in leakage of intracellular components. This ITC derivative showed synergistic action with erythromycin against *S. pyogenes* and with streptomycin against *P. aeruginosa* and *E. coli* (Khameneh et al. 2019).

8.4.4 Antidiabetic and Hypolipidemic Activities

The role of the alkaloid berberine in regulating glucose homeostasis, lipid metabolism, inflammation, and cancer growth is well recognized. Berberine exerts a potent hypoglycemic action comparable to that of metformin with less side effects. This action is mediated by AMPK and MAPK pathways (Li et al. 2020). Moreover, the role of this alkaloid in polycystic ovary syndrome is currently under investigation (Och et al. 2020). Berberine was extensively studied for its protective effect on the cardiovascular system. For instance, the cholesterol lowering action of berberine is attributed to numerous mechanisms, reduction of the intestinal absorption of cholesterol, stimulation of cholesterol excretion, reduction of fatty acid synthesis through activation of AMPK pathway, and enhancement of hepatic exchange of cholesterol. In hyperlipidemic patients, administration of berberine for 3 months improved the lipid profile; reduced total cholesterol, triglycerides, and LDL-cholesterol; and increased high-density lipoprotein cholesterol (Och et al. 2020). Additionally, berberine is a potentially antiobesity drug that acts through the downregulation of adipogenesis and lipogenesis. By modulation of PPARγ and C/EBPα pathways, berberine suppresses the differentiation of adipocytes and preadipocytes. Furthermore, in adipocytes, berberine stimulates triglycerides basal lipolysis state by upregulating the AMPK-mediated ATGL expression leading to long-term weight loss. However, the potential role of berberine in metabolic syndrome has yet to be clinically studied (Och et al. 2020). Berberine was also found to lower alanine and aspartate transaminase levels in indirect clinical trials (Wang et al. 2017).

Hyperlipidemia is a complex state of lipid dysregulation, making drug development a challenge and emphasizing the potential of natural products as multitarget agents. Ginseng (genus *Panax*) is one of the most popular plants in traditional folk remedies. It has a long history of use in traditional Chinese medicine and throughout other parts of Asia. The triterpenoid saponins known as ginsenosides constitute the principal bioactive ingredients of ginseng root responsible for its health benefits (Jin et al. 2023). Extensive studies proved the hypolipidemic and antioxidant effects of ginsenosides by modulating several pathways such as peroxisome proliferator-activated receptor alpha (PPARα), AMPK/sirtuin-1 (SIRT1), MAPK, SIRT3/Forkhead Box O3 (FOXO3)/SOD, and Nrf2, with the ginsenosides Rg1 and Rb1 being the most promising. Ginsenosides counteract oxidative stress by upregulating the endogenous antioxidant enzymes SOD, CAT, and glutathione peroxidase (GPx), taking into consideration the vicious cycle of oxidative stress and hyperlipidemia. Ginsenosides like Rg1 proved to upregulate the expression of PPARα, which facilitates β-oxidation of free fatty acids in mitochondria leading to their decomposition and downregulates triglycerides synthesis. Ginsenosides also induce autophagy and inhibit apoptosis. Interestingly, they reduce the aging of atherosclerotic endothelium and the formation of atherosclerotic plaques. Regulation of these events is useful in management of hyperlipidemia and related metabolic disorders, for example, atherosclerosis, diabetes, and nonalcoholic fatty liver. Moreover, saponins from ginseng were found to regulate the intestinal microflora, increase short chain fatty acids, and improve the response to antioxidants. While preclinical studies approve their role in alleviating oxidative stress and improving lipid metabolism, clinical evidence remains limited. Further rigorous experimental and clinical studies are recommended to clarify mechanisms and specific pathophysiological processes (Jin et al. 2023).

The broad range of biological actions of polyphenolic compounds holds immense promise for addressing metabolic disorders including type 2 diabetes mellitus (T2DM) which is marked by complex origins, disrupted cellular pathways, and an unclear distinction between primary causes and consequences. Polyphenols may exert both pancreatic and extrapancreatic effects. By preventing amyloid buildup, reducing oxidative and inflammatory stress, and protecting against mitochondrial and endoplasmic reticulum damage, polyphenolics can help protect β-cell function. At the same time, their anti-inflammatory properties, lipid-lowering capacity, and insulin-like influence may ameliorate insulin sensitivity and contribute to improved glucose regulation (Nie and Cooper 2021).

Soluble human amylin (hA) oligomers are cytotoxic to pancreatic islet β-cells, contributing to the pathogenesis of T2DM. Through their contribution to oxidative stress and inflammation, hA oligomers can induce β-cell apoptosis. Accumulating evidence indicates that the aromatic molecular structure and the adjacent phenolic hydroxyl groups of polyphenolic compounds are crucial for their ability to prevent and destabilize hA monomers assembly. Many polyphenolics, for example, EGCG, rutin, oleuropein, baicalein, and rosmarinic acid, successfully inhibited amyloid formation from hA in vitro and displayed cytoprotective effects on β-cells in different insulinoma cell lines (Nie and Cooper 2021). In transgenic animal models expressing hA, the human form of amylin, rutin, rosmarinic acid, and EGCG delayed the onset and progression of diabetes, alleviated its severity, reduced pancreatic amyloid fibril deposition, and increased serum insulin levels (Nie and Cooper 2021; Shahidi and Danielski 2024). The overall evidence from clinical trials suggests that polyphenols can lower blood sugar levels and enhance insulin sensitivity. However, more studies are recommended considering the good safety profile of polyphenols which are consumed in fairly high amounts in human diet. Indeed, the results of some clinical trials are controversial due to limitations in sample size, variations in type, dose and duration of intervention, health condition of participants, the intake of other medications, and the multifactorial nature of T2DM.

Interestingly, the pathogenesis of T2DM shares similarities with other diseases such as AD, indicating that polyphenols may serve as multitarget therapeutics in many disorders being general inhibitors of amyloid. They act against amyloid-β (AD), α-synuclein (PD), and prion proteins (Creutzfeldt-Jakob Disease). With their strong anti-inflammatory and antioxidant activities, which are nondisease specific, polyphenols are especially useful for complex-etiology diseases and ongoing studies explore their potential in preventing and managing cancer, cardiovascular, and neurodegenerative disorders (Shahidi and Danielski 2024; Nie and Cooper 2021; Li et al. 2023).

8.4.5 Anticancer Activity

Most of the synthetic chemotherapeutic agents in clinical cancer therapy are targeted therapies which focus on particular targets such as certain proteins, nucleic acids, or specific oncogenic pathways. For instance, many of the synthetic anticancer drugs target tyrosine kinases, nucleotide synthesis, or angiogenesis. Most natural anticancer products, on the other hand, are multitargeted. They can simultaneously affect numerous oncogenic targets and pathways controlling cell proliferation, differentiation, growth, survival, invasion, apoptosis, metastasis, and angiogenesis (Naeem et al. 2022). Natural products can stop the cell cycle and regulate various cell signaling pathways, for example, matrix metalloproteinases (MMP), phosphoinositide 3-kinase (PI3K), and MAPK pathway. They can aid in DNA damage response by upregulating tumor suppressor gene p53 and p21, leading to induction of pro-apoptotic proteins such as Bax and Bak. This event will provoke the synthesis of crucial apoptotic enzymes, for example, caspases 3 and 7. Natural products can also exert anticancer effects by modulating endogenous antioxidants, for example, GSH, glutathione peroxidase (GPx), and glutathione S-transferase. Moreover, natural products are less toxic, more tolerable, readily available, more affordable, structurally complex, and chemically diverse with a wide spectrum of scaffolds, compared to synthetic drugs that cover a narrower chemical space. Many anticancer drugs in clinical use are of plant origin, such as the diterpenoid taxane alkaloids, vinca terpenoid indole alkaloids, and podophyllotoxin derivatives (Naeem et al. 2022).

Many alkaloids were documented for their outstanding anticancer activity. Vinca alkaloids, for example, vinblastine and vincristine, are obtained from *Catharanthus roseus*. These terpenoid indole alkaloids target microtubules where they bind to the β-tubulin heterodimers, damaging the mitotic spindle, disrupting mitosis, and inducing apoptosis. Vinca alkaloids have broad clinical applications in treating breast cancer, ovarian cancer, nonsmall cell lung cancer, leukemia, and lymphoma. Due to their several side effects, they are pharmacomodulated, used in combinations, or as liposomal or other vector delivery systems (Lichota and Gwozdzinski 2018).

Paclitaxel is a potent anticancer taxane alkaloid isolated from *Taxus* spp (yew trees). Taxanes act, partially similar to vinca alkaloids, as mitosis inhibitors. However, paclitaxel binds with high affinity to microtubules leading to microtubule stabilization and polymerization and mitosis inhibition. Paclitaxel is widely used alone or in combination for the management of breast, ovarian, and prostate cancers, nonsmall cell lung carcinoma, and Kaposi's sarcoma. Despite their potency, taxane therapy is associated with undesirable side effects such as cardiotoxicity and oxidative stress, especially when used with drugs like doxorubicin (Lichota and Gwozdzinski 2018).

Camptothecin is a potent cytotoxic pentacyclic quinoline alkaloid isolated from the fruit, cortex, and wood of *Camptotheca acuminata*. Camptothecin and derivatives act at the S phase of the cell cycle. Their mechanism of anticancer action involves DNA intercalation and binding to the DNA–topoisomerase I complex, causing DNA breaks during replication and hindering the reassembly of the DNA strands (Lichota and Gwozdzinski 2018). Additionally, camptothecin significantly inhibits Nrf2 expression and transcription in various cancer cell lines and can be used to sensitize cancer cells to other antineoplastic agents particularly in high Nrf2-expressing cancers (Chen et al. 2017). A plethora of semisynthetic derivatives were developed from camptothecin to overcome its poor solubility and lactone ring instability, such as topotecan, irinotecan, and diflomotecan which hold great promise for clinical application in metastatic pancreatic, metastatic lung, colorectal, and cervical cancers. New derivatives such as exatecan show improved potency, but toxicity remains a concern, especially in combination therapies (Lichota and Gwozdzinski 2018).

Berberine cytotoxic actions were reported in numerous cancer cell lines, for example, human colon adenocarcinoma HCA-7, breast cancer MDA-MB-231, uterine cancer HeLa, and hepatoma HepG2, and in different leukemia cell lines. A plethora of mechanisms mediate the anticancer action of this alkaloid (Och et al. 2020). Berberine causes cell cycle arrest in the G0/G1, G1, and G2/M phases according to the concentration and the cell line. Additionally, berberine induces apoptosis via modulating gene expression, upregulating proapoptotic and downregulating antiapoptotic genes. Berberine upregulated the expression of proapoptotic genes BAX, BAK1, and BIK and the caspase genes including CASP3, CASP8, and CASP9 and downregulated the antiapoptotic genes BNIP1, BNIP3, BCL2, and BCL2L2 (Och et al. 2020). Meanwhile, it increased many key proapoptotic proteins, for example, p53, and increased ROS levels. Moreover, berberine activates MAPK signaling pathways in diverse human tumor cell lines. Berberine can also halt metastasis through its effect on MMP. Some MMPs degrade collagen in the extracellular matrix. This is a key event in cancer cell migration allowing metastases formation. Other mechanisms of berberine anticancer facets include the inhibition of angiogenesis which in turn prevents metastasis and the anti-inflammatory activity via inhibition of the expression of COX_2 and PGE_2 (Och et al. 2020; Ayati et al. 2017; Singh et al. 2024). The multimechanistic action of berberine makes it a viable option in cancer therapy.

Podophyllotoxin is an aryl tetralin lactone lignan isolated from the resin of *Podophyllum* species (*P. peltatum* and *P. emodi*) with strong cytostatic, antiviral, and anticancer activities. This lignan inhibits tubulin polymerization, disrupts karyokinesis spindle, forms stable DNA–topoisomerase II complex, and causes cell cycle arrest at the G2-phase. Additionally, podophyllotoxins provoke single and double strand DNA breaks and display promising antimultidrug resistance potential. Podophyllotoxin's high toxicity, however, limits its clinical use. Other semisynthetic derivatives such as etoposide and teniposide were developed and are widely employed for various neuroblastomas, lymphomas, and sarcomas. These compounds constitute key combination therapies. Deoxypodophyllotoxin, obtained from *Juniperus communis*, shows promising anticancer properties in preclinical studies (Lichota and Gwozdzinski 2018; Shah et al. 2021).

Among the various pharmacological actions of organosulfur compounds are their anticancer activities. Since allicin is redox active, this thiosulfinate causes a redox shift in cancerous cells and induces apoptosis by influencing Nrf2 and ERK1/2 MAP kinases pathways (Borlinghaus et al. 2014). Nonetheless, the clinical application of allicin in a pharmaceutical context is problematic due to its poor stability. Once in the body, allicin reacts rapidly with available thiols especially glutathione and becomes inactivated. Meanwhile, the use of *Allium* and *Brassica* as nutraceuticals remains valid.

Curcumin suppresses cancer progression by simultaneously targeting multiple pathways including inhibition of Akt/PI3K/mTOR and Wnt/ß-catenin pathways, stimulation of p38 MAPK and p53 pathways, and modulation of JAK/STAT and NF-kB signaling pathways (Naeem et al. 2022). Additionally, this phenolic compound improves the immune response by inhibiting the immunosuppressive function of Treg cells in cancer microenvironment. Natural products such as curcumin, resveratrol, and thymoquinone were reported to chemosensitize cancer cells to traditional drugs, for example, doxorubicin and 5-fluorouracil in several cancer cell lines including human colorectal carcinoma HCT116, neuroblastoma SH-SY5Y, and glioblastoma DBTRG cells. This adjuvant therapy approach proved to enhance anticancer efficacy, limit the dose and toxicity of conventional chemotherapeutic agents, tackle the hurdle of chemoresistance, and avoid cancer recurrence and metastasis, with substantial success in preclinical trials (Naeem et al. 2022).

8.5 Case Studies

The therapeutic application of plant-derived bioactive compounds has received significant attention in both preclinical and clinical investigations. This section highlights three extensively studied compounds, curcumin, resveratrol, and quercetin, with emphasis on their plant sources, molecular mechanisms, and evidence supporting their role in disease prevention and treatment.

8.5.1 Curcumin

Plant Source Curcumin is the principal curcuminoid found in the rhizome of *Curcuma longa* (turmeric), a spice widely used in traditional medicine systems such as Ayurveda and Traditional Chinese Medicine (Tian et al. 2025).

Clinical Evidence Curcumin exhibits anti-inflammatory, antioxidant, hepatoprotective, anticancer, and neuroprotective activities (Islam et al. 2024). At the molecular level, it modulates signaling pathways such as NF-κB, STAT3, and MAPK, thereby suppressing pro-inflammatory cytokines alongside oxidative stress (Hewlings and Kalman 2017). Preclinical studies have demonstrated that curcumin can suppress tumor initiation, promotion, and metastasis in diverse experimental models of cancer (Gupta et al. 2013). Clinical trials have reported potential benefits in conditions such as metabolic syndrome, osteoarthritis, and inflammatory bowel disease, although challenges with bioavailability remain a significant limitation (Daily et al. 2016). Novel formulations, such as curcumin nanoparticles, liposomes, and phospholipid complexes, are being developed to overcome this barrier.

8.5.2 *Resveratrol*

Plant Source Resveratrol is a natural stilbene polyphenol most abundant in the skin of grapes (*Vitis vinifera*), berries, peanuts, and red wine (Thirumalaisamy et al. 2022).

Clinical Evidence Resveratrol exhibits cardioprotective, anticancer, antiaging, and neuroprotective effects. It activates AMP-activated protein kinase (AMPK) and sirtuin-1 (SIRT1), both of which are implicated in longevity, metabolic energy processes, and mitochondrial performance (Baur and Sinclair 2006). Preclinical studies suggest protective roles in CVDs by improving endothelial function and reducing oxidative stress (Xia et al. 2017). Clinical trials report modest improvements in lipid profile, insulin sensitivity, and vascular health in metabolic syndrome and type 2 diabetic patients (Hausenblas et al. 2015). However, variability in clinical outcomes is linked to dosage, formulation, and individual metabolic differences.

8.5.3 *Quercetin*

Plant Source Quercetin is a flavonol present in high concentrations in onions (*Allium cepa*), apples, berries, grapes, and leafy greens (Anand David et al. 2016).

Clinical Evidence Quercetin exerts anti-inflammatory, antiviral, antioxidant, and anticancer effects. Mechanistically, it modulates inflammatory mediators (e.g., IL-6 and TNF-α), scavenges reactive oxygen species, and influences signaling pathways including PI3K/Akt and MAPK (Li et al. 2016). Preclinical studies demonstrate protective effects against cardiovascular disorders, hypertension, and viral infections, including influenza and SARS-CoV-2 (Colunga Biancatelli et al. 2020). Clinical trials suggest that quercetin supplementation may reduce blood pressure, improve endothelial function, and enhance exercise performance, although bioavailability again poses limitations (Serban et al. 2016).

8.5.4 Comparative Insights

While each of these bioactive compounds demonstrates promising health benefits, their clinical translation faces common challenges: poor solubility, rapid metabolism, and low systemic bioavailability. Innovations in delivery technologies like nanoparticle systems, phytosomal formulations, and encapsulation approaches are essential for utilizing their full therapeutic potential. Furthermore, rigorous clinical trials with larger cohorts and standardized formulations are necessary to confirm efficacy and safety across diverse populations.

8.6 Challenges and Future Prospects

Despite the remarkable therapeutic benefits of botanically derived bioactive constituents (e.g., polyphenols, flavonoids, and alkaloids) for disease prevention and management, their translational journey from bench to bedside remains hindered by scientific, technical, and regulatory obstacles. In this section, we systematically address these challenges and propose a roadmap for future research and innovation.

8.6.1 Low Bioavailability and Stability

Phytochemicals often suffer from poor water solubility, rapid metabolic degradation, and limited gastrointestinal absorption, which markedly reduce systemic bioavailability and therapeutic efficacy. Moreover, many bioactive compounds are chemically labile and susceptible to degradation under physiological and processing conditions (Kartah and El Monfalouti 2024; Nsairat et al. 2023).

Recent advances in nanotechnology provide effective strategies to overcome these limitations. Nanoformulations, including lipid-based and polymeric nanoparticles, liposomes, micelles, and emulsions, have been shown to enhance solubility, protect against degradation, and allow controlled or targeted release of bioactives (Yang et al. 2020; Kamil et al. 2015; Nsairat et al. 2023). For example, nanoencapsulation of tea polyphenols has been demonstrated to significantly improve their stability in the gastrointestinal tract and enhance their systemic delivery (Yin et al. 2020).

Polymer-based carriers like polylactic acid (PLA), and poly (lactic-co-glycolic acid) (PLGA) nanoparticles offer biocompatibility, biodegradability, and controlled release, thereby enhancing the functional activity of encapsulated polyphenols, including anticancer and anti-inflammatory activities (Yang et al. 2020). Nevertheless, concerns regarding safety, toxicity, scale-up challenges, and cost remain critical and necessitate rigorous evaluation (Abdul Rahim et al. 2019).

8.6.2 Extraction and Standardization of Plant Extracts

The chemical composition of plant-derived extracts is highly variable due to factors such as geographical origin, seasonal variation, soil conditions, and postharvest processing. This variability hinders quality control, reproducibility, and reliability of therapeutic outcomes (Parveen et al. 2015; Mssusa et al. 2023).

Innovative green extraction techniques, such as ultrasound-assisted extraction, supercritical fluid extraction, and microwave-assisted extraction, can offer greater efficiency and sustainability. Simultaneously, advanced analytical platforms like metabolomic profiling, chromatographic fingerprinting, and chemometric tools are essential for ensuring consistency and establishing standardized protocols for extract quality.

8.6.3 Clinical Validation and Regulatory Concerns

Although preclinical investigations frequently report promising effects of plant-derived bioactives, well-designed clinical trials remain limited. Challenges include difficulty in standardizing doses, lack of placebo comparability due to unique sensory attributes of herbal preparations, and insufficient pharmacokinetic and safety data (Parveen et al. 2015).

Regulatory frameworks for herbal and botanical products vary widely globally. For instance, the European Union's Traditional Herbal Medicinal Products Directive (2004/24/EC) streamlines approval for long-used therapeutics, but excludes novel entities and imposes strict manufacturing and documentation requirements (Gatt et al. 2024). In many developing regions, regulatory infrastructure is underdeveloped. One study highlighted that approval processes for herbal trials in Zimbabwe ranged from 8 to 72 weeks, accompanied with lack of dedicated guidelines (Monera-Penduka et al. 2017; Usure et al. 2024).

Quality control is another issue where many submissions for herbal product approval omit essential testing such as heavy metal, pesticide, mycotoxin, or microbial contamination analysis. This omission undermines safety assessments and regulatory approval processes (Mssusa et al. 2023).

8.6.4 Future Research and Innovation Directions

To facilitate clinical translation and regulatory acceptance, research must address multiple aspects:

- *Advanced Delivery Systems*: Continuous development of nanocarriers (e.g., polymeric nanoparticles, nanoemulsions, and liposomes) to improve solubility, stability, targeted release, and therapeutic index of phytochemicals.

- *Omics and Mechanistic Insights*: Multiomics techniques (e.g., genomics, metabolomics), systems biology, and bioinformatics can elucidate mechanisms of action, reveal synergism among compounds, and support personalized medicinal approaches.
- *Integrative Therapies*: Evaluating phytochemical combinations with conventional drugs may yield synergistic effects, reduce required dosages, and minimize adverse effects, especially relevant in combating drug resistance.
- *Standardization and Quality Assurance*: Establish standardized extraction protocols, reference materials, and validated analytical methods to ensure batch-to-batch consistency, quality, and safety.
- *Regulatory Harmonization and Infrastructure*: Promote global alignment of regulatory guidelines, improve regulatory infrastructure, and foster communication between researchers and policymakers to speed up ethical and scientific approval processes.

8.6.5 Conclusion

In summary, plant-derived bioactive compounds offer significant promise for disease prevention and management. Yet, their full potential remains untapped due to barriers in bioavailability, standardization, clinical validation, and regulatory acceptance. Overcoming these challenges will require multidisciplinary collaboration among phytochemists, pharmacologists, toxicologists, clinicians, and regulatory agencies, together with technological innovation and rigorous validation.

References

Abate L, Tadesse MG, Bachheti A, Bachheti RK (2022) Traditional and phytochemical bases of herbs, shrubs, climbers, and trees from Ethiopia for their anticancer response. Biomed Res Int 2022:1589877

Abbas MM, Kandil Yİ, Abbas MA (2020) R-(−)-carvone attenuated doxorubicin induced cardiotoxicity in vivo and potentiated its anticancer toxicity in vitro. Balkan Med J 37:98

Abdul Rahim R, Jayusman PA, Muhammad N, Ahmad F, Mokhtar N, Naina Mohamed I, Mohamed N, Shuid AN (2019) Recent advances in nanoencapsulation systems using PLGA of bioactive phenolics for protection against chronic diseases. Int J Environ Res Public Health 16:4962

Abu-Izneid T, Rauf A, Shariati MA, Khalil AA, Imran M, Rebezov M, Uddin MS, Mahomoodally MF, Rengasamy KR (2020) Sesquiterpenes and their derivatives-natural anticancer compounds: an update. Pharmacol Res 161:105165

Achan J, Talisuna AO, Erhart A, Yeka A, Tibenderana JK, Baliraine FN, Rosenthal PJ, D'alessandro U (2011) Quinine, an old anti-malarial drug in a modern world: role in the treatment of malaria. Malar J 10:144

Adefegha SA, Oboh G, Oluokun OO (2022) Food bioactives: the food image behind the curtain of health promotion and prevention against several degenerative diseases. Stud Nat Prod Chem 72:391–421

Ahmed LI, Ibrahim N, Abdel-Salam AB, Fahim KM (2021) Potential application of ginger, clove and thyme essential oils to improve soft cheese microbial safety and sensory characteristics. Food Biosci 42:101177

Akinboye ES, Bakare O (2011) Biological activities of emetine. Open Nat Prod J 4:8–15

Alov P, Al Sharif M, Najdenski H, Pencheva T, Tsakovska I, Zaharieva MM, Pajeva I (2022) New potential pharmacological targets of plant-derived hydroxyanthraquinones from Rubia spp. Molecules 27:3274

Al-Shahwany AW, Al-Hemiri AA, Abed KM (2013) Comparative evaluation of alkaloids extraction methods from the root bark of *Punica granatum* Linn. Adv Biores 4:33–39

Altemimi A, Lakhssassi N, Baharlouei A, Watson DG, Lightfoot DA (2017) Phytochemicals: extraction, isolation, and identification of bioactive compounds from plant extracts. Plants 6:42

Ames-Sibin AP, Barizão CL, Castro-Ghizoni CV, Silva FM, Sá-Nakanishi AB, Bracht L, Bersani-Amado CA, Marçal-Natali MR, Bracht A, Comar JF (2018) β-Caryophyllene, the major constituent of copaiba oil, reduces systemic inflammation and oxidative stress in arthritic rats. J Cell Biochem 119:10262–10277

Anand David AV, Arulmoli R, Parasuraman S (2016) Overviews of biological importance of quercetin: a bioactive flavonoid. Pharmacogn Rev 10:84–89

Aniszewski T (2015) Alkaloids: chemistry, biology, ecology, and applications. Elsevier

Anwar S, Ahmad B, Sultan M, Gul W, Islam N (2003) Biological and pharmacological properties of *Aconitum chasmanthum*. J Biol Sci 3:989–993

Arens AM, Shah K, Al-Abri S, Olson KR, Kearney T (2018) Safety and effectiveness of physostigmine: a 10-year retrospective review. Clin Toxicol 56:101–107

Ashidate K, Kawamura M, Mimura D, Tohda H, Miyazaki S, Teramoto T, Yamamoto Y, Hirata Y (2005) Gentisic acid, an aspirin metabolite, inhibits oxidation of low-density lipoprotein and the formation of cholesterol ester hydroperoxides in human plasma. Eur J Pharmacol 513:173–179

Ashrafizadeh M, Ahmadi Z, Mohammadinejad R, Kaviyani N, Tavakol S (2020) Monoterpenes modulating autophagy: a review study. Basic Clin Pharmacol Toxicol 126:9–20

Ayati SH, Fazeli B, Momtazi-Borojeni AA, Cicero AF, Pirro M, Sahebkar A (2017) Regulatory effects of berberine on microRNome in cancer and other conditions. Crit Rev Oncol Hematol 116:147–158

Ayoub IM, Eldahshan OA, Roxo M, Zhang S, Wink M, Singab ANB (2024) Stress resistance, anti-aging, and neuroprotective activities of baicalein 5,6-dimethyl ether and *Alnus rugosa* extract in *Caenorhabditis elegans* model. Arch Pharm 357:2400464

Azeez S, Narayana C, Oberoi H (2017) Extraction and utilisation of bioactive compounds from agricultural waste. In: Utilisation of bioactive compounds from agricultural and food production waste. CRC Press

Bahramsoltani R, Ebrahimi F, Farzaei MH, Baratpourmoghaddam A, Ahmadi P, Rostamiasrabadi P, Rasouli Amirabadi AH, Rahimi R (2019) Dietary polyphenols for atherosclerosis: a comprehensive review and future perspectives. Crit Rev Food Sci Nutr 59:114–132

Bai X, Tang J (2020) Myrcene exhibits antitumor activity against lung cancer cells by inducing oxidative stress and apoptosis mechanisms. Nat Prod Commun 15:1934578X20961189

Bai L, Li J, Li H, Song J, Zhou Y, Lu R, Liu B, Pang Y, Zhang P, Chen J (2019) Renoprotective effects of artemisinin and hydroxychloroquine combination therapy on IgA nephropathy via suppressing NF-κB signaling and NLRP3 Inflammasome activation by exosomes in rats. Biochem Pharmacol 169:113619

Ballal M, Srujan D, Bhat K, Shirwaikar A, Shivananda P (2000) Antibacterial activity of *Holarrhena antidysenterica* (Kurchi) against the enteric pathogens. Indian J Pharmacol 33:392–393

Battelli MG, Polito L, Bolognesi A (2014) Xanthine oxidoreductase in atherosclerosis pathogenesis: not only oxidative stress. Atherosclerosis 237:562–567

Baur JA, Sinclair DA (2006) Therapeutic potential of resveratrol: the in vivo evidence. Nat Rev Drug Discov 5:493–506

Benowitz NL (2009) Pharmacology of nicotine: addiction, smoking-induced disease, and therapeutics. Annu Rev Pharmacol Toxicol 49:57–71

Bimonte S, Cascella M, Barbieri A, Arra C, Cuomo A (2020) Current shreds of evidence on the anticancer role of EGCG in triple negative breast cancer: an update of the current state of knowledge. Infect Agents Cancer 15:2

Blevins LK, Bach AP, Crawford RB, Zhou J, Henriquez JE, Rizzo MD, Sermet S, Khan DIO, Turner H, Small-Howard AL (2022) Evaluation of the anti-inflammatory effects of selected cannabinoids and terpenes from *Cannabis sativa* employing human primary leukocytes. Food Chem Toxicol 170:113458

Borlinghaus J, Albrecht F, Gruhlke MC, Nwachukwu ID, Slusarenko AJ (2014) Allicin: chemistry and biological properties. Molecules 19:12591–12618

Bravo-Chaucanés CP, Chitiva LC, Vargas-Casanova Y, Diaz-Santoyo V, Hernández AX, Costa GM, Parra-Giraldo CM (2023) Exploring the potential mechanism of action of piperine against *Candida albicans* and targeting its virulence factors. Biomolecules 13:1729

Brewer M (2011) Natural antioxidants: sources, compounds, mechanisms of action, and potential applications. Compr Rev Food Sci Food Saf 10:221–247

Brinza I, Boiangiu RS, Mihasan M, Gorgan DL, Stache AB, Abd-Alkhalek A, El-Nashar H, Ayoub I, Mostafa N, Eldahshan O (2024) Rhoifolin, baicalein 5,6-dimethyl ether and agathisflavone prevent amnesia induced in scopolamine zebrafish (Danio rerio) model by increasing the mRNA expression of BDNF, NPY, EGR-1, NFR2A, and CREB1 genes. Eur J Pharmacol 984:177013

Câmara JS, Perestrelo R, Ferreira R, Berenguer CV, Pereira JA, Castilho PC (2024) Plant-derived terpenoids: a plethora of bioactive compounds with several health functions and industrial applications—a comprehensive overview. Molecules 29:3861

Canedo-Santos JC, Carrillo-Garmendia A, Mora-Martinez I, Gutierrez-Garcia IK, Ramirez-Romero MG, Regalado-Gonzalez C, Nava GM, Madrigal-Perez LA (2022) Resveratrol shortens the chronological lifespan of *Saccharomyces cerevisiae* by a pro-oxidant mechanism. Yeast 39:193–207

Cao Y, True AD, Chen J, Xiong YL (2016) Dual role (anti-and pro-oxidant) of gallic acid in mediating myofibrillar protein gelation and gel in vitro digestion. J Agric Food Chem 64:3054–3061

Cappelletti S, Daria P, Sani G, Aromatario M (2015) Caffeine: cognitive and physical performance enhancer or psychoactive drug? Curr Neuropharmacol 13:71–88

Carbonell-Capella JM, Barba FJ, Esteve MJ, Frigola A (2014) Quality parameters, bioactive compounds and their correlation with antioxidant capacity of commercial fruit-based baby foods. Food Sci Technol Int 20:479–487

Cermanova J, Kadova Z, Zagorova M, Hroch M, Tomsik P, Nachtigal P, Kudlackova Z, Pavek P, Dubecka M, Ceckova M, Staud F, Laho T, Micuda S (2015) Boldine enhances bile production in rats via osmotic and Farnesoid X receptor dependent mechanisms. Toxicol Appl Pharmacol 285:12–22

Chambial S, Dwivedi S, Shukla KK, John PJ, Sharma P (2013) Vitamin C in disease prevention and cure: an overview. Indian J Clin Biochem 28:314–328

Chandimali N, Bak SG, Park EH, Lim H-J, Won Y-S, Kim E-K, Park S-I, Lee SJ (2025) Free radicals and their impact on health and antioxidant defenses: a review. Cell Death Discov 11:19

Chatzivasileiou AO, Ward V, Edgar SM, Stephanopoulos G (2019) Two-step pathway for isoprenoid synthesis. Proc Natl Acad Sci 116:506–511

Chen J, Qu Y, Wang D, Peng P, Cai H, Gao Y, Chen Z, Cai B (2014) Pharmacological evaluation of total alkaloids from nux vomica: effect of reducing strychnine contents. Molecules 19:4395–4408

Chen W, Ma X, Lin Y, Xiong Y, Zheng C, Hu Y, Yu D, Jiang Z (2016) Dietary supplementation with a high dose of daidzein enhances the antioxidant capacity in swine muscle but experts pro-oxidant function in liver and fat tissues. J Anim Sci Biotechnol 7:43

Chen F, Wang H, Zhu J, Zhao R, Xue P, Zhang Q, Bud Nelson M, Qu W, Feng B, Pi J (2017) Camptothecin suppresses NRF2–are activity and sensitises hepatocellular carcinoma cells to anticancer drugs. Br J Cancer 117:1495–1506

Chen M, Qin Y, Ma H, Zheng X, Zhou R, Sun S, Huang Y, Duan Q, Liu W, Wu P (2019) Downregulating NF-κB signaling pathway with triterpenoids for attenuating inflammation: in vitro and in vivo studies. Food Funct 10:5080–5090

Chen X, Lan W, Xie J (2024) Natural phenolic compounds: antimicrobial properties, antimicrobial mechanisms, and potential utilization in the preservation of aquatic products. Food Chem 440:138198

Chen M, Nie Y-K, Liu X-Y, Liu Y, Guo D-Y (2025) Antiviral properties of the natural product eugenol: a review. Fitoterapia 185:106674

Cho KS, Lim Y-R, Lee K, Lee J, Lee JH, Lee I-S (2017) Terpenes from forests and human health. Toxicol Res 33:97–106

Chu M, Ding R, Chu Z-Y, Zhang M-B, Liu X-Y, Xie S-H, Zhai Y-J, Wang Y-D (2014) Role of berberine in anti-bacterial as a high-affinity LPS antagonist binding to TLR4/MD-2 receptor. BMC Complement Altern Med 14:89

Colunga Biancatelli RML, Berrill M, Catravas JD, Marik PE (2020) Quercetin and vitamin C: an experimental, synergistic therapy for the prevention and treatment of SARS-COV-2 related disease (COVID-19). Front Immunol 11:1451

Coricovac D, Dehelean CA, Pinzaru I, Mioc A, Aburel O-M, Macasoi I, Draghici GA, Petean C, Soica C, Boruga M (2021) Assessment of betulinic acid cytotoxicity and mitochondrial metabolism impairment in a human melanoma cell line. Int J Mol Sci 22:4870

Corpas R, Griñán-Ferré C, Rodríguez-Farré E, Pallàs M, Sanfeliu C (2019) Resveratrol induces brain resilience against Alzheimer neurodegeneration through proteostasis enhancement. Mol Neurobiol 56:1502–1516

Cosín-Tomàs M, Senserrich J, Arumí-Planas M, Alquézar C, Pallàs M, Martín-Requero Á, Suñol C, Kaliman P, Sanfeliu C (2019) Role of resveratrol and selenium on oxidative stress and expression of antioxidant and anti-aging genes in immortalized lymphocytes from Alzheimer's disease patients. Nutrients 11:1764

Daily JW, Yang M, Park S (2016) Efficacy of turmeric extracts and curcumin for alleviating the symptoms of joint arthritis: a systematic review and meta-analysis of randomized clinical trials. J Med Food 19:717–729

Daniel OO, Adeoye AO, Ojowu J, Olorunsogo OO (2018) Inhibition of liver mitochondrial membrane permeability transition pore opening by quercetin and vitamin E in streptozotocin-induced diabetic rats. Biochem Biophys Res Commun 504:460–469

Das B, Mandal D, Dash SK, Chattopadhyay S, Tripathy S, Dolai DP, Dey SK, Roy S (2016) Eugenol provokes ROS-mediated membrane damage-associated antibacterial activity against clinically isolated multidrug-resistant staphylococcus aureus strains. Infect Dis 9:11–19

Das S, Paul P, Dastidar DG, Chakraborty P, Chatterjee S, Sarkar S, Maiti D, Tribedi P (2023) Piperine exhibits potential antibiofilm activity against *Pseudomonas aeruginosa* by accumulating reactive oxygen species, affecting cell surface hydrophobicity and quorum sensing. Appl Biochem Biotechnol 195:3229–3256

Davoodvandi A, Shabani Varkani M, Clark CC, Jafarnejad S (2020) Quercetin as an anticancer agent: focus on esophageal cancer. J Food Biochem 44:E13374

De Costa C (2002) St Anthony's fire and living ligatures: a short history of ergometrine. Lancet 359:1768–1770

De Morais Oliveira-Tintino CD, Pessoa RT, Fernandes MNM, Alcântara IS, Da Silva BAF, De Oliveira MRC, Martins AOBPB, Da Silva MDS, Tintino SR, Rodrigues FFG (2018) Anti-inflammatory and anti-edematogenic action of the Croton campestris A. St.-Hil (Euphorbiaceae) essential oil and the compound β-caryophyllene in in vivo models. Phytomedicine 41:82–95

Debnath B, Singh WS, Das M, Goswami S, Singh MK, Maiti D, Manna K (2018) Role of plant alkaloids on human health: a review of biological activities. Mater Today Chem 9:56–72

Del Prado-Audelo ML, Cortés H, Caballero-Florán IH, González-Torres M, Escutia-Guadarrama L, Bernal-Chávez SA, Giraldo-Gomez DM, Magaña JJ, Leyva-Gómez G (2021) Therapeutic applications of terpenes on inflammatory diseases. Front Pharmacol 12:704197

Dereli FTG (2023) Plant-based bioactive components: phytochemicals: a review. In: Thakur M, Belwal T (eds) Bioactive components: a sustainable system for good health and well-being. Springer Nature Singapore, Singapore

Di Petrillo A, Orrù G, Fais A, Fantini MC (2022) Quercetin and its derivates as antiviral potentials: a comprehensive review. Phytother Res 36:266–278

Ding H-W, Huang A-L, Zhang Y-L, Li B, Huang C, Ma T-T, Meng X-M, Li J (2017) Design, synthesis and biological evaluation of hesperetin derivatives as potent anti-inflammatory agent. Fitoterapia 121:212–222

Dominguez Avila JA, Rodrigo Garcia J, Gonzalez Aguilar GA, De La Rosa LA (2017) The anti-diabetic mechanisms of polyphenols related to increased glucagon-like peptide-1 (GLP1) and insulin signaling. Molecules 22:903

Duringer J, Lewis R, Kuehn L, Fleischmann T, Craig A (2005) Growth and hepatic in vitro metabolism of ergotamine in mice divergently selected for response to endophyte toxicity. Xenobiotica 35:531–548

Eid HM, Haddad PS (2017) The antidiabetic potential of quercetin: underlying mechanisms. Curr Med Chem 24:355–364

El Khodary YA, Ayoub IM, Pério P, Bourgeade-Delmas S, Ibrahim N, El-Ahmady SH (2024) A comparative untargeted metabolomic analysis and assessment of antiplasmodial potential of nine *Albizia* species. Arch Pharm 357:2300543

Estévez-Sarmiento F, Saavedra E, Ruiz-Estévez M, León F, Quintana J, Brouard I, Estévez F (2020) Chlorinated guaiane-type sesquiterpene lactones as cytotoxic agents against human tumor cells. Int J Mol Sci 21:9767

Fahey JW, Stephenson KK, Wade KL, Talalay P (2013) Urease from helicobacter pylori is inactivated by sulforaphane and other isothiocyanates. Biochem Biophys Res Commun 435:1–7

Falsaperla M, Morgia G, Tartarone A, Ardito R, Romano G (2005) Support ellagic acid therapy in patients with hormone refractory prostate cancer (HRPC) on standard chemotherapy using vinorelbine and estramustine phosphate. Eur Urol 47:449–455

Fan D, Guo Q, Shen J, Zheng K, Lu C, Zhang G, Lu A, He X (2018) The effect of triptolide in rheumatoid arthritis: from basic research towards clinical translation. Int J Mol Sci 19:376

Fayez S, Ayoub IM, Mostafa NM, Moussa AY, Gamal Eldin MI, El-Shazly M (2022) Nutraceuticals in cancer therapy: challenges and opportunities. In: Handbook of oxidative stress in cancer: therapeutic aspects. Springer, pp 189–208

Fernández IS, Cuevas P, Angulo J, López-Navajas P, Canales-Mayordomo A, González-Corrochano R, Lozano RM, Valverde S, Jiménez-Barbero J, Romero A (2010) Gentisic acid, a compound associated with plant defense and a metabolite of aspirin, heads a new class of in vivo fibroblast growth factor inhibitors. J Biol Chem 285:11714–11729

Fujisawa S, Atsumi T, Kadoma Y, Sakagami H (2002) Antioxidant and prooxidant action of eugenol-related compounds and their cytotoxicity. Toxicology 177:39–54

Gandy KAO, Zhang J, Nagarkatti P, Nagarkatti M (2019) Resveratrol (3,5,4′-trihydroxy-trans-stilbene) attenuates a mouse model of multiple sclerosis by altering the miR-124/sphingosine kinase 1 axis in encephalitogenic T cells in the brain. J Neuroimmune Pharmacol 14:462–477

García-Conesa M-T, Larrosa M (2020) Polyphenol-rich foods for human health and disease. Nutrients 12:400

Gatt AR, Vella Bonanno P, Zammit R (2024) Ethical considerations in the regulation and use of herbal medicines in the European Union. Front Med Technol 6:1358956

Ge J, Liu Z, Zhong Z, Wang L, Zhuo X, Li J, Jiang X, Ye X-Y, Xie T, Bai R (2022) Natural terpenoids with anti-inflammatory activities: potential leads for anti-inflammatory drug discovery. Bioorg Chem 124:105817

George MY, Farag N, Menze ET, Elezaby RS, Salem NA, Elrahmany SK, Adel N, Elkhatib RM, Galal M, Assem H, Elkhouly R, Wesa MA, Hesham N, Hossam N, Ayoub IM (2025) Intranasal delivery of blackberry-loaded chitosan nanoparticles for antipsychotic potential in ketamine-induced schizophrenia in rats. Sci Rep 15:16707

Ghiringhelli F, Rebe C, Hichami A, Delmas D (2012) Immunomodulation and anti-inflammatory roles of polyphenols as anticancer agents. Anticancer Agents Med Chem 12:852–873

Goleniowski M, Bonfill M, Cusido R, Palazón J (2013) Phenolic acids. In: Natural products: phytochemistry, botany and metabolism of alkaloids, phenolics and terpenes. Springer, Berlin, Heidelberg

González-Cofrade L, De Las Heras B, Ticona LA, Palomino OM (2019) Molecular targets involved in the neuroprotection mediated by terpenoids. Planta Med 85:1304–1315

Griffin C, Karnik A, Mcnulty J, Pandey S (2011) Pancratistatin selectively targets cancer cell mitochondria and reduces growth of human colon tumor xenografts. Mol Cancer Ther 10:57–68

Grzesik M, Naparło K, Bartosz G, Sadowska-Bartosz I (2018) Antioxidant properties of catechins: comparison with other antioxidants. Food Chem 241:480–492

Guo Y, Wu G, Su X, Yang H, Zhang J (2009) Antiobesity action of a daidzein derivative on male obese mice induced by a high-fat diet. Nutr Res 29:656–663

Gupta SC, Patchva S, Aggarwal BB (2013) Therapeutic roles of curcumin: lessons learned from clinical trials. AAPS J 15:195–218

Gutiérrez-Grijalva EP, Contreras-Angulo LA, Emus-Medina A, Heredia JB (2021) Plant alkaloids with antidiabetic potential. In: Structure and health effects of natural products on diabetes mellitus. Springer

Gyawali R, Ibrahim SA (2014) Natural products as antimicrobial agents. Food Control 46:412–429

Haines DD, Cowan FM, Tosaki A (2024) Evolving strategies for use of phytochemicals in prevention and long-term management of cardiovascular diseases (CVD). Int J Mol Sci 25:6176

Hameed A, Ijaz S, Mohammad IS, Muhammad KS, Akhtar N, Khan HMS (2017) Aglycone solanidine and solasodine derivatives: a natural approach towards cancer. Biomed Pharmacother 94:446–457

Hancke JL, Srivastav S, Cáceres DD, Burgos RA (2019) A double-blind, randomized, placebo-controlled study to assess the efficacy of *Andrographis paniculata* standardized extract (ParActin®) on pain reduction in subjects with knee osteoarthritis. Phytother Res 33:1469–1479

Haq IU, Imran M, Nadeem M, Tufail T, Gondal TA, Mubarak MS (2021) Piperine: a review of its biological effects. Phytother Res 35:680–700

Hausenblas HA, Schoulda JA, Smoliga JM (2015) Resveratrol treatment as an adjunct to pharmacological management in type 2 diabetes mellitus--systematic review and meta-analysis. Mol Nutr Food Res 59:147–159

Heil M, Bostock RM (2002) Induced systemic resistance (ISR) against pathogens in the context of induced plant defences. Ann Bot 89:503–512

Hewlings SJ, Kalman DS (2017) Curcumin: a review of its effects on human health. Foods 6:92

Himesh Soni HS, Jitender Malik JM, Singhai A, Sarvesh Sharma SS (2013) Antimicrobial and anti-inflammatory activity of the hydrogels containing rutin delivery. Asian J Chem 25:8371–8373

Huang W-Y, Davidge ST, Wu J (2013) Bioactive natural constituents from food sources—potential use in hypertension prevention and treatment. Crit Rev Food Sci Nutr 53:615–630

Husen A (2021) Traditional herbal therapy for the human immune system. CRC Press, Boca Raton. https://doi.org/10.1201/9781003137955

Husen A (2022) Herbs, shrubs and trees of potential medicinal benefits. CRC Press, Boca Raton. https://doi.org/10.1201/9781003205067

Husen A (2023a) Exploring poisonous plants: medicinal values, toxicity response and other potential uses. CRC Press, Boca Raton. https://doi.org/10.1201/b23017

Husen A (2023b) Antidiabetic medicinal plants and herbal treatments. CRC Press, Boca Raton. https://doi.org/10.1201/b23347

Husen A (2024a) Antimalarial medicinal plants. Taylor & Francis Group, LLC. https://doi.org/10.1201/9781003378396

Husen A (2024b) Medicinal spices and condiment crops. Taylor & Francis Group, LLC. https://doi.org/10.1201/9781003387046

Husen H, Pant M (2024a) Exploring medicinal orchids. Taylor & Francis Group, LLC. https://doi.org/10.1201/b23375

Husen H, Pant M (2024b) Tissue culture techniques and medicinal plants: enhancing propagation and production. Taylor & Francis Group, LLC. https://doi.org/10.1201/b23374

Ibrahim A, Nasr M, El-Sherbiny IM (2022) Baicalin as an emerging magical nutraceutical molecule: emphasis on pharmacological properties and advances in pharmaceutical delivery. J Drug Delivery Sci Technol 70:103269

Imran M, Saeed F, Hussain G, Imran A, Mehmood Z, Gondal TA, El-Ghorab A, Ahmad I, Pezzani R, Arshad MU (2021) Myricetin: a comprehensive review on its biological potentials. Food Sci Nutr 9:5854–5868

Islam MR, Rauf A, Akash S, Trisha SI, Nasim AH, Akter M, Dhar PS, Ogaly HA, Hemeg HA, Wilairatana P, Thiruvengadam M (2024) Targeted therapies of curcumin focus on its therapeutic benefits in cancers and human health: molecular signaling pathway-based approaches and future perspectives. Biomed Pharmacother 170:116034

Itani WS, El-Banna SH, Hassan SB, Larsson RL, Bazarbachi A, Gali-Muhtasib HU (2008) Anti colon cancer components from Lebanese sage (*Salvia libanotica*) essential oil: mechanistic basis. Cancer Biol Ther 7:1765–1773

Ibrahim N, Moussa AY, Taher EM, Luo J, Xu B, & Ahmed LI (2026) Chemical profiling, molecular docking, and harnessing Manuka honey's antimicrobial and prebiotic potential in probiotic milk and yoghurt. Food Microbiol 137:105045. https://doi.org/10.1016/j.fm.2026.105045

Ibrahim N, & Moussa AY (2021) A comparative volatilomic characterization of Florence fennel from different locations: antiviral prospects. Food & Funct, 12(4):1498–1515

Jankowska M, Rogalska J, Wyszkowska J, Stankiewicz M (2018) Molecular targets for components of essential oils in the insect nervous system—a review. Molecules 23:34

Janssen B, Schäfer B (2017) Galantamine. ChemTexts 3:7

Jin W, Li C, Yang S, Song S, Hou W, Song Y, Du Q (2023) Hypolipidemic effect and molecular mechanism of ginsenosides: a review based on oxidative stress. Front Pharmacol 14:1166898

Jourdes M, Pouységu L, Deffieux D, Teissedre P-L, Quideau S (2013) Hydrolyzable tannins: gallotannins and ellagitannins. In: Ramawat KG, Mérillon J-M (eds) Natural products: phytochemistry, botany and metabolism of alkaloids, phenolics and terpenes. Springer, Berlin, Heidelberg

Kamil A, Chen C-YO, Blumberg JB (2015) The application of nanoencapsulation to enhance the bioavailability and distribution of polyphenols. In: Nanotechnology and functional foods. Wiley, Hoboken

Karimi A, Azar PS, Kadkhodayi M, Tandorost A, Vajdi M, Shoorei H, Farhangi MA (2022) A comprehensive insight into effects of resveratrol on molecular mechanism in rheumatoid arthritis: a literature systematic review. Int J Rheum Dis 25:827–843

Kartah BE, El Monfalouti H (2024) Enhancing polyphenol bioavailability through nanotechnology: current trends and challenges. In: Gouvinhas I, Novo Barros A (eds) Exploring natural phenolic compounds – recent progress and practical applications. IntechOpen, Rijeka

Khameneh B, Iranshahy M, Soheili V, Fazly Bazzaz BS (2019) Review on plant antimicrobials: a mechanistic viewpoint. Antimicrob Resist Infect Control 8:118

Khodavandi A, Alizadeh F, Harmal NS, Sidik SM, Othman F, Sekawi Z, Jahromi MAF, Ng K-P, Chong PP (2011) Comparison between efficacy of allicin and fluconazole against *Candida albicans* in vitro and in a systemic candidiasis mouse model. FEMS Microbiol Lett 315:87–93

Khusnutdinova E, Petrova A, Zileeva Z, Kuzmina U, Zainullina L, Vakhitova Y, Babkov D, Kazakova O (2021) Novel A-ring chalcone derivatives of oleanolic and ursolic amides with anti-proliferative effect mediated through ROS-triggered apoptosis. Int J Mol Sci 22:9796

Kim T, Song B, Cho KS, Lee I-S (2020) Therapeutic potential of volatile terpenes and terpenoids from forests for inflammatory diseases. Int J Mol Sci 21:2187

Kłos P, Chlubek D (2022) Plant-derived terpenoids: a promising tool in the fight against melanoma. Cancer 14:502

Koche D, Shirsat R, Kawale M (2016) An overerview of major classes of phytochemicals: their types and role in disease prevention. Hislopia J 9:1–11

Kovacic P, Somanathan R (2009) Novel, unifying mechanism for mescaline in the central nervous system: electrochemistry, catechol redox metabolite, receptor, cell signaling and structure activity relationships. Oxidative Med Cell Longev 2:181–190

Kumar A (2016) Vincristine and vinblastine: a review. IJMPS 6:23–30

Kumar N, Awasthi A, Kumari A, Sood D, Jain P, Singh T, Sharma N, Grover A, Chandra R (2022) Antitussive noscapine and antiviral drug conjugates as arsenal against COVID-19: a comprehensive chemoinformatics analysis. J Biomol Struct Dyn 40:101–116

Lang Y, Gao N, Zang Z, Meng X, Lin Y, Yang S, Yang Y, Jin Z, Li B (2024) Classification and antioxidant assays of polyphenols: a review. J Future Foods 4:193–204

Lanzotti V, Scala F, Bonanomi G (2014) Compounds from *Allium* species with cytotoxic and antimicrobial activity. Phytochem Rev 13:769–791

Lee S-H, Lee Y-J (2021) Synergistic anticancer activity of resveratrol in combination with docetaxel in prostate carcinoma cells. Nutr Res Pract 15:12–25

Lee S, Jo C, Choi H-Y, Lee K (2021) Effect of co-administration of curcumin with amlodipine in hypertension. Nutrients 13:2797

Leis K, Kulczyńska A, Racinowski M, Kaczor P, Gołębiewski J, Januszko-Giergielewicz B (2021) Genistein–a supplement improving efficiency of the human body: a review. Sci Sports 36:359–367

Leopoldini M, Russo N, Toscano M (2011) The molecular basis of working mechanism of natural polyphenolic antioxidants. Food Chem 125:288–306

Li W, Beta T (2013) Food sources of phenolics compounds. In: Natural products: phytochemistry, botany and metabolism of alkaloids, phenolics and terpenes. Springer, Berlin, Heidelberg

Li C, Schluesener H (2017) Health-promoting effects of the citrus flavanone hesperidin. Crit Rev Food Sci Nutr 57:613–631

Li Y, Yao J, Han C, Yang J, Chaudhry MT, Wang S, Liu H, Yin Y (2016) Quercetin, inflammation and immunity. Nutrients 8:167

Li C, Guan X-M, Wang R-Y, Xie Y-S, Zhou H, Ni W-J, Tang L-Q (2020) Berberine mitigates high glucose-induced podocyte apoptosis by modulating autophagy via the mTOR/P70S6K/4EBP1 pathway. Life Sci 243:117277

Li Z, Zhao T, Shi M, Wei Y, Huang X, Shen J, Zhang X, Xie Z, Huang P, Yuan K (2023) Polyphenols: natural food grade biomolecules for treating neurodegenerative diseases from a multi-target perspective. Front Nutr 10:1139558

Lichota A, Gwozdzinski K (2018) Anticancer activity of natural compounds from plant and marine environment. Int J Mol Sci 19:3533

Lin H, Hu J, Mei F, Zhang Y, Ma Y, Chen Q, Wang C, Fu J, Yang M, Wen Z (2022) Anti-microbial efficacy, mechanisms and druggability evaluation of the natural flavonoids. J Appl Microbiol 133:1975–1988

Linden M, Brinckmann C, Feuereisen MM, Schieber A (2020) Effects of structural differences on the antibacterial activity of biflavonoids from fruits of the Brazilian peppertree (Schinus terebinthifolius Raddi). Food Res Int 133:109134

Liu RH (2013) Dietary bioactive compounds and their health implications. J Food Sci 78:A18–A25

Liu H-T, Ho Y-S (2018) Anticancer effect of curcumin on breast cancer and stem cells. Food Sci Human Wellness 7:134–137

Liu Y-J, Peng W, Hu M-B, Xu M, Wu C-J (2016) The pharmacology, toxicology and potential applications of arecoline: a review. Pharm Biol 54:2753–2760

Lv Q-Z, Long J-T, Gong Z-F, Nong K-Y, Liang X-M, Qin T, Huang W, Yang L (2021) Current state of knowledge on the antioxidant effects and mechanisms of action of polyphenolic compounds. Nat Prod Commun 16:1934578X211027745

Ma G, Bavadekar SA, Davis YM, Lalchandani SG, Nagmani R, Schaneberg BT, Khan IA, Feller DR (2007) Pharmacological effects of ephedrine alkaloids on human A1- and A2-adrenergic receptor subtypes. J Pharmacol Exp Ther 322:214–221

Ma K, Yuen M, Yuen T, Yuen H, Peng Q (2024) Protective mechanism of Sea buckthorn proanthocyanidins against hydrogen peroxide-introduced oxidative damage in adult retinal pigment epithelial-19. Antioxidants 13:1352

Malikova J, Zdarilova A, Hlobilkova A (2006) Effects of sanguinarine and chelerythrine on the cell cycle and apoptosis. Biomed Pap 150:5–12

Malonga H, Neault J, Diamantoglou S, Tajmir-Riahi H (2005) Taxol anticancer activity and DNA binding. Mini Rev Med Chem 5:307–311

Mandal SM, Chakraborty D, Dey S (2010) Phenolic acids act as signaling molecules in plant-microbe symbioses. Plant Signal Behav 5:359–368

Manzoor MF, Ahmad N, Ahmed Z, Siddique R, Zeng XA, Rahaman A, Muhammad Aadil R, Wahab A (2019) Novel extraction techniques and pharmaceutical activities of Luteolin and its derivatives. J Food Biochem 43:E12974

Marchese A, Barbieri R, Coppo E, Orhan IE, Daglia M, Nabavi SF, Izadi M, Abdollahi M, Nabavi SM, Ajami M (2017) Antimicrobial activity of eugenol and essential oils containing eugenol: a mechanistic viewpoint. Crit Rev Microbiol 43:668–689

Martín Ortega AM, Segura Campos MR (2019) Bioactive compounds as therapeutic alternatives. In: Bioactive compounds: health benefits and potential applications. Woodhead Publishing

Martino E, Della Volpe S, Terribile E, Benetti E, Sakaj M, Centamore A, Sala A, Collina S (2017) The long story of camptothecin: from traditional medicine to drugs. Bioorg Med Chem Lett 27:701–707

Masyita A, Sari RM, Astuti AD, Yasir B, Rumata NR, Emran TB, Nainu F, Simal-Gandara J (2022) Terpenes and terpenoids as main bioactive compounds of essential oils, their roles in human health and potential application as natural food preservatives. Food Chem X 13:100217

Mayo B, Vázquez L, Flórez AB (2019) Equol: a bacterial metabolite from the daidzein isoflavone and its presumed beneficial health effects. Nutrients 11:2231

Mazzantini C, El Bourji Z, Parisio C, Davolio PL, Cocchi A, Pellegrini-Giampietro DE, Landucci E (2024) Anti-inflammatory properties of cannabidiol and beta-caryophyllene alone or combined in an in vitro inflammation model. Pharmaceuticals 17:467

Medina I, Gallardo J, González MJ, Lois S, Hedges N (2007) Effect of molecular structure of phenolic families as hydroxycinnamic acids and catechins on their antioxidant effectiveness in minced fish muscle. J Agric Food Chem 55:3889–3895

Meng X, Dong X, Wang W, Yang L, Zhang X, Li Y, Chen T, Ma H, Qi D, Su J (2018) Natural borneol enhances paclitaxel-induced apoptosis of ESCC cells by inactivation of the PI3K/AKT. J Food Sci 83:1436–1443

Mittal M, Mehta P, Rajput S, Rajender S, Chattopadhyay N (2021) The pharmacological assessment of resveratrol on preclinical models of rheumatoid arthritis through a systematic review and meta-analysis. Eur J Pharmacol 910:174504

Monera-Penduka TG, Maponga CC, Morse GD, Nhachi CFB (2017) Capacity for ethical and regulatory review of herbal trials in developing countries: a case study of Moringa oleifera research in HIV-infected patients. J Pharm Policy Pract 10:9

Moudgil KD, Venkatesha SH (2022) The anti-inflammatory and immunomodulatory activities of natural products to control autoimmune inflammation. Int J Mol Sci 24:95

Mssusa AK, Holst L, Kagashe G, Maregesi S (2023) Safety profile of herbal medicines submitted for marketing authorization in Tanzania: a cross-sectional retrospective study. J Pharm Policy Pract 16:149

Munteanu IG, Apetrei C (2021) Analytical methods used in determining antioxidant activity: a review. Int J Mol Sci 22:3380

Naeem A, Hu P, Yang M, Zhang J, Liu Y, Zhu W, Zheng Q (2022) Natural products as anticancer agents: current status and future perspectives. Molecules 27:8367

Naeimi AF, Alizadeh M (2017) Antioxidant properties of the flavonoid fisetin: an updated review of in vivo and in vitro studies. Trends Food Sci Technol 70:34–44

Nath S, Choudhury MD, Roychoudhury S, Talukdar AD, Sirotkin AV, Baková Z, Kádasi A, Maruniaková N, Kolesárová A (2011) Restorative aspect of castor plant on mammalian physiology: a review. J Microbiol Biotechnol Food Sci 1:236–246

Negreiros HA, De Moura KG, Barreto Do Nascimento ML, Do Nascimento Rodrigues DC, Ferreir PM, Braz DC, De Farias MG, De Sousa Corrêia L, Pereira AR, Santos LK (2021) Alpha-terpineol as antitumor candidate in pre-clinical studies. Anticancer Agents Med Chem 21:2023–2031

Nie T, Cooper GJ (2021) Mechanisms underlying the antidiabetic activities of polyphenolic compounds: a review. Front Pharmacol 12:798329

Nordin N, Yeap SK, Rahman HS, Zamberi NR, Abu N, Mohamad NE, How CW, Masarudin MJ, Abdullah R, Alitheen NB (2019) In vitro cytotoxicity and anticancer effects of citral nanostructured lipid carrier on MDA MBA-231 human breast cancer cells. Sci Rep 9:1614

Nsairat H, Lafi Z, Al-Sulaibi M, Gharaibeh L, Alshaer W (2023) Impact of nanotechnology on the oral delivery of phyto-bioactive compounds. Food Chem 424:136438

Och A, Podgórski R, Nowak R (2020) Biological activity of berberine—a summary update. Toxins 12:713

Okhunov I, Mirzaev YR, Bobakulov KM, Abdullaev N, Aripova S (2016) Constituents of *Convolvulus fruticosus* and pharmacological activity of the main alkaloid cuscohygrine. Chem Nat Compd 52:558–559

Oliveira J, Mateus N, De Freitas V (2013) Flavanols: catechins and proanthocyanidins. In: Ramawat KG, Mérillon J-M (eds) Natural products: phytochemistry, botany and metabolism of alkaloids, phenolics and terpenes. Springer, Berlin, Heidelberg

Oliveira-Costa JF, Meira CS, Neves MVGD, Dos Reis BPZC, Soares MBP (2022) Anti-inflammatory activities of betulinic acid: a review. Front Pharmacol 13:883857

Olufunmilayo EO, Gerke-Duncan MB, Holsinger RD (2023) Oxidative stress and antioxidants in neurodegenerative disorders. Antioxidants 12:517

Pant M, Husen A (2025) Exploring medicinal bamboos. Taylor & Francis Group, LLC. https://doi.org/10.1201/9781003485933

Pant M, Sohrab SS, Husen A (2025) Genetic improvement and conservation practice of medicinal plants. Taylor & Francis Group, LLC. https://doi.org/10.1201/9781003482659

Parveen A, Parveen B, Parveen R, Ahmad S (2015) Challenges and guidelines for clinical trial of herbal drugs. J Pharm Bioallied Sci 7:329–333

Passie T, Seifert J, Schneider U, Emrich HM (2002) The pharmacology of psilocybin. Addict Biol 7:357–364

Passie T, Halpern JH, Stichtenoth DO, Emrich HM, Hintzen A (2008) The pharmacology of lysergic acid diethylamide: a review. CNS Neurosci Ther 14:295–314

Pelletier SW, Pelletier SW (1983) Alkaloids: chemical and biological perspectives. Springer

Peng Y, Ao M, Dong B, Jiang Y, Yu L, Chen Z, Hu C, Xu R (2021) Anti-inflammatory effects of curcumin in the inflammatory diseases: status, limitations and countermeasures. Drug Des Devel Ther 15:4503–4525

Perreca E, Rohwer J, González-Cabanelas D, Loreto F, Schmidt A, Gershenzon J, Wright LP (2020) Effect of drought on the methylerythritol 4-phosphate (MEP) pathway in the isoprene emitting conifer Picea glauca. Front Plant Sci 11:546295

Petrovska BB (2012) Historical review of medicinal plants' usage. Pharmacogn Rev 6:1

Phan HT, Yoda T, Chahal B, Morita M, Takagi M, Vestergaard MDC (2014) Structure-dependent interactions of polyphenols with a biomimetic membrane system. Biochim Biophys Acta 1838:2670–2677

Pinheiro-Neto FR, Lopes EM, Acha BT, Da Silva Gomes L, Dias WA, Dos Reis Filho AC, De Sousa Leal B, Do Nascimento Rodrigues DC, Do Nascimento Silva J, Dittz D (2021) α-Phellandrene exhibits antinociceptive and tumor-reducing effects in a mouse model of oncologic pain. Toxicol Appl Pharmacol 418:115497

Platzer M, Kiese S, Herfellner T, Schweiggert-Weisz U, Miesbauer O, Eisner P (2021) Common trends and differences in antioxidant activity analysis of phenolic substances using single electron transfer based assays. Molecules 26:1244

Platzer M, Kiese S, Tybussek T, Herfellner T, Schneider F, Schweiggert-Weisz U, Eisner P (2022) Radical scavenging mechanisms of phenolic compounds: a quantitative structure-property relationship (QSPR) study. Front Nutr 9:882458

Priya A, Pandian SK (2022) Biofilm and hyphal inhibitory synergistic effects of phytoactives piperine and cinnamaldehyde against *Candida albicans*. Med Mycol 60:Myac039

Prolo C, Piacenza L, Radi R (2024) Peroxynitrite: a multifaceted oxidizing and nitrating metabolite. Curr Opin Chem Biol 80:102459

Pugazhendhi D, Pope G, Darbre P (2005) Oestrogenic activity of *p*-hydroxybenzoic acid (common metabolite of paraben esters) and methylparaben in human breast cancer cell lines. J Appl Toxicol 25:301–309

Raheja S, Girdhar A, Lather V, Pandita D (2018) Biochanin a: a phytoestrogen with therapeutic potential. Trends Food Sci Technol 79:55–66

Ramalingam M, Kim H, Lee Y, Lee Y-I (2018) Phytochemical and pharmacological role of liquiritigenin and isoliquiritigenin from *Radix glycyrrhizae* in human health and disease models. Front Aging Neurosci 10:348

Reyes-Escogido MDL, Gonzalez-Mondragon EG, Vazquez-Tzompantzi E (2011) Chemical and pharmacological aspects of capsaicin. Molecules 16:1253–1270

Sabogal-Guáqueta AM, Osorio E, Cardona-Gómez GP (2016) Linalool reverses neuropathological and behavioral impairments in old triple transgenic Alzheimer's mice. Neuropharmacology 102:111–120

Salihović M, Huseinović S, Špirtović-Halilović S, Osmanović A, Dedić A, Ašimović Z, Završnik D (2014) DFT study and biological activity of some methylxanthines. Bull Chem Technol Bosnia Herzegovina 42:31–36

Santora VJ, Covel JA, Hayashi R, Hofilena BJ, Ibarra JB, Pulley MD, Weinhouse MI, Sengupta D, Duffield JJ, Semple G (2008) A new family of H3 receptor antagonists based on the natural product Conessine. Bioorg Med Chem Lett 18:1490–1494

Santos AP, Moreno PRH (2004) Pilocarpus spp.: a survey of its chemical constituents and biological activities. Rev Bras Ciênc Farm 40:116–137

Semwal R, Joshi SK, Semwal RB, Semwal DK (2021) Health benefits and limitations of rutin – a natural flavonoid with high nutraceutical value. Phytochem Lett 46:119–128

Serban MC, Sahebkar A, Zanchetti A, Mikhailidis DP, Howard G, Antal D, Andrica F, Ahmed A, Aronow WS, Muntner P, Lip GY, Graham I, Wong N, Rysz J, Banach M (2016) Effects of quercetin on blood pressure: a systematic review and meta-analysis of randomized controlled trials. J Am Heart Assoc 5:e002713

Shah Z, Gohar UF, Jamshed I, Mushtaq A, Mukhtar H, Zia-Ui-Haq M, Toma SI, Manea R, Moga M, Popovici B (2021) Podophyllotoxin: history, recent advances and future prospects. Biomolecules 11:603

Shahidi F, Danielski R (2024) Review on the role of polyphenols in preventing and treating type 2 diabetes: evidence from in vitro and in vivo studies. Nutrients 16:3159

Shamima AR, Fakurazi S, Hidayat MT, Hairuszah I, Moklas MAM, Arulselvan P (2012) Antinociceptive action of isolated mitragynine from *Mitragyna speciosa* through activation of opioid receptor system. Int J Mol Sci 13:11427–11442

Shanbhag SM, Kulkarni HJ, Gaitonde BB (1970) Pharmacological actions of berberine on the central nervous system. Jpn J Pharmacol 20:482–487

Shimizu K, Yoshihara E, Takahashi M, Gotoh K, Orita S, Urakawa N, Nakajyo S (2000) Mechanism of relaxant response to papaverine on the smooth muscle of non-pregnant rat uterus. J Smooth Muscle Res 36:83–91

Simonetto M, Infante M, Sacco RL, Rundek T, Della-Morte D (2019) A novel anti-inflammatory role of omega-3 PUFAs in prevention and treatment of atherosclerosis and vascular cognitive impairment and dementia. Nutrients 11:2279

Singh B, Singh JP, Kaur A, Singh N (2020) Phenolic composition, antioxidant potential and health benefits of citrus peel. Food Res Int 132:109114

Singh K, Gupta JK, Jain D, Chaudhary AN, Kumar S, Sharma MC, Singh B (2024) Unveiling the therapeutic potential of traditional Chinese medicine Rhizome coptidis: a comprehensive review. Pharmacol Res Mod Chin Med 13:100522

Slobodnick A, Shah B, Pillinger MH, Krasnokutsky S (2015) Colchicine: old and new. Am J Med 128:461–470

Somani SJ, Modi KP, Majumdar AS, Sadarani BN (2015) Phytochemicals and their potential usefulness in inflammatory bowel disease. Phytother Res 29:339–350

Stacchiotti A, Corsetti G (2020) Natural compounds and autophagy: allies against neurodegeneration. Front Cell Dev Biol 8:555409

Stanca E, Serviddio G, Bellanti F, Vendemiale G, Siculella L, Giudetti AM (2013) Down-regulation of LPCAT expression increases platelet-activating factor level in cirrhotic rat liver: potential antiinflammatory effect of silybin. Biochim Biophys Acta 1832:2019–2026

Stansbury J (2013) Lobelia's mechanisms of action on the pulmonary system. Naturopathic Doctor News & Review, 9

Strawbridge R, Javed RR, Cave J, Jauhar S, Young AH (2023) The effects of reserpine on depression: a systematic review. J Psychopharmacol 37:248–260

Stuart A, Brooks C, Prescott R, Blackwell A (2000) Repellent and antifeedant activity of salicylic acid and related compounds against the biting midge, Culicoides impunctatus (Diptera: Ceratopogonidae). J Med Entomol 37:222–227

Sung H-Y, Hong C-G, Suh Y-S, Cho H-C, Park J-H, Bae J-H, Park W-K, Han J, Song D-K (2010) Role of (−)-epigallocatechin-3-gallate in cell viability, lipogenesis, and retinol-binding protein 4 expression in adipocytes. Naunyn Schmiedebergs Arch Pharmacol 382:303–310

Tabakam TG, Makhafola TJ (2024) Plant-derived alkaloids as a potential source of treatment for colorectal cancer over the past five years: a comprehensive review. Plants 13:2723

Taiwo B, Akinkunmi E, Omisore N (2013) Antimicrobial and antiplasmodial activities of a quaternary compound from *Ritchiea capparoides* var. *longipedicellata*. Afr J Tradit Complement Altern Med 10:528–531

Tanaka M, Zhao J, Suyama A, Matsui T (2012) Epigallocatechin gallate promotes the vasorelaxation power of the antiatherosclerotic dipeptide Trp-His in contracted rat aorta. J Agric Food Chem 60:9048–9054

Tasso B, Mattioli LB, Tonelli M, Boido V, Chiarini A, Sparatore F, Budriesi R (2023) Further quinolizidine derivatives as antiarrhythmic agents-3. Molecules 28:6916

Thakur K, Zhu Y-Y, Feng J-Y, Zhang J-G, Hu F, Prasad C, Wei Z-J (2020) Morin as an imminent functional food ingredient: an update on its enhanced efficacy in the treatment and prevention of metabolic syndromes. Food Funct 11:8424–8443

Thilakarathna WW, Langille MG, Rupasinghe HV (2018) Polyphenol-based prebiotics and synbiotics: potential for cancer chemoprevention. Curr Opin Food Sci 20:51–57

Thirumalaisamy R, Bhuvaneswari M, Haritha S, Jeevarathna S, Janani KSS, Suresh K (2022) Curcumin, naringenin and resveratrol from natural plant products hold promising solutions for modern world diseases—a recent review. S Afr J Bot 151:567–580

Tian WW, Liu L, Chen P, Yu DM, Li QM, Hua H, Zhao JN (2025) *Curcuma longa* (turmeric): from traditional applications to modern plant medicine research hotspots. Chin Med 20:76

Ting Y, Chang W-T, Shiau D-K, Chou P-H, Wu M-F, Hsu C-L (2018) Antiobesity efficacy of quercetin-rich supplement on diet-induced obese rats: effects on body composition, serum lipid profile, and gene expression. J Agric Food Chem 66:70–80

Tu Y (2011) The discovery of artemisinin (qinghaosu) and gifts from Chinese medicine. Nat Med 17:1217–1220

Turner N, Li J-Y, Gosby A, To SWC, Cheng Z, Miyoshi H, Taketo MM, Cooney GJ, Kraegen EW, James DE, Hu L-H, Li J, Ye J-M (2008) Berberine and its more biologically available derivative, dihydroberberine, inhibit mitochondrial respiratory complex I: a mechanism for the action of berberine to activate AMP-activated protein kinase and improve insulin action. Diabetes 57:1414–1418

Ulanowska M, Olas B (2021) Biological properties and prospects for the application of eugenol—a review. Int J Mol Sci 22:3671

Usure RE, Kebebe D, Mekasha YT, Hasen G, Chura Waritu N, Dubale S, Suleman S (2024) Traditional herbal medicine regulatory implementation in Ethiopia: a qualitative study. Front Pharmacol 15:1392330

Vayalil PK (2012) Date fruits (*Phoenix dactylifera* Linn): an emerging medicinal food. Crit Rev Food Sci Nutr 52:249–271

Velez Z, Campinho MA, Guerra ÂR, García L, Ramos P, Guerreiro O, Felício L, Schmitt F, Duarte M (2012) Biological characterization of *Cynara cardunculus* L. methanolic extracts: antioxidant, anti-proliferative, anti-migratory and anti-angiogenic activities. Agriculture 2:472–492

Viesca FS, Gómez RG (2023) The mechanism of the oxidation of narceine by means of selenous acid-sulphuric acid. Magna Sci Adv Res Rev 9:30

Vigerelli H, Sciani JM, Eula MAC, Sato LA, Antoniazzi MM, Jared C, Pimenta DC (2018) Biological effects and biodistribution of Bufotenine on mice. Biomed Res Int 2018:1032638

Viveiros MMH, Silva MG, Da Costa JGM, De Oliveira AG, Rubio C, Padovani CR, Rainho CA, Schellini SA (2022) Anti-inflammatory effects of α-humulene and β-caryophyllene on pterygium fibroblasts. Int J Ophthalmol 15:1903

Volke DC, Rohwer J, Fischer R, Jennewein S (2019) Investigation of the methylerythritol 4-phosphate pathway for microbial terpenoid production through metabolic control analysis. Microb Cell Factories 18:192

Wang H, Zhu C, Ying Y, Luo L, Huang D, Luo Z (2017) Metformin and berberine, two versatile drugs in treatment of common metabolic diseases. Oncotarget 9:10135

Wang Z-Z, Li H, Maskey AR, Srivastava K, Liu C, Yang N, Xie T, Fu Z, Li J, Liu X (2024) The efficacy & molecular mechanisms of a terpenoid compound ganoderic acid C1 on corticosteroid-resistant neutrophilic airway inflammation: in vivo and in vitro validation. J Inflamm Res 17:2547–2561

Williamson G (2017) The role of polyphenols in modern nutrition. Nutr Bull 42:226–235

Xia N, Daiber A, Förstermann U, Li H (2017) Antioxidant effects of resveratrol in the cardiovascular system. Br J Pharmacol 174:1633–1646

Xu H, Dickschat JS (2020) Germacrene A – a central intermediate in sesquiterpene biosynthesis. Chemistry 26:17318–17341

Xu B, Huang S, Chen Y, Wang Q, Luo S, Li Y, Wang X, Chen J, Luo X, Zhou L (2021) Synergistic effect of combined treatment with baicalin and emodin on DSS-induced colitis in mouse. Phytother Res 35:5708–5719

Xu B, Bai L, Chen L, Tong R, Feng Y, Shi J (2022) Terpenoid natural products exert neuroprotection via the PI3K/Akt pathway. Front Pharmacol 13:1036506

Xue H, Xu M, Gong D, Zhang G (2023) Mechanism of flavonoids inhibiting xanthine oxidase and alleviating hyperuricemia from structure – activity relationship and animal experiments: a review. Food Front 4:1643–1665

Yang Y, Zhang T (2019) Antimicrobial activities of tea polyphenol on phytopathogens: a review. Molecules 24:816

Yang L, Xian D, Xiong X, Lai R, Song J, Zhong J (2018) Proanthocyanidins against oxidative stress: from molecular mechanisms to clinical applications. Biomed Res Int 2018:8584136

Yang B, Dong Y, Wang F, Zhang Y (2020) Nanoformulations to enhance the bioavailability and physiological functions of polyphenols. Molecules 25:4613

Yehia SM, Ayoub IM, Watanabe M, Devkota HP, Singab ANB (2023) Metabolic profiling, antioxidant, and enzyme inhibition potential of Iris pseudacorus L. from Egypt and Japan: a comparative study. Sci Rep 13:5233

Yin C, Cheng L, Zhang X, Wu Z (2020) Nanotechnology improves delivery efficiency and bioavailability of tea polyphenols. J Food Biochem 44:E13380

Zhang Y-J, Gan R-Y, Li S, Zhou Y, Li A-N, Xu D-P, Li H-B (2015) Antioxidant phytochemicals for the prevention and treatment of chronic diseases. Molecules 20:21138–21156

Zhang J, Zhang M, Zhang W-H, Zhu Q-M, Ning J, Huo X-K, Xiao H-T, Sun C-P (2022a) Total terpenoids of Inula japonica activated the Nrf2 receptor to alleviate the inflammation and oxidative stress in LPS-induced acute lung injury. Phytomedicine 107:154377

Zhang Z, Li X, Sang S, Mcclements DJ, Chen L, Long J, Jiao A, Jin Z, Qiu C (2022b) Polyphenols as plant-based nutraceuticals: health effects, encapsulation, nano-delivery, and application. Foods 11:2189

Zhang L, Liu X, Wang J, Li Z, Wang S, Yang W, Hai Y, Liu D (2025) Kaempferol protects against doxorubicin-induced myocardial damage by inhibiting mitochondrial ROS-dependent ferroptosis. Redox Rep 30:2503130

Zhou X, Wang F, Zhou R, Song X, Xie M (2017) Apigenin: a current review on its beneficial biological activities. J Food Biochem 41:E12376

Zhou X, Xie D, Huang J, Lu A, Wang R, Jin Y, Zhang R, Chang C, Xu L, Xu L (2021) Therapeutic effects of (5r)-5-hydroxytriptolide on fibroblast-like synoviocytes in rheumatoid arthritis via lncRNA WAKMAR2/miR-4478/E2F1/p53 axis. Front Immunol 12:605616

Chapter 9
Nutraceuticals: Plant-Based Nutrition for Health and Wellness

Salma Sameh, Ashaimaa Y. Moussa, Nada M. Mostafa, and Abdel Nasser B. Singab

Abstract Neurological and psychiatric problems, such as depression, insomnia, anxiety, Parkinson's, and Alzheimer's diseases, are considered huge health challenges all over the world. Conventional therapies are found to be ineffective in some cases owing to resistance development. Moreover, they could elicit a wide range of side effects. Currently, there is increased demand for therapeutically beneficial plant-derived nutraceuticals as natural alternatives to address these conditions. The current chapter explores the effectiveness and therapeutic activity of natural phytometabolites of terrestrial and marine origin, such as *Hypericum perforatum*, turmeric, and saffron, against a range of disorders. The key neurobiological pathways by which these herbs and compounds exert their actions are stated in this chapter. Notably, the use of these nutraceuticals is associated with critical challenges, as discussed in this chapter, including regulatory obstacles, safety concerns, and bioavailability issues. This comprehensive chapter emphasizes the role of herbal nutraceuticals as a promising and valuable component in the development of sustainable strategies for mental and neurological health.

Keywords Neurological · Depression · Nutraceuticals · *Hypericum perforatum* · Neurobiological pathways

---

S. Sameh · A. Y. Moussa · N. M. Mostafa
Department of Pharmacognosy, Faculty of Pharmacy, Ain Shams University, Cairo, Egypt

A. N. B. Singab (✉)
Department of Pharmacognosy, Faculty of Pharmacy, Ain Shams University, Cairo, Egypt

Center for Drug Discovery Research and Development, Faculty of Pharmacy, Ain Shams University, Cairo, Egypt
e-mail: abdelnassersingab@pharma.asu.edu.eg; vpr.nassersingab@asu.edu.eg

L. A. Worku et al. (eds.), *Plant-Based Solutions for Sustainable Agriculture and Healthcare*, Sustainability Sciences in Asia and Africa,
https://doi.org/10.1007/978-981-95-8910-4_9

9.1 Introduction

Hundreds of millions of people worldwide are affected by a profound and persistent debilitating mental health condition, namely, depression. According to the World Health Organization, over 280 million people suffer from this disease (World Health Organization 2023). This disease is considered a significant global drain on personal well-being and public health resources. Additionally, anxiety, insomnia, Parkinson's disease, and Alzheimer's disease are among other debilitating disorders. Conventional therapies are available to treat these conditions; however, they possess significant limitations. For example, the conventional drugs, namely, selective serotonin reuptake inhibitors (SSRIs) and SNRIs, were considered as the primary defence against depression for decades (Nawaz et al. 2024). Unfortunately, these medications were proven to be ineffective, with approximately one-third of patients suffering from depression due to the development of resistance (Al-Harbi 2012). On the other hand, severe side effects arise from the employment of these drugs, leading to patients' premature discontinuation of treatment. Emotional blunting, sexual dysfunction, and weight gain are among the side effects (Kutzer et al. 2020; Ma et al. 2021).

Dysregulated stress response, chronic neuroinflammation, and oxidative stress represent multi-system dysfunctions associated with neurological and mental health conditions. Thus, there is a great demand for alternative therapies targeting multiple biological pathways rather than single-target synthetic drugs (Bottaccioli et al. 2025; Dantzer et al. 2008). Due to the limitations of conventional drugs, there has been a growing interest in herbal nutraceuticals, which exhibit potential therapeutic activities and nutritional value. These compounds are derived from terrestrial and marine sources and are composed of a diverse array of bioactive components that act synergistically (Peng et al. 2022; Singab et al. 2024; Swati et al. 2025).

Nutraceuticals could exhibit protective effects against neurological disorders (Moussa et al. 2024). Herbal nutraceuticals are widely employed in traditional medicine. Currently, these plant-derived substances are being extensively studied by modern science to determine the mechanisms by which they exert their effects (Husen 2021, 2022, 2023a, b, 2024a, b; Husen and Pant 2024a, b; Pant et al. 2025; Pant and Husen 2025). These agents could relieve the inflammatory storms which are responsible for depressive symptoms and neurological conditions. Also, they can protect the brain cells from damage by acting as antioxidants. In addition, they influence the release of neurotransmitters (Sarris et al. 2011). The protective potential of natural products of terrestrial and marine origin against neurological and psychiatric problems is explored in the confines of this comprehensive chapter. Interestingly, herbal nutraceuticals offer a promising, sustainable path to hope and healing, as the future of mental health and neurological strategies may depend on them.

9.2 A Natural Approach to Brain Health

Targeting key neurobiological pathways involved in depression and in a wide array of neurological and psychiatric problems, such as insomnia, anxiety, Parkinson's, and Alzheimer's diseases, represents a promising therapeutic approach offered by herbal nutraceuticals.

9.2.1 Monoamine Hypothesis

The classic monoamine theory assumes that depressive behavior is elicited by a lack of monoamine neurotransmitters, primarily serotonin (5-HT), norepinephrine, and dopamine (Delgado 2000). Selective serotonin reuptake inhibitors (SSRIs) are conventional antidepressants, acting by increasing the level of monoamines in the synapses. These drugs lead to a rapid elevation in the monoamine levels; however, the therapeutic actions of these drugs take place slowly. It was observed that herbal nutraceuticals can act on monoamine systems in different ways. They can either inhibit serotonin reuptake or regulate monoamine oxidase enzymes that break down these neurotransmitters (Sarris et al. 2011).

The development of depression and anxiety is associated with the dysregulation of these neurotransmitters. Notably, reduced amounts of norepinephrine and serotonin result in depression, lack of motivation, and fatigue. On the other hand, anxiety is characterized by a reversed condition. To the best of our knowledge, the stress response originates in brain regions such as the amygdala and the cerebral cortex. In stress- and fear-related conditions, enhanced norepinephrine release is observed in these areas. This increase in the firing of the locus coeruleus is directly connected to anxiety-type behaviors. Previous studies showed that drugs that can decrease this firing can alleviate anxiety (Bremner et al. 1996). Thus, high norepinephrine causes panic symptoms of anxiety. Similarly, an unbalanced serotonin system can result in heightened anxiety. The serotonin action is more complicated in the case of anxiety. It was observed that selective serotonin reuptake inhibitors (SSRIs) are effective in the long term; however, at the beginning of treatment, they exacerbate anxiety. Acute changes in certain brain circuits, such as the bed nucleus of the stria terminalis and amygdala, which are pivotal for dealing with fear and anxiety, are associated with this paradoxical effect. By increasing activity within the BNST, SSRIs can enhance fear and anxiety (Ravinder et al. 2013).

In addition, reports proved that social anxiety disorder (SAnD) is observed in cases of reduced dopamine function. This results in social withdrawal and avoidance owing to the reduced reward from social interaction. SAnD and Parkinson's disease are strongly connected, as the symptoms of SAnD frequently appear before the decline in motor activity (Robinson et al. 2006).

Herbal nutraceuticals offer a unique approach by targeting multiple pathways simultaneously. They can regulate the reuptake and breakdown of monoamines, making them a promising alternative for achieving neurochemical balance in cases of anxiety and depression (Urbanska et al. 2025).

9.2.2 Neuroinflammation and Oxidative Stress

Neuroinflammation and oxidative stress are key factors in several neurological and psychiatric disorders. Pro-inflammatory cytokines are released owing to chronic stress and inflammatory processes. This could result in cognitive and emotional symptoms of depression due to disruption in neurotransmitter synthesis, reduction in neurogenesis, and increase in oxidative damage in the brain (Nobis et al. 2020). Additionally, previous in vivo studies proved that anxiety-related behaviors could be reversed through inhibiting the enzymes related to oxidative stress. Thus, it was deduced that targeting oxidative stress may be more effective for some anxiety symptoms than conventional drugs such as diazepam (Bouayed et al. 2009). In addition, oxidative imbalance was reported as a cause of insomnia and sleep disruption (Heyat et al. 2022).

Moreover, dopaminergic neurons in the condition of Parkinson's disease are degenerated owing to oxidative stress. Initial insults caused by mitochondrial dysfunction and altered dopamine metabolism, along with the influence of PD-related genes such as alpha-synuclein, lead to oxidative stress, which in turn results in further stress (Dias et al. 2013). In the same context, Alzheimer's disease is characterized by the formation of (Aβ) plaques and tangles of the tau protein. Oxidative stress is responsible for their formation (Huang et al. 2016). It was observed that herbal products can reduce the incidence of degenerative diseases by controlling oxidative stress (El-Nashar et al. 2023; Mostafa et al. 2022; Zengin et al. 2023). Several herbal nutraceuticals can modulate key inflammatory pathways, thereby protecting neurons from oxidative damage by increasing antioxidant defence systems, as they represent a rich source of anti-inflammatory and antioxidant compounds, such as polyphenols and flavonoids (Figueiredo Godoy et al. 2025).

9.2.3 HPA Axis, Neurotrophic Factors, and Neurogenesis

It was observed that chronic stress and depression lead to over-activation of the hypothalamic-pituitary-adrenal axis (HPA), which represents the primary response system to stress. Hippocampus and other brain regions are damaged owing to high cortisol levels resulting from HPA axis stimulation (Bottaccioli et al. 2025). Long-lasting changes in the HPA axis caused by early stressful situations could lead to a wide range of psychiatric problems, comprising all types of anxiety (Faravelli et al. 2012). Additionally, chronic insomnia is associated with high cortisol levels. Short-term insomnia could be developed owing to initial stressful conditions; however, the stress of not sleeping could turn the condition into a long-term problem, leading to overactivity of HPA (Elder et al. 2023). The response of the HPA axis can be normalized by using adaptogens, such as *Rhodiola rosea* (Vasanthi et al. 2025). Furthermore, reduced amounts of brain-derived neurotrophic factor contribute to the symptoms of depression, Parkinson's disease, and Alzheimer's disease. Certain herbal nutraceuticals could exert protective effects against depression by increasing BDNF expression, enhancing neuronal survival and development, and reversing the structural damage associated with this disease (Numakawa and Kajihara 2025; Palasz et al. 2020; Gao et al. 2022). In addition, BDNF contributes to the regeneration of dopamine-producing neurons destroyed in PD (Palasz et al. 2020).

9.2.4 *Excitotoxicity*

The neurodegenerative conditions are characterized by progressive neuronal loss owing to excitotoxicity—the neurotransmitter glutamate can damage and kill cells by overactivating neurons. Furthermore, Alzheimer's disease is caused by glutamate, which causes excessive stimulation of NMDA receptors, leading to a cascade of events resulting in this disease. In the same context, excitotoxic neurodegeneration could be caused by the tau protein. Modulation of the sigma-1 receptor (σ1R) could target this process, thereby conferring protective effects on neurons. Furthermore, Parkinson's pathology is linked to the overstimulation of neurons by excessive glutamate, leading to a toxic influx of calcium ions, promoting oxidative stress and cell death (Wu et al. 2025).

9.3 Therapeutic Effects of Herbal Extracts and Natural Compounds Obtained from Terrestrial and Marine Origins Against Psychiatric and Neurological Disorders

Herbal nutraceuticals employ a multi-target approach, making them well-suited for treating several conditions and modulating biological pathways, including neurotransmitter levels, the inflammatory response, and cellular health. Figure 9.1 shows the mechanisms by which the active compounds elicit their antidepressant effect. Table 9.1 lists the natural bioactive compounds exhibiting protective effects against psychiatric and neurological diseases, and their chemical structures are illustrated in Fig. 9.2.

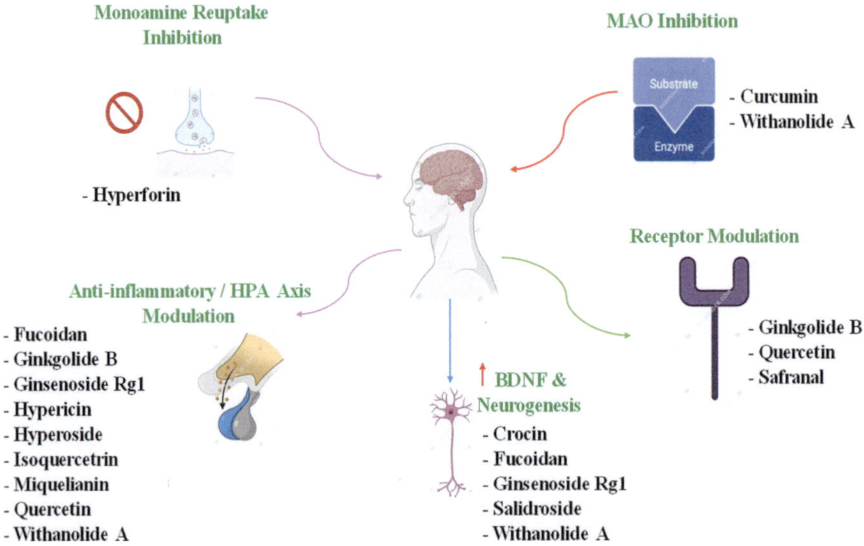

Fig. 9.1 Overview of the multifaceted mechanisms underlying the antidepressant effects of natural compounds

9.3.1 Hypericum perforatum

Teas and tinctures were prepared from *Hypericum perforatum* and utilized in the management of depression and anxiety (Klemow et al. 2011). Its use dates back to ancient Greece. Hippocrates first documented it for its therapeutic effects. The common name, "St. John's Wort," refers to its blooming period around June 24th, St. John the Baptist's Day (Istikoglou et al. 2010). Nobakht et al. (2022) stated that the plant was employed to alleviate melancholia and nervous problems.

Serotonin, norepinephrine, and dopamine are neurotransmitters that modulate mood. Reports revealed that the plant compound hyperforin is a potent inhibitor of the reuptake of these neurotransmitters. Chatterjee et al. (1998) showed that the compound exerted its inhibitory action, with IC_{50} values of 0.05–0.10 μg/mL. When a standardized hydroalcoholic extract containing 4.5% hyperforin prepared from *Hypericum* extract was administered to rats at doses of 3.12 and 6.25 mg/kg, it was observed that the immobility was reduced. A significant reduction in body weight and locomotor activity was observed when the extract was administered at a higher dose. Thus, it was deduced that the components of other extracts exhibited toxic and sedative effects. When isolated hyperforin, as its DCHA salt, was tested, it produced a similar antidepressant effect at much lower doses (0.19 and 0.38 mg/kg), without reducing locomotor activity or body weight. Pharmacokinetic analysis revealed that adequate doses of both the extract and the hyperforin salt resulted in micromolar plasma concentrations. Thus, it was concluded that hyperforin represents the key active component. However, its antidepressant action may not be solely due to direct

Table 9.1 Protective mechanisms of bioactive compounds against neurological and psychiatric disorders

Compound	Chemical class	Plant/marine source	Mechanism of action	Doses/IC_{50}	References
Hyperforin	Phloroglucinol derivative	*Hypericum perforatum*	Inhibits neurotransmitter reuptake (serotonin, norepinephrine, dopamine)	IC_{50}: 0.05–0.10 μg/mL (for reuptake inhibition). Doses: 0.19 and 0.38 mg/kg (isolated DCHA salt)	Chatterjee et al. (1998); Cervo et al. (2002)
			Reduces Aβ deposit formation and disaggregates existing Aβ fibrils, reduces reactive oxygen species (ROS), and alleviates oxidative stress	–	Dinamarca et al. (2006)
Hypericin	Naphthodianthrone		Modulates HPA axis	Dose: 0.2 mg/kg	Butterweck et al. (2004)
Hyperoside	Flavonoid glycoside			Dose: 0.6 mg/kg	
Isoquercitrin					
Miquelianin					
Curcumin	Curcuminoid	*Curcuma longa*	Modulates central monoaminergic system by inhibiting MAO-A and MAO-B, increasing serotonin and dopamine	Doses: 10–80 mg/kg (ip)	Kulkarni et al. (2008)
			Mixed-type inhibitor of monoamine oxidase-B	IC_{50}: 500.46 nM	Khatri and Juvekar (2016)
			Inhibits astrocytic activation (GFAP) and NFκB, reducing pro-inflammatory cytokines, inhibits NADPH oxidase, restores glutathione, prevents iron accumulation, and regulates Bax, Bcl-2, and caspases.	Dose: 40 mg/kg daily for 21 days	Sharma and Nehru (2018)
			Activates BDNF-ERK signaling pathway	Dose: 100 and 200 mg/kg	Zhang et al. (2015)

(continued)

Table 9.1 (continued)

Compound	Chemical class	Plant/marine source	Mechanism of action	Doses/IC_{50}	References
Crocin	Carotenoid	*Crocus sativus*	Regulates Wnt/*β*-catenin signaling pathway and promotes neurogenesis of the adult hippocampus	Doses: 12.5 and 25 mg/kg (oral)	Tao et al. (2023)
Safranal	Monoterpene aldehyde		Modulates noradrenergic system	Doses: 0.15–0.5 mg/kg	Hosseinzadeh et al. (2003)
			Facilitates the GABAergic inhibitory system and binds selectively to specific subtypes of the $GABA_A$-benzodiazepine receptor complex.	Dose: 0.15 and 0.35 mL/kg	Hosseinzadeh and Noraei (2009)
Crocetin	Carotenoid		Exhibits binding affinity for NMDA and (σ1) receptors.	Tested at concentrations up to 10 μM Ki = 1.6 μM at (σ1) receptor	D'Onofrio et al. (2021); Lechtenberg et al. (2008)
Ginsenoside Rg1	Triterpene saponin	*Panax ginseng*	Promotes hippocampal neurogenesis via CREB/BDNF pathway, modulates HPA axis, and reduces IL-6	Doses: 40 mg/kg (i.p.), 10 and 30 mg/kg	Liu et al. (2016); Zheng et al. (2014)
			Interrupts apoptotic cascade, reduces ROS and nitric oxide, prevents the release of cytochrome c, and blocks caspase-3 activation	Concentrations: 5–20 μM	Chen et al. (2003)
			Facilitates clearance of Aβ and p-Tau, upregulates BDNF and p-TrkB protein levels, restores LTP	Dose: 10 mg/kg daily for 30 consecutive days	Li et al. (2016)

Quercetin	Flavonoid	*Ginkgo biloba*	Dual mechanism: antioxidant effect (reduces LOOH), inhibits NMDA and/or nitric oxide pathway	Dose: 25 mg/kg	Holzmann et al. (2015)
Ginkgolide B	Terpene lactone		Modulates neurochemical and inflammatory pathways (restores 5-HT, inhibits IL-1β), possibly via the STAT3 pathway	Dose: 3.5 mg/kg	Ge et al. (2020)
			Reduces pro-inflammatory cytokines, promotes M1 to M2 microglial phenotype shift, and inactivates the NLRP3 inflammasome via autophagy	Concentrations: 20 and 40 μM Dose: 50 and 100 mg/kg, daily	Shao et al. (2022)
Withanolide A	Withanolide (steroid lactone)	*Withania somnifera*	Dual mechanism: inhibits MAO activity and reduces oxidative stress (ROS, MDA). Upregulates TrkB-BDNF signaling pathway	Concentrations: 1.56–3.12 μg/mL (in vitro)	Kim et al. (2025)
Salidroside	Phenylpropanoid glycoside	*Rhodiola rosea*	Promotes hippocampal neurogenesis via the SIRT1/PGC-1α signaling pathway	Doses: 25 and 50 mg/kg	Xing et al. (2024)
			Decreases concentrations of MDA and nitrate, increases SOD activity and GSH concentration, suppresses IL-6 and TNF-α, and suppresses the rate of hippocampal neuronal apoptosis.	Dose: 30 mg/kg	Li et al. (2018)
Fucoidan	Sulfated polysaccharide	Brown seaweed	Inhibits hippocampal inflammation via caspase-1, promoting BDNF-dependent synaptic plasticity	Doses: 25, 50, and 100 mg/kg	Li et al. (2020)
			Protects dopaminergic neurons and fibers, restores striatal dopamine and its metabolites, alleviates mitochondrial damage, restores mitochondrial respiratory function, reduces oxidative stress, and restores the PGC-1α and NRF2 expression	Doses: 35, 70, and 140 mg/kg	Li et al. (2018)

(continued)

Table 9.1 (continued)

Compound	Chemical class	Plant/marine source	Mechanism of action	Doses/IC_{50}	References
Aaptamine	Alkaloid	Marine sponges of genus *Aaptos*	–	Doses: 1, 10, and 20 mg/kg	Diers et al. (2008)
			Directly inhibits AChE and BuChE via a mixed-type inhibition mechanism	IC_{50}: 16.0 μM AChE and 4.6 μM BuChE; Ki = 6.96 μM AChE and 6.35 μM BuChE	Miao et al. (2022)
			Promotes recovery from dyskinesia	Dose: 5, 10, and 20 μM	

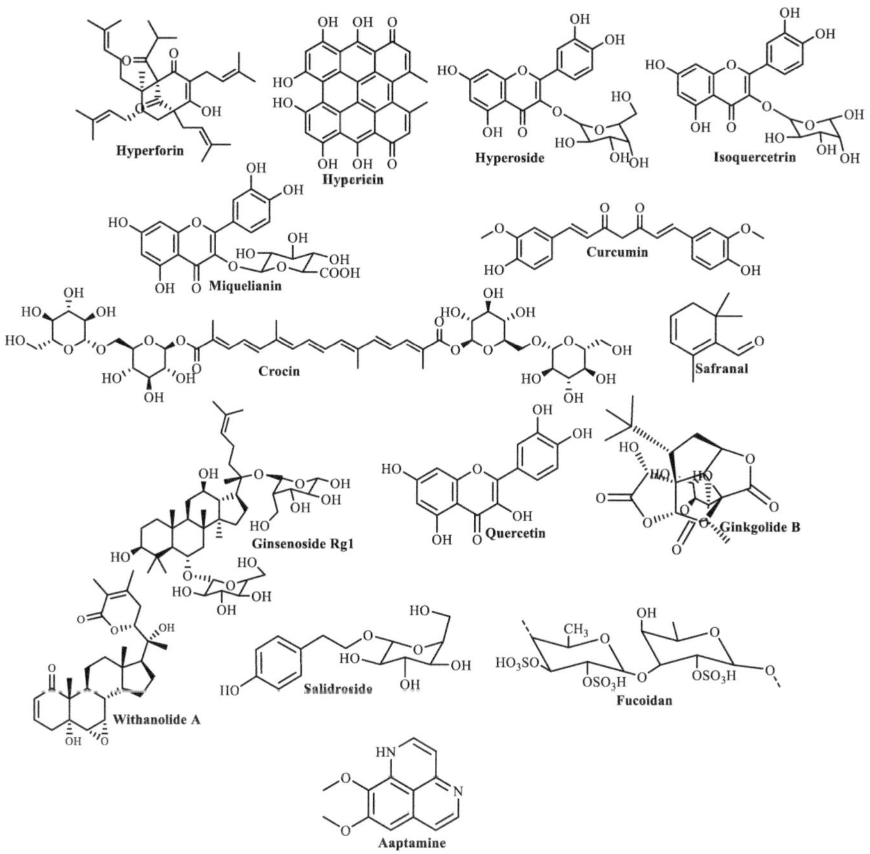

Fig. 9.2 Reported nutraceutical possessing protective effects against psychiatric and neurological conditions

neurotransmitter reuptake inhibition, as the therapeutic doses result in very low brain concentrations (Cervo et al. 2002).

Additionally, the plant's antidepressant effects are related to its ability to modulate the HPA axis, leading to a reduced stress response. Previous research showed that a methanol extract prepared from the plant and its component, hypericin, delayed the HPA axis. However, hyperforin exhibited no activity in this model. The plant's therapeutic properties are also linked to its flavonoid content. Imipramine (15 mg/kg), hypericin (0.2 mg/kg), and the flavonoids hyperoside (0.6 mg/kg), miquelianin (0.6 mg/kg), and isoquercitrin (0.6 mg/kg) were administered every day for 14 days in a study investigating this relationship. This treatment regimen significantly reduced plasma ACTH levels and decreased corticosterone levels by 40–70%. Furthermore, it was observed that with long-term treatment, the levels of corticosterone and ACTH were not influenced (Butterweck et al. 2004). Based on the findings of Butterweck et al. (2004), it was deduced that the HPA axis is modulated by hypericin, along with flavonoids. Moreover, the various constituents of the plant exert their antidepressant effects by modulating several biological pathways.

Preclinical studies have shown mixed but generally positive results, with animal models revealing anxiolytic-like effects. Interestingly, some of these effects were observed in extracts lacking hyperforin, suggesting that other compounds contribute to its anti-anxiety action. It is important to note that clinical evidence remains limited and often contradictory (Zirak et al. 2019). It was observed that in patients suffering from co-morbid depression, the symptoms of anxiety were reduced in open-label trials. On the other hand, a failure in efficacy determination was observed in larger, well-controlled studies. Zirak et al. (2019) indicated that *Hypericum perforatum* could not be used alone for the management of anxiety disorders, as the extract prepared from it, when tested against obsessive-compulsive disorder and social anxiety disorder, showed comparable results to placebo in placebo-controlled trials.

Mohanasundari and Sabesan (2007) showed that the extract from *Hypericum perforatum* exhibited antioxidant activity and inhibited monoamine oxidase-B, leading to neuroprotective effects against Parkinson's disease. Administration of the methanol extract at a dose of 300 mg/kg in a mouse model of Parkinson's disease led to 31% inhibition of striatal monoamine oxidase-B, preventing the formation of the neurotoxic metabolite MPP+. Additionally, astrogliosis was reduced, and GFAP-positive astrocytes showed a significant decrease. The neuroprotective effects were attributed to flavonoids, especially quercetin and hyperoside, owing to their potent antioxidant properties, which help scavenge free radicals.

Kraus et al. (2007) demonstrated that the flavanols (+)-catechin and (−)-epicatechin present in the plant extract exhibited neuroprotective effects against Alzheimer's disease and could serve as promising leads for future drug development. They exert their actions by counteracting the structural damage and formation of toxic ion channels caused by Aβ peptides and by restoring cellular membrane fluidity. Moreover, the antioxidant properties of the extract help protect microglial cells from amyloid-beta (Aβ) toxicity. Furthermore, Dinamarca et al. (2006) demonstrated the neuroprotective effect of hyperforin against Alzheimer's disease in a rat model of amyloidosis. It alleviates oxidative stress, disaggregates existing amyloid fibrils, and reduces the production of new Aβ aggregates in a time- and dose-dependent manner.

9.3.2 Curcuma longa

Yu et al. (2002) showed that *Curcuma longa* is a potent inhibitor of monoamine oxidase A, leading to an antidepressant effect, when administered at a dose of 140 mg/kg and above for 14 days to mice. The action was observed as reduced immobility in the tail suspension and forced swimming tests. By employing a dose (560 mg/kg), MAO-B activity inhibition was observed, and the extract's action was more potent than that of fluoxetine. Fluoxetine showed only a weak ability to inhibit MAO-A and MAO-B under the same conditions. Neither *C. longa* extract nor fluoxetine affected locomotor activity at the tested doses, indicating that the observed effects were antidepressant rather than sedative (Yu et al. 2002).

Reports revealed that curcumin exhibits antidepressant effects without affecting locomotor activity in animal models by modulating the central monoaminergic system. It was observed that intraperitoneal administration of the compound at doses ranging from 10 to 80 reduced immobility dose-dependently in the Forced Swim Test (FST) and in reserpine-induced immobility. After 60 min of administration of a dose of 40 mg/kg, the maximal anti-immobility effect in the FST was observed. At doses of 20–80 mg/kg, curcumin effectively increased brain serotonin (5-HT) levels. Additionally, at higher doses, it increased dopamine levels. However, norepinephrine levels remained unchanged. This activity is supported by its potent inhibition of monoamine oxidase A (MAO-A) in a dose-dependent manner and MAO-B at higher, non-dose-dependent levels. To the best of our knowledge, these findings were supported by combination studies. Curcumin (20 mg/kg) potentiated the antidepressant effects of monoaminergic drugs, including venlafaxine (4 mg/kg), fluoxetine (5 mg/kg), and bupropion (5 mg/kg). In contrast, no substantiation of the effects of norepinephrine-focused antidepressants like imipramine or desipramine (5 mg/kg) was observed (Kulkarni et al. 2008).

In addition, reports revealed that *Curcuma longa* possesses promising therapeutic activity in alleviating the neurodegeneration in Parkinson's disease. A water extract prepared from the plant exhibited antioxidant and anti-apoptotic activities through reducing the production of reactive oxygen species and down-regulating the expression of pro-apoptotic genes, such as p53, Bax, and caspase 3, resulting in significant neuroprotective effects against toxicity induced by the endogenous neurotoxin, namely, salsolinol (Ma and Guo 2017).

Khatri and Juvekar (2016) demonstrated that curcumin exhibited potent inhibitory activity against monoamine oxidase-B, owing to its structural similarity to MAO inhibitors, with an IC_{50} of 500.46 nM in a preclinical rat brain model. The study's kinetic analysis revealed that curcumin acts as a mixed-type inhibitor of MAO-B, binding to multiple sites to inhibit enzyme activity. This inhibition increases the availability of dopamine in the brain, prevents the activation of neurotoxins, and prevents the formation of damaging free radicals, thereby enhancing curcumin's therapeutic activity against Parkinson's disease.

Another study by Sharma and Nehru (2018) demonstrated that curcumin exhibits significant neuroprotective effects against Parkinson's disease by targeting several pathways. Administration of 40 mg/kg of curcumin every day for 21 days in an LPS-induced rat model of Parkinson's disease proved its neuroprotective action. It acts by preventing the aggregation of alpha-synuclein (α-syn). In addition, it reduces pro-inflammatory cytokines and iNOS by inhibiting astrocytic activation (GFAP) and the transcription factor NFκB. Furthermore, it inhibits NADPH oxidase, prevents neurotoxic iron accumulation, and restores the glutathione system, resulting in strong antioxidant effects. Also, the compound modulates the intrinsic apoptotic pathway, thereby blocking programmed cell death.

Curcumin elicited protective effects against chemotherapy-induced cognitive impairment (Rabie et al. 2023a). Zhang et al. (2015) demonstrated that chronic administration of curcumin at 100 and 200 mg/kg in an amyloid-beta (Aβ)-induced rat model of Alzheimer's disease alleviates cognitive dysfunction associated with Alzheimer's disease (AD) by modulating the BDNF-ERK signaling cascade. It demonstrates its activity by significantly reversing Aβ-induced deficits in working and spatial memory, without affecting the rats' locomotor activity. Moreover, it activates the ERK pathway and increases hippocampal BDNF levels.

9.3.3 Crocus sativus

Crocus sativus, a perennial herb, is used in traditional medicine for its antidepressant activity, owing to its crocin and safranal content (Siddiqui et al. 2018). Tao et al. (2023) performed an in vivo study to test the antidepressant effect of crocin when orally administered to male mice at doses of 12.5 or 25 mg/kg for 2 weeks in the tail suspension test (TST) and the Forced Swim Test (FST). Immobility times were reduced significantly in both tests. Notably, it was observed that the administration of the standard antidepressant fluoxetine at a dose of 20 mg/kg, i.p., resulted in adverse effects such as impaired male sexual function; however, crocin exhibited a safer profile.

In addition, Tao et al. (2023) demonstrated that crocin combats long-term stress and exhibits its antidepressant action by activating the Wnt/*β*-catenin signaling pathway. Moreover, when used at a dose of 25 mg/kg every day, it increases the proliferation and differentiation of neural progenitor cells, leading to increased dendritic length and spine density in new neurons, resulting in enhanced transmission in the synapse and potentiation for long term. Also, it reverses stress-induced behaviors such as immobility and anhedonia. Moreover, safranal exhibited antidepressant action. In the Forced Swim Test, an increase in climbing behavior is often associated with the induction of dopaminergic and noradrenergic systems. Also, increased swimming time indicates serotonergic modulation. Particular enhancement of climbing behavior by safranal at the dose (0.5 mg/kg) suggests potential involvement of the noradrenergic system in its antidepressant effects (Hosseinzadeh et al. 2003).

Hosseinzadeh and Noraei (2009) conducted an in vivo study to test the anxiolytic and hypnotic effects of the aqueous extract of *Crocus sativus*. The anxiolytic effect was most remarkable at low doses (56 and 80 mg/kg). At the same time, muscle-relaxant, sedative, and hypnotic properties were observed at higher doses up to 560 mg/kg, with results similar to those of diazepam. The anxiolytic and hypnotic effects were primarily caused by safranal and were observed at doses of 0.15 and 0.35 mL/kg. The whole extract and benzodiazepines exhibited motor incoordination and muscle relaxation, which were not observed with safranal. No anxiolytic, hypnotic, or muscle-relaxant activities were observed with crocin at doses reaching up to 600 mg/kg. The study proved the facilitation of the GABAergic inhibitory system. The unique therapeutic profile of safranal suggests the potential for selective binding to specific subtypes of the $GABA_A$-benzodiazepine receptor complex (Hosseinzadeh and Noraei 2009). Additionally, the compound crocetin could have potential applications in the management of Alzheimer's disease. Crocetin exhibited significant in vitro affinity for both NMDA and sigma-1 (σ1) receptors. The compound was tested at concentrations up to 10 μM and exhibited a Ki of 1.6 μM at the σ1 receptor, indicating strong binding affinity (D'Onofrio et al. 2021; Lechtenberg et al. 2008).

9.3.4 Panax ginseng

The potency of *Panax ginseng* as an antidepressant, anxiolytic, and cognitive enhancer was first systematically recorded in the Ming Dynasty by Shi-Zhen Li in the Ben Cao Gang Mu. Preclinical and clinical studies have proved these traditional applications. It is documented that various formulations containing the plant, including Gincosan, Sho-ju-sen, and Kai Xin San aqueous extract, can alleviate depressive symptoms in human and animal studies (Choi et al. 2018). Reports revealed that when the aqueous root extract was administered orally, it exhibited antidepressant activity via several mechanisms. Choi et al. (2018) showed that the extract inhibited the expression of nuclear factor erythroid 2-related factor 2 and heme oxygenase-1. Moreover, the administration of the extract decreased the chronic restraint stress (CRS)-induced increase in serum adrenocorticotropic hormone and corticosterone levels, thereby regulating the hypothalamic-pituitary-adrenal axis.

Furthermore, the extract increased the neuroplasticity, which is attributed to the increased expression of (BDNF) mRNA. Also, it down-regulated the levels of inducible nitric oxide synthase and cyclooxygenase-2, resulting in decreased neuroinflammation and reversed microglial activation. An in vivo study employing the CRS model revealed that the oral administration of the aqueous extract at doses of 75, 150, and 300 mg/kg pronouncedly reduced the immobility time in the forced swim and tail suspension tests, demonstrating a potent antidepressant effect. Based on the findings, it was concluded that the 150 mg/kg dose was optimal for eliciting a therapeutic effect (Choi et al. 2018).

Previous reports have shown that ginsenosides obtained from *P. ginseng* exhibit neuroprotective and antioxidant activities (Moussa et al. 2025). Liu et al. (2016) performed an in vivo study in a rat model of chronic unpredictable mild stress (CUMS) and showed that ginsenoside Rg1 at a dose of 40 mg/kg (i.p.) administered for 5 weeks exhibited antidepressant action by increasing BDNF expression in the amygdala. Another in vivo study by Zheng et al. (2014) showed that the compound elicited an antidepressant effect at 10 and 30 mg/kg in a rat model of LPS-induced neuroinflammation. It exerted its action by lowering the release of pro-inflammatory mediators, such as iNOS and myeloperoxidase, reducing microglial activity, regulating the HPA axis, and inhibiting the elevation of interleukin-6.

An in vivo study showed that *Panax ginseng* G115 root extract inhibited apoptosis, ER stress, and neuroinflammation. It prevented the accumulation of insoluble alpha-synuclein (α-syn), exhibiting neuroprotective effects in a model of Parkinson's condition activated by BSSG intoxication when administered at a dose of 100 mg/kg. It prevented motor impairment and the underlying neurodegeneration. It was observed that the extract preserved dopamine transporter immunolabeling in the striatum and protected the dopaminergic neurons located in the substantia nigra pars compacta (Van Kampen et al. 2014).

Chen et al. (2003) proved that ginsenoside Rg1, when tested at concentrations of 5–20 μM, alleviated neurotoxicity induced by dopamine in PC12 cells and protected the cells from the destruction caused by 0.15–0.45 mM dopamine through interrupting the apoptotic cascade at multiple levels, resulting in attenuating DNA fragmentation and maintaining cell viability. Also, it reduces the release of reactive oxygen species and nitric oxide and prevents subsequent cytochrome c release from the mitochondria.

Li et al. (2016) demonstrated that ginsenoside Rg1 exerted neuroprotective effects and improved memory performance when administered at 10 mg/kg for 30 consecutive days in a transgenic mouse model of Alzheimer's disease (AD) overexpressing APP and PS1. The compound exerted its action by increasing p-TrkB protein levels and upregulating BDNF, thereby restoring synaptic plasticity. Also, it inhibited the accumulation of phosphorylated (p)-Tau and β-amyloid 1–42.

9.3.5 Ginkgo biloba

Rojas et al. (2011) demonstrated that EGb 761 (standardized *Ginkgo biloba* extract) exhibited significant antidepressant action by modulating several key neurobiological pathways, including inhibiting the generation of the superoxide radical and decreasing lipid peroxidation. Also, it regulates serotonergic and dopaminergic neurotransmission in the prefrontal cortex, midbrain, and hippocampus. The activity of the extract is attributed to its flavonoid and terpene content, which are small molecules able to penetrate the blood-brain barrier effectively, resulting in the observed neuroprotective therapeutic activities. Pretreatment of male BALB/c mice with EGb 761 at 10 mg/kg (i.p.) daily for 17 days resulted in a significant reduction by 39% in immobility time during the forced swimming test.

In the same context, Woelk et al. (2007) showed that EGb 761 is an effective anxiolytic, particularly in patients where anxiety is co-morbid with cognitive decline. In a randomized, placebo-controlled clinical trial involving 107 patients suffering from generalized anxiety disorder (GAD) or adjustment disorder with anxious mood (ADWAM), the extract (240 and 480 mg) was administered daily for 4 weeks. A significant reduction in anxiety symptoms was observed, with a decrease in the Hamilton Anxiety Rating Scale (HAM-A) total scores in the high- and low-dose groups, which was significantly superior to the placebo group's decrease.

Holzmann et al. (2015) performed an in vivo study and stated that chronic treatment with quercetin (25 mg/kg) for 2 weeks exhibited antidepressant action in an olfactory bulbectomy (OB) induced depression model by reversing the anhedonia-like symptoms and behavioral hyperactivity. Additionally, it reduced immobility time in the forced swimming and tail suspension tests, without affecting locomotor activity, owing to its antioxidant effects, thereby inhibiting the OB-induced increase in lipid hydroperoxide (LOOH) levels in the hippocampus. Moreover, the compound inhibited the nitric oxide pathway or NMDA receptors.

Moreover, ginkgolide B has been proven to treat depression symptoms following myocardial infarction. This therapeutic effect of 3.5 mg/kg ginkgolide B was attributed to a modulation of neurochemical and inflammatory pathways in the median raphe nucleus and cortex, where the compound restored reduced serotonin (5-HT) levels and inhibited the increase of the pro-inflammatory cytokine interleukin 1 beta (IL-1β). In addition, the compound modulated the STAT3 pathway and reduced IL-1β production (Ge et al. 2020).

In addition, ginkgolide B exhibited powerful anti-inflammatory and cognitive-enhancing properties in both cellular and animal models. Shao et al. (2022) demonstrated that the compound effectively improves learning and memory by promoting autophagy and inhibiting the NLRP3 inflammasome via autophagic degradation. The study revealed that the compound, when tested at concentrations of 20 and 40 μM in LPS-challenged BV2 microglial cells, pronouncedly inhibited pro-inflammatory cytokines and promoted a conversion from the pro-inflammatory M1 to the anti-inflammatory M2 microglial phenotype. Furthermore, *an* in vivo study showed that the compound exhibited cognitive protection in SAMP8 mice when tested at doses of 50 and 100 mg/kg daily in both the Novel Object Recognition and Morris Water Maze tests by attenuating NLRP3 inflammasome activation and promoting autophagy in the mice's brains. Also, it reduces neuroinflammation and shifts the microglial phenotype in the hippocampus (Shao et al. 2022).

9.3.6 Withania somnifera

Kim et al. (2025) conducted an in vitro study demonstrating that ashwagandha root extract (ARE) and the compound withanolide A possess antidepressant potential. In an HT-22 neuronal cell corticosterone (CORT) exposure model, both ARE and withanolide A reversed the CORT-induced reduction in cell viability. Moreover, both (100–200 μg/mL) ARE and (100–200 μg/mL) withanolide A increased the availability of monoamine neurotransmitters as they inhibited the monoamine oxidase activity induced by CORT treatment. Moreover, they reduced the production of reactive oxygen species and malondialdehyde, resulting in potent antioxidant effects. Notably, the TrkB-BDNF signaling pathway is vital for promoting neuronal regeneration and neural plasticity by phosphorylating AKT and ERK and stimulating CREB. The results revealed that ARE, when employed at a concentration of 200 μg/mL, promoted BDNF, TrkB, p-AKT, p-ERK, and p-CREB levels compared with those of CORT-treated controls. Moreover, withanolide A increased protein levels at a concentration of 3.12 μg/mL. Furthermore, it was found that the chronic administration of ARE at a dose of 100 mg/kg in a mouse model of unpredictable chronic mild stress (UCMS) elicited significant antidepressant and anxiolytic properties, which were comparable to the positive control, fluoxetine, by reversing the conditions of weight loss and adrenal hypertrophy, which are the physiological indicators of chronic stress. (Kim et al. 2025).

Langade et al. (2019) showed that ARE represents a promising sleep-inducing and anxiolytic agent for patients suffering from insomnia and anxiety. Participants in a randomized, double-blind, placebo-controlled clinical trial received either a 300 mg ARE capsule twice daily or a placebo for 10 weeks. A significant improvement in various sleep parameters was observed. Notably, after 10 weeks, sleep efficiency (SE) increased from 75.63% to 83.48% compared with placebo. Overall sleep quality, Pittsburgh Sleep Quality Index (PSQI) scores, and Hamilton Anxiety Rating Scale (HAM-A) scores improved.

9.3.7 Rhodiola rosea

Rhodiola rosea is well tolerated in short-term evaluations and exhibits a favorable safety profile. It is considered an adaptogen. It was observed that *R. rosea* extracts and their key constituent, salidroside, when investigated in preclinical studies, exhibited multi-target effects through interactions with the neuroendocrine-immune system and the neurotransmitter receptor system, which are implicated in depressive symptoms (Amsterdam and Panossian 2016).

In a corticosterone (CORT)-induced depression animal study, salidroside showed both molecular and behavioral actions and has appeared as a good treatment for depression. A significant reduction in the lack of mobility time in different tests was observed when administering 25 and 50 mg/kg of the compound. Also, the treatment restored usual sucrose preference, a key indicator of anhedonia, which CORT had reduced. Additionally, significant neurobiological changes in the hippocampus accompanied these behavioral improvements. The compound promoted hippocampal neurogenesis. Moreover, key neurogenesis markers, such as NeuN, Ki-67, and BrdU, were reduced in this model. However, treatment reversed these conditions by enhancing SIRT1/PGC-1α signaling and restoring the expression of SIRT1, PGC-1α, and BDNF. Notably, the antidepressant effects of salidroside were abolished by pharmacological inhibition of SIRT1 signaling. This suggests that the promotion of neurogenesis by salidroside and its subsequent antidepressant action depend on the induction of the SIRT1/PGC-1α signaling pathway (Xing et al. 2024).

Li et al. (2018) showed that the compound can attenuate the effects of AD-associated cognitive impairment by exhibiting several mechanisms. It exerts its action by suppressing the rate of hippocampal neuronal apoptosis. Moreover, it enhances superoxide dismutase activity and glutathione level. Also, it suppresses the generation of pro-inflammatory cytokines (IL-6) and tumor necrosis factor-α (TNF-α), resulting in anti-inflammatory activity. Furthermore, it reduces malondialdehyde and nitrate levels, thereby inhibiting brain oxidative/nitrosative damage.

9.3.8 *Brown Algae*

Brown seaweed is rich with a sulfated polysaccharide, namely, fucoidan. Li et al. (2020) showed that the compound possesses antidepressant activity by promoting BDNF-dependent synaptic plasticity and targeting hippocampal inflammation, as demonstrated in studies using lipopolysaccharide (LPS) and chronic restraint stress (CRS) animal models. The results indicated that no effects were observed in the case of acute administration of fucoidan, while the chronic administration of 25, 50, and 100 mg/kg of the compound remarkably decreased the stress-induced depressive behaviors in the tail suspension test (TST), novelty-suppressed feeding test (NSFT), Forced Swim Test (FST), and sucrose preference test (SPT). Fucoidan down-regulated caspase-1-mediated inflammation in the hippocampus, resulting in alleviating the down-regulation of BDNF-dependent synaptic plasticity and inhibiting hippocampal inflammation.

It is important to note that rotenone elicits behavioral and biochemical alterations in rats (El-Nashar et al. 2021). In an animal model in which Parkinson's is induced by rotenone, fucoidan exhibited significant neuroprotective actions. The findings revealed that the compound's cataleptic effect at a dose of 140 mg/kg was superior to that of rasagiline. Moreover, doses of 70 and 140 mg/kg effectively reversed locomotor deficits. Fucoidan prevented the decrease in striatal dopamine and its metabolites at doses of 35, 70, and 140 mg/kg and prevented the rotenone-induced loss of dopaminergic neurons and fibers. Also, it protected nigral neurons. Protection of the striatal fibers was observed at the two higher doses. It also alleviated mitochondrial damage and restored mitochondrial respiratory function in a dose-dependent manner. In particular, the 140 mg/kg dose fully restored mitochondrial respiration and ATP production. The compound significantly reduced oxidative stress markers, such as MDA, 3-NT, and 8-OHdG, and at 140 mg/kg completely blocked the increases in 3-NT and 8-OHdG. In addition, it restored the release of PGC-1α and NRF2 (key mitochondrial-related proteins) at all doses (Zhang et al. 2018).

9.3.9 *Marine Sponge*

Reports revealed that the alkaloid aaptamine is present in the genus *Aaptos* and possesses promising antidepressant activity. An in vivo study showed that 1, 10, and 20 mg/kg of aaptamine decreased immobility time in the Forced Swim Test. Notably, 20 mg/kg of the compound markedly reduced immobility time by 36% compared with the control group (Diers et al. 2008).

Furthermore, aaptamine has shown significant potential as a management for Alzheimer's condition. An in vitro study showed that the compound exerted potent inhibitory activity against both acetylcholinesterase and butyrylcholinesterase and elicited IC_{50} (16.0 and 4.6 μM), respectively. Kinetic analysis showed that the compound exhibited Ki values equal to 6.96 and 6.35 μM for AChE and BuChE, respectively. It was found to act as a mixed-type inhibitor. AChE inhibitors are among the most effective drugs for protecting against Alzheimer's disease, as they increase the availability of acetylcholine at the synapse (Rabie et al. 2023b). Also, Miao et al. (2022) reported that the compound showed a dose-dependent restoration of dyskinesia in an in vivo zebrafish study. The dyskinesia recovery rates were 34.8% at 5 μM, increasing to 58.8% and 60.0% at 10 μM and 20 μM, respectively. A recovery rate of 63.7% was observed with the standard drug donepezil, which was comparable to aaptamine. Thus, aaptamine is considered a promising structural scaffold from which a future drug able to directly inhibit cholinesterases and thereby confer protective effects against Alzheimer's could be developed.

9.4 Challenges, Safety, and Regulatory Considerations

The use of herbal nutraceuticals is hampered by several obstacles that warrant careful study. Standardization, bioavailability, and drug interactions are among the challenges which complicate their use. Notably, they should be addressed to ensure efficacy and safety when utilized by patients.

9.4.1 Standardization and Quality Control

Herbal extracts differ from conventional drugs, as they are composed of a mixture of varying natural bioactive metabolites. These wide composition variabilities render the issue of reaching consistent pharmacological actions and reproducible clinical results very challenging. To overcome this, a defined concentration of the bioactive components must be achieved in each batch of the product through standardization (Calixto 2000). Reports showed that the standardized *Ginkgo biloba* extract, namely, EGb 76, composed of 24% flavonoid glycosides and 6% terpene lactones, was used clinically (Rojas et al. 2011). Similarly, the standardized curcuminoid extract, namely, BCM-95, had enhanced bioavailability and exhibited consistent therapeutic effects (Barrett 2019).

In the same context, quality control measures should be implemented to ensure the safety of herbal products. Several sources, such as heavy metals, mycotoxins, microbial pathogens, and pesticides, could contaminate these products (Ovuru et al. 2023). Thus, rigorous testing must be performed at every step of the production process.

9.4.2 Pharmacokinetics and Bioavailability

Poor bioavailability, which reduces the ability of herbal compounds to reach therapeutic levels in the bloodstream, is considered a pharmacokinetic challenge. To the best of our knowledge, the natural compound, namely, curcumin, suffers from poor bioavailability. Previous reports revealed that its bioavailability could be enhanced through co-administering 2.5 mg/kg piperine with 20 mg/kg curcumin. This combination pronouncedly increased curcumin's anti-immobility activity in the FST, and the results were comparable to 40 mg/kg curcumin when administered alone. The overall antidepressant effect was improved, owing to piperine's potential to increase brain curcumin and serotonin levels (Kulkarni et al. 2008).

9.4.3 Drug Interactions and Side Effects

Clinically significant drug-herb interactions arise owing to the complex chemical profile of plants. Despite the favorable safety profile of *Hypericum perforatum* when used alone, remarkable interactions are linked to its use. Previous studies revealed that it enhances the activities of drugs that are metabolized by enzymes and transported by transporters, including the cytochrome P450 system and P-glycoprotein, leading to adverse effects. Several interactions were documented. It was observed that the plasma levels of the drugs, namely, cyclosporin, antiretrovirals, and anticancer agents, were reduced. In the same context, co-administration of serotonin reuptake inhibitors with the herb leads to serotonin syndrome. Also, reports revealed that this herb reduces the effectiveness of oral contraceptives. Therefore, careful patient screening and monitoring should be performed to prevent herbal-drug interactions (Borrelli and Izzo 2009).

9.4.4 Regulatory Landscape

No regulatory definition exists for the term nutraceuticals. Treatment and prevention of diseases are performed by nutraceutical and pharmaceutical compounds. However, governmental sanctions exist only for pharmaceutical compounds (Singab et al. 2024). Different sets of sales rules regulate herbal products and supplements in the USA and in the European Union. A law, namely, the Dietary Supplement Health and Education Act (DSHEA), passed in 1994, classifies these products as a specific type of compound distinct from drugs and regulates them in the United States by the FDA. Based on this, companies do not need premarket approval before selling their products. This represents a post-market surveillance system. The manufacturer is responsible for safety and accurate but non-medical labeling. In contrast, products are categorized as food supplements or medicines in the European Union.

It is important to note that no singular, centralized approval from the EU is required for supplements that fall under the category of foods. Individual member states control this. In addition, any ingredient that was not widely utilized in the EU prior to 1997 must be evaluated for safety by the European Food Safety Authority (EFSA). Moreover, a product approved in one EU member state can be marketed across all member states under the principle of mutual recognition. However, a country may prohibit the use of a product if it is demonstrated to pose a health hazard. Thus, the US regulatory framework is more reactive, while the EU's approach is more proactive and systematically organized (Gamil et al. 2025).

9.5 Future Research

The beneficial effects of several herbal nutraceuticals on various neurological and psychiatric conditions have been demonstrated by numerous studies. However, extra studies are required to consolidate their usage and incorporate them into clinical practice. More long term, comprehensive, and carefully planned randomized controlled trials (RCTs) are highly recommended. Most available data are derived from small-scale research or animal models. Further studies must be conducted with larger patient cohorts to demonstrate effectiveness, determine optimal dosing schedules, and evaluate long-term safety and adverse effects (Ng et al. 2017; Swati et al. 2025).

In the same context, upcoming research should focus on investigating the effects of combining herbal compounds with conventional treatments, as well as exploring the effects of using herbal compounds together. Additionally, drug metabolism alterations, interactions, and adverse effects elicited by these combinations must be explored (Gamil et al. 2025). Moreover, the ways by which nutraceuticals impact brain circuits, gene expression, and neuroplasticity should be evaluated.

9.6 Conclusion

Considerable interest is directed toward herbal nutraceuticals to be used in the management of several psychiatric and neurological problems, including anxiety, depression, insomnia, Parkinson's, and Alzheimer's. The neuroprotective effects of herbal nutraceuticals from various origins, including terrestrial and marine sources, are supported by previous reports. These nutraceuticals target key neurobiological pathways. Modulation of neurotransmitter systems is one mechanism by which these compounds exert their actions. Moreover, promotion of hippocampal neurogenesis, regulation of the HPA axis, and the reduction of inflammation and oxidative stress are also observed. Unfortunately, many obstacles to the employment of these nutraceuticals include standardization issues, poor bioavailability, safety concerns, and possible interactions. Further research is needed to address these challenges. Regulatory landscapes should adopt consistent evidence-based approaches to ensure the safety and quality of these substances. In conclusion, herbal nutraceuticals are considered a beneficial complementary or alternative approach to alleviating neurological conditions.

References

Al-Harbi KS (2012) Treatment-resistant depression: therapeutic trends, challenges, and future directions. Patient Prefer Adherence 6:369–388

Amsterdam JD, Panossian AG (2016) Rhodiola rosea L. as a putative botanical antidepressant. Phytomedicine 23(7):770–783

Barrett ML (2019) BCM-95®/CURCUGREEN™. Proprietary botanical ingredient scientific and clinical monograph for BCM-95®/CURCUGREEN™ turmeric rhizome/curcumin extract preparation. American Botanical Council

Borrelli F, Izzo AA (2009) Herb–drug interactions with St John's wort (*Hypericum perforatum*): an update on clinical observations. AAPS J 11(4):710–727

Bottaccioli AG, Bologna M, Bottaccioli F (2025) Rethinking depression—beyond neurotransmitters: an integrated psychoneuroendocrineimmunology framework for depression's pathophysiology and tailored treatment. Int J Mol Sci 26(6):2759

Bouayed J, Rammal H, Soulimani R (2009) Oxidative stress and anxiety: relationship and cellular pathways. Oxidative Med Cell Longev 2(2):63–67

Bremner JD, Krystal JH, Southwick SM, Charney DS (1996) Noradrenergic mechanisms in stress and anxiety: I. Preclinical studies. Synapse 23(1):28–38

Butterweck V, Hegger M, Winterhoff H (2004) Flavonoids of St. John's Wort reduce HPA axis function in the rat. Planta Med 70(10):1008–1011

Calixto J (2000) Efficacy, safety, quality control, marketing and regulatory guidelines for herbal medicines (phytotherapeutic agents). Braz J Med Biol Res 33:179–189

Cervo L, Rozio M, Ekalle-Soppo C, Guiso G, Morazzoni P, Caccia S (2002) Role of hyperforin in the antidepressant-like activity of *Hypericum perforatum* extracts. Psychopharmacology 164(4):423–428

Chatterjee S, Bhattacharya S, Wonnemann M, Singer A, Müller W (1998) Hyperforin as a possible antidepressant component of hypericum extracts. Life Sci 63(6):499–510

Chen X-C, Zhu Y-G, Zhu L-A, Huang C, Chen Y, Chen L-M, Fang F, Zhou Y-C, Zhao C-H (2003) Ginsenoside Rg1 attenuates dopamine-induced apoptosis in PC12 cells by suppressing oxidative stress. Eur J Pharmacol 473(1):1–7

Choi JH, Lee MJ, Jang M, Kim H-J, Lee S, Lee SW, Kim YO, Cho I-H (2018) *Panax ginseng* exerts antidepressant-like effects by suppressing neuroinflammatory response and upregulating nuclear factor erythroid 2 related factor 2 signaling in the amygdala. J Ginseng Res 42(1):107–115

D'Onofrio G, Nabavi SM, Sancarlo D, Greco A, Pieretti S (2021) *Crocus sativus* L. (Saffron) in Alzheimer's disease treatment: bioactive effects on cognitive impairment. Curr Neuropharmacol 19(9):1606–1616

Dantzer R, O'Connor JC, Freund GG, Johnson RW, Kelley KW (2008) From inflammation to sickness and depression: when the immune system subjugates the brain. Nat Rev Neurosci 9(1):46–56

Delgado PL (2000) Depression: the case for a monoamine deficiency. J Clin Psychiatry 61(6):7–11

Dias V, Junn E, Mouradian MM (2013) The role of oxidative stress in Parkinson's disease. J Parkinsons Dis 3(4):461–491

Diers JA, Ivey KD, El-Alfy A, Shaikh J, Wang J, Kochanowska AJ, Stoker JF, Hamann MT, Matsumoto RR (2008) Identification of antidepressant drug leads through the evaluation of

marine natural products with neuropsychiatric pharmacophores. Pharmacol Biochem Behav 89(1):46–53
Dinamarca M, Cerpa W, Garrido J, Hancke J, Inestrosa N (2006) Hyperforin prevents β-amyloid neurotoxicity and spatial memory impairments by disaggregation of Alzheimer's amyloid-β-deposits. Mol Psychiatry 11(11):1032–1048
Elder GJ, Altena E, Palagini L, Ellis JG (2023) Stress and the hypothalamic–pituitary–adrenal axis: how can the COVID-19 pandemic inform our understanding and treatment of acute insomnia? J Sleep Res 32(4):e13842
El-Nashar HA, Mostafa NM, Abd El-Ghffar EA, Eldahshan OA, Singab ANB (2021) The genus *Schinus* (Anacardiaceae): a review on phytochemicals and biological aspects. Nat Prod Res 36(18):4833–4851
El-Nashar HA, Sayed AM, El-Sherief HA, Rateb ME, Akil L, Khadra I, Majrashi TA, Al-Rashood ST, Binjubair FA, El Hassab MA (2023) Metabolomic profile, anti-trypanosomal potential and molecular docking studies of *Thunbergia grandifolia*. J Enzyme Inhib Med Chem 38(1):2199950
Faravelli C, Sauro CL, Godini L, Lelli L, Benni L, Pietrini F, Lazzeretti L, Talamba GA, Fioravanti G, Ricca V (2012) Childhood stressful events, HPA axis and anxiety disorders. World J Psychiatry 2(1):13
Figueiredo Godoy AC, Frota FF, Araújo LP, Valenti VE, Pereira E d SBM, Detregiachi CRP, Galhardi CM, Caracio FC, Haber RS, Fornari Laurindo L (2025) Neuroinflammation and natural antidepressants: balancing fire with flora. Biomedicine 13(5):1129
Gamil NM, Elsayed HA, Hamed RM, Salah ET, Ahmed AM, Mostafa HA, El-Shiekh RA, Abou-Hussein D (2025) Insights from herb interactions studies: a foundational report for integrative medicine. Futur J Pharm Sci 11(1):46
Gao L, Zhang Y, Sterling K, Song W (2022) Brain-derived neurotrophic factor in Alzheimer's disease and its pharmaceutical potential. Transl Neurodegener 11(1):4
Ge Y, Xu W, Zhang L, Liu M (2020) Ginkgolide B attenuates myocardial infarction-induced depression-like behaviors *via* repressing IL-1β in central nervous system. Int Immunopharmacol 85:106652
Heyat MBB, Akhtar F, Sultana A, Tumrani S, Teelhawod BN, Abbasi R, Kamal MA, Muaad AY, Lai D, Wu K (2022) Role of oxidative stress and inflammation in insomnia sleep disorder and cardiovascular diseases: herbal antioxidants and anti-inflammatory coupled with insomnia detection using machine learning. Curr Pharm Des 28(45):3618–3636
Holzmann I, da Silva LM, da Silva JAC, Steimbach VMB, de Souza MM (2015) Antidepressant-like effect of quercetin in bulbectomized mice and involvement of the antioxidant defenses, and the glutamatergic and oxidonitrergic pathways. Pharmacol Biochem Behav 136:55–63
Hosseinzadeh H, Noraei NB (2009) Anxiolytic and hypnotic effect of *Crocus sativus* aqueous extract and its constituents, crocin and safranal, in mice. Phytother Res 23(6):768–774
Hosseinzadeh H, Karimi G, Niapoor M (2003) Antidepressant effect of *Crocus sativus* L. stigma extracts and their constituents, crocin and safranal, in mice. Proceedings of the Ist international symposium on saffron biology and biotechnology. Acta Hort 650:435–445
Huang WJ, Zhang X, Chen WW (2016) Role of oxidative stress in Alzheimer's disease. Biomed Rep 4(5):519–522
Husen A (2021) Traditional herbal therapy for the human immune system. CRC Press, Boca Raton. https://doi.org/10.1201/9781003137955
Husen A (2022) Herbs, shrubs and trees of potential medicinal benefits. CRC Press, Boca Raton. https://doi.org/10.1201/9781003205067
Husen A (2023a) Exploring poisonous plants: medicinal values, toxicity response and other potential uses. CRC Press, Boca Raton. https://doi.org/10.1201/b23017
Husen A (2023b) Antidiabetic medicinal plants and herbal treatments. CRC Press, Boca Raton. https://doi.org/10.1201/b23347
Husen A (2024a) Antimalarial medicinal plants. Taylor & Francis Group, LLC. https://doi.org/10.1201/9781003378396

Husen A (2024b) Medicinal spices and condiment crops. Taylor & Francis Group, LLC. https://doi.org/10.1201/9781003387046

Husen H, Pant M (2024a) Exploring medicinal orchids. Taylor & Francis Group, LLC. https://doi.org/10.1201/b23375

Husen H, Pant M (2024b) Tissue culture techniques and medicinal plants: enhancing propagation and production. Taylor & Francis Group, LLC. https://doi.org/10.1201/b23374

Istikoglou C, Mavreas V, Geroulanos G (2010) History and therapeutic properties of *Hypericum perforatum* from antiquity until today. Psychiatriki 21(4):332–338

Khatri DK, Juvekar AR (2016) Kinetics of inhibition of monoamine oxidase using curcumin and ellagic acid. Pharmacogn Mag 12(Suppl 2):S116

Kim H, Choi H-S, Han K, Sim W, Suh HJ, Ahn Y (2025) Ashwagandha (*Withania somnifera* (L.) dunal) root extract containing withanolide a alleviates depression-like behavior in mice by enhancing the brain-derived neurotrophic factor pathway under unexpected chronic mild stress. J Ethnopharmacol 340:119224

Klemow KM, Bartlow A, Crawford J, Kocher N, Shah J, Ritsick M (2011) Medical attributes of St. John's Wort (Hypericum perforatum). In: Herbal medicine: biomolecular and clinical aspects, 2nd edn. CRC Press, pp 211–228

Kraus B, Wolff H, Heilmann J, Elstner EF (2007) Influence of *Hypericum perforatum* extract and its single compounds on amyloid-β mediated toxicity in microglial cells. Life Sci 81(11):884–894

Kulkarni SK, Bhutani MK, Bishnoi M (2008) Antidepressant activity of curcumin: involvement of serotonin and dopamine system. Psychopharmacology 201(3):435–442

Kutzer T, Dick M, Scudamore T, Wiener M, Schwartz T (2020) Antidepressant efficacy and side effect burden: an updated guide for clinicians. Drugs Context 9:1

Langade D, Kanchi S, Salve J, Debnath K, Ambegaokar D, Langade DG (2019) Efficacy and safety of Ashwagandha (*Withania somnifera*) root extract in insomnia and anxiety: a double-blind, randomized, placebo-controlled study. Cureus 11(9):e5797

Lechtenberg M, Schepmann D, Niehues M, Hellenbrand N, Wünsch B, Hensel A (2008) Quality and functionality of saffron: quality control, species assortment and affinity of extract and isolated saffron compounds to NMDA and σ1 (sigma-1) receptors. Planta Med 74(7):764–772

Li F, Wu X, Li J, Niu Q (2016) Ginsenoside Rg1 ameliorates hippocampal long-term potentiation and memory in an Alzheimer's disease model. Mol Med Rep 13(6):4904–4910

Li Q, Wang J, Li Y, Xu X (2018) Neuroprotective effects of salidroside administration in a mouse model of Alzheimer's disease. Mol Med Rep 17(5):7287–7292

Li M, Sun X, Li Q, Li Y, Luo C, Huang H, Chen J, Gong C, Li Y, Zheng Y (2020) Fucoidan exerts antidepressant-like effects in mice *via* regulating the stability of surface AMPARs. Biochem Biophys Res Commun 521(2):318–325

Liu Z, Qi Y, Cheng Z, Zhu X, Fan C, Yu S (2016) The effects of ginsenoside Rg1 on chronic stress induced depression-like behaviors, BDNF expression and the phosphorylation of PKA and CREB in rats. Neuroscience 322:358–369

Ma XW, Guo RY (2017) Dose-dependent effect of *Curcuma longa* for the treatment of Parkinson's disease. Exp Ther Med 13(5):1799–1805

Ma H, Cai M, Wang H (2021) Emotional blunting in patients with major depressive disorder: a brief non-systematic review of current research. Front Psych 12:792960

Miao S, He Q, Li C, Wu Y, Liu M, Chen Y, Qi S, Gong K (2022) Aaptamine – a dual acetyl – and butyrylcholinesterase inhibitor as potential anti-Alzheimer's disease agent. Pharm Biol 60(1):1502–1510

Mohanasundari M, Sabesan M (2007) Modulating effect of *Hypericum perforatum* extract on astrocytes in MPTP induced Parkinson's disease in mice. Eur Rev Med Pharmacol Sci 11(1):17–20

Mostafa NM, Edmond MP, El-Shazly M, Fahmy HA, Sherif NH, Singab ANB (2022) Phytoconstituents and renoprotective effect of *Polyalthia longifolia* leaves extract on radiation-induced nephritis in rats via TGF-*β*/smad pathway. Nat Prod Res 36(16):4187–4192

Moussa AY, Alanzi AR, Riaz M, Fayez S (2024) Could mushrooms' secondary metabolites ameliorate Alzheimer disease? A computational flexible docking investigation. J Med Food 27(8):775–796

Moussa AY, Alanzi AR, Luo J, Wang J, Cheang WS, Xu B (2025) Role of saponins from Platycodon grandiflorum in Alzheimer's disease: DFT, molecular docking, and simulation studies in key enzymes. Molecules 30(8):1812

Nawaz A, Mamoon B, Batool TE, Khattak MI, Amir F, Akbar A, Khan S (2024) Advances in antidepressant therapy: comparing the efficacy of selective serotonin reuptake inhibitors (SSRIs), serotonin-norepinephrine reuptake inhibitors (SNRIs), and novel agents. Cureus 16(12):e76318

Ng QX, Koh SSH, Chan HW, Ho CYX (2017) Clinical use of curcumin in depression: a meta-analysis. J Am Med Dir Assoc 18(6):503–508

Nobakht SZ, Akaberi M, Mohammadpour AH, Moghadam AT, Emami SA (2022) *Hypericum perforatum*: traditional uses, clinical trials, and drug interactions. Iran J Basic Med Sci 25(9):1045

Nobis A, Zalewski D, Waszkiewicz N (2020) Peripheral markers of depression. J Clin Med 9(12):3793

Numakawa T, Kajihara R (2025) The role of brain-derived neurotrophic factor as an essential mediator in neuronal functions and the therapeutic potential of its mimetics for neuroprotection in neurologic and psychiatric disorders. Molecules 30(4):848

Ovuru KF, Izah SC, Yasmin H, Enerijiofi KE, Das M, Ogwu MC (2023) Microbial contaminants of herbal remedies: health risks and sustainable quality control strategies. In: Herbal medicine phytochemistry: applications and trends. Springer, pp 1–30

Palasz E, Wysocka A, Gasiorowska A, Chalimoniuk M, Niewiadomski W, Niewiadomska G (2020) BDNF as a promising therapeutic agent in Parkinson's disease. Int J Mol Sci 21(3):1170

Pant M, Husen A (2025) Exploring medicinal bamboos. Taylor & Francis Group, LLC. https://doi.org/10.1201/9781003485933

Pant M, Sohrab SS, Husen A (2025) Genetic improvement and conservation practice of medicinal plants. Taylor & Francis Group, LLC. https://doi.org/10.1201/9781003482659

Peng S, Zhou Y, Lu M, Wang Q (2022) Review of herbal medicines for the treatment of depression. Nat Prod Commun 17(11):1934578X221139082

Rabie O, El-Nashar HA, George MY, Majrashi TA, Al-Warhi T, Hassan FE, Eldehna WM, Mostafa NM (2023a) Phytochemical profiling and neuroprotective activity of *Callistemon subulatus* leaves against cyclophosphamide-induced chemobrain. Biomed Pharmacother 167:115596

Rabie O, El-Nashar HA, Majrashi TA, Al-Warhi T, El Hassab MA, Eldehna WM, Mostafa NM (2023b) Chemical composition, seasonal variation and antiaging activities of essential oil from *Callistemon subulatus* leaves growing in Egypt. J Enzyme Inhib Med Chem 38(1):2224944

Ravinder S, Burghardt NS, Brodsky R, Bauer EP, Chattarji S (2013) A role for the extended amygdala in the fear-enhancing effects of acute selective serotonin reuptake inhibitor treatment. Transl Psychiatry 3(1):e209

Robinson HM, Hood SD, Bell CJ, Nutt DJ (2006) Dopamine and social anxiety disorder. Braz J Psychiatry 28:263–264

Rojas P, Serrano-García N, Medina-Campos ON, Pedraza-Chaverri J, Ögren SO, Rojas C (2011) Antidepressant-like effect of a *Ginkgo biloba* extract (EGb761) in the mouse forced swimming test: role of oxidative stress. Neurochem Int 59(5):628–636

Sarris J, Panossian A, Schweitzer I, Stough C, Scholey A (2011) Herbal medicine for depression, anxiety and insomnia: a review of psychopharmacology and clinical evidence. Eur Neuropsychopharmacol 21(12):841–860

Shao L, Dong C, Geng D, He Q, Shi Y (2022) Ginkgolide B inactivates the NLRP3 inflammasome by promoting autophagic degradation to improve learning and memory impairment in Alzheimer's disease. Metab Brain Dis 37(2):329–341

Sharma N, Nehru B (2018) Curcumin affords neuroprotection and inhibits α-synuclein aggregation in lipopolysaccharide-induced Parkinson's disease model. Inflammopharmacology 26(2):349–360

Siddiqui MJ, Saleh MS, Basharuddin SNB, Zamri SHB, bin Mohd Najib MH, bin Che Ibrahim MZ, Mazha HNB, Hassan NM, Khatib A (2018) Saffron (Crocus sativus L.): as an antidepressant. J Pharm Bioallied Sci 10(4):173–180

Singab ANB, Elhawary EA, Elkhawas YA, Fawzy IM, Moussa AY, Mostafa NM (2024) Role of nutraceuticals in obesity management: a mechanism and prospective supported by molecular docking studies. J Med Food 27(2):176–197

Swati K, Sharma R, Singh A (2025) Herbal antidepressants: a comprehensive review of plant–based therapeutic agents. Int J Pharm Sci 3(6):3248–3265. https://doi.org/10.5281/zenodo.15716771

Tao W, Ruan J, Wu R, Zhao M, Zhao T, Qi M, Yau SS, Yao G, Zhang H, Hu Y (2023) A natural carotenoid crocin exerts antidepressant action by promoting adult hippocampal neurogenesis through Wnt/β-catenin signaling. J Adv Res 43:219–231

Urbanska N, Ashaolu TJ, Mattova S, Simko P, Kiskova T (2025) The potential of selected plants and their biologically active molecules in the treatment of depression and anxiety disorders. Int J Mol Sci 26(5):2368

Van Kampen JM, Baranowski DB, Shaw CA, Kay DG (2014) *Panax ginseng* is neuroprotective in a novel progressive model of Parkinson's disease. Exp Gerontol 50:95–105

Vasanthi AV, Gayatri BM, Sreeja C, Nitiashwarya G, Sruthi C, Thanmai A, Kumar PR (2025) The role of adaptogenic herbs in combating stress: a review of scientific evidence. Int J Adv Res Sci Commun Technol 5(3):11–27

Woelk H, Arnoldt K, Kieser M, Hoerr R (2007) *Ginkgo biloba* special extract EGb 761® in generalized anxiety disorder and adjustment disorder with anxious mood: a randomized, double-blind, placebo-controlled trial. J Psychiatr Res 41(6):472–480

World Health Organization (2023) Depressive disorder (depression). World Health Organization, Geneva. https://www.who.int/news-room/fact-sheets/detail/depression

Wu W-L, Gong X-X, Qin Z-H, Wang Y (2025) Molecular mechanisms of excitotoxicity and their relevance to the pathogenesis of neurodegenerative diseases—an update. Acta Pharmacol Sin 46:3129–3142

Xing S, Xu S, Wang L, Guo L, Zhou X, Wu H, Wang W, Liu L (2024) Salidroside exerts antidepressant-like action by promoting adult hippocampal neurogenesis through SIRT1/PGC-1α signalling. Acta Neuropsychiatr 36(6):446–456

Yu Z, Kong L, Chen Y (2002) Antidepressant activity of aqueous extracts of *Curcuma longa* in mice. J Ethnopharmacol 83(1–2):161–165

Zengin G, El-Raey M, El-Kashak W, Batiha GE-S, Althumairy D, Alamer S, Mostafa NM, Eldahshan OA (2023) Sweroside: an iridoid glycoside of potential neuroprotective, antidiabetic, and antioxidant activities supported by molecular docking. Amino Acids 55(12):1765–1774

Zhang L, Fang Y, Xu Y, Lian Y, Xie N, Wu T, Zhang H, Sun L, Zhang R, Wang Z (2015) Curcumin improves amyloid β-peptide (1-42) induced spatial memory deficits through BDNF-ERK signaling pathway. PLoS One 10(6):e0131525

Zhang L, Hao J, Zheng Y, Su R, Liao Y, Gong X, Liu L, Wang X (2018) Fucoidan protects dopaminergic neurons by enhancing the mitochondrial function in a rotenone-induced rat model of Parkinson's disease. Aging Dis 9(4):590

Zheng X, Liang Y, Kang A, Ma S-J, Xing L, Zhou Y-Y, Dai C, Xie H, Xie L, Wang G-J (2014) Peripheral immunomodulation with ginsenoside Rg1 ameliorates neuroinflammation-induced behavioral deficits in rats. Neuroscience 256:210–222

Zirak N, Shafiee M, Soltani G, Mirzaei M, Sahebkar A (2019) *Hypericum perforatum* in the treatment of psychiatric and neurodegenerative disorders: current evidence and potential mechanisms of action. J Cell Physiol 234(6):8496–8508

Chapter 10
Innovations in Plant-Based Nanomaterials for Healthcare Applications

Heba A. S. El-Nashar, Ayman M. Al-Qaaneh, Eyad Ayman, Abdallah M. Ali, Omar Nabil, Hamza Elkoshairy, Omar Ibrahim, Mohammed Hamdy, and Abdel Nasser B. Singab

Abstract In recent years, the focus on nanoparticles has been steadily rising due to their increasing potential in health care applications, while plants have proved to be a great asset to synthesize stable and safe nanoparticles, as they are rich in phytochemicals such as polyphenols, flavonoids, and alkaloids, which act as natural reducing and stabilizing agents, thus synthesizing nanoparticles without the need for hazardous chemicals, rendering them eco-friendly and safe for further exploration for their therapeutic use. This chapter outlines the green synthesis of nanoparticles using plant extracts as a superior form of nanoparticle synthesis, which is eco-friendly, cost-effective, and biocompatible; the characterization techniques that emphasize the influence of size and morphology of nanoparticles on the biological activity; and potential therapeutic activities such as wound healing, neuroprotective, antimicrobial, antioxidant, antiparasitic, and anticancer with some of the formulation advancing to clinical trials. It highlights their applications in medicine as diagnostic tools or as innovative drug delivery systems that enhance solubility, bioavailability, and targeted delivery of therapeutic agents. By integrating the advantages of nanotechnology and phytochemistry, this chapter provides an overview of current knowledge and identifies

H. A. S. El-Nashar (✉)
Department of Pharmacognosy, Faculty of Pharmacy, Ain Shams University, Cairo, Egypt
e-mail: heba_pharma@pharma.asu.edu.eg

A. M. Al-Qaaneh
Faculty of Allied Medical Sciences, Al-Balqa Applied University (BAU), Al-Salt, Jordan

E. Ayman · A. M. Ali · O. Nabil · H. Elkoshairy · O. Ibrahim · M. Hamdy
Faculty of Pharmacy, Ain Shams University, Abbassia, Cairo, Egypt

A. N. B. Singab
Department of Pharmacognosy, Faculty of Pharmacy, Ain Shams University, Cairo, Egypt

Center for Drug Discovery Research and Development, Faculty of Pharmacy, Ain Shams University, Cairo, Egypt

L. A. Worku et al. (eds.), *Plant-Based Solutions for Sustainable Agriculture and Healthcare*, Sustainability Sciences in Asia and Africa,
https://doi.org/10.1007/978-981-95-8910-4_10

future research directions to further advance the clinical and industrial translation of plant-based nanoparticles.

Keywords Nanoparticles · Plant extracts · Phytochemistry · Health care · Flavonoids · Neuroprotective

10.1 Introduction

The relentless ascent of global health challenges—from the silent epidemic of neurodegenerative disorders like Alzheimer's disease to the critical threat of antimicrobial resistance—demands a paradigm shift in our therapeutic and diagnostic arsenals (Eldahshan et al. 2025). In recent decades, conventional nanoparticles have emerged as a cornerstone of nanomedicine, offering unprecedented opportunities to overcome major therapeutic challenges such as poor solubility, limited bioavailability, systemic toxicity, and lack of target specificity associated with traditional pharmaceuticals (Abd-elaty et al. 2025; Askar et al. 2022). Their unique physicochemical properties—particularly their nanoscale size, high surface-to-volume ratio, and tunable surface chemistry—enable them to act as efficient carriers for drugs, genes, and biomolecules and improve therapeutic outcomes (El-Nashar et al. 2022). Despite these advances, they are often associated with significant drawbacks, including high energy consumption, reliance on hazardous chemicals, and the generation of toxic by-products (Masood et al. 2024).

To overcome these limitations, green synthesis approaches have emerged as a sustainable alternative. Green synthesis of nanoparticles (NPs) is based on three fundamental principles: the use of environmentally friendly solvents (preferably water, ethanol, or their mixtures), non-toxic reducing agents, and safe stabilizers (Banjara et al. 2024). Within this context, plant-based synthesis has gained particular attention for its numerous advantages over conventional methods, as plant materials contain a diverse range of bioactive metabolites that can serve as reducing and stabilizing agents (Villagrán et al. 2024). Such metabolites are sourced from different parts of the plant including leaves, flowers, stems, bark, seeds, roots, peels, and latex and encompass both primary metabolites (proteins, amino acids, organic acids, and vitamins) and secondary metabolites (flavonoids, alkaloids, terpenoids, polyphenols, heterocyclic compounds, and polysaccharides) (El-Nashar et al. 2025). In this way, plant-based green synthesis effectively bridges the gap between the efficiency of conventional nanoparticle methods and the growing demand for safe, sustainable, and environmentally responsible nanotechnologies (Shahzadi et al. 2025).

Expanding on benefits, plant-based nanoparticles (PBNPs) have attracted growing attention not only as environmentally sustainable alternatives to chemically synthesized nanomaterials but also as multifunctional agents with inherent therapeutic potential (Liyana and Jasmine 2025). The phytochemicals involved in their synthesis provide PBNPs with unique bioactivities that extend beyond simple drug delivery (Codreanu et al. 2025). As a result, PBNPs have demonstrated a wide spectrum of therapeutic effects, including anticancer properties through the induction of apoptosis and inhibition of tumor proliferation, neuroprotective effects by attenuating oxidative stress and modulating neuronal signaling pathways, and wound healing capabilities via enhanced antimicrobial activity and promotion of tissue regeneration (Li et al. 2022). Notably, PBNPs have also shown promising antiparasitic effects, with several studies reporting their efficacy against protozoan parasites, including malaria, leishmaniasis, schistosomiasis, and helminthic infections (Rajpal et al. 2024).

Furthermore, emerging clinical trials are beginning to validate their biomedical promise, particularly in oncology and infectious disease management (Mohite et al. 2024). Thus, PBNPs represent a confluence of green chemistry and nanomedicine, offering both sustainability and efficacy while overcoming critical challenges in modern therapeutics (Ifijen et al. 2025).

Integrating nanotechnology into medicine has fostered the emergence of new fields of study and provided fresh perspectives on understanding biological systems (Abdelazim et al. 2024; Hashad et al. 2025). These advances enable specific parameters, such as improved solubility, controlled drug retention in the bloodstream, targeted drug delivery, and controlled release under various conditions. The ability to manipulate substances at the nanometer scale also allows for the modification of their physicochemical properties and biological activity (Kumar et al. 2024). Drug delivery systems utilizing nanoparticles can enhance the effectiveness of medications, increase absorption, and target specific tissues (Bonifácio et al. 2014). Plant-based nanoparticles (PBNPs) have also emerged as promising candidates for drug delivery (Liyana and Jasmine 2025). Their small size, high surface-to-volume ratio, and natural phytochemical coating enable them to encapsulate, protect, and transport bioactive compounds with improved stability and bioavailability (Qin et al. 2018). Unlike conventional nanocarriers, PBNPs offer enhanced biocompatibility and reduced systemic toxicity (Forgách et al. 2020). These advantages position PBNPs as versatile nanocarriers capable of addressing the limitations of traditional drug delivery systems, particularly in cancer therapy, neurodegenerative diseases, and infectious disorders (Wang et al. 2024b). In addition to drug delivery, PBNPs are playing a growing role in diagnostic applications, where their unique optical, magnetic, and electronic properties are leveraged for sensitive biosensing and advanced bioimaging (Karnwal et al. 2024). Conventional diagnostic tools, while effective, often face challenges such as high cost, limited sensitivity, and potential toxicity of contrast agents (Ibrahim et al. 2021). In contrast, PBNPs provide green, low-cost, and biocompatible alternatives, enabling fluorescence imaging and real-time biomarker detection. This dual role—both as therapeutic carriers and diagnostic tools—underscores their versatility and positions them as a transformative platform for advancing precision and personalized medicine (Ibrahim et al. 2021). This book chapter focuses on the synthesis of plant-based nanoparticles, their therapeutic activities, and their diagnostic applications. Further, we illustrated the potential role of plant-based nanoparticles in drug delivery systems.

10.2 Synthesis of Plant-Based Nanoparticles

Green synthesis of nanoparticles (NPs) requires three criteria: the use of sustainable green solvents (preferably water, ethanol, or their mixture), non-toxic reducing agents, and non-harmful stabilizing agents (Poh Yan et al. 2022). Plant-based synthesis offers many benefits over conventional physical and chemical techniques (Husen and Siddiqi 2014; Husen 2019, 2020, 2023). They include, but are not limited to, availability, cost-effectiveness, lower energy requirements, reduced production of harmful by-products, and overall environmental sustainability (Shahzadi et al. 2025). Plant extracts are fascinating because they contain numerous natural metabolites that have the potential to serve as reducing and stabilizing agents (Singh et al. 2023a). The metabolites are derived from the whole plant, including leaves, flowers, stems, bark, seeds, roots, peels, and latex. The extracts contain primary metabolites (proteins, amino acids, organic acids, and vitamins) and secondary metabolites (flavonoids, alkaloids, terpenoids, polyphenols, heterocyclic compounds, and polysaccharides) (Baky et al. 2024). Collectively, these compounds facilitate metal-ion reduction, promote nanoparticle formation, and stabilize the resulting structures (Yin et al. 2012; Bachheti and Bachheti 2023; Husen et al. 2023). Most plant species contain multiple phytochemicals and antioxidative molecules, which are potent natural reductants (Bhuia et al. 2024). These antioxidative molecules work very efficiently to enhance the reduction process in nanoparticle synthesis. Flavonoids and terpenoids can reduce Ag^+ ions to form silver nanoparticles; in this process, the hydroxyl functional groups of flavonoids such as quercetin and myricetin are oxidized to carbonyl functional groups by the bioreduction of metal ions (Anwar et al. 2019; Bachheti et al. 2024). Similarly, naturally occurring vitamins like B1, B2, and C are biocompatible reducing agents that minimize or even eliminate the formation of toxic by-products. In addition, pigments such as chlorophyll stabilize nanoparticles by preventing aggregation and extending shelf life (Cortez et al. 2017). Stabilization is necessary because phytochemicals also prevent agglomeration, thereby improving dispersion and surface activity of nanoparticles.

Various parameters influence the synthesis and properties of plant-mediated nanoparticles, including the concentration and release type of plant extract, pH of the reaction medium, concentration of metal salt, reaction time, and temperature (Berta et al. 2021). Such parameters could regulate the size, morphology, yield, and stability of the nanoparticles formed. Synthesis is usually achieved by mixing a plant extract with an aqueous solution of a metal salt under mild conditions, typically at ambient temperature. This is a low-energy, straightforward process that is easy to scale up and does not require sophisticated laboratory hardware, making it an appealing method for large-scale nanoparticle synthesis (Nguyen et al. 2022). The extraction technique of plant-based nanoparticles is illustrated in Fig. 10.1.

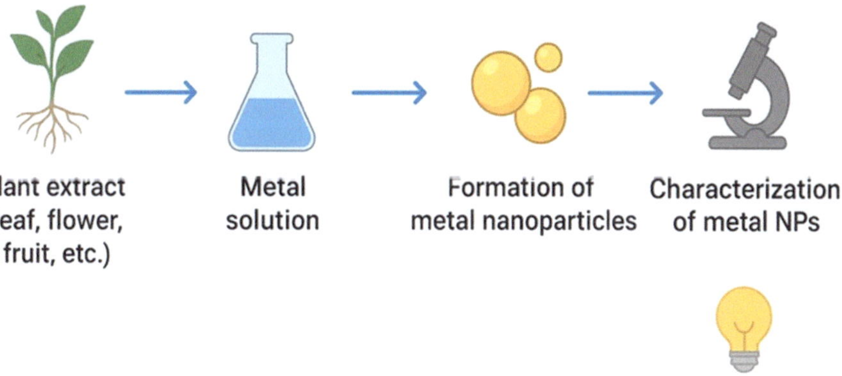

Fig. 10.1 The extraction technique of plant-based nanoparticles

10.3 Potential Therapeutic Effects of Plant-Based Nanoparticles

10.3.1 Neuroprotective Effects

Plant-based nanoparticles are widely used for the management of neurological conditions; they can protect and treat nerve cells through different mechanisms, such as increasing the passage of potent drugs across the blood-brain barrier (BBB), which previously had poor brain bioavailability (Bhattacharya et al. 2022). Also, PBNPs exhibit antioxidant, anti-inflammatory, and anti-apoptotic effects, thereby improving outcomes (Wang et al. 2023a). Plant-based nanoparticles (PBNPs) can be used in various ways to combat dementia, including their direct anti-acetylcholinesterase activity, which can slow the progression of Alzheimer's disease, as demonstrated by Youssif et al. (2019). It was found that *Lampranthus coccineus* silver nanoparticles exhibited high anti-acetylcholinesterase activity in adult male demented albino rats, comparable to that of Rivastigmine (Youssif et al. 2019).

Another example of PBNPs with high neuroprotective effects is biosynthesized Selenium nanoparticles using epigallocatechin gallate (EGCG-SeNPs), which protected against acute epileptic seizures in mice and restored the oxidant/antioxidant balance better than valproic acid, a well-established anti-epileptic drug. EGCG--SeNPs also exhibited an anti-inflammatory effect by restoring the levels of tumor necrosis factor-α (TNF-α), interleukin-1β (IL-1β), and nitric oxide synthase 2 (NOS2). Moreover, EGCG-SeNPs reduced the levels of pro-apoptotic proteins (Bax and caspase-3) while raising the level of the anti-apoptotic protein (Bcl-2) (Alrashdi et al. 2023).

Also, PBNPs can be used as delivery systems, thereby drastically increasing the bioavailability of otherwise vapid agents, such as asiatic acid, which has great neuroprotective potential but is limited by its poor brain bioavailability. The work of Raval et al. (2015) showed a tenfold increase in the brain bioavailability of asiatic acid and a 627% increase in targeting efficiency using bovine serum albumin nanoparticles (BSANP) as a delivery system and glutathione as a brain-targeting ligand (Raval et al. 2015). Recently, the silver nanoparticles of *Lepidagathis pungens* ethanolic extract reduced the levels of malondialdehyde (MDA) and reactive oxygen species (ROS) by 45% and 50%, respectively, and increased the levels of antioxidant enzymes superoxide dismutase (SOD) and catalase (CAT) by 40% and 35%, respectively, in a lipopolysaccharide-induced neuroinflammation in male mice (Amtaghri et al. 2024). Moreover, the use of plant-based nanoparticles as a neuroprotective was challenged by pretreating human neuroblastoma SH-SY5 cells with biosynthesized silver nanoparticles using pomegranate extract before exposing the cells to methamphetamine. The results were promising, showing a reduction in the methamphetamine-induced elevation of malondialdehyde (MDA) and intracellular reactive oxygen species (ROS) while heavily preventing DNA fragmentation (Khorrami et al. 2023).

10.3.2 Antimicrobial Effects

Bacterial resistance has been on the rise due to abuse and poor adherence to antibiotic regimens; the development of new antimicrobials is of utmost importance (El-Nashar et al. 2021). While generating new molecules is challenging, implementing plant-based nanoparticles continues to show encouraging results (El-Nashar et al. 2024). Plant-mediated synthesis of nanoparticles has been accelerated due to their antimicrobial potential (Akbar et al. 2020). Increased inhibitory potential of phytochemical-capped nanoparticles has been reported against a variety of microorganisms (Amini 2019). Among all metal nanoparticles, silver and gold nanoparticles derived from plant extracts are best studied for their antimicrobial activity (Sher et al. 2022). Owing to their extremely small size and extensive surface area, they can readily interact with microbial membranes, killing and destroying cells. Mechanisms of action of plant-based nanoparticles against antimicrobials include cell wall lysis, generation of reactive oxygen species (ROS), protein-DNA interactions, and inhibition of enzyme activity (Ali et al. 2024). These working combinations make them bactericidal and fungicidal. Vegetal silver nanoparticles have been shown in some studies to be highly effective against drug-resistant microbial strains—so effective that they are likely candidates to replace conventional antibiotics in attack operations against resistant microbial ministrains (Rajak et al. 2023).

Plant nanoparticles exhibit high antimicrobial activity because phytochemicals also act as capping and stabilizing agents (Nguyen et al. 2022). Plant nanoparticles exhibit increased interactions with microbial cell walls and membranes via biomolecules (Asghar and Asghar 2020). Silver nanoparticles' antimicrobial activity is attributed to their ability to release silver ions and induce oxidative stress in microbial cells (Quinteros et al. 2016). Plant-gold nanoparticles have been shown to suppress disease-causing microbes by interfering with microbial metabolism and cell membrane permeability. Green synthesis using medicinal plants not only reduces toxicity but also confers additional antimicrobial activity through the synergistic effects of the phytoconstituents (Mohamed et al. 2024).

Copper oxide nanoparticles synthesized from extracts and fractions of *Colebrookea oppositifolia* were shown to have promising antibacterial activity against *Pseudomonas aeruginosa* with a zone of inhibition of ~13.67 mm and antifungal activity against *Rosellinia necatrix* and *Fusarium oxysporum* (58% inhibition) (Bhardwaj et al. 2024). Another example demonstrating the antimicrobial activity of PBNPs is the selenium nanoparticles (SeNPs) synthesized using *Dahlia pinnata* L. tuber extract, which exhibited potent efficacy against *E. coli* isolated from different patients with various infections (urinary tract infection, bloodstream infection, respiratory tract infection), including multidrug-resistant strains, nine of which were completely resistant to penicillin, amoxicillin-clavulanic acid, clindamycin, tetracycline, and vancomycin (Mekky et al. 2025).

Moreover, silver nanoparticles (AgNPs) prepared from *Magnolia alba* leaf extract have shown antibacterial, antifungal, and antiviral effects. AgNPs bind to bacterial cell walls, leading to their destruction. Once inside, they bind to proteins, prevent ATP generation, and generate ROS that induce apoptosis. AgNPs exhibit strong antibacterial activity, with zones of inhibition greater than 15 mm with *E. coli*, *P. aeruginosa*, *E. faecalis*, methicillin-sensitive *Staphylococcus aureus* (MSSA), and methicillin-resistant *Staphylococcus aureus* (MRSA). Also, AgNPs biosynthesized from *Magnolia alba* leaf extracts exhibited significant antifungal activity against the nosocomial pathogen *Candida albicans*, compromising the fungal membrane integrity. Furthermore, the tested AgNPs displayed antiviral activity against T7 bacteriophage by binding to the viral capsid, causing structural damage that kills the phages (De Mel et al. 2025).

10.3.3 Antioxidant Effects

Antioxidants are vital in neutralizing free radicals, thereby protecting cells from oxidative stress. Plant-based nanoparticles, often synthesized through "green synthesis" methods using plant extracts, offer a biocompatible, eco-friendly, and sustainable alternative to chemically or physically produced nanoparticles (Radulescu et al. 2023).

An example of plant-based nanoparticles being used as antioxidants is lignin (Barhoum et al. 2020). Because lignin contains persistent organic radicals in its π-conjugated phenolic framework, some researchers expect to utilize it as an antioxidant in the pharmaceutical business (Qu et al. 2025). Because of its low molecular weight, lignin promotes cell and mitochondrial infiltration, thereby inhibiting biological activity (i.e., antimicrobial activity). Because of their biological toxicity, other synthetic low-molecular-weight antioxidants have also proven challenging to employ in medical research (Alzagameem et al. 2019). This has led to the

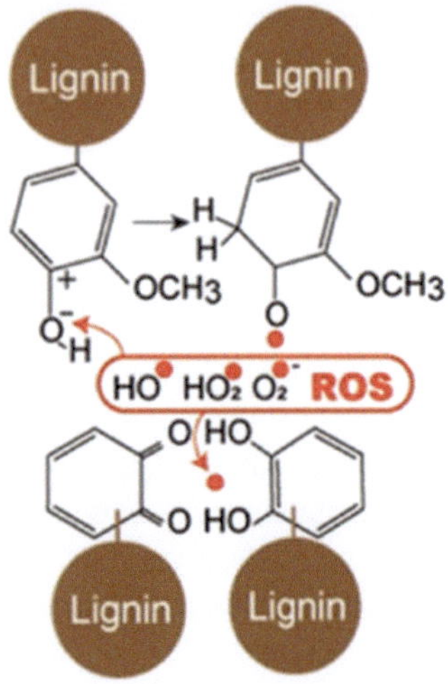

Fig. 10.2 The radical scavenging activity of lignin (Barhoum et al. 2020)

development of antioxidant nanoparticles, macromolecules with antioxidant functional groups, such as nitroxyl radicals, that act as novel antioxidants without biological toxicity because they do not penetrate cells or mitochondria. The lignin nanoparticle, which is negatively charged in pure water, has a diameter of many tens of nanometers. The water-dispersed lignin nanoparticle's antioxidant properties are demonstrated by its ability to absorb oxygen radicals. The lignin nanoparticle isolated from Japanese cedar has a Trolox equivalent antioxidant capacity (TEAC) of 25.3 μmol TE g^{-1}. Because it has the chemical structure depicted in Fig. 10.2, it exhibits a distinct absorption property for reactive oxygen species (ROS) that is comparable to that of low-molecular-weight antioxidants, including its macromolecular (i.e., nanoparticulated) character (Shikinaka et al. 2018). Another example is the green synthesis of silver nanoparticles using a polar extract of taro corms (*Colocasia esculenta*) and the evaluation of its antioxidant properties and wound-healing applications (Qubtia et al. 2024). Taro is an important tropical root crop grown for its starchy corms, or underground stems. A variety of carbohydrates, nutrients, and mineral substances, including iron, calcium, and potassium, that may undergo rapid oxidation are present in taro corms and are responsible for their increased antioxidant activity (Ribeiro Pereira et al. 2020).

10.3.4 Antiparasitic Effects

Due to their natural origin, biocompatibility, and reduced risk of side effects or environmental toxicity, plant-based nanoparticles hold great promise for use against a wide range of parasitic diseases, including malaria, leishmaniasis, schistosomiasis, and helminth infections (Tiwari et al. 2023). One of the parasites that can potentially be treated using this technology is malaria (Pham et al. 2018). The prescribed antimalarial medications exhibit low-to-no selectivity for the host cell and extensive volume of distribution, even though the causal pathogen (*Plasmodium* spp.) is an intracellular parasite (Lokole et al. 2024). Serious adverse effects result from this, which impair patient adherence and encourage the development of drug-resistant strains.

Green nanotechnology-based approaches to malaria control hold promise, as they can eradicate malaria vectors (*Anopheles* sp.) and encapsulate new and existing antimalarial drugs to deliver them to various target sites (Barati et al. 2024). These carriers have demonstrated the potential to improve the targeting of antimalarials to host cells and to reduce the adverse effects of medication therapy. Additionally, the discovery of novel molecular targets within the malaria parasite, including aminopeptidases and histone deacetylases, has created opportunities to develop novel antimalarial drugs (Wei et al. 2021).

Another parasite, leishmaniasis, is a group of diseases spread by mosquito species in the genera Phlebotomus (New World) and Lutzomyia (Old World) and causes health, social, and economic issues. Sodium stibogluconate (Pentostam®) and meglumine antimony (Glucantime®) are the primary medications used to treat leishmaniasis (Solomon et al. 2024). Amphotericin B (AmB) and pentamidine are other medications that can be utilized, although their therapeutic effects are limited, frequently because microorganisms are resistant to them (Vyas and Gupta 2006). These medications have issues with cost, effectiveness, and serious patient toxicity (Gupta et al. 2010). They also need lengthy treatment regimens, which complicates treatment.

A nanoemulsion was formulated and added to a gel for transdermal topical application to enhance the skin's absorption of the active ingredients in *Citrus sinensis* essential oil (Ghanbariasad et al. 2021). In addition to increasing the product's viscosity, the formulation promoted greater accumulation and enhanced penetration at the application site compared to the free essential oil. The encapsulated product demonstrated greater in vitro antileishmanial activity against promastigote forms of *L. major* and *L. tropica*, the species responsible for cutaneous leishmaniasis.

10.3.5 Anticancer Effects

Cancer is among the leading causes of mortality worldwide and is highly prevalent (Yadav et al. 2025). It is defined as the rapid proliferation of specific cells relative to others and manifests in the lung, prostate, thyroid, breast, and colorectal tissues, among others (Sachan et al. 2024). It was stated by the International Agency for Research on Cancer (IARC) that one in every five people becomes ill with cancer (Sachan et al. 2024). Therefore, this worrisome level of spread, in addition to the severity of the side effects of available drugs, such as pain, fatigue, vomiting, and diarrhea, prompts us to search for alternatives in nature and plants continuously (Katta et al. 2023; Sachan et al. 2024).

The anticancer effect of plant-based NPs was highlighted by using molecules such as paclitaxel, vincristine, and topotecan plant-derived and formulating them as liposomes, which have undergone clinical trials, and with doxorubicin (DOX) liposomal form (AmBisome) getting approved by the FDA for cancer, leishmaniasis, and fungal infections (Chang and Yeh 2012; Meyerhoff 1999; Sritharan and Sivalingam 2021). Liver, colon, lung, and cervical cancers were all responsive to treatment with naturally derived NPs, mainly due to reactive oxygen species generated within cancer cells, leading to cell cycle arrest (Javed et al. 2021).

Chemotherapy and radiotherapy, typically used for people living with cancer, suffer from side effects due to a lack of targeting; therefore, NPs were used to enhance the selectivity of drugs to cancerous cells and pack them as a means to protect normal cells (Javed et al. 2021). Taxifolin, a flavonoid, was used in zinc oxide (ZnO) NPs for this purpose, along with the aforementioned DOX (Mishra et al. 2022; Sundraraman and Jayakumari 2020). Such particles, whether loaded with antitumor drugs or not, enter cancerous cells by endocytosis and intoxicate them, leading to apoptosis. Their surface can also be modified to alter drug release, which is beneficial against cancer (Anjum et al. 2021).

Plant extracts are also the preferred agents for the biosynthesis of metal NPs (Javed et al. 2021). This can be explained as a result of certain chemical groups in secondary metabolites, such as carboxylic and amine groups, which play a role in the reduction of metal ions into NPs as silver NPs, as well as the stabilization and capping of said NPs, such as *Eucalyptus procera* extract and *Butea monosperma* extract used to form silver NPs loaded with DOX, inhibiting cancer cells more intensely than DOX alone (Gomes et al. 2021). Silver NPs also contained epigallocatechin gallate (EGCG), a potent antioxidant with potential for treating tumors (Ali et al. 2022). EGCG NPs harvested from green tea have antioxidant and anti-inflammatory properties and have been tested as a cancer therapy (Karnwal et al. 2024). Moreover, curcumin- and limonin-containing citrus fruits have been used to form nanocurcumin and copper NPs, respectively, enhancing cytotoxicity, with the former improving the solubility and bioavailability of curcumin (Hettiarachchi et al. 2021; Singh et al. 2023b).

ZnO NPs were also used for anticancer therapy with plant bioactive compounds, sometimes surpassing synthetic NPs, as in colon cancer targeting, acting in the exact mechanism and carrying curcumin, enhancing its activity against gastric cancer cells while being safe (Anjum et al. 2021). Trans-resveratrol, when coupled with ZnO NPs, induces the apoptotic pathway and can be combined with curcumin and other antitumor drugs, thereby reducing their chemotoxicity (Annaji et al. 2021). The activity of Resveratrol NPs is currently being validated in a clinical trial evaluating their efficacy against pathogenic human amoebas (NCT07148752).

10.3.6 Wound Healing Effects

Wound healing is a complex process that aims to resolve several problems that arise alone or together. Wounds are prone to colonization by bacteria that form persistent biofilms; these biofilms protect microbes from host defenses and antimicrobials and are a significant cause of chronic, non-healing wounds (Zhao et al. 2013). Another challenge is oxidative stress, which is an imbalance between reactive oxygen species (ROS) production and antioxidant defenses. It is a key factor in impaired wound healing, particularly in diabetic wounds where excessive ROS causes cellular dysfunction and delayed repair (Deng et al. 2021). Also, inflammation at the wound site, driven by persistent immune activation, impairs progression through the normal healing stages and contributes to tissue breakdown rather than repair (Versey et al. 2021). Moreover, effective re-epithelialization, a keratinocyte-driven process that restores the epidermal barrier, is slowed in chronic wounds due to altered keratinocyte migration, dysregulated growth factor signaling, and a hostile wound microenvironment (Pastar et al. 2014).

Plant-based nanoparticles continue to show promising results when applied to promote wound healing due to a multitude of synergistic factors, such as their ability to reduce inflammation (Lakkim et al. 2023), stimulate cell migration and proliferation (Sezen et al. 2023), enhance collagen deposition (Vijapur et al. 2025), and their antimicrobial and antioxidant activities (Aydin Acar et al. 2024). According to recent research, producing silver nanoparticles from plants offers a safe and effective method for wound treatment. For instance, silver nanoparticles were biosynthesized from a *Cyperus rotundus* rhizome extract and then incorporated into a carbopol-based hydrogel (Sharma et al. 2025a). Key clinical issues in wound care, such as bacterial infection, oxidative stress, and chronic inflammation, were successfully handled by this formulation. The hydrogel demonstrated strong antibacterial activity against both Gram-positive and Gram-negative pathogens, notable anti-inflammatory and antioxidant properties, and good cytocompatibility with human cells. Furthermore, in vivo studies using a rat excision wound model showed that the AgNP-hydrogel significantly expedited wound contraction, promoted re-epithelialization, and facilitated skin appendage regeneration, with healing outcomes on par with conventional therapy (Sharma et al. 2025a, b).

Chronic inflammation, oxidative stress, poor angiogenesis, and persistent infection all contribute to delayed healing and a high risk of complications, making diabetic wounds a significant clinical issue. Wei et al. (2025) developed a novel herbal-based hydrogel by self-assembling mangiferin with rhein-functionalized silver nanoparticles (Rh@Ag NPs) to produce multipurpose wound dressings (Wei et al. 2025). The main obstacles to diabetic wound healing were intended to be addressed concurrently by this bioengineered system: mangiferin provided strong antioxidative effects. It promoted the recruitment of cells necessary for tissue regeneration, while Rh@Ag NPs offered robust, broad-spectrum antibacterial activity. In infected diabetic wound models, preclinical studies showed that this dual-component hydrogel was not only biocompatible but also highly effective at accelerating wound closure. In addition to beneficial modulation of signaling pathways associated with wound healing, treatment led to improved re-epithelialization, increased collagen deposition, increased angiogenesis, and decreased inflammatory responses. All of these results pointed to the potential of plant-derived hydrogels, coupled with nanoparticles, as sophisticated, multipurpose treatments that may successfully navigate the intricate pathological milieu of chronic diabetic wounds (Wei et al. 2025).

Another notable example is the work of Jin et al. (2025), who integrated plant-derived lemon exosomes into a GelMA-DAS hydrogel to address several of the constraints in diabetic wound healing, including chronic inflammation, inadequate vascularization, and insufficient sustained delivery. In diabetic rat wound models, this hybrid dressing increased angiogenesis and collagen deposition, improved fibroblast and endothelial cell activity, altered macrophage polarization, and demonstrated good biocompatibility (Jin et al. 2025).

Although the use of plant-based nanoparticles in wound healing is promising, future literature should address remaining challenges, such as cytotoxicity and biocompatibility, as silver and other metallic NPs can still cause oxidative stress, mitochondrial dysfunction, or apoptosis in mammalian cells at higher concentrations. Moreover, the therapeutic window is not always clear (AshaRani et al. 2009). Also, the dose-dependent effects of NPs are not well standardized. Higher doses may enhance antibacterial action but increase the risk of cytotoxicity, and the persistence and accumulation of NPs in tissues, especially metallic ones, may raise long-term safety concerns (e.g., inflammation, fibrosis) (Choi and Hu 2008). While preclinical evidence is promising, there is a lack of well-designed clinical studies testing safety, efficacy, and dosage in human patients. Future research must bridge this gap before clinical translation (Mayegowda et al. 2023) (Table 10.1).

10.4 Role of Plant-Based Nanoparticles in Diagnostic Applications

Diagnostic applications are among the most surprising uses of plant nanoparticles (PNPs) (Karnwal et al. 2024). PNPs are biocompatible and stable and exhibit novel chemical and physical properties that enhance the sensing and monitoring of disease (Song et al. 2021). PNPs provide green, low-cost, and rapid alternatives to traditional diagnostic techniques with sensitive imaging and precise biosensing (Fig. 10.3).

The most auspicious PNP application in diagnosis is bioimaging. Their design in the optical range enhances imaging of disease mechanisms and biological structures (Sarasati et al. 2023). Gold nanoparticles are extensively used as SPR probes for imaging (Chen et al. 2023). They offer high-contrast fluorescent microscopy and computer tomography signals (Ibrahim et al. 2021). Silver nanoparticles enable signal amplification and detection at lower limits, as well as real-time detection of disease biomarkers. Zinc oxide nanoparticles exhibit intense luminescence in the

Table 10.1 Summary of different biological activities of some plant-based nanoparticles

Activity	Plant/component responsible	Mechanism of action	Reference
Neuroprotective	Silver nanoparticles *of Lampranthus coccineus*	Direct anti--acetylcholinesterase activity	Youssif et al. (2019)
	Selenium nanoparticles biosynthesized with epigallocatechin gallate (EGCG-SeNPs)	Restore oxidant/antioxidant balance, anti-inflammatory (↓TNF-α, IL-1β, NOS2), anti-apoptotic (↑Bcl-2, ↓Bax, caspase-3)	Alrashdi et al. (2023)
	Asiatic acid with BSANP (bovine serum albumin nanoparticles) + glutathione ligand	Nanoparticle delivery system improves bioavailability and targeting	Raval et al. (2015)
	Lepidagathis pungens ethanolic extract silver nanoparticles	Reduce MDA (45%) and ROS (50%), increase SOD (40%) and CAT (35%) antioxidant enzymes	Amtaghri et al. (2024)
	Pomegranate extract silver nanoparticles	Reduce MDA and ROS, prevention of DNA fragmentation in SH-SY5 cells	Khorrami et al. (2023)
Antimicrobial	Copper oxide nanoparticles from *Colebrookea oppositifolia* extract	Antibacterial activity of *Pseudomonas aeruginosa*, (13.67% inhibition) antifungal activity against *Rosellinia necatrix*, *Fusarium oxysporum* (58% inhibition)	Bhardwaj et al. (2024)
	Selenium nanoparticles *Dahlia pinnata* L. tuber extract	Effective against multidrug--resistant *E. coli* strains	Mekky et al. (2025)
	Silver nanoparticles *Magnolia alba* leaf extract	Antibacterial disrupts bacterial wall, inhibits ATP generation, generates ROS, antifungal compromises fungal membrane, antiviral damages viral capsid	De Mel et al. (2025)
	Lignin nanoparticles (from Japanese cedar)	π-conjugated phenolic framework with persistent organic radicals → absorbs oxygen radicals; macromolecular nanoparticles prevent mitochondrial/cell penetration	Otsuka et al. (2026)

(continued)

Table 10.1 (continued)

Activity	Plant/component responsible	Mechanism of action	Reference
Antioxidant capacity	AgNPs from *Carissa carandas* leaf extract	–	Singh et al. (2021)
	Gold NPs and Silver NPs from *Clerodendrum inerme* leaf extract	–	Khan et al. (2020)
	Lignin nanoparticles	π-conjugated phenolic radicals absorb ROS	Shikinaka et al. (2018)
	Silver nanoparticles *Colocasia esculenta* (taro) corms extract	Carbohydrates and minerals with antioxidant properties	Ribeiro Pereira et al. (2020)
Antiparasitic	*Citrus sinensis* essential oil (encapsulated in NE gel)	Improved absorption, penetration, and accumulation at skin application site; enhanced in vitro activity vs. *L. major* and *L. tropica* promastigotes	Ghanbariasad et al. (2021)
Anticancer	Paclitaxel, Vincristine, Topotecan (plant-derived molecules); Doxorubicin (liposomal form, AmBisome)	Formulated as liposomes; FDA-approved; enhanced delivery for cancer, leishmaniasis, and fungal infections	Chang and Yeh (2012)
	Taxifolin (flavonoid) in ZnO NPs; Doxorubicin in ZnO NPs	Enhanced drug selectivity, protection of normal cells; entry by endocytosis → apoptosis; modified surface for drug release.	Sundraraman and Jayakumari (2020)
	Silver NPs biosynthesized with *Eucalyptus procera* and *Butea monosperma* extracts, loaded with DOX	Secondary metabolites (carboxylic and amine groups) reduce and stabilize AgNPs; DOX-loaded AgNPs inhibit cancer cells more strongly than DOX alone	Gomes et al. (2021)
	Silver NPs housing epigallocatechin gallate (EGCG); EGCG NPs from green tea	Antioxidant and anti--inflammatory activity; tested as cancer therapy	Ali et al. (2022)
	Nanocurcumin from curcumin; copper NPs with limonin from (*citrus* fruit)	Curcumin: improved solubility and bioavailability; Copper NPs: enhanced cytotoxicity	Hettiarachchi et al. (2021)
	(*E*)-resveratrol with ZnO NPs	Induces apoptosis; can combine with curcumin and other drugs → reduces chemotoxicity	Annaji et al. (2021)

(continued)

Table 10.1 (continued)

Activity	Plant/component responsible	Mechanism of action	Reference
Wound healing	ZnO nanoparticles synthesized with *Capparis spinosa* extract	Antimicrobial effects + wound healing support	Hettiarachchi et al. (2021)
	Silver nanoparticles from *Glycyrrhiza glabra* root extract	Enhanced wound healing properties	Vijapur et al. (2025)
	ZnO nanoparticles synthesized with *Calendula officinalis* flower extract	Antioxidant + wound healing promotion	Aydin Acar et al. (2024)
	Silver nanoparticles from *Cyperus rotundus* rhizome extract (AgNP-hydrogel)	Antibacterial, antioxidant, anti-inflammatory, cytocompatible, accelerates wound contraction + skin regeneration	Sharma et al. (2025a, b)
	Herbal-based hydrogel (mangiferin + rhein--functionalized AgNPs, Rh@AgNPs)	Antioxidant (mangiferin) + antibacterial (Rh@Ag NPs) → improved re-epithelialization, collagen deposition, angiogenesis, and reduced inflammation	Wei et al. (2025)

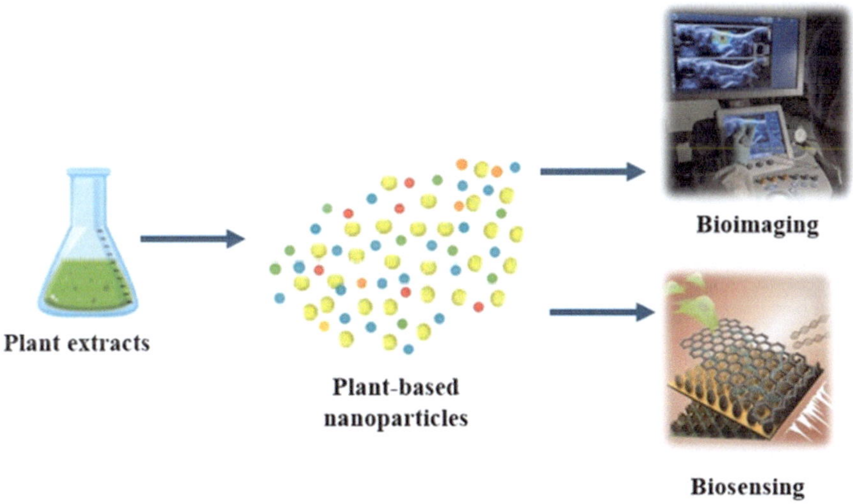

Fig. 10.3 The diagnostic role of plant-based nanoparticles

visible and near-UV regions of the spectrum and are therefore widely used in fluorescence imaging. Selenium nanoparticles are fluorescent probes that can be used for imaging and antioxidant activity (Dhir et al. 2023; Sharma et al. 2025a, b).

Plant-based nanoparticles are also being investigated in new imaging modalities such as MRI, CT, and optical imaging. Plant-origin biocompatibility renders them harmless for applications in biological systems (Nazli et al. 2018). Targeted nanoparticles can be tailored to specific cancer biomarkers, infectious agents, or diseased cells, thereby enhancing sensitivity and specificity (Gu and Minko 2024). It has the potential to provide real-time, non-invasive imaging of biological processes at the cellular and molecular levels, which is particularly appropriate for disease detection and early-stage treatment monitoring (Mai and Inkielewicz-Stepniak 2024).

In addition, "smart" nanostructures such as quantum dots and protein nanoparticles can be used for plant nanomaterials in imaging (Mai and Inkielewicz-Stepniak 2024). The green, fluorescent quantum dots in plants can fluoresce when excited by UV light and can be used for tumor imaging or tracing brain function. The nanoparticles of the smart protein labeled with imaging moiety can be programmed to migrate and target a specific site for imaging and co-delivery therapy (Chen et al. 2025). The multimodal facilitates the theragnostic application of PNPs, combining diagnosis and treatment into a single device (Chen et al. 2025).

Apart from imaging, PNPs are used on a gigantic scale in biosensing (Kucherenko et al. 2019). PNPs are part of diagnostic sensors for pathogen and toxin detection that exhibit high sensitivity and specificity. PNPs possess a very high surface-to-volume ratio, and functionalization with molecules such as antibodies, enzymes, and nucleic acids is highly effective for generating analyte-specific surfaces. There are a whole lot of applications that demonstrate their versatility.

PNPs are widely used in electrochemical and optical biosensors. PNPs facilitate rapid electron transfer, signal amplification, and detection, thereby reducing the limit of detection. This renders them suitable for point-of-care diagnostics and poor populations, where they must be fast, cheap, and effective at detection (Kucherenko et al. 2019). The application of a PNP-based biosensor is also reported for tracking environmental and food safety (Mocan et al. 2017). Microbial toxins and pollutants are detected by PNP-based biosensors using biosensor-assisted detection (Zaimy et al. 2017). As they can produce fluorescent, electrical, or visible signals, they can diagnose a wide range of conditions without compromising integrity. Further, PNPs' applications in nanochips and micro-robots improve sensitivity and throughput in diagnosis, offering intelligent diagnosis systems in the future with new, promising opportunities.

Plant-based nanoparticles are widely used as green, biocompatible, and sensitive diagnostic reagents, serving as an alternative to conventional reagents. PNPs' potential to enable instant detection, in situ real-time monitoring, and even therapeutics or antimicrobial activity indicates their potential as future multifunctional drugs. Due to a surge in research, PNP-based diagnostics will be at the forefront of next-generation theragnostic platform development, revolutionizing disease detection, monitoring, and personalized medicine.

10.5 Role of Plant-Based Nanoparticles in the Drug Delivery Systems

The dawn of the era of nanoparticles (NPs) is a remarkable advancement in science. Nanomedicine has disadvantages, including immunogenicity, toxicity, and difficulty in loading nanoparticles (Ilinskaya and Dobrovolskaia 2016). Despite that, it has been used to deliver many drugs extensively such as anticancer drugs (Pelaz et al. 2017). Nano-systems, such as liposomes and exosomes, are used to protect drugs from degradation, reduce side effects, and perhaps even alter their pharmacokinetic properties (dos Santos et al. 2022). Leaf extracts, stems, roots, flowers, and seeds are good sources for nanoparticles (Griffin et al. 2018; Rey-Méndez et al. 2022). These natural-origin NPs, called Plant-Derived Nanoparticles (PDNPs), are favored over synthetic NPs due to their low immunogenicity and toxicity (Yang et al. 2018). This has pushed us toward harnessing plants to create nanoparticles that can carry drugs, serving as new drug delivery systems (Iravani and Varma 2019). The normal size distribution of natural nanoparticles ranges from 30 to 500 nm. If smaller than 30 nm, they become difficult to pack with hydrophobic material, and their capacity for loading decreases (Yang et al. 2018). Vegetable proteins harvested from plants are widely used as nanocarriers. Lectin, gliadin, zein, and soy proteins are among examples of plant proteins that are formulated as nanoparticles. They have been used to contain drugs such as anticancer drugs successfully (Kianfar 2021).

Many edible plants release nano-sized vesicles named exosomes, which contain proteins, lipids, and nucleic acids and have been utilized in drug delivery. This content is used for their characterization in addition to other bioactive compounds, such as naringenin in grapefruits and gingerol in ginger (Alfieri et al. 2021; Viršilė et al. 2022). They emerge from plant cells after combining with the plasma membrane. They are commonly isolated by ultracentrifugation, but immunoaffinity, polymer precipitation, and size-exclusion chromatography are also used (Wang et al. 2023b; Yang et al. 2018). They enter cells again by direct fusion with the cell membrane or by receptor-mediated endocytosis, pinocytosis, or phagocytosis (Wang et al. 2023b). Ginger, lemon, grapefruit, broccoli, and grapes are all examples of plants; plant exosome-like nanoparticles (PELNVs) were isolated from them (Wang et al. 2024a). They were found to be synthesized by several biogenesis pathways; the most common method was their budding from the endosomal membrane, as mentioned above, which is called the multivesicular body (MVB) pathway (Wang et al. 2024a). We can find many proteins, such as receptors, cytokines, and transport proteins, which perform many functions, and lipids, such as glycerol lipids and ceramide, which provide structural rigidity (Wang et al. 2023b, Yang et al. 2018), in addition to DNA, mRNA, noncoding RNA, and miRNA, which were used as potential treatments for colon cancer (Iravani and Varma 2019; Wang et al. 2023b). Ginseng PELNVs and PELNVs in general were found to be highly stable (Wang et al. 2024a; Yang et al. 2018). Ginseng-derived NPs were tested and found to promote angiogenesis and tissue and bone regeneration as well (Karnwal et al. 2024). Ginger-derived NPs were found to affect the gut microbiota positively and are beneficial for the treatment and prevention of ulcerative colitis (Alfieri et al. 2021; Yang et al. 2018), while lemon ones had antitumor activity, reducing progression and growth (Alfieri et al. 2021). Grapefruit NPs also played a role in fighting tumors and ulcerative colitis, and they were potential carriers for anticancer drugs, as they were tested along with ginger NPs to deliver doxorubicin, which showed stronger activity than doxorubicin alone (Alfieri et al. 2021). The lipids within PDNPs were also isolated and used to form liposomes that deliver doxorubicin to tumor cells (Yang et al. 2018). Grapefruit-derived NPs were selectively taken up by intestinal macrophages and ameliorated DSS-induced mouse colitis, allowing their use for oral delivery of methotrexate to reduce its side effects (Mohammadinejad et al. 2016).

Ultimately, more research should be directed toward the abilities PDNPs possess alone and as drug carrier systems for other drugs, and their performance in clinical trials needs to be evaluated (Wang et al. 2024a). Moreover, the important role of plants and their extracts as environmentally friendly reducing agents for the synthesis of metal nanoparticles cannot be overstated, as they serve diverse purposes in drug delivery (Bao et al. 2021).

10.6 Conclusion and Future Remarks

Positioned as the future of nanomedicine, plant-based nanoparticles (PBNPs) have emerged as a promising and sustainable alternative to conventional nanomaterials. This is due to their green synthesis, wide range of applications, and biocompatibility. By leveraging the diversity of plants and their phytochemicals, scientists have produced stable, safe, and biologically functional nanoparticles. Results presented previously in this chapter highlight the vast array of biomedical applications of PBNPs. They possess neuroprotective, antimicrobial, antioxidant, antiparasitic, anticancer, wound healing, and diagnostic capabilities, in addition to serving as effective drug delivery systems. Notably, plant-origin nanoparticles not only enhance the therapeutic targeting and bioavailability of drugs but also demonstrate synergistic pharmacological activity. Their ability to penetrate biological barriers, reduce side effects, and enhance pharmacokinetic profiles makes them valued in modern therapy.

While these advantages exist, large-scale production, protocol-standardized synthesis, and the complete characterization of their pharmacodynamics, along with long-term safety profiles, remain ongoing challenges. Further preclinical and clinical investigations will be needed to take these experimental observations to regulatory-approved treatments. This requires further collaboration among biotechnology, pharmacology, and nanotechnology to realize PBNP's full potential. In conclusion, plant-based nanoparticles are a milestone toward greener, safer, and more effective healthcare solutions. Their unique capacity to enhance drug delivery and confer intrinsic bioactivity makes them an exceptional tool for combating complex diseases and addressing global health challenges. With future continued research efforts, PBNPs can shape the future of personalized medicine and the next generation of nanotherapeutics.

References

Abd-elaty DM, Al-Qaaneh AM, Al-Azzawi MA, Ishaq AR, Alamri A, Al-Qahtani WH, El-Nashar HA (2025) Development of a novel allspice essential oil-loaded cream against skin wrinkles: GC-MS metabolic profiling, characterization and enzyme inhibition assays. J Essent Oil Bear Plants 28:708–721

Abdelazim EB, Abed T, Goher SS, Alya SH, El-Nashar HA, El-Moslamy SH, El-Fakharany EM, Abdul-Baki EA, Shakweer MM, Eissa NG (2024) In vitro and in vivo studies of Syzygium cumini-loaded electrospun PLGA/PMMA/collagen nanofibers for accelerating topical wound healing. RSC Adv 14(1):101–117

Akbar S, Tauseef I, Subhan F, Sultana N, Khan I, Ahmed U, Haleem KS (2020) An overview of the plant-mediated synthesis of zinc oxide nanoparticles and their antimicrobial potential. Inorg Nano-Metal Chem 50(4):257–271

Alfieri M, Leone A, Ambrosone A (2021) Plant-derived nano and microvesicles for human health and therapeutic potential in nanomedicine. Pharmaceutics 13(4):498

Ali SG, Jalal M, Ahmad H, Sharma D, Ahmad A, Umar K, Khan HM (2022) Green synthesis of silver nanoparticles from Camellia sinensis and its antimicrobial and antibiofilm effect against clinical isolates. Materials (Basel) 15(19):6978

Ali A, Ambrose S, Hussain D, Hafeez F, Asghar T, Shah SNA, Ahsan M, Shakir M, Ali A, Khan AMA (2024) Antimicrobial resistance and antimicrobial activity of plant-based antimicrobial peptides against bacteria: plant-based antimicrobial peptides. Lett Anim Biol 4(2):19–27

Alrashdi BM, Fehaid A, Kassab RB, Rizk S, Habotta OA, Abdel Moneim AE (2023) Biosynthesized selenium nanoparticles using epigallocatechin gallate protect against pentylenetetrazole-induced acute epileptic seizures in mice via antioxidative, anti-inflammatory, and anti-apoptotic activities. Biomedicines 11(7):1955

Alzagameem A, Klein SE, Bergs M, Do XT, Korte I, Dohlen S, Hüwe C, Kreyenschmidt J, Kamm B, Larkins M, Schulze M (2019) Antimicrobial activity of lignin and lignin-derived cellulose and chitosan composites against selected pathogenic and spoilage microorganisms. Polymers (Basel) 11(4):670

Amini SM (2019) Preparation of antimicrobial metallic nanoparticles with bioactive compounds. Mater Sci Eng C 103:109809

Amtaghri S, Slaoui M, Eddouks M (2024) Mentha pulegium: a plant with several medicinal properties. Endocr Metab Immune Disord Drug Targets 24(3):302–320

Anjum S, Hashim M, Malik SA, Khan M, Lorenzo JM, Abbasi BH, Hano C (2021) Recent advances in zinc oxide nanoparticles (ZnO NPs) for cancer diagnosis, target drug delivery, and treatment. Cancers (Basel) 13(18):4570

Annaji M, Poudel I, Boddu SHS, Arnold RD, Tiwari AK, Babu RJ (2021) Resveratrol-loaded nanomedicines for cancer applications. Cancer Rep (Hoboken) 4(3):e1353

Anwar A, Masri A, Rao K, Rajendran K, Khan NA, Shah MR, Siddiqui R (2019) Antimicrobial activities of green synthesized gums-stabilized nanoparticles loaded with flavonoids. Sci Rep 9(1):3122

Asghar MA, Asghar MA (2020) RETRACTED: green synthesized and characterized copper nanoparticles using various new plants extracts aggravate microbial cell membrane damage after interaction with lipopolysaccharide. Int J Biol Macromol 160:1168

AshaRani P, Low Kah Mun G, Hande MP, Valiyaveettil S (2009) Cytotoxicity and genotoxicity of silver nanoparticles in human cells. ACS Nano 3(2):279–290

Askar MA, El-Nashar HA, Al-Azzawi MA, Rahman SSA, Elshawi OE (2022) Synergistic effect of quercetin magnetite nanoparticles and targeted radiotherapy in treatment of breast cancer. Breast Cancer 16:11782234221086728

Aydin Acar C, Gencer MA, Pehlivanoglu S, Yesilot S, Donmez S (2024) Green and eco-friendly biosynthesis of zinc oxide nanoparticles using Calendula officinalis flower extract: wound healing potential and antioxidant activity. Int Wound J 21(1):e14413

Bachheti RK, Bachheti A (eds) (2023) Secondary metabolites from medicinal plants: nanoparticles synthesis and their applications. CRC Press, Boca Raton. https://doi.org/10.1201/9781003213727

Bachheti A(J), Bachheti RK, Husen A (eds) (2024) Carbon-based nanomaterials: synthesis, agricultural, biomedical, and environmental interventions. Springer Nature Singapore. https://doi.org/10.1007/978-981-97-0240-4

Baky MH, Elkenawy NM, El-Nashar HAS, Abib B, Farag MA (2024) Comparison of autoclaving and γ-radiation impact on four spices aroma profiles and microbial load using HS-SPME GC-MS and chemometric tools. Sci Rep 14(1):5752

Banjara RA, Kumar A, Aneshwari RK, Satnami ML, Sinha S (2024) A comparative analysis of chemical vs green synthesis of nanoparticles and their various applications. Environ Nanotechnol Monit Manage 22:100988

Bao Y, He J, Song K, Guo J, Zhou X, Liu S (2021) Plant-extract-mediated synthesis of metal nanoparticles. J Chem 2021(1):6562687

Barati A, Huseynzade A, Imamova N, Shikhaliyeva I, Keles S, Alakbarli J, Akgul B, Bagirova M, Allahverdiyev AM (2024) Nanotechnology and malaria: evaluation of efficacy and toxicity of green nanoparticles and future perspectives. J Vector Borne Dis 61(3):340–356

Barhoum A, Jeevanandam J, Rastogi A, Samyn P, Boluk Y, Dufresne A, Danquah MK, Bechelany M (2020) Plant celluloses, hemicelluloses, lignins, and volatile oils for the synthesis of nanoparticles and nanostructured materials. Nanoscale 12(45):22845–22890

Berta L, Coman NA, Rusu A, Tanase C (2021) A review on plant-mediated synthesis of bimetallic nanoparticles, characterisation and their biological applications. Materials (Basel) 14(24):7677

Bhardwaj N, Puri S, Kumari A, Chauhan A, Kumar A (2024) Phyto-mediated synthesis of copper oxide nanoparticles: an investigation of antioxidant, antimicrobial, anti-inflammatory, and protective neuropsychiatric potential. Bionanoscience 14:5377

Bhattacharya T, Soares G, Chopra H, Rahman MM, Hasan Z, Swain SS, Cavalu S (2022) Applications of phyto-nanotechnology for the treatment of neurodegenerative disorders. Materials (Basel) 15(3):804

Bhuia MS, Chowdhury R, Sonia FA, Biswas S, Ferdous J, El-Nashar HAS, El-Shazly M, Islam MT (2024) Efficacy of rotundic acid and its derivatives as promising natural anticancer triterpenoids: a literature-based study. Chem Biodivers 21(2):e202301492

Bonifácio BV, da Silva PB, Ramos MA d S, Negri KMS, Bauab TM, Chorilli M (2014) Nanotechnology-based drug delivery systems and herbal medicines: a review. Int J Nanomedicine 9:1–15

Chang HI, Yeh MK (2012) Clinical development of liposome-based drugs: formulation, characterization, and therapeutic efficacy. Int J Nanomedicine 7:49–60

Chen Y, Xiong X, Chen Y, Chen L, Liu G, Xiao W, Shi J, Chen Z, Luo Y (2023) MoS(2)-nanoflower and nanodiamond co-engineered surface plasmon resonance for biosensing. Biosensors (Basel) 13(5):506

Chen Q, Sun J, Ling S, Yang H, Li T, Yang X, Li M, Du M, Zhang Y, Li C, Wang Q (2025) Tumor microenvironment-responsive nano-immunomodulaTors for enhancing chimeric antigen receptor-T cell therapy in lung cancer. ACS Nano 19(8):8212–8226

Choi O, Hu Z (2008) Size dependent and reactive oxygen species related nanosilver toxicity to nitrifying bacteria. Environ Sci Technol 42(12):4583–4588

Codreanu S, Cepoi L, Rudi L, Chiriac T (2025) Photosynthetic microorganisms and biogenic synthesis of nanomaterials for sustainable agriculture. Nanomaterials 15(13):990

Cortez R, Luna-Vital DA, Margulis D, Gonzalez de Mejia E (2017) Natural pigments: stabilization methods of anthocyanins for food applications. Compr Rev Food Sci Food Saf 16(1):180–198

De Mel S, Gruenler J, Khoury L, Heynes A, Fazekas J, Damaske K, Galbadage T, Gunasekera RS, Anderson RS (2025) Green synthesis of silver nanoparticles using Magnolia alba leaf extracts and evaluating their antimicrobial, anticancer, antioxidant, and photocatalytic properties. Sci Rep 15(1):23709

Deng L, Du C, Song P, Chen T, Rui S, Armstrong DG, Deng W (2021) The role of oxidative stress and antioxidants in diabetic wound healing. Oxidative Med Cell Longev 2021(1):8852759
Dhir R, Chauhan S, Subham P, Kumar S, Sharma P, Shidiki A, Kumar G (2023) Plant-mediated synthesis of silver nanoparticles: unlocking their pharmacological potential-a comprehensive review. Front Bioeng Biotechnol 11:1324805
dos Santos DB, Lemos JA, Miranda SEM, Di Filippo LD, Duarte JL, Ferreira LAM, Barros ALB, Oliveira AEMFM (2022) Current applications of plant-based drug delivery nano systems for leishmaniasis treatment. Pharmaceutics 14(11):2339
Eldahshan OA, Nilofar N, Zengin G, El-Nashar HA (2025) Chemical profile of essential oil from Pimenta racemosa leaves, antioxidant potential, and its enzyme inhibitory properties. Sci Rep 15(1):30705
El-Nashar HAS, Mostafa NM, El-Badry MA, Eldahshan OA, Singab ANB (2021) Chemical composition, antimicrobial and cytotoxic activities of essential oils from Schinus polygamus (Cav.) cabrera leaf and bark grown in Egypt. Nat Prod Res 35(23):5369–5372
El-Nashar H, Abbas H, Zewail M, Noureldin MH, Ali MM, Shamaa MM, Khattab MA, Ibrahim N (2022) Neuroprotective effect of artichoke-based nanoformulation in sporadic Alzheimer's disease mouse model: focus on antioxidant, anti-inflammatory, and amyloidogenic pathways. Pharmaceuticals 15(10):1202
El-Nashar HAS, Abbas AM, Al-Qaaneh AM, Abdel-Maksoud MA, El-Tayeb MA, Almutairi S, Hafeez Kiani B, Elhawary EA (2024) Chemical composition and antimicrobial investigation of the essential oil isolated from Montanoa bipinnatifida (Kunth) K. Koch leaves. J Essent Oil Bear Plants 27(5):1428–1438
El-Nashar HA, Zengin G, Tarek A, Eldahshan OA (2025) Comparative metabolic profiling of Cordia myxa leaves and fruits based on UPLC-ESI/MS-MS analysis and investigation of antioxidant activities and enzyme inhibitory properties. Sci Rep 15(1):19296
Forgách L, Hegedűs N, Horváth I, Kiss B, Kovács N, Varga Z, Jakab G, Kovács T, Padmanabhan P, Szigeti K (2020) Fluorescent, prussian blue-based biocompatible nanoparticle system for multimodal imaging contrast. Nanomaterials 10(9):1732
Ghanbariasad A, Amoozegar F, Rahmani M, Zarenezhad E, Osanloo M (2021) Impregnated nanofibrous mat with nanogel of citrus sinensis essential oil as a new type of dressing in cutaneous leishmaniasis. Biointerface Res Appl Chem 11(4):11066–11076
Gomes HIO, Martins CSM, Prior JAV (2021) Silver nanoparticles as carriers of anticancer drugs for efficient target treatment of cancer cells. Nanomaterials (Basel) 11(4):964
Griffin S, Masood MI, Nasim MJ, Sarfraz M, Ebokaiwe AP, Schäfer K-H, Keck CM, Jacob C (2018) Natural nanoparticles: a particular matter inspired by nature. Antioxidants 7(1):3
Gu X, Minko T (2024) Targeted nanoparticle based diagnostic and treatment options for pancreatic cancer. Cancers (Basel) 16(8):1589
Gupta S, Pal A, Vyas SP (2010) Drug delivery strategies for therapy of visceral leishmaniasis. Expert Opin Drug Deliv 7(3):371–402
Hashad IM, Aly SH, Saleh DO, Abo El-Nasr NM, Shabana ME, El-Tokhy FS e, El-Nashar HA, Abdelmohsen UR, Mostafa NM, Mostafa AM (2025) Mechanistic wound healing of Ficus trijuja leaf extract and its lipid nanocapsule supported by metabolomic profiling and in vivo studies. Int J Mol Sci 26(3):928
Hettiarachchi SS, Dunuweera SP, Dunuweera AN, Rajapakse RMG (2021) Synthesis of curcumin nanoparticles from raw turmeric rhizome. ACS Omega 6(12):8246–8252
Husen A (2019) Natural product-based fabrication of zinc oxide nanoparticles and their application. In: Husen A, Iqbal M (eds) N*anomaterials* and plant potential. Springer International Publishing AG, Cham, pp 193–291. https://doi.org/10.1007/978-3-030-05569-1_7
Husen A (2020) Introduction and techniques in nanomaterials formulation. In: Husen A, Jawaid M (eds) Nanomaterials for agriculture and forestry applications. Elsevier, Cambridge, MA, pp 1–14. https://doi.org/10.1016/B978-0-12-817852-2.00001-9
Husen A (2023) Secondary metabolites based green synthesis of nanomaterials and their applications. Springer Nature Singapore Pte Ltd., Singapore. https://doi.org/10.1007/978-981-99-0927-8

Husen A, Siddiqi KS (2014) Phytosynthesis of nanoparticles: concept, controversy and application. Nanoscale Res Lett 9(229):1–24. https://doi.org/10.1186/1556-276X-9-229
Husen A, Bachheti RK, Bachheti A (2023) Current trends in green nano-emulsions (food, agriculture and biomedical sectors). Springer Nature Singapore Pte Ltd., Singapore. https://doi.org/10.1007/978-981-99-5398-1
Ibrahim N, Jamaluddin ND, Tan LL, Mohd Yusof NY (2021) A review on the development of gold and silver nanoparticles-based biosensor as a detection strategy of emerging and pathogenic RNA virus. Sensors 21:5114
Ifijen IH, Awoyemi RF, Faderin E, Akobundu UU, Ajayi AS, Chukwu JU, Lekan OK, Asiriuwa OD, Maliki M, Ikhuoria EU (2025) Protein-based nanoparticles for antimicrobial and cancer therapy: implications for public health. RSC Adv 15(19):14966–15016
Ilinskaya AN, Dobrovolskaia MA (2016) Understanding the immunogenicity and antigenicity of nanomaterials: past, present and future. Toxicol Appl Pharmacol 299:70–77
Iravani S, Varma RS (2019) Plant-derived edible nanoparticles and miRNAs: emerging frontier for therapeutics and targeted drug-delivery. ACS Sustain Chem Eng 7(9):8055–8069
Javed B, Ikram M, Farooq F, Sultana T, Mashwani ZU, Raja NI (2021) Biogenesis of silver nanoparticles to treat cancer, diabetes, and microbial infections: a mechanistic overview. Appl Microbiol Biotechnol 105(6):2261–2275
Jin E, Yang Y, Cong S, Chen D, Chen R, Zhang J, Hu Y, Chen W (2025) Lemon-derived nanoparticle-functionalized hydrogels regulate macrophage reprogramming to promote diabetic wound healing. J Nanobiotechnol 23(1):68
Karnwal A, Jassim AY, Mohammed AA, Sharma V, Al-Tawaha A, Sivanesan I (2024) Nanotechnology for healthcare: plant-derived nanoparticles in disease treatment and regenerative medicine. Pharmaceuticals 17(12):1711
Katta B, Vijayakumar C, Dutta S, Dubashi B, Nelamangala Ramakrishnaiah VP (2023) The incidence and severity of patient-reported side effects of chemotherapy in routine clinical care: a prospective observational study. Cureus 15(4):e38301
Khan SA, Shahid S, Lee C-S (2020) Green synthesis of gold and silver nanoparticles using leaf extract of Clerodendrum inerme; characterization, antimicrobial, and antioxidant activities. Biomolecules 10(6):835
Khorrami S, Dogani M, Mahani SE, Moghaddam MM, Taheri RA (2023) Neuroprotective activity of green synthesized silver nanoparticles against methamphetamine-induced cell death in human neuroblastoma SH-SY5Y cells. Sci Rep 13(1):11867
Kianfar E (2021) Protein nanoparticles in drug delivery: animal protein, plant proteins and protein cages, albumin nanoparticles. J Nanobiotechnol 19(1):159
Kucherenko IS, Soldatkin OO, Kucherenko DY, Soldatkina OV, Dzyadevych SV (2019) Advances in nanomaterial application in enzyme-based electrochemical biosensors: a review. Nanoscale Adv 1(12):4560–4577
Kumar A, Jain SK, Mishra DK, Gautam R (2024) Influence of drug properties and routes of drug administration on design of sustained and controlled release systems. In: Novel carrier systems for targeted and controlled drug delivery. Springer, pp 1–46
Lakkim V, Reddy MC, Lekkala VV, Lebaka VR, Korivi M, Lomada D (2023) Antioxidant efficacy of green-synthesized silver nanoparticles promotes wound healing in mice. Pharmaceutics 15(5):1517
Li D, Liu M, Li W, Fu Q, Wang L, Lai E, Zhao W, Zhang K (2022) Synthesis of Prussian blue nanoparticles and their antibacterial, antiinflammation and antitumor applications. Pharmaceuticals 15(7):769
Liyana E, Jasmine JS (2025) Transformative potential of plant-based nanoparticles in cancer diagnosis and treatment: bridging traditional medicine and modern therapy. Naunyn Schmiedebergs Arch Pharmacol 398:12869–12890
Lokole PB, Byamungu GG, Mutwale PK, Ngombe NK, Mudogo CN, Krause RWM, Nkanga CI (2024) Plant-based nanoparticles targeting malaria management. Front Pharmacol 15:1440116
Mai S, Inkielewicz-Stepniak I (2024) Graphene oxide nanoparticles and organoids: a prospective advanced model for pancreatic cancer research. Int J Mol Sci 25(2):1066

Masood M, Albayouk T, Saleh N i, El-Shazly M, El-Nashar HA (2024) Carbon nanotubes: a novel innovation as food supplements and biosensing for food safety. Front Nutr 11:1381179

Mayegowda SB, Sarma G, Gadilingappa MN, Alghamdi S, Aslam A, Refaat B, Almehmadi M, Allahyani M, Alsaiari AA, Aljuaid A (2023) Green-synthesized nanoparticles and their therapeutic applications: a review. Green Process Synth 12(1):20230001

Mekky AE, Abdo AM, Haggag MI, Elhaw MH, Kadry MM, Ghanem SM, Salama MM, Soliman AM, Mahmoud NN (2025) Biosynthesis of selenium nanoparticles from Dahlia pinnata tuberous roots with antibacterial, antidiabetic, and erythrocyte membrane protective activities. Sci Rep 15(1):27177

Meyerhoff A (1999) U.S. Food and Drug Administration approval of AmBisome (liposomal amphotericin B) for treatment of visceral leishmaniasis. Clin Infect Dis 28(1):42–48; discussion 49–51

Mishra P, Ahmad A, Al-Keridis LA, Alshammari N, Alabdallah NM, Muzammil K, Saeed M, Ansari IA (2022) Doxorubicin-conjugated zinc oxide nanoparticles, biogenically synthesised using a fungus Aspergillus niger, exhibit high therapeutic efficacy against lung cancer cells. Molecules 27(8):2590

Mocan T, Matea CT, Pop T, Mosteanu O, Buzoianu AD, Puia C, Iancu C, Mocan L (2017) Development of nanoparticle-based optical sensors for pathogenic bacterial detection. J Nanobiotechnol 15(1):25

Mohamed A, Dayo M, Alahmadi S, Ali S (2024) Anti-inflammatory and antimicrobial activity of silver nanoparticles green-synthesized using extracts of different plants. Nanomaterials 14(17):1383

Mohammadinejad R, Karimi S, Iravani S, Varma RS (2016) Plant-derived nanostructures: types and applications. Green Chem 18(1):20–52

Mohite P, Yadav V, Pandhare R, Maitra S, Saleh FM, Saleem RM, Al-Malky HS, Kumarasamy V, Subramaniyan V, Abdel-Daim MM (2024) Revolutionizing cancer treatment: unleashing the power of viral vaccines, monoclonal antibodies, and proteolysis-targeting chimeras in the new era of immunotherapy. ACS Omega 9(7):7277–7295

Nazli A, Baig MW, Zia M, Ali M, Shinwari ZK, Haq IU (2018) Plant-based metallic nanoparticles as potential theranostics agents: bioinspired tool for imaging and treatment. IET Nanobiotechnol 12(7):869–878

Nguyen NTT, Nguyen LM, Nguyen TTT, Nguyen TT, Nguyen DTC, Tran TV (2022) Formation, antimicrobial activity, and biomedical performance of plant-based nanoparticles: a review. Environ Chem Lett 20(4):2531–2571

Otsuka Y, Saito Y, Suzuki Y, Araki T, Shikinaka K, Inoue H (2026) Chemical properties of lignin and lignin-polysaccharide complexes obtained by simultaneous enzymatic saccharification and comminution (SESC) from lignocellulosic biomass of Japanese cedar (Cryptomeria japonica). Biomass Bioenergy 204:108428

Pastar I, Stojadinovic O, Yin NC, Ramirez H, Nusbaum AG, Sawaya A, Patel SB, Khalid L, Isseroff RR, Tomic-Canic M (2014) Epithelialization in wound healing: a comprehensive review. Adv Wound Care 3(7):445–464

Pelaz B, Alexiou C, Alvarez-Puebla RA, Alves F, Andrews AM, Ashraf S, Balogh LP, Ballerini L, Bestetti A, Brendel C, Bosi S, Carril M, Chan WCW, Chen C, Chen X, Chen X, Cheng Z, Cui D, Du J, Dullin C, Escudero A, Feliu N, Gao M, George M, Gogotsi Y, Grünweller A, Gu Z, Halas NJ, Hampp N, Hartmann RK, Hersam MC, Hunziker P, Jian J, Jiang X, Jungebluth P, Kadhiresan P, Kataoka K, Khademhosseini A, Kopeček J, Kotov NA, Krug HF, Lee DS, Lehr C-M, Leong KW, Liang X-J, Ling Lim M, Liz-Marzán LM, Ma X, Macchiarini P, Meng H, Möhwald H, Mulvaney P, Nel AE, Nie S, Nordlander P, Okano T, Oliveira J, Park TH, Penner RM, Prato M, Puntes V, Rotello VM, Samarakoon A, Schaak RE, Shen Y, Sjöqvist S, Skirtach AG, Soliman MG, Stevens MM, Sung H-W, Tang BZ, Tietze R, Udugama BN, VanEpps JS, Weil T, Weiss PS, Willner I, Wu Y, Yang L, Yue Z, Zhang Q, Zhang Q, Zhang X-E, Zhao Y, Zhou X, Parak WJ (2017) Diverse applications of nanomedicine. ACS Nano 11(3):2313–2381

Pham NM, Karlen W, Beck HP, Delamarche E (2018) Malaria and the "last" parasite: how can technology help? Malar J 17(1):260

Poh Yan L, Gopinath SCB, Subramaniam S, Chen Y, Velusamy P, Chinni SV, Gobinath R, Lebaka VR (2022) Greener synthesis of nanostructured iron oxide for medical and sustainable agro-environmental benefits. Front Chem 10:984218

Qin Z, Li Y, Gu N (2018) Progress in applications of Prussian blue nanoparticles in biomedicine. Adv Healthc Mater 7(20):1800347

Qu L, Lin Z, Liu F, Kong F, Zhang Y, Ni X, Zhang X, Zhao Y, Lu Q, Zou B (2025) Research progress on the application of metal porphyrin electrochemical sensors in the detection of phenolic antioxidants in food. Polymers (Basel) 17(6):789

Qubtia M, Ghumman SA, Noreen S, Hameed H, Noureen S, Kausar R, Irfan A, Akhtar Shah P, Afzal H, Hameed M, Raish M, Rana M, Ahmad A, Kotwica-Mojzych K, Bin Jardan YA (2024) Evaluation of plant-based silver nanoparticles for antioxidant activity and promising wound-healing applications. ACS Omega 9(10):12146–12157

Quinteros M, Aristizábal VC, Dalmasso PR, Paraje MG, Páez PL (2016) Oxidative stress generation of silver nanoparticles in three bacterial genera and its relationship with the antimicrobial activity. Toxicol In Vitro 36:216–223

Radulescu DM, Surdu VA, Ficai A, Ficai D, Grumezescu AM, Andronescu E (2023) Green synthesis of metal and metal oxide nanoparticles: a review of the principles and biomedical applications. Int J Mol Sci 24(20):15397

Rajak KK, Pahilani P, Patel H, Kikani B, Desai R, Kumar H (2023) Green synthesis of silver nanoparticles using Curcuma longa flower extract and antibacterial activity. arXiv preprint arXiv:2304.04777

Rajpal VR, Dhingra Y, Khungar L, Mehta S, Minkina T, Rajput VD, Husen A (2024) Current research in biotechnology. Curr Res Biotechnol 7:100219

Raval N, Mistry T, Acharya N, Acharya S (2015) Development of glutathione-conjugated asiatic acid-loaded bovine serum albumin nanoparticles for brain-targeted drug delivery. J Pharm Pharmacol 67(11):1503–1511

Rey-Méndez R, Rodríguez-Argüelles MC, González-Ballesteros N (2022) Flower, stem, and leaf extracts from Hypericum perforatum L. to synthesize gold nanoparticles: effectiveness and antioxidant activity. Surf Interfaces 32:102181

Ribeiro Pereira P, Bertozzi de Aquino Mattos É, Nitzsche Teixeira Fernandes Corrêa AC, Afonso Vericimo M, Margaret Flosi Paschoalin V (2020) Anticancer and immunomodulatory benefits of taro (Colocasia esculenta) corms, an underexploited tuber crop. Int J Mol Sci 22(1):265

Sachan R, Sachan K, Singh A, Karnwal A, Shidiki A, Kumar G (2024) Plant-mediated gold nanoparticles in cancer therapy: exploring anticancer mechanisms, drug delivery applications, and future prospects. Front Nanotechnol 6:1490980

Sarasati A, Syahruddin MH, Nuryanti A, Ana ID, Barlian A, Wijaya CH, Ratnadewi D, Wungu TD, Takemori H (2023) Plant-derived exosome-like nanoparticles for biomedical applications and regenerative therapy. Biomedicines 11:1053

Sezen S, Ertuğrul MS, Balpınar Ö, Bayram C, Özkaraca M, Okkay IF, Hacımüftüoğlu A, Güllüce M (2023) Assessment of antimicrobial activity and in vitro wound healing potential of ZnO nanoparticles synthesized with Capparis spinosa extract. Environ Sci Pollut Res 30(55):117609–117623

Shahzadi S, Fatima S, Shafiq Z, Janjua MRSA (2025) A review on green synthesis of silver nanoparticles (SNPs) using plant extracts: a multifaceted approach in photocatalysis, environmental remediation, and biomedicine. RSC Adv 15(5):3858–3903

Sharma S, Bose A, Biswas S, Sen S, Roy I (2025a) Cyperus rotundus mediated green synthesis of silver nanoparticles for antibacterial wound dressing applications. Sci Rep 15(1):18394

Sharma R, Tyagi S, Bachheti RK, Bachheti A, Worku LA (2025b) Biosynthesis of zinc oxide nanoparticles using aqueous leaf extract of Alternanthera philoxeroides for antioxidant and antibacterial activities. Nanomater Nanotechnol 2025(1):5594031

Sher N, Alkhalifah DHM, Ahmed M, Mushtaq N, Shah F, Fozia F, Khan RA, Hozzein WN, Aboul-Soud MA (2022) Comparative study of antimicrobial activity of silver, gold, and silver/gold bimetallic nanoparticles synthesized by green approach. Molecules 27(22):7895

Shikinaka K, Nakamura M, Navarro RR, Otsuka Y (2018) Plant-based antioxidant nanoparticles without biological toxicity. ChemistryOpen 7(9):709–712

Singh R, Hano C, Nath G, Sharma B (2021) Green biosynthesis of silver nanoparticles using leaf extract of Carissa carandas L. and their antioxidant and antimicrobial activity against human pathogenic bacteria. Biomolecules 11(2):299

Singh H, Desimone MF, Pandya S, Jasani S, George N, Adnan M, Aldarhami A, Bazaid AS, Alderhami SA (2023a) Revisiting the green synthesis of nanoparticles: uncovering influences of plant extracts as reducing agents for enhanced synthesis efficiency and its biomedical applications. Int J Nanomedicine 18:4727–4750

Singh S, Prasad AS, Rajeshkumar S (2023b) Cytotoxicity, antimicrobial, anti-inflammatory and antioxidant activity of camellia sinensis and citrus mediated copper oxide nanoparticle-an in vitro study. J Int Soc Prev Community Dent 13(6):450–457

Solomon M, Ollech A, Pavlotsky F, Barzilai A, Schwartz E, Baum S, Astman N (2024) Comparison of intralesional sodium stibogluconate versus intralesional meglumine antimoniate for the treatment of leishmania major cutaneous leishmaniasis. Acta Derm Venereol 104:adv35089

Song L, Chen J, Xu BB, Huang Y (2021) Flexible plasmonic biosensors for healthcare monitoring: progress and prospects. ACS Nano 15(12):18822–18847

Sritharan S, Sivalingam N (2021) A comprehensive review on time-tested anticancer drug doxorubicin. Life Sci 278:119527

Sundraraman G, Jayakumari L (2020) Meticulous taxifolin releasing performance by the zinc oxide nanoparticles: as a short road to drug delivery system for cancer therapeutics. J Clust Sci 31:241

Tiwari R, Gupta RP, Singh VK, Kumar A, Rajneesh, Madhukar P, Sundar S, Gautam V, Kumar R (2023) Nanotechnology-based strategies in parasitic disease management: from prevention to diagnosis and treatment. ACS Omega 8(45):42014–42027

Versey Z, da Cruz Nizer WS, Russell E, Zigic S, DeZeeuw KG, Marek JE, Overhage J, Cassol E (2021) Biofilm-innate immune interface: contribution to chronic wound formation. Front Immunol 12:648554

Vijapur LS, Shalavadi M, Desai AR, Hiremath JN, Gudigennavar AS, Shidramshettar SL, Hiremath SR, Peram MR, Kittur BS (2025) Wound healing potential of green synthesized silver nanoparticles of Glycyrrhiza glabra linn root extract: a preclinical study. J Trace Elem Miner 11:100214

Villagrán Z, Anaya-Esparza LM, Velázquez-Carriles CA, Silva-Jara JM, Ruvalcaba Gómez JM, Aurora-Vigo EF, Rodríguez-Lafitte E, Rodríguez-Barajas N, Balderas-León I, Martínez-Esquivias F (2024) Plant-based extracts as reducing, capping, and stabilizing agents for the green synthesis of inorganic nanoparticles. Resources 13(6):70

Viršilė A, Samuolienė G, Laužikė K, Šipailaitė E, Balion Z, Jekabsone A (2022) Species-specific plant-derived nanoparticle characteristics. Plants 11(22):3139

Vyas SP, Gupta S (2006) Optimizing efficacy of amphotericin B through nanomodification. Int J Nanomedicine 1(4):417–432

Wang Y, Cheng H, Wang T, Zhang K, Zhang Y, Kang X (2023a) Oxidative stress in intervertebral disc degeneration: molecular mechanisms, pathogenesis and treatment. Cell Prolif 56(9):e13448

Wang Y, Wei Y, Liao H, Fu H, Yang X, Xiang Q, Zhang S (2023b) Plant exosome-like nanoparticles as biological shuttles for transdermal drug delivery. Bioengineering 10(1):104

Wang X, Xin C, Zhou Y, Sun T (2024a) Plant-derived vesicle-like nanoparticles: the next-generation drug delivery nanoplatforms. Pharmaceutics 16(5):588

Wang X, Yin X, Li Y, Zhang S, Hu M, Wei M, Li Z (2024b) Novel insight and perspectives of nanoparticle-mediated gene delivery and immune-modulating therapies for pancreatic cancer. J Nanobiotechnol 22(1):771

Wei L, Adderley J, Leroy D, Drewry DH, Wilson DW, Kaushansky A, Doerig C (2021) Host-directed therapy, an untapped opportunity for antimalarial intervention. Cell Rep Med 2(10):100423

Wei S, Wang Y, Su S, Hao M, Wang Y (2025) Herbal-based hydrogel fabricated by self-assembly of mangiferin loading with rhein-functionalized silver nanoparticles for treating infected diabetic wound. Colloids Surf B Biointerfaces 255:114890

Yadav R, Chawra HS, Dubey G, Alam MS, Kumar V, Sharma P, Upadhayay NK, Yadav T (2025) Herbal based nanoparticles as a possible and potential treatment of cancer: a review. Explor Target Antitumor Ther 6:1002285

Yang C, Zhang M, Merlin D (2018) Advances in plant-derived edible nanoparticle-based lipid nano-drug delivery systems as therapeutic nanomedicines. J Mater Chem B 6(9):1312–1321

Yin Y, Liu J, Jiang G (2012) Sunlight-induced reduction of ionic Ag and Au to metallic nanoparticles by dissolved organic matter. ACS Nano 6(9):7910–7919

Youssif KA, Haggag EG, Elshamy AM, Rabeh MA, Gabr NM, Seleem A, Salem MA, Hussein AS, Krischke M, Mueller MJ, Abdelmohsen UR (2019) Anti-Alzheimer potential, metabolomic profiling and molecular docking of green synthesized silver nanoparticles of Lampranthus coccineus and Malephora lutea aqueous extracts. PLoS One 14(11):e0223781

Zaimy MA, Saffarzadeh N, Mohammadi A, Pourghadamyari H, Izadi P, Sarli A, Moghaddam LK, Paschepari SR, Azizi H, Torkamandi S, Tavakkoly-Bazzaz J (2017) New methods in the diagnosis of cancer and gene therapy of cancer based on nanoparticles. Cancer Gene Ther 24(6):233–243

Zhao G, Usui ML, Lippman SI, James GA, Stewart PS, Fleckman P, Olerud JE (2013) Biofilms and inflammation in chronic wounds. Adv Wound Care 2(7):389–399

Chapter 11
Aromatic and Essential Oil Plants in Physical and Mental Well-Being

Orchid A. Mahmoud, Jumana Haider Al-Anssari, Naglaa S. Ashmawy, Nada M. Mostafa, Omayma A. Eldahshan, and Abdel Nasser B. Singab

Abstract Maintaining physical and mental well-being is essential to enhancing the quality of life. Essential oils (EOs) have been used for centuries in different cultures, serving as therapeutic, cosmetic, aromatic and even spiritual agents. EOs are extracted from various parts of the plant, such as leaves, flowers, barks and fruits, using methods such as steam and hydrodistillation. The main chemical components of volatile oils are classified into terpenes and phenylpropanoids. They are considered resourceful, possessing various pharmacological and psychological properties, including antibacterial, antiviral, antioxidant, anti-inflammatory, immunity--boosting, calming and memory- and alertness-improving effects. This chapter discusses the potential impact of essential oils on physical and mental well-being, focusing on mechanisms and their chemistry. The chapter also provides insight into their use in clinical and pre-clinical trials. In addition, safety precautions are also covered. The data are collected from databases such as PubMed, Google Scholar, Scopus, Science Direct and Web of Science.

O. A. Mahmoud · N. M. Mostafa
Department of Pharmacognosy, Faculty of Pharmacy, Ain Shams University, Cairo, Egypt

J. H. Al-Anssari
Department of Pharmaceutical Sciences, College of Pharmacy, Dubai Medical University, Dubai, United Arab Emirates

N. S. Ashmawy
Department of Pharmacognosy, Faculty of Pharmacy, Ain Shams University, Cairo, Egypt

Department of Pharmaceutical Sciences, College of Pharmacy, Dubai Medical University, Dubai, United Arab Emirates

O. A. Eldahshan (✉) · A. N. B. Singab (✉)
Department of Pharmacognosy, Faculty of Pharmacy, Ain Shams University, Cairo, Egypt

Center for Drug Discovery Research and Development, Faculty of Pharmacy, Ain Shams University, Cairo, Egypt
e-mail: Oeldahshan@pharma.asu.edu.eg; vpr.nassersingab@asu.edu.eg

L. A. Worku et al. (eds.), *Plant-Based Solutions for Sustainable Agriculture and Healthcare*, Sustainability Sciences in Asia and Africa,
https://doi.org/10.1007/978-981-95-8910-4_11

Keywords Essential oil • Aromatic plants • Mental well-being • Depression • Insomnia • Anxiety

11.1 Introduction

The term 'essential oil' (EO) is derived from 'quinta essentia' in earliest Latin, which basically means the fifth element. The fifth element is the spirit connected with the other main elements: fire, air, earth and water. Essential oil extraction was believed to be the separation of the spirit from the plant and used as a healing promoter (Sattayakhom et al. 2023). Essential oils (EOs) are scented products extracted from aromatic plant parts. They have a hydrophobic nature and complex composition that differ according to the plant species from which they are extracted (Aziz et al. 2018; Pezantes-Orellana et al. 2024; Abate et al. 2021). EOs are composed of hydrocarbons and oxygenated hydrocarbons, namely, alcohols, aldehydes and ketones, which may produce characteristic odours (Ali et al. 2015; Aziz et al. 2018; Bachheti 2015). Terpenoids and phenylpropanoids are considered the main classes in essential oils (Aziz et al. 2018). The biosynthesis of terpenoids is mainly through mevalonate and mevalonate-independent (deoxyxylulose phosphate); meanwhile, the biosynthesis of phenylpropanoids happens through the shikimate pathway (Sattayakhom et al. 2023; Gonfa et al. 2022; Chandra et al. 2017).

Several ancient civilisations, including Egypt, India and China, have used aromatherapy. This therapeutic approach utilises essential oils as a complementary and alternative medicine, to enhance well-being and promote holistic health (Caballero-Gallardo et al. 2025). The ancient Egyptians used frankincense (*Boswellia sacra*) and myrrh (*Commiphora myrrha*) in religious rituals and embalming. The ancient Chinese and Indians employed aromatic plants for treating ailments (El-Nashar et al. 2023; Vora et al. 2024). Nowadays, essential oils are widely used in several industries, including cosmetics, perfumes, food flavouring, and food preservation. They have been resourceful, possessing various biological activities including anti-inflammatory, antioxidant, antimicrobial, antifungal and antidepressant (Sattayakhom et al. 2023; Pezantes-Orellana et al. 2024). In this chapter, a summary of essential oils is provided, focusing on their influence on physical and mental well-being, their chemistry and possible underlying mechanisms. Clinical studies are also presented in this chapter.

11.2 Methodology

This chapter explores the potential efficacy of aromatic and essential oil plants in promoting mental and physical well-being. It also discusses different extraction methods, the chemistry of volatile oils and the possible underlying mechanisms for their pharmacological actions. Clinical and pre-clinical trials are also discussed.

Several search engines were used, including Google Scholar, PubMed, MDPI and Science Direct. The words used were 'mental well-being', 'essential oils' and 'physical well-being'.

11.3 Methods of Extraction of Essential Oils

Different extraction methods for essential oils from plants have been utilised, including conventional methods such as hydrodistillation, steam distillation, hydrodiffusion and solvent extraction, as well as modern methods such as supercritical fluid extraction and microwave-assisted extraction (Aziz et al. 2018; Baptista-Silva et al. 2020). Hydrodistillation is considered the oldest, simplest technique: the plant material is placed in a flask with water, heated to boiling, and the mixture is then condensed to collect the essential oil and water. It is usually used for hydrophobic plant material that is insoluble in boiling water. Steam distillation is considered the most widely used technique, in which plant material is heated with steam, causing plant oil glands to burst and releasing their contents. The difference between steam distillation and hydrodiffusion is that the steam is supplied from above in the latter. In solvent extraction methods, delicate plant materials that cannot be exposed to higher temperatures can be mixed with different solvents such as acetone, hexane, ethanol and methanol. Then it is filtered and evaporated, and the resulting resin is mixed with alcohol and then evaporated at a low temperature. This method is considered time-consuming and more expensive (Aziz et al. 2018). Modern methods have an advantage over traditional methods, such as a lower oxidation potential of compounds. For example, supercritical fluid extraction uses CO_2, which is easily separated from the plant extract. Using CO_2 also has many advantages, as it is non--flammable, inexpensive and available at high purity (Baptista-Silva et al. 2020).

11.4 Introduction to Essential Oils and Physical Well-Being

11.4.1 Overview of Aromatherapy's Role in Physical Wellness

Aromatherapy, a complementary medicine that improves health and well-being, involves the therapeutic use of essential oils (EOs), highly concentrated extracts from fragrant plants. Aromatherapy has been used for both physical and psychological treatment with roots in the ancient Egyptian, Chinese, Indian and Arab traditions.

This treatment has been revived by recent clinical interest and is now part of supportive care regimens aimed at improving immunity, controlling inflammation, and providing analgesia (Dhifi et al. 2016; Vora et al. 2024). Terpenes and phenolic derivatives are the main volatile compounds found in essential oils, which exhibit pharmacological activities, including anti-inflammatory, analgesic, antibacterial, and immunomodulatory properties (Sarmento-Neto et al. 2015; Dhifi et al. 2016). When essential oils are used topically or through inhalation, their effects are usually produced through a combination of mechanisms. These include stimulation of olfactory pathways, absorption into the systemic circulation, and direct local interactions within tissues (Vora et al. 2024). A wide range of studies has reported that essential oils help mitigate various physical conditions such as cancer-related discomfort, muscle stiffness and infections. Essential oils are crucial in integrative oncology settings because a thorough review and meta-analysis showed that they significantly reduced pain intensity in cancer patients (Sarmento-Neto et al. 2015; Corasaniti et al. 2023). Similarly, oils such as lavender, eucalyptus and peppermint have analgesic (antinociceptive) properties via both central and peripheral mechanisms. Essential oils, characterised by their varied chemical compositions and methods of administration (inhalation, topical application or supervised oral intake), are increasingly recognised as versatile natural agents in physical healthcare. This chapter examines the pivotal functions of essential oils in physical health, utilising both ancient wisdom and contemporary scientific research (Sarmento-Neto et al. 2015; Corasaniti et al. 2023; Vora et al. 2024).

11.5 Mechanisms of Action of Essential Oils Related to Physical Well-Being

11.5.1 Anti-inflammatory Mechanisms of Essential Oils

Essential oils (EOs) are complex mixtures of volatile bioactive molecules that exert strong anti-inflammatory effects by modulating innate and adaptive immune pathways in distinct ways. Controlling cytokine production, lowering the activity of inflammatory enzymes and blocking intracellular signalling pathways linked to the inflammatory response are the main ways these activities are made easier.

To better understand how essential oils exert these effects, the following key mechanisms have been identified. These mechanisms and their associated essential oils are also summarised in Table 11.1.

11.5.1.1 Changing Cytokines

Many essential oils lower the levels of pro-inflammatory cytokines, including tumour necrosis factor-α (TNF-α), interleukin (IL)-1β, IL-6 and IL-8, while raising the levels of anti-inflammatory cytokines like IL-10. For example, thyme (*Thymus vulgaris*) essential oil lowers TNF-α and IL-1β levels by inhibiting cytokine transcription at the gene level, with its primary ingredients, thymol and p-cymene (Ocaña and Reglero 2012). Also, 1,8-cineole (eucalyptol), present in *Eucalyptus globulus*, strongly inhibits the release of cytokines by monocytes and lymphocytes, suggesting a strong systemic immunomodulatory effect (Juergens 2014).

Table 11.1 Anti-inflammatory mechanisms of essential oils: Key actions and representative compounds

Mechanism	Key actions	Essential oils involved	References
Cytokine modulation	↓ TNF-α, IL-1β, IL-6, IL-8; ↑ IL-10 via thymol, p-cymene, 1,8-cineole	Thyme, eucalyptus	Ocaña and Reglero (2012), Juergens (2014)
Enzyme and transcription factor inhibition	Inhibition of COX-2, iNOS, NF-κB, AP-1, MAPKs by thymol, carvacrol, p-cymene, eugenol	Thyme, clove, eucalyptus	Cha et al. (2009) and Gholijani et al. (2016)
Cellular-level effects	Prevention of monocyte/neutrophil adhesion; modulation of TREM-1, NLRP3 via α-pinene, bornyl acetate, eucalyptol	Eucalyptus, pine	Kim et al. (2015), Yadav and Chandra (2017)
Wound healing modulation	↑ TGF-β1, IL-10; wound healing and inflammation control by tea tree oil (terpinen-4-ol), *Artemisia* oil	Tea tree, *Artemisia*	Mohammed et al. (2022)
Chemical synergy	Minor compounds (e.g., citronellal) enhance the effects of primary ones like thymol (synergistic interaction)	*E. citriodora*, *Thymus vulgaris*	Pérez-Recalde et al. (2018) and Mohammed et al. (2022)

11.5.1.2 Inhibition of Key Enzymes and Transcription Factors

Essential oils contain various chemicals that demonstrate anti-inflammatory properties by inhibiting key enzymes, including cyclooxygenase-2 (COX-2) and inducible nitric oxide synthase (iNOS), which are involved in the synthesis of pro-inflammatory mediators. Thymol and carvacrol inhibit COX-2 activity, thus reducing the production of Prostaglandin E2 (PGE-2) (Gholijani et al. 2016).

In addition to this effect, essential oils can regulate several important transcription factors, including NF-κB, AP-1 and MAPKs, which are closely involved in triggering and sustaining inflammatory responses. By blocking NF-κB signalling activation and nuclear translocation, substances such as p-cymene, eugenol and 1,8-cineole reduce the expression of downstream pro-inflammatory cytokines and chemokines (Cha et al. 2009).

11.5.1.3 Cellular-Level Effects

Essential oils can help calm inflammatory processes by slowing the entry of immune cells into tissues and their activation. Research shows that molecules such as α-pinene and bornyl acetate interfere with the ability of monocytes and neutrophils to stick to surfaces and migrate, which ultimately helps protect tissues from excessive damage (Kim et al. 2015). Eucalyptol also contributes to its anti-inflammatory effect by enhancing the production of MAPK phosphatase-1 (MKP-1), an enzyme that attenuates MAPK signalling. In addition, it influences key pattern recognition receptors, including TREM1 and the NLRP3 inflammasome, whose regulatory actions are particularly valuable in the respiratory epithelium, where they help prevent an overactive inflammatory response (Yadav and Chandra 2017).

Immunoregulatory Effects in Wound Healing

Essential oils can support wound healing by helping the body regulate its inflammatory response. The oil from *Artemisia judaica*, for instance, has been shown to boost the levels of TGF-β1 and the IL-10 at the site of injury, two signalling molecules that help keep inflammation under control while supporting the repair of damaged tissues. Terpinen-4-ol, one of the major active compounds in tea tree oil, also plays a restorative role. It reduces the expression of pro-inflammatory cytokines such as IL-1β and IL-6 while promoting faster wound closure, giving adults anti-inflammatory and regenerative benefits (Pérez-Recalde et al. 2018; Mohammed et al. 2022).

11.5.1.4 Chemical Synergy

Synergy among the various compounds in an essential oil often drives its anti-inflammatory potency. Research on the *oils of Eucalyptus citriodora* and *Thymus vulgaris* shows that even the less abundant molecules can influence the behaviour of major constituents such as citronellal and thymol. These subtle interactions can boost or fine-tune the activity of the primary compounds, ultimately intensifying the oil's overall biological impact (Stojanović et al. 2024).

11.5.2 Antimicrobial Mechanisms of Essential Oils

Essential oils exhibit antimicrobial effects through multiple mechanisms that target both the structure and the function of microbial cells. Their activity can include disturbing the integrity of the cell membrane, interfering with essential intracellular pathways, or blocking the communication systems microbes use to coordinate their behaviour. The specific mechanisms and the oils associated with them are listed in Table 11.2, and the major modes of action are summarised as follows.

11.5.2.1 Disruption of Microbial Cell Membranes

Essential oils are strongly hydrophobic, allowing them to slip into the lipid layers of microbial cell membranes. Once inside, they disrupt the membrane's structure, making it more permeable. As a result, essential oils and ions such as ATP, K^+ and

Table 11.2 Antimicrobial mechanisms of essential oils: Key actions and representative compounds

Mechanism	Key actions	Essential oils involved	References
Disruption of microbial membranes	Increased membrane permeability, ion leakage, and cell lysis	Various (general membrane disruptors)	Moghrovyan and Sahakyan (2024)
Gram-negative membrane Penetration	Penetration of the outer membrane; disruption of energy metabolism	Eugenol, carvacrol, thymol	Andrade-Ochoa et al. (2021)
Intracellular targeting	Inhibition of DNA/ ATP-synthesising enzymes by compounds like eugenol	Eugenol, cinnamaldehyde	Yammine et al. (2022)
Induction of oxidative stress	ROS generation leads to oxidative damage to cell structures	Monoterpenes, sesquiterpenes	Chueca et al. (2014) and Kong et al. (2022)
Inhibition of quorum sensing	Blocks quorum sensing and biofilm formation without killing microbes directly	Multiple EOs affecting N-acyl homoserine lactones	Maggio et al. (2025)
Efflux pump inhibition	Blocks efflux pumps in resistant strains, restoring antibiotic sensitivity	Geraniol	Aelenei et al. (2016)
Synergy with antibiotics	Lowers MICs of antibiotics; enhances clearance of MDR bacteria	Various oils with aminoglycosides	Sena et al. (2024)
Nano--encapsulation delivery	Improves EO stability, solubility, and release for enhanced efficacy	All major EOs (nano-formulated)	Arora et al. (2025)
Broad-spectrum antimicrobial activity	Effective against Gram-positive and Gram-negative bacteria (e.g., *E. coli*, *S. aureus*)	*Cymbopogon citratus*, *Cinnamomum zeylanicum*, *Origanum vulgare*	Angane et al. (2022)

Ca^{2+} begin to leak out of the cell. Losing these components disrupts the cell's internal equilibrium, hampers nutrient transport across the membrane and destabilises the membrane's electrical potential. When this system fails, the cell can no longer maintain homeostasis, leading ultimately to its death (Moghrovyan and Sahakyan 2024).

11.5.2.2 Penetration and Action Within Gram-Negative Bacteria

Gram-negative bacteria possess an outer membrane packed with lipopolysaccharides (LPS), which typically forms a formidable protective barrier. Even so, certain essential constituents such as eugenol, carvacrol and thymol can penetrate this layer. After crossing the membrane, these molecules weaken its structural integrity, thereby increasing permeability. This increased permeability allows antimicrobial substances to enter the cell more rapidly, where they interfere with energy generation and compromise essential cellular structure (Andrade-Ochoa et al. 2021).

11.5.2.3 Targeting Intracellular Structures and Enzymes

Some components of essential oils can move beyond the cell membrane and reach the inner machinery of microbial cells. Compounds like eugenol and cinnamaldehyde can alter protein structure and inhibit the activity of enzymes essential to processes such as DNA replication and ATP generation. When this internal interference occurs alongside neighbouring damage, the combined stress greatly enhances the antimicrobial action (Yammine et al. 2022).

11.5.2.4 Induction of Oxidative Stress

Essential oils can harm microbes by triggering oxidative stress within their cells. Some monoterpenes and sesquiterpenes stimulate the production of reactive oxygen species (ROS), which then react with and damage vital cellular components. As the ROS level rises, cell structures and functions become increasingly compromised, eventually leading to microbial death and further strengthening the overall antimicrobial action (Chueca et al. 2014; Kong et al. 2022).

11.5.2.5 Inhibition of Quorum Sensing and Biofilm Formation

Several components of essential oils can disrupt quorum sensing, the communication network that microbes rely on to coordinate biofilm formation and control their virulence. These compounds interfere with signalling molecules such as N-acyl homoserine lactones, weakening pathogens' ability to activate harmful behaviours without directly killing cells. Because this approach targets microbial communication instead of growth, essential oils place less evolutionary pressure on microbes, reducing the likelihood that resistance will develop (Maggio et al. 2025).

11.5.2.6 Efflux Pump Inhibition and Resistance Modulation

Some essential oil constituents, such as geraniol, can interfere with the efflux pumps used by Gram-negative bacteria, such as *Escherichia coli* and *Pseudomonas aeruginosa*. These pumps typically expel antibiotics from the cell but when their function is inhibited, the drugs remain inside the bacteria for a longer time and exert stronger effects. This renewed intracellular accumulation of antibiotics increases the bacterial susceptibility and helps reverse certain forms of antimicrobial resistance (Aelenei et al. 2016).

11.5.2.7 Synergistic Effects with Antibiotics

Essential oils can also work alongside antibiotics in a complementary way, strengthening the overall antimicrobial effect. Research indicates that when specific essential oils are paired with antibiotics, such as aminoglycoside amikacin, the required minimum inhibitory concentrations can drop significantly, allowing bacteria to be cleared more quickly. These energetic combinations have shown promise even against multidrug-resistant strains, highlighting their potential to address difficult-to-treat conditions (Sena et al. 2024).

11.5.2.8 Nano-Encapsulation and Antimicrobial Potency

Nano-encapsulation of essential oils enhances their performance by increasing solubility and shielding their active constituents from degradation. This approach also enables a slow, controlled release of the oil, prolonging the antibacterial activity. In addition, encapsulation minimises the quick loss of volatile compounds through evaporation, resulting in a more stable product that is better suited for therapeutic applications as well as food preservation (Arora et al. 2025).

11.5.2.9 Broad-Spectrum Activity

Essential oils are known for their broad-spectrum antimicrobial properties, acting against both Gram-positive and Gram-negative bacteria. *Cymbopogon citratus* (lemongrass) oil, for example, has been shown to suppress the growth of *Staphylococcus aureus* and *E. coli* at concentrations as low as 0.6 μL/mL. *Cinnamomum zeylanicum* (cinnamon) oil also demonstrated strong activity against *S. aureus* and *Klebsiella pneumoniae* at concentrations of 2–8 μL/mL. Essential oils derived from *Origanum vulgare* (oregano) exhibit notable antimicrobial activity, with minimum inhibitory concentration (MIC) values of 800–1800 ppm against both *E. coli* and *S. aureus* (Angane et al. 2022).

11.5.3 Analgesic Mechanisms of Essential Oils

Essential oils can exert a pain-relieving effect through multiple pathways that influence both the central nervous system and peripheral pain responses. Their analgesic activity largely stems from bioactive constituents, including monoterpenes, sesquiterpenes and phenylpropanoids. These compounds can modulate inflammatory signalling, interact with the key neurotransmitter systems and alter the function of various ion channels involved in pain perception. The mechanism described in the next section highlights how these actions collectively contribute to the pain-reducing properties of essential oils.

11.5.3.1 Modulation of Inflammatory Pathways

Essential oils can influence pain through several pathways, acting on the neurotransmitter system, inflammatory mediators, and a range of receptor types involved in both central and peripheral pain processing. Their analgesic activity is often associated with the presence of monoterpenes, sesquiterpenes and phenylpropanoid compounds, which are known to modulate inflammation and pain signalling. One of the key ways essential oils reduce pain is by modifying inflammatory pathways. Studies show that constituents such as linalool, limonene and myrcene can reduce the production of prostaglandins and other inflammatory mediators, which, in turn, decrease nociceptive input from related tissues (Lenardão et al. 2015). An experimental research using chemical and heat induced pain modules in mice has demonstrated this effect clearly, for example, *Citrus limon* which is contained in the 50% limonene and *Cymbopogon citratus* rich in myrcene significantly reduced pain reaction result from the acetic acid writhing test the formalin essay and hot plate test all support the idea that these oils act on both peripheral and central mechanisms to produce analgesia (Rahman et al. 2012; Scuteri et al. 2021).

11.5.3.2 Biphasic Inhibition in the Formalin Test

The formalin test is commonly used model for studying pain as it produces an early phase of nociception caused by direct activation of pain receptor followed by a later phase driven by inflammation essential oils from *Bunium persicum* and *Eugenia caryophyllata* have been shown to reduce pain response in both phases suggesting they act on both neurogenic and inflammatory component of pain in particular the oil from *Eugenia caryophyllata* which contain a high proportion of eugenol 87.34% produced strong inhibitory effect which when naloxone reduced the activity it indicated that the opioid receptor system may play a role in the oil analgesic mechanism (Halder et al. 2012).

11.5.3.3 Opioid-Like Activity

Some essential oil appear to produce analgesic effect similar to those opioids either by acting directly on opioid-related receptor or by influencing the body natural of your system, for example, the pain relieving action of *Citrus limon* essential oil was reduced when naloxone when was administered suggesting involvement of central opioid receptors oils from *Ocimum gratissimum* (rich in eugenol at 67.17%) and *Ocimum basilicum* (containing 69.54% linalool) have also shown opioid-based central analgesic effect in animal studies using hot plate and tail flick test (Paula-Freire et al. 2013; Paula-Freire et al. 2015).

11.5.3.4 Multimodal Mechanisms: Transient Potential Vanilloid 1 (TRPV1), Adrenergic, and More

Essential oils can influence several pain-related signalling systems, including opioid pathways and receptors, as well as TRPV1, adrenergic, cholinergic, nitrergic and glutamatergic activity. For instance, the oil from *Vanillosmopsis arborea*, which contains a high concentration of alpha-bisabolol, has shown pain-relieving effects by acting on TRPV1 receptors and on adrenergic, serotonergic and cholinergic pathways. Likewise, *Zingiber zerumbet*, whose main active component is zerumbone, appears to reduce pain by inhibiting TRPV1 glutamatergic and nitrergic signalling and ATP-sensitive potassium channels. These findings support the idea that essential oils may act through multiple mechanisms simultaneously (Kamatou and Viljoen 2010).

11.5.3.5 Peripheral Action and Ion Channel Modulation

Eucalyptus citriodora, which contains a high level of citronellal 83.5%, and *Cymbopogon winterianus*, rich in geraniol (40.06%), have both been shown to reduce pain-related behaviour in animal models, such as the acetic acid writhing test and the formalin assay. These effects are believed to stem from decreased prostaglandin synthesis and from changes in ion channels involved in the transmission of pain signals (Júnior et al. 2017).

11.5.3.6 Immune-Modulatory Mechanisms of Essential Oils

Essential oils contain a diverse range of volatile components, including terpenes and phenylpropanoids, which contribute to their biological activity. Many of these effects involved the immune system influencing both innate and adaptive responses They may act by adjusting cytokine levels, stimulating phagocytic cell activity and controlling oxidative stress, among other pathways.

11.5.3.7 Activation of Innate Immunity

Essential oil can have a strong impact on innate immunity by boosting the activity of important defensive cells such as neutrophils, dendritic cells, macrophages and natural killer cells for example a human monocyte derived macrophages become more active when exposed to *Eucalyptus globulus* essential oil showing increased phagocytic capacity without signs of toxicity this activation was evident through noticeable cytoskeletal changes and altered cell morphology including greater surface ruffling and the formation of pseudopodia. Macrophages internalised significantly more fluorescent beads per cell, with a maximum of 64.8 beads, compared to untreated controls, which internalised only 11 beads per cell.

Essential oils such as eucalyptus oil may enhance immune surveillance while minimising excessive inflammation, as indicated by the reduced production of pro-inflammatory cytokines (e.g., IL-6, IL-1β) associated with this phagocytic enhancement. The necessity of an intact microtubule network for the EO's stimulatory effects was demonstrated, suggesting that complement receptor-mediated endocytosis may underlie this mechanism (Zonfrillo et al. 2022).

11.5.3.8 Cytokine Regulation and Th1/Th2 Modulation

Essential oils can modulate immune activity by altering cytokine production, thereby shifting the balance between Th1- and Th2-type responses. In a mouse model of *Toxoplasma gondii* infection, *Myrtus communis* essential oil was shown to markedly elevate the expression of IFN-γ and IL-12. These two cytokines play central roles in activating macrophages and strengthening Th1-driven cytotoxic immunity. IL-12 guides naïve T cells towards a Th1 phenotype, boosting the killing capacity of $CD8^+$ T cells and natural killer cells. IFN-γ further amplifies this defense by enhancing macrophage function and improving antigen presentation (Shaapan et al. 2021).

Chamaecyparis obtuse essential oil has been shown to lower the production of IL-25 and IL-33 in bronchial epithelial cells and, in a dose-dependent fashion, the cytokines are key drivers of Th-2-type allergic inflammation and contribute to eosinophilic responses. This decrease in their expression appears to result from the oil's ability to suppress transcription factors such as NF-κB and AP-1, both of which are central regulators of inflammatory pathways and cytokine gene activation (Shin et al. 2021).

Enhancement of Humoral and Cell-Mediated Immunity

A study examining the immunomodulatory effect of essential oil from *Zingiber officinale*, *Salvia officinalis* and *Syzygium aromaticum* found that clove oil produced an increase in total white blood cells count and strengthened both humoral and cell-mediated immunity in immunocompromised mice. These benefits were largely linked to eugenol the major bioactive compound in clove oil which appears to enhance the antigen presentation and simulate the lymphocyte proliferation. The same study also reported that ginger essential oil helped restore humoral immune response in mice with suppressed immunity, although its influence on cell-mediated pathways was less pronounced (Carrasco et al. 2009).

11.6 Key Therapeutic Essential Oils Supporting Physical Health

Essential oils, obtained from various aromatic plants, are valued not only for their fragrance but also for their therapeutic potential. Their biological activity arises from complex chemical composition and the interactions among their constituents, which may influence immune activity, inflammation, mental well-being and the progression of certain chronic conditions. This section reviews essential oils that have demonstrated meaningful benefits for physical and mental health, particularly in relation to their immunomodulatory and anti-inflammatory properties. A summary of these essential oils, their major constituents and associated therapeutic effects is presented in Table 11.3.

11.6.1 Nigella sativa *(Black Cumin)*

Black cumin oil (*Nigella sativa* seed oil) is rich in thymoquinone, a major active compound that contributes to a wide range of immunomodulatory and anti-inflammatory effects. Research shows that the oil can regulate several key inflammatory mediators—such as IL-1β, IL-6 and TNF-α—particularly in macrophages activated by lipopolysaccharide (LPS) in modules related to asthma. It has been shown to influence IL-2 and IL-17 production and alter the expression of miRNA-133a. These molecular shifts help stimulate other projects through the SIRT1 pathway and affect cellular energy balance via the AMPK–SIRT1–PGC-1α signalling axis.

Beyond its immunological effect, black cumin oil has demonstrated neuroprotective potential studies suggesting that it can reduce neuroinflammation and suppress TLR- and NF-κB dependent signalling mechanism that may contribute to protection against Alzheimer-related pathology. When applied to the skin, the oil exhibit mild anaesthetic and antiseptic actions supporting wound healing and reducing localised. In addition, black cumin has been shown to enhance activity of chemotherapeutic agents such as 5-fluorouracil and piperine by modulating cytokine profiles and promoting apoptosis in cancer cells (Amin and Hosseinzadeh 2015).

11.6.2 Thymus vulgaris *(Thyme)*

Thyme essential oil is rich in phenolic constituents, particularly thymol and carvacrol, which account for much of its biological activity. These compounds contribute to the oil's antioxidant, antimicrobial, and anticancer properties. High levels of thymol (up to 36.5%) and carvacrol (around 29.8%) enable the oil to neutralise free radicals and effectively limit lipid peroxidation. Experimental studies also show

Table 11.3 Therapeutic essential oils and their major components: Evidence-based benefits for physical health

Essential oil	Major components	Therapeutic effects	References
Nigella sativa (Black cumin)	Thymoquinone	Anti-inflammatory, immunomodulatory, anticancer	Amin and Hosseinzadeh (2015)
Thymus vulgaris (Thyme)	Thymol, carvacrol	Antioxidant, anti-inflammatory, antimicrobial, anticancer	Gholijani et al. (2016)
Chamaecyparis obtusa	Unknown (likely terpenes)	Anti-allergic, anti-inflammatory (respiratory)	Shin et al. (2021)
Myrtus communis (Myrtle)	1,8-Cineole, linalool, myrtenol	Immunostimulatory, antibacterial, antioxidant	Alipour et al. (2014)
Elettaria cardamomum (Cardamom)	1,8-Cineole	Anti-inflammatory, GI health, Respiratory health	Heimesaat et al. (2021)
Pogostemon cablin (Patchouli)	Patchouli alcohol	Immunomodulation, anticancer, gut microbiota	Leong et al. (2022)
Origanum vulgare (Oregano)	Carvacrol, thymol	Antibacterial (Gram ±), membrane disruption	Moghrovyan and Sahakyan (2024)
Cinnamomum zeylanicum (Cinnamon)	Eugenol, cinnamaldehyde	Antimicrobial, antifungal, Detoxification	Guo et al. (2024)
Lavandula officinalis (Lavender)	Linalool	Anti-inflammatory, cytotoxic (anti-cancer), immune-enhancing	Fahmy et al. (2022)
Salvia officinalis (Sage)	Camphor, 1,8-cineole	Antioxidant, anti-inflammatory, anticancer	Ribeiro Junior (2024)
Foeniculum vulgare (Fennel)	Trans-anethole, fenchone	Antioxidant, antimicrobial, detoxifying	El-Kersh et al. (2022) and Naaz (2022)
Mentha piperita (Peppermint)	Menthol, menthone	Digestive aid, antimicrobial, cognitive enhancer	Zhao et al. (2022)
Curcuma longa (Turmeric)	Ar-turmerone, curcuminoids	Anticancer, antifungal	Hu et al. (2017)
Eucalyptus globulus	1,8-Cineole	Antibacterial, antiviral, respiratory support	Mieres-Castro et al. (2021) and Sutrisna et al. (2025)
Zingiber officinale (Ginger)	Zingiberene, citral	Humoral immunity enhancer, antioxidant	Peterfalvi et al. (2019)
Syzygium aromaticum (Clove)	Eugenol	Immune stimulation, antibacterial, antiviral	Bachiega et al. (2012)

that both molecules can suppress the growth of cancer cells in both colon and breast cancer models. Their actions include reducing mitochondrial membrane potential, triggering caspase activation, and ultimately causing DNA fragmentation.

Beyond antioxidant and anti-cancer effects, thyme essential oil exhibits a strong anti-inflammatory activity. Thymol mode and carvacrol downregulate pro-inflammatory cytokines such as TNF-α, IL-1β, and IL-6 while enhancing the expression of interleukin IL-10, a key anti-inflammatory cytokine mediator, by inhibiting the NF-κB and MAPK signalling pathways. The oil reduces COX enzyme production and induces nitric oxide synthase. Thymol also directly inhibits elastase and enzymes released by activated neutrophils, which contribute to tissue damage during inflammation.

Thyme oil further influences immune regulation by modulating IL-2 and IFN-γ production, thereby affecting T cell-mediated response. The immunomodulatory effect may benefit individuals with autoimmune or allergic conditions, where balancing inflammatory and immune activity is essential (Gholijani et al. 2016).

11.6.3 Chamaecyparis obtusa *(Japanese Cypress)*

Chamaecyparis obtusa essential oil exhibits significant anti-inflammatory and anti-allergic effects. It inhibits epithelial cells from producing IL-25 and IL-33, which are crucial for initiating Th2-mediated allergic responses. It inhibits the NF-κB and AP-1 pathways in bronchial epithelial cells and reduces the expression of TNF and prostaglandin genes. This oil may help reduce airway inflammation associated with conditions such as asthma or allergic rhinitis (Shin et al. 2021).

11.6.4 Myrtus communis *(Myrtle)*

Myrtle essential oil has a historical application in the treatment of skin and digestive issues. Elevated levels of IL-12 and IFN-γ mRNA are believed to contribute to its immunological effects, particularly in toxoplasmosis models. Cytokines enhance macrophage activity and increase nitric oxide production, both of which are crucial for pathogen elimination. Linalool, 1,8-cineole and myrtenol are constituents of myrtle oil that contribute to its anti-inflammatory, antibacterial and antioxidant properties (Alipour et al. 2014).

11.6.5 Elettaria cardamomum *(Cardamom)*

Cardamom essential oil, commonly known as the 'Queen of Spices', improves digestive and respiratory health. It reduces levels of nitric oxide, TNF and IFN-γ, and inhibits the infiltration of macrophages, monocytes and T lymphocytes into the intestinal mucosa. Cardamom oil inhibited Th1 cytokine production in mice and improved symptoms of enterocolitis. A primary component, 1,8-cineole, is responsible for numerous associated benefits (Heimesaat et al. 2021).

11.6.6 Pogostemon cablin *(Patchouli)*

Patchouli oil has a meaningful impact on the immune system by promoting a shift in macrophages from the M1 (pro-inflammatory) to the M2 (anti-inflammatory) state. This shifts the levels of inflammatory cytokines, enhancing the functionality of CD4+ and CD8+ T cells within immune tissues. Patchouli alcohol is the active compound in the oil and is thought to play a role in cancer prevention and the regulation of the gut microbiome (Leong et al. 2022).

11.6.7 Origanum vulgare *(Oregano)*

Oregano essential oil owes much of its antimicrobial potency to its phenolic constituents, particularly carvacrol and thymol. These compounds exhibit vigorous germicidal activity and are effective against a broad range of bacterial species. The reported minimum inhibitory concentration for *Escherichia coli* is 1600–1800 ppm, while *S. aureus* is inhibited at 800–900 ppm. As bacterial cells are exposed to these compounds, their membranes become increasingly permeable, leading to disrupted protein synthesis and loss of essential cellular functions. The oil's hydrophobicity allows its components to embed in the phospholipid bilayer, leading to ion leakage and, ultimately, cell death. Carvacrol is especially potent and important because it alters membrane structure and fluidity, which contributes to its effectiveness against both Gram-positive and Gram-negative bacteria (Moghrovyan and Sahakyan 2024).

11.6.8 Cinnamomum zeylanicum *(Cinnamon)*

Cinnamon essential oil is rich in bioactive compounds such as eugenol and cinnamaldehyde, both of which exhibit strong antimicrobial activity. These molecules compromise bacterial cell membrane causing cytoplasmic leakage and breakdown of cellular integrity. Beyond antibacterial effect, the oil also effective against various fungal species research further suggests that cinnamon essential oil may play a role in cancer prevention by boosting the activity of detoxifying enzymes like glutathione S-transferase and by influencing cytochrome P450 enzyme involved in the metabolism of carcinogens (Guo et al. 2024).

11.6.9 Lavandula officinalis *(Lavender)*

Lavender oil has antibacterial activity, but its cytotoxic and anti-inflammatory effects are especially significant. Linalool, a main component, has been shown to reduce drug resistance in specific cancer cells, including MCF7 breast adenocarcinoma cells. Lavender oil contains antioxidants and may support immune function by helping reduce stress through aromatherapy (Fahmy et al. 2022).

11.6.10 Salvia officinalis *(Sage)*

Sage essential oil has been shown to possess strong antioxidant properties relative to ascorbic acid and butylated hydroxytoluene (BHT). It is a key component. Camphor and 1,8-cineole contribute to its anti-inflammatory effects and help reduce cellular proliferation (Sidiropoulou et al. 2022). These compounds suppress the synthesis of inflammatory cytokines and induce apoptosis in cancer cells, making them suitable candidates for integrative cancer therapy (Ribeiro Junior 2024).

11.6.11 Foeniculum vulgare *(Fennel)*

Fennel essential oil has demonstrated the ability to disrupt the bacterial cell membrane particularly in pathogens such as *Shigella dysenteriae*. In addition to its antimicrobial effect the oil exhibit notable anti-inflammatory activity and strong free radical scavenging capacity Much of this bioactivity is linked to key consistent like trans-anethole and fenchone which contributes to it therapeutic profile. The oil is also rich in phenylpropanoids; these compounds not only enhance its antimicrobial action but also support the body natural detoxification process (El-Kersh et al. 2022; Naaz 2022).

11.6.12 Mentha piperita *(Peppermint)*

Peppermint essential oil exhibit antibacterial activity against wide range of microorganism includes both Gram-positive and Gram-negative bacteria. It is effective both when applied in its pure form and delivered as capsule. Its primary constituents, menthol and menthone, are capable of distributing bacterial membranes and have demonstrated activity even against drug resistant strains beyond its antimicrobial properties peppermint oils also associated with digestive support immune enhancement and increased mental alertness (Zhao et al. 2022).

11.6.13 Curcuma longa *(Turmeric)*

Turmeric essential oil is rich in bioactive compounds, including Ar-turmerone and various curcuminoids, both of which are recognised for their anticancer potential. Ar-turmerone has been shown to trigger apoptosis in HepG2 liver cancer cells by activating caspase-3 and engaging mitochondrial-dependent pathways. In addition to its anticancer effect, Ar-turmerone can disrupt ergosterol synthesis in *Aspergillus flavus*, suggesting that turmeric essential oil may also offer therapeutic value in managing fungal infections (Hu et al. 2017).

11.6.14 *Eucalyptus*

Eucalyptus oil, mainly derived from *Eucalyptus globulus*, has been widely used to treat respiratory conditions such as bronchitis, coughs, the common cold and sinusitis. However, the predominant active chemical, 1,8-Cineole (eucalyptol), accounts for about 90% of the total composition of Eucalyptus essential oil and is mainly responsible for its various activities. This essential oil has been shown to possess potent antibacterial activity, particularly against respiratory bacteria such as *Haemophilus influenzae*, *Streptococcus pneumoniae* and *Stenotrophomonas maltophilia*. However, its efficacy against the bacteria *Klebsiella pneumoniae* has been less potent. In addition, this essential oil demonstrates its ability to act against viral organisms, mainly enveloped ones.

Some studies have found that dispersing eucalyptus oil in the vapour phase is the most effective, suggesting its efficacy in inhalation therapy and air purification. In addition to its antimicrobial properties, the essential oil also appears to have immunoenhancing effects. In vitro experiments have shown that eucalyptus essential oil can enhance phagocytosis and decrease the production of inflammatory cytokines in immune cells. Animal experiments have also found that eucalyptus essential oil can stimulate monocytes and granulocytes and partially alleviate immunosuppression induced by chemotherapy.

Eucalyptus oil, especially 1,8-cineole, is effective in asthma, chronic obstructive pulmonary disease (COPD), bronchitis, and sinusitis. In inhalation tests, significant improvements in lung function parameters, such as forced vital capacity (FVC) and forced expiratory volume (FEV1), have been observed. This can be attributed largely to the disruption of cell membranes of bacteria, which increases the ability of the immune system to engulf the bacteria, as well as stimulation of sensory ion channels in the airway passages, which produces a decrease in the cough reflex and the sensation of airflow (Mieres-Castro et al. 2021; Sutrisna et al. 2025).

11.6.15 Zingiber officinale *(Ginger)*

Ginger essential oil (GEO) is mainly composed of monoterpenes, with key constituents such as ar-curcumene, β-myrcene, 1,8-cineole, citral and zingiberene. Studies using mouse immune models have shown that GEO can restore humoral immunity in animals whose immune system was suppressed by cyclophosphamide. This restoration was reflected by an increase in antibodies directed against sheep red blood cells (SRBCs), suggesting improved B-cell-mediated immune response. However, GEO did not stimulate T-cell activity, as evidenced by its lack of effect on total white blood cell count (Peterfalvi et al. 2019).

11.6.16 Syzygium aromaticum *(Clove)*

The main constituent of Clove is eugenol. Studies have shown that administering clove oil increased white blood cell counts, which, in turn, improved humoral and cell-mediated immune responses. In vivo testing with clove oil showed a noticeable increase in paw volume, indicating increased T-cell activity. Clove oil administration also restored antibody levels and white blood cells back to normal in mice with weak immunity. Both T and B cells have demonstrated enhanced functional performance. Clove oil shows immune-boosting effects similar to those of levamisole, suggesting it may support immune function, haematopoiesis and effector cells such as T cells, B cells and macrophages. These findings suggest the potential use of clove essential oil as a natural adjuvant for individuals with compromised immunity (Bachiega et al. 2012).

11.7 Therapeutic Applications of Natural Agents

11.7.1 Pain Relief Uses

Essential and aromatic oils have been shown to help relieve various types of pain, including cancer-related pain, tension headaches and primary dysmenorrhea. Evidence from a systematic review and meta-analysis indicates that essential oils can significantly reduce cancer-related discomfort, with a very strong statistical support ($p < 0.00001$). Among the oils studied most frequently are lavender, lemon, peppermint, chamomile, frankincense and bergamot. Several clinical reports have effectively reduced pain in clinical settings.

Although one clinical trial did not show significance ($p = 0.368$), aroma lymphatic tressage with frankincense oil was still observed to lessen pain among breast cancer patients receiving taxane-based chemotherapy. Another study found that aromatherapy massage with a mixture of peppermint, chamomile and rosemary oils significantly improved chemotherapy-induced peripheral neuropathic pain (CIPN). By week six, the proportion of patients experiencing neuropathic pain dropped from 45.8% to 18.2% ($p = 0.046$) (Corasaniti et al. 2023).

Essential oils have been studied for their role in managing primary dysmenorrhea, a condition marked by painful menstrual cramps. In a randomised, double-masked clinical trial, an abdominal massage with a blend of lavender, clary sage and marjoram essential oils significantly reduced the intensity and duration of menstrual cramps, shortening the average duration from 2.4 to 1.8 days ($p < 0.001$). The analgesic effect of this mixture comes from major components such as linalyl acetate, linalool, eucalyptol and β-caryophyllene, which together make up more than 79% of the blend (Ou et al. 2012).

Valeriana officinalis has shown promise in easing tension-type headaches. Valerian capsules were shown to significantly reduce headache pain, as evidenced by lower VAS scores in patients compared with placebo (Azizi et al. 2020).

In addition, citrus and basil oils demonstrated analgesic effects in several in vivo studies, including the formalin and hot-plate tests. It is suggested that their effects are mediated by pain signalling pathways (Sarmento-Neto et al. 2015).

11.7.2 Immune System Modulation Uses

Eucalyptus essential oil is one of the oils that regulate immune activity. The main active constituent is 1,8-cineole. The oil exerts its effect by increasing macrophage phagocytosis and reducing cytokine levels. Studies have shown that macrophages treated with eucalyptus oil displayed higher phagocytic activity than untreated cells. Also, reductions in TNF-α, IL-6, and IL-4 levels were observed. An in vivo study showed that administration of eucalyptus oil reversed immunosuppression induced by 5-fluorouracil by activating phagocytosis (Zonfrillo et al. 2022).

Black cumin essential oil (*Nigella sativa*) is another example of an oil that regulates the immune system. It is reported that the oil activates NF-κB and TLR signalling pathways, suppresses pro-inflammatory cytokines such as IL-6, TNF and IL-1β, and regulates autophagy via SIRT1 signalling. Studies have shown that the oil also enhances macrophage activity without altering normal leukocyte counts (Ciesielska-Figlon et al. 2023).

Another essential oil reported to exhibit immunomodulatory and anti-inflammatory effects is thyme oil, which is rich in thymol and carvacrol. Studies have shown that the oil may decrease cytokines such as TNF and IL-1β, while increasing IL-10 levels (Gholijani et al. 2016).

A study reported that *Chamaecyparis obtusa* oil may reduce cytokines involved in allergies, such as IL-25 and IL-33. It also demonstrated that NF-κB and AP-1 levels may change. This suggests the potential effect of this oil in allergy management (Shin et al. 2021).

11.7.3 Antimicrobial Uses

Essential oils have been reported to exert antimicrobial activity owing to their bioactive constituents, including terpenes, terpenoids and phenylpropenes. For instance, *Cymbopogon flexuosus* oil displayed antimicrobial activity against *S. aureus* and *Candida albicans* (Gao et al. 2020). *Origanum vulgare* and *Lavandula officinalis* oils also displayed antibacterial activity against *S. aureus* and *E. coli*, with minimum inhibitory concentrations (MICs) ranging from 1600 to 1800 ppm for oregano oil, respectively. Meanwhile, lavender oil showed activity with MIC values equal to 2000 and (1000–1200) ppm, respectively. This activity is attributed to thymol and carvacrol (Martucci et al. 2015).

Moreover, cinnamon oil exhibited noticeable antibacterial activity, with MIC values ranging from 0.5 to 8 mg/mL against *Enterococcus faecalis*, *Klebsiella pneumoniae* and *S. aureus*. The major oil component, cinnamaldehyde, was associated with antibacterial activity (Saki et al. 2020).

Eucalyptus oil was also reported to display antibacterial effect against *Haemophilus influenzae*, *H. parainfluenzae*, and *Stenotrophomonas maltophilia* with MIC values of approximately 1.25 μL/mL (de Mendonça Braga et al. 2018).

Applying nano-encapsulation may improve stability and ensure controlled release of the oil, thus enhancing their antimicrobial activity. A mixture of peppermint, thyme, and clove essential oils in water reduced *E. coli* levels within 30 min (Sharma et al. 2022).

The findings indicate that essential oils possess significant potential as natural antibacterial agents, both independently and in conjunction with contemporary pharmaceuticals, to combat drug-resistant microbes.

11.7.4 Respiratory Health Uses

Essential oils have long been used for respiratory conditions due to their antibacterial, antiviral and anti-inflammatory properties, as well as their ability to help clear mucus and widen airways. Allowing them to reach both the upper and lower respiratory pathways.

Many essential oils have been traditionally used to relieve respiratory problems. Eucalyptus oil, which contains a high amount of 1,8-cineole, is often used for coughs, bronchitis and catarrh. Its effectiveness via inhalation is due to its rapid action on the lungs' mucous membranes. Menthol, present in peppermint oil, alleviates cold symptoms when inhaled.

Eucalyptus oil shows mild antiviral activity against the mumps virus, suggesting its potential use in viral respiratory infections. However, it has shown effectiveness against adenovirus (Mieres-Castro et al. 2021).

In addition to their antimicrobial effects, essential oils influence the transient receptor potential channels in the airways. Thymol and cinnamaldehyde function as agonists for TRPA1, while menthol and 1,8-cineole act as agonists for TRPM8. These chemicals promote airway opening, lessen coughing and help reduce airway irritation. In one study, guinea pigs exposed to menthol experienced a 56% drop in cough frequency triggered by external stimuli (Stinson et al. 2023) (Stinson et al. 2023).

Human studies support these findings. A spray containing *Eucalyptus citriodora*, *E. globulus*, *Mentha × piperita*, *Origanum syriacum* and *Rosmarinus officinalis* provided short-term relief to patients with upper respiratory tract infections after a 20-min application. A post-market study of over 3000 patients found that an ointment containing eucalyptus oil, pine needle oil, and menthol improved bronchitis and hoarseness, with good tolerability reported (Smith and Matthews 2022).

Thyme, cinnamon, and lemongrass oils exhibited antibacterial activity against *H. influenzae* and *S. pneumoniae* in the vapour phase test (VPT), indicating their potential for the management of respiratory tract infections (Ács et al. 2016).

11.7.5 Skin and Dental Health Uses

11.7.5.1 Dental Health Applications

One of the most widely used essential oils in dentistry is clove oil, which is rich in eugenol as its main component. The oil is well known for its antibacterial and antifungal activities, especially against *C. albicans* and *Staphylococcus* species (Didehdar et al. 2022).

Another oil has shown promising activity against biofilm formation, particularly *against C. albicans*, making it effective against oral candidiasis. The oil is tea tree oil, and the main component, terpinene-4-ol, is of interest for combating candidiasis (Francisconi et al. 2020).

Other oils, such as cinnamon and lemon oils, showed promising antibacterial and antifungal effects. These oils can be incorporated into formulations for the management of oral candidiasis (Yanakiev 2020; Białoń et al. 2014). It is also worth noting that not only oils can be used as bactericidal agents, but also to reduce anxiety, especially during dental procedures, such as lavender oil (Hugar et al. 2022).

Eucalyptus oil was studied for its ability to combat oral pathogens, particularly Lactobacillus acidophilus. The oil also exhibited an immunomodulatory effect, which might help strengthen oral immunity (Ishnava et al. 2013).

11.7.5.2 Skin Health Applications

Several essential oils can be used to promote skin health, as they possess various activities. For instance, clove oil possesses antioxidant activity and can thus be used to treat skin ageing. Also, the oil showed antifungal activity, suggesting it can be used to treat fungal infections (Huang et al. 2013).

Another oil is tea tree oil, which is famous in skin preparations for its antimicrobial and anti-inflammatory properties and can be used to treat acne. Terpinen-4-ol, the main oil component, has been reported to be safer than the oil itself (Pazyar et al. 2013). Eucalyptus oil has been reported to exhibit antioxidant and anti-inflammatory activities (Insuan et al. 2021). Lemon oil has been reported to exhibit antifungal activity (Abd Rashed et al. 2021).

11.8 Clinical and Pre-clinical Evidence on the Therapeutic Effects of Essential Oils

Clinical studies on the use of essential oils have also been reported. For instance, lavender, chamomile and peppermint oils have been studied in a randomised cross-over study in patients with leukaemia. The oils significantly decreased depression and anxiety symptoms, but they did not show any significant reduction in pain (Corasaniti et al. 2023).

Studies have shown that aromatherapy combined with reflexology significantly reduced pain and anxiety in patients with advanced cervical cancer undergoing brachytherapy (Blackburn et al. 2021). Other studies involving aromatherapy showed that a reduction in pain scores is observed when used before, during, and after breast cancer therapy. The study also showed that using aromatherapy lowered IL-6 levels (Deng et al. 2022).

Another clinical trial in which lavender, sage and marjoram oils were used during abdominal massage for participants. This combination significantly decreased the severity and duration of pain in dysmenorrhea patients (Ou et al. 2012).

A double-masked, placebo-controlled study of *Valeriana officinalis* demonstrated a reduction in the frequency of tension-type headaches. Participants taking valerian showed lower pain levels (mean VAS: 3.5 compared to 5.1, $p < 0.001$), headache related disability, and difficulties with daily activities after 1 month of treatment (Azizi et al. 2020).

Many of these clinical outcomes are supported by earlier pharmacological studies and by animal experiments using established pain models, including acetic acid-induced writhing, formalin, hot-plate and tail flick tests, which showed that essential oils from *Cymbopogon citratus*, *Ocimum basilicum* and *Eugenia caryophyllata* exhibit analgesic effects. These actions involve both central and peripheral mechanisms, often involving changes in opioid receptors, suppression of inflammatory mediators and interactions with ion channels (Sarmento-Neto et al. 2015).

Essential oils extracted from *Hyptis pectinata* and *Lippia gracilis* exhibited anti-inflammatory and analgesic effects in pre-clinical investigations. This was accomplished by modulating nitric oxide pathways, cholinergic modulation, and possibly the synergistic effects of monoterpenes including carvacrol, linalool and thymol (Raymundo et al. 2011).

Clinical and pre-clinical studies support the use of essential oils in integrative approaches for managing physical pain and mental distress.

11.9 Introduction to Essential Oils and Mental Well-Being

Among mental health problems, stress, mood swings, anxiety, depression, sleeping issues, and cognitive dysfunction are considered the most prevalent. Maintaining mental well-being is gaining interest nowadays, especially the use of natural resources, such as essential oils, to enhance mental health by harnessing their power to improve mood and promote relaxation (Pferschy-Wenzig et al. 2022; Rabie et al. 2023; Chavda et al. 2025).

11.10 Mechanisms of Action of Essential Oils Related to Mental Well-Being

Anxiety is a disorder that emerges as a result of dangerous or stressful events, arising from the interaction of biological, psychological and environmental aspects. The gamma aminobutyric acid (GABA), norepinephrine and serotonin 5-HT systems play crucial roles in modulating anxiety and depression (Cui et al. 2022). The first line of anxiety treatment is using selective serotonin reuptake inhibitors (SSRIs) and serotonin-norepinephrine reuptake inhibitors (SNRIs) (Chen et al. 2022; Algarzae et al. 2025). Depression is characterised by low serotonin and norepinephrine levels, which could be due to high stress levels or hypothalamic–pituitary–adrenal (HPA) hyperactivation that leads to an increase in cortisol levels, which in turn causes an increase in serotonin reuptake. Inflammation also plays a role in mental problems. Stress can lead to increased release of adrenaline and noradrenaline, which in turn triggers a cascade of events, starting with the activation of nuclear factor kappa-b and subsequently the release of pro-inflammatory cytokines, including interleukin 1 (IL-1), interleukin 6 (IL-6), and tumour necrosis factor α (TNF-α). Those mediators affect the afferent vagus nerve, which, in turn, influences mood regulation or, upon entering the brain, alters neurotransmitter metabolism (Lizarraga-Valderrama 2021). Insomnia is a complex disorder that involves many factors; one of the hypotheses of its pathogenesis is HPA dysfunction and endocrine system dysregulation. The other hypothesis is dysregulation of neurotransmitters involved in the wake-sleep system (Chen et al. 2022).

EOs may enhance their mental health through different mechanisms. For instance, they may activate the GABA system, such as lavender oil (Chavda et al. 2025). López et al. have studied the effect of lavender EO at a molecular level and found that its anxiolytic activity may be attributed to NMDA receptor antagonism and serotonin transporter (SERT) inhibition (López et al. 2017).

It was also reported that certain oil constituents, for example, linalool, citronellol, geranyl acetate and linalyl acetate, may have anxiolytic activity (Harada et al. 2018; Qneibi et al. 2019). Basil leaf oil was reported to act through HPA as it significantly increases serotonin and decreases cortisol levels in a depression rat model (Sentari et al. 2019). Limonene, the main component of *Citrus sinensis* EO, significantly restored HPA hyperactivity and restored monoamine neurotransmitter levels (Zhang et al. 2019). Ylang ylang essential oil has been reported to increase serotonin levels in the hippocampus region of rats (Fung et al. 2021). The oil has also been reported to modulate the HPA axis and regulate the MAPK pathway, relieving anxiety (Zhang et al. 2018).

11.11 Key Therapeutic Essential Oils Supporting Mental Health

Several essential oils have been used in maintaining mental health. Their chemistry, mode of action, in vivo studies and clinical studies are discussed. The structure of the main components is illustrated in Table 11.4.

*11.11.1 Lavender (*Lavandula *sp.)*

Lavender is considered the most studied plant in aromatherapy studies. The main components of lavender oil are linalyl acetate, linalool, and lavandulyl acetate (Lizarraga-Valderrama 2021). It has anxiolytic properties due to its potential effect on the limbic system. Possible mechanisms of action include inhibition of voltage-gated calcium channels (VGCCs), reduction of $5HT_{1A}$ receptor activity, and increased parasympathetic tone (Malcolm and Tallian 2017; da Silva et al. 2023). Studies have also shown that lavender oil may enhance GABA activation, thus relieving symptoms of anxiety. Previous reports have shown that lavender oil, administered orally and by inhalation in mice, can significantly reduce pain induced by chemical and thermal stimuli (Chavda et al. 2025). Another in vivo study showed that inhalation of lavender oil in rats led to enhanced sleep and had anxiolytic activity as measured by electroencephalogram (EEG) (Sattayakhom et al. 2023). A double-blind trial using standardised lavender oil orally (80 mg) found that it decreased hamilton anxiety scale (HAMA) scores compared to placebo (Malcolm and Tallian 2017). Another randomised controlled trial showed that inhalation of *Lavandula stoechas* reduced anxiety scores in the lavender group (Ebrahimi et al. 2022).

Table 11.4 Main essential oil plants' components to maintain mental well-being and their chemical structures

Name of the plant	Chemical structure
Lavender	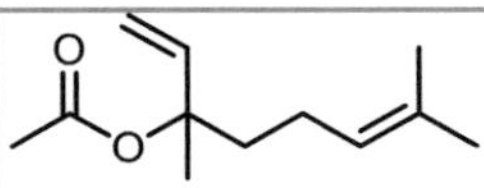 Linalyl acetate 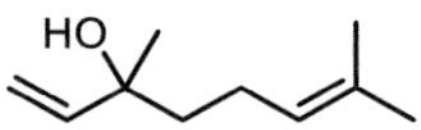 Linalool 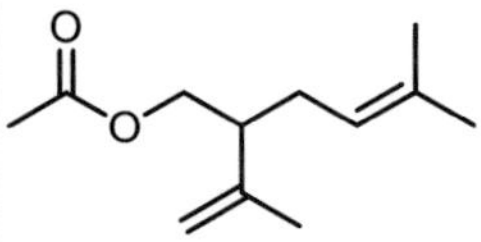 Lavandulyl acetate
Bergamot	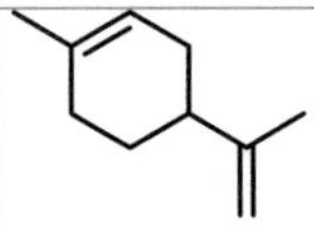 Limonene 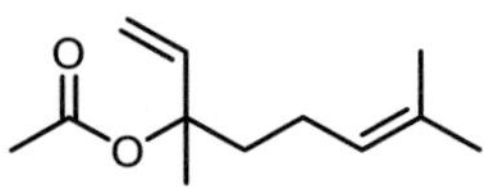 Linalyl acetate 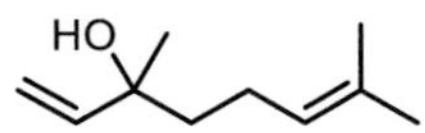 Linalool
Roman Chamomile	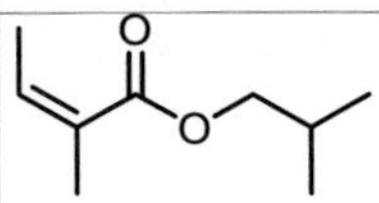 Isobutyl angelate 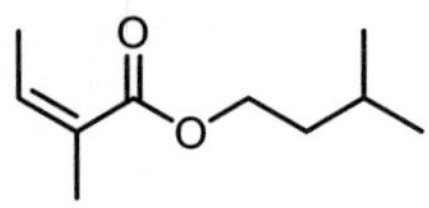 Isoamyl angelate 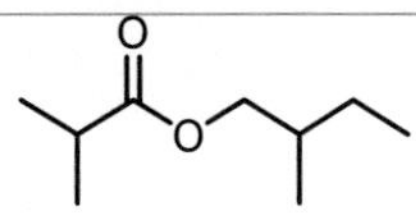 2-Methylbutyl isobutyrate 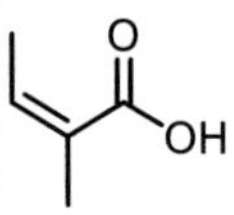 Angelic acid 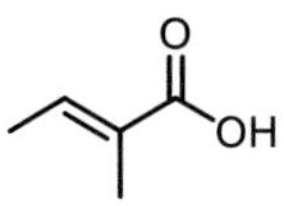 Tiglic acid
Lemongrass	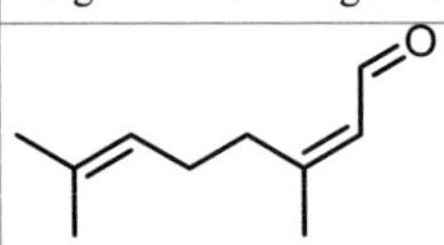 Neral 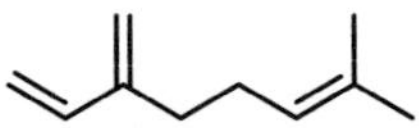 β-myrcene 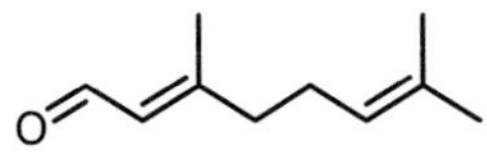 Geranial
Frankincense	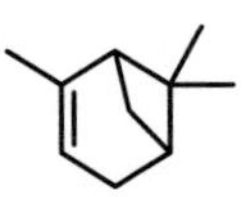 α-pinene 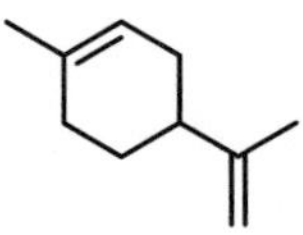 Limonene 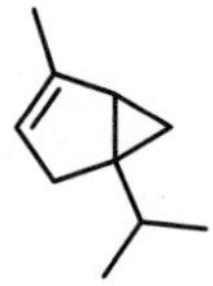 α-Thujene 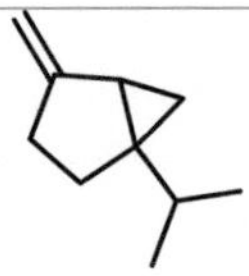 Sabinene

(continued)

Table 11.4 (continued)

Name of the plant	Chemical structure
Rose	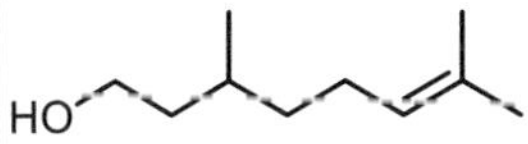 Citronellol 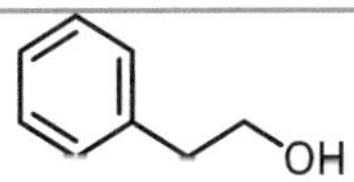 2-Phenethyl alcohol 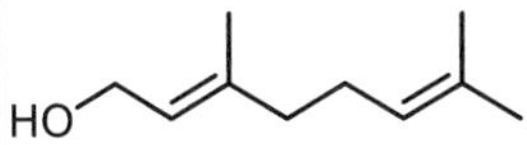 Geraniol
Rosemary	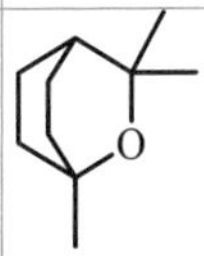 Cineole 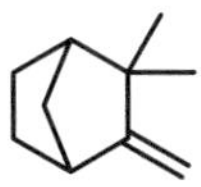 Camphene 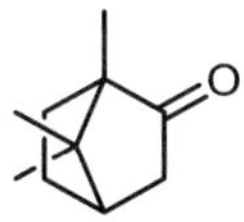 Camphor 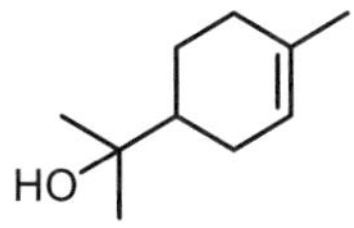 α-Terpineol 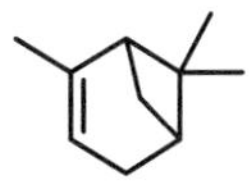 α-pinene
Valerian	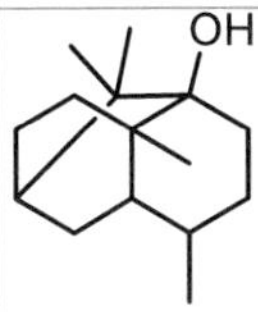 Patchouli alcohol 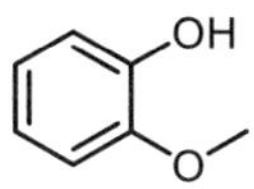 δ-Guaiacole 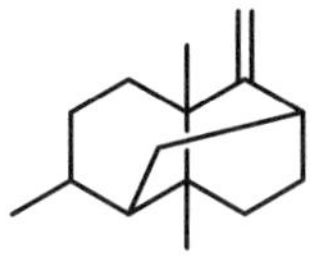 Seychellene 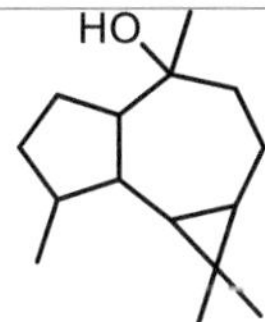 Viridiflorol

*11.11.2 Bergamot (*Citrus bergamia*)*

Bergamot has demonstrated positive results in people with anxiety (da Silva et al. 2023). Principal components are limonene, linalyl acetate and linalool (Lizarraga-Valderrama 2021). In vivo studies have demonstrated that the bergamot oil may affect synaptic transmission and regulate electroencephalography (EEG) activity (Cui et al. 2022). The possible mechanism of action is reported in a study in which bergamot oil administration (100 μL/kg, i.p.) in rats led to a significant increase in extracellular amino acids aspartate, glycine and taurine in a Ca^{2+}-dependent manner (Lizarraga-Valderrama 2021). Another study showed that bergamot oil reduced anxiety-like behaviour in rats as evidenced through the forced swimming test (FST), open field test (OFT) and elevated maze test (EMT) (Rombolà et al. 2017). In a randomised crossover human study, inhalation of bergamot oil significantly slowed anxiety-induced tachycardia, decreased salivary cortisol levels, and enhanced mood (Watanabe et al. 2015). Another study showed that inhaling bergamot for 15 min also enhanced positive feelings by 17% compared with the control group (Han et al. 2017).

11.11.3 *Roman Chamomile (***Anthemis nobilis***)*

Roman chamomile has been used as an anxiolytic, to calm nerves and mitigate insomnia (Chavda et al. 2025). The major compounds identified in the oil are isobutyl angelate, isoamyl angelate and 2-methylbutyl isobutyrate (Lizarraga-Valderrama 2021). Other components identified included angelic acid, tiglic acid, α-pinene and 2-methylbutanoic acid. It was reported that the possible mode of action of Roman chamomile is to increase mitochondrial function, increase parvalbumin mRNA expression in the hippocampus and increase 5-hydroxytryptamine in synapses (Kong et al. 2017; Jia et al. 2021). A study found that inhalation of Roman chamomile essential oil reduced depressive-like behaviour in rats during the FST (Kong et al. 2017). Another group studied the effect of combining inhalation of Roman chamomile oil with chlorpromazine and found that this combination reduced drug-resistant depression-like behaviour in mice (Hashikawa-Hobara et al. 2019).

11.11.4 *German Chamomile (***Matricaria chamomilla***)*

German chamomile oil has long been used in aromatherapy (Ebrahimi et al. 2022). A randomised study showed that there was a significant enhancement in depression, anxiety, and stress levels after using chamomile oil compared to the control group ($p < 0.01$) (Ebrahimi et al. 2022)—another study, German chamomile to prevent generalised anxiety disorder (GAD) relapse. Chamomile extract (1500 mg/day) exhibited significant activity in reducing GAD symptoms over 8 weeks in 179 patients (Keefe et al. 2016).

11.11.5 *Lemongrass (***Cymbopogon citratus***)*

Lemongrass has been reported to treat insomnia as well as possessing anxiolytic, sedative, and anticonvulsant activities (da Silva et al. 2023). Neral, geranial and *β*-myrcene are the main oil components (Lizarraga-Valderrama 2021). Costa et al. studied the mechanism of action by which lemongrass can cause anxiolytic effects in rats. They suggested that this is due to interaction with the GABA-A receptor–benzodiazepine (BDZ) complex (Rodrigues de Almeida Costa et al. 2011). To evaluate its anxiolytic activity in humans, it has been shown that exposure to the oil after inducing anxiety has significantly decreased tension and anxiety compared to the control group. This was assessed using the Social Phobia Inventory (SPIN) questionnaire (Goes et al. 2015).

*11.11.6 Frankincense (***Boswellia** *sp.)*

Frankincense is derived from the resins of various *Boswellia* species. The oil's main components have been identified to be α-pinene, limonene, α-thujene, myrcene and sabinene (Lizarraga-Valderrama 2021). It has stress-reducing qualities as it possesses anti-inflammatory activity, thus relieving the neuroinflammation which is related to anxiety and depression (Chavda et al. 2025). An in vivo study reported that topical application of frankincense oil reduced corticosterone levels, a stress marker, in sleep-deprived rats (Okano et al. 2019). Another study revealed that administration of a frankincense ethyl acetate extract significantly decreased the inflammatory cytokines interleukin-6 (IL-6) and tumour necrosis factor-α (TNF-α) in lipopolysaccharide (LPS)-induced depression and anxiety-like behaviour in rats (Shahidpour et al. 2021).

*11.11.7 Rose (***Rosa damascena***)*

Rose oil has been shown to have anxiolytic effects, helping create a sense of relaxation and comfort (Vora et al. 2024). Its main phytoconstituents include citronellol, 2-phenethyl alcohol and geraniol. The anxiolytic action of rose oil is thought to involve the hypothalamic–pituitary–adrenal (HPA) axis and the serotonergic system (Lizarraga-Valderrama 2021). Inhalation studies in rats demonstrated apparent anxiolytic activity, with treated animals showing more frequent entries and longer time spent on the open arms of the EPM than diazepam-treated animals (de Almeida et al. 2004). Clinical research has also reported that rose oil inhalation produced a sedative effect and improved sleep quality in patients (Hajibagheri et al. 2014). Another study found that inhaling the oil reduced both anxiety and pain in burn patients. (Bikmoradi et al. 2016).

*11.11.8 Rosemary (***Rosmarinus officinalis***)*

Rosemary essential oil has been shown to have anxiolytic effects and to improve both memory and mood. The main components of the oil are pinene, cineole, camphene, camphor and α-terpineol (Lizarraga-Valderrama 2021). α-pinene, the main component, is believed to be involved in the HPA axis (Lizarraga-Valderrama 2021). In a rat model, inhalation of rosemary essential oil has significantly decreased the mobility time in the TST, decreased serum corticosterone levels, and increased brain dopamine levels (Villareal et al. 2017). In a human trial, the impact of rosemary essential oil on cognitive functions, viz., memory, attention, and mood, was evaluated using computerised cognitive assessment and mood questionnaires, respectively. The study concluded that rosemary oil significantly improved memory quality and other memory-related factors compared with the control group (no-odour group) (Moss et al. 2003).

*11.11.9 Agarwood (*Aquilaria *sp.)*

It is a famous aromatic plant in Chinese culture due to its characteristic odour. Sesquiterpenes and chromones are the primary key components of the essential oil. Agarwood essential oil has been reported to enhance sleep quality and to attenuate depression and anxiety through various mechanisms, including inhibiting the HPA axis, increasing serotonin levels and exerting antioxidant and anti-inflammatory activities (Chen et al. 2022).

*11.11.10 Valerian (*Valeriana *Spp.)*

Valerian has been used as a sedative to improve sleep quality. Patchouli alcohol, δ-guaiacole, seychellene, 8-acetyl patchouli alcohol and viridiflorol are the main key constituents of valerian EO. In an animal study, the root essential oil of *Valeriana wallichi* has been reported to act as an antidepressant by increasing noradrenaline and 5-hydroxytryptamine. It is suggested that its mechanism of action is through the L-arginine-NO-cGMP pathway (Sah et al. 2011). In a randomised, double-blind study, the standardised root extract of *Valeriana officinalis* significantly improved sleep quality and reduced anxiety in patients with insomnia (Chandra Shekhar et al. 2024).

11.12 Challenges and Future Considerations

Essential oils (EOs) offer valuable therapeutic potential; however, several challenges still limit their broader use. Although extensive research has shown that aromatic plants can support both physical and mental well-being (El-Nashar et al. 2021; Elkousy et al. 2022; Mahmoud et al. 2024).

One challenge is their nature and complexity, which depend on many factors, including plant origin, harvest timing, plant maturity and extraction methods. This variability renders standardisation difficult (Aziz et al. 2018).

Although aromatic plants and essential oils are safe, they can cause several side effects, especially when misused or used in large amounts. Among these side effects are allergy, oral toxicity, eye irritation, and phototoxic reaction. One of the significant adverse effects observed is contact dermatitis, mainly when the oil is used undiluted; for example, peppermint oil can cause skin burns (Vora et al. 2024; Parys

1983). In addition, lavender and tea tree oil caused erythema in some cases (Varma et al. 2000). Another significant effect is gastrointestinal tract (GIT) irritation when the oils are ingested in high doses (Lemmens-Gruber 2020). Other problems can be related to active constituents in the oil, such as thujone, which is found in wormwood and sage oil. It has been reported that when used in large amounts, it may cause liver and brain damage (Al-Harrasi et al. 2022).

One of the main challenges is the stability of the oils as they degrade, especially when exposed to heat or light. Another problem that hinders their use is their hydrophobic nature, which reduces their solubility and bioavailability (Severino et al. 2015). To solve these problems, novel drug delivery systems must be applied; for instance, nanocapsulation has many advantages over conventional methods, such as stability enhancement, volatility reduction, and controlled release (Nair et al. 2022).

In addition, the interactions between EOs and drugs, or among the components of EOs, must be investigated, as this may help develop combined therapies with greater efficacy (Owen and Laird 2018).

11.13 Conclusion and Future Remarks

Aromatic plants and essential oils have been used since ancient times to promote physical and mental well-being. This chapter provides an intensive review of essential oils, their possible applications in the treatment of various disorders and the mechanisms underlying their actions. The challenges that hinder the use of essential oils are numerous and require further studies. Essential oils may act on multiple targets via different pathways. Thus, it is imperative to explore further their mechanism of action, toxicities and their implementation in clinical studies. Another important factor to be considered is their bioavailability; hence, novel drug delivery systems research is essential to obtain the maximum benefit from these valuable plants.

References

Abate L, Bachheti A, Bachheti RK, Husen A, Getachew M, Pandey DP (2021) Potential role of forest-based plants in essential oil production: an approach to cosmetic and personal health care applications. In: Husen A, Bachheti RK, Bachheti A (eds) Non-timber forest products: food, healthcare and industrial applications. Springer, Cham, pp 1–18

Abd Rashed A, Rathi D-NG, Ahmad Nasir NAH, Abd Rahman AZ (2021) Antifungal properties of essential oils and their compounds for application in skin fungal infections: conventional and nonconventional approaches. Molecules 26(4):1093

Ács K, Bencsik T, Böszörményi A, Kocsis B, Horváth G (2016) Essential oils and their vapors as potential antibacterial agents against respiratory tract pathogens. Nat Prod Commun 11(11):1709–1712

Aelenei P, Miron A, Trifan A, Bujor A, Gille E, Aprotosoaie A (2016) Essential oils and their components as modulators of antibiotic activity against gram-negative bacteria. Medicine 3(3):19

Algarzae NK, Saleem U, Alsharif I, Alsulami FT, Mostafa NM, El-Nashar HA, Eldahshan OA, Chauhdary Z, Farrukh M, Taiba AE (2025) Unlocking the neuroprotective potential of Citrus japonica peel oil: gas chromatography-mass spectrometry analysis and anti-Parkinsonian activity in a paraquat rat model. Ital J Food Sci 37(2):391–412

Al-Harrasi A, Bhatia S, Behl T, Kaushik D, Mittal V, Sharma A (2022) Toxicity associated with essential oils. In: Role of essential oils in the management of COVID-19. CRC Press, Boca Raton, pp 391–399

Ali B, Al-Wabel NA, Shams S, Ahamad A, Khan SA, Anwar F (2015) Essential oils used in aromatherapy: a systemic review. Asian Pac J Trop Biomed 5(8):601–611

Alipour G, Dashti S, Hosseinzadeh H (2014) Review of pharmacological effects of Myrtus communis L. and its active constituents. Phytother Res 28(8):1125–1136

Amin B, Hosseinzadeh H (2015) Black cumin (Nigella sativa) and its active constituent, thymoquinone: an overview on the analgesic and anti-inflammatory effects. Planta Med 82(01/02):8–16

Andrade-Ochoa S, Chacón-Vargas KF, Sánchez-Torres LE, Rivera-Chavira BE, Nogueda-Torres B, Nevárez-Moorillón GV (2021) Differential antimicrobial effect of essential oils and their main components: insights based on the cell membrane and external structure. Membranes 11(6):405–405

Angane M, Swift S, Huang K, Butts CA, Quek SY (2022) Essential oils and their major components: an updated review on antimicrobial activities, mechanism of action and their potential application in the food industry. Foods 11(3):464–464

Arora S, Chhetri S, Bhandari P (2025) Nanoencapsulation of essential oil: a tailored system of green therapeutic potential. Curr Top Med Chem 25(25). https://doi.org/10.2174/0115680266342801250303004033

Aziz ZA, Ahmad A, Setapar SHM, Karakucuk A, Azim MM, Lokhat D, Rafatullah M, Ganash M, Kamal MA, Ashraf GM (2018) Essential oils: extraction techniques, pharmaceutical and therapeutic potential-a review. Curr Drug Metab 19(13):1100–1110

Azizi H, Shojaii A, Hashem-Dabaghian F, Noras M, Boroumand A, Haghani BE, Ghods R (2020) Effects of Valeriana officinalis (Valerian) on tension-type headache: a randomized, placebo-controlled, double-blind clinical trial. Avicenna J Phytomed 10:297–304

Bachheti RK (2015) Chemical composition and antibacterial activity of the essential oil from the leaves of Eucalyptus globulus collected from Haramaya University, Ethiopia. Der Pharma Chem 7:209–214

Bachiega TF, de Sousa JPB, Bastos JK, Sforcin JM (2012) Clove and eugenol in noncytotoxic concentrations exert immunomodulatory/anti-inflammatory action on cytokine production by murine macrophages. J Pharm Pharmacol 64(4):610–616

Baptista-Silva S, Borges S, Ramos OL, Pintado M, Sarmento B (2020) The progress of essential oils as potential therapeutic agents: a review. J Essent Oil Res 32(4):279–295

Białoń M, Krzyśko-Łupicka T, Koszałkowska M, Wieczorek PP (2014) The influence of chemical composition of commercial lemon essential oils on the growth of Candida strains. Mycopathologia 177(1–2):29–39

Bikmoradi A, Harorani M, Roshanaei G, Moradkhani S, Falahinia GH (2016) The effect of inhalation aromatherapy with damask rose (Rosa damascena) essence on the pain intensity after dressing in patients with burns: a clinical randomized trial. Iran J Nurs Midwifery Res 21(3):247

Blackburn L, Hill C, Lindsey A, Sinnott L, Thompson K, Quick A (2021) Effect of foot reflexology and aromatherapy on anxiety and pain during brachytherapy for cervical cancer. Oncol Nurs Forum 48(3):265–276

Caballero-Gallardo K, Quintero-Rincón P, Olivero-Verbel J (2025) Aromatherapy and essential oils: holistic strategies in complementary and alternative medicine for integral wellbeing. Plants 14(3):400

Carrasco FR, Schmidt G, Romero AL, Sartoretto JL, Caparroz-Assef SM, Bersani-Amado CA, Cuman RKN (2009) Immunomodulatory activity of *Zingiber officinale* Roscoe, *Salvia offi-*

cinalis L. and *Syzygium aromaticum* L. essential oils: evidence for humor- and cell-mediated responses. J Pharm Pharmacol 61(7):961–967
Cha J-D, Moon S-E, Kim H-Y, Lee J-C, Lee K-Y (2009) The essential oil isolated from *Artemisia capillaris* prevents LPS-induced production of NO and PGE_2 by inhibiting MAPK-mediated pathways in RAW 264.7 macrophages. Immunol Investig 38(6):483–497
Chandra M, Prakash O, Kumar R, Bachheti RK, Bhushan B, Kumar M, Pant AK (2017) β-Selinene-rich essential oils from the parts of Callicarpa macrophylla and their antioxidant and pharmacological activities. Medicine 4:52
Chandra Shekhar H, Joshua L, Thomas JV (2024) Standardized extract of Valeriana officinalis improves overall sleep quality in human subjects with sleep complaints: a randomized, double-blind, placebo-controlled, clinical study. Adv Ther 41(1):246–261
Chavda V, Balar P, Apostolopoulos V (2025) A review on essential oils: a potential tonic for mental wellbeing in the aging population? Maturitas 192:108158
Chen X, Wang C, He Q, Feng J, Chen D, Wei J, Liu Y (2022) Chemical composition and potential properties in mental illness (anxiety, depression and insomnia) of agarwood essential oil: a review. Molecules 27(14):4528
Chueca B, Pagán R, García-Gonzalo D (2014) Oxygenated monoterpenes citral and carvacrol cause oxidative damage in Escherichia coli without the involvement of tricarboxylic acid cycle and Fenton reaction. Int J Food Microbiol 189:126–131
Ciesielska-Figlon K, Wojciechowicz K, Wardowska A, Lisowska KA (2023) The immunomodulatory effect of Nigella sativa. Antioxidants 12(7):1340–1340
Corasaniti MT, Bagetta G, Morrone LA, Tonin P, Hamamura K, Hayashi T, Guida F, Maione S, Scuteri D (2023) Efficacy of essential oils in relieving cancer pain: a systematic review and meta-analysis. Int J Mol Sci 24(8):7085–7085
Cui J, Li M, Wei Y, Li H, He X, Yang Q, Li Z, Duan J, Wu Z, Chen Q (2022) Inhalation aromatherapy via brain-targeted nasal delivery: natural volatiles or essential oils on mood disorders. Front Pharmacol 13:860043
da Silva CC, Berbert LV, Sakalem M (2023) Essential oils as a resource to support the mental health of academics and health-care professionals. Id on Line Rev Psic:81–93. https://doi.org/10.14295/idonline.v17i69.3872
de Almeida RN, Motta SC, de Brito Faturi C, Catallani B, Leite JR (2004) Anxiolytic-like effects of rose oil inhalation on the elevated plus-maze test in rats. Pharmacol Biochem Behav 77(2):361–364
de Mendonça Braga NS, da Costa Silva M, Cunha AL, Sant'Ana AEG, Pires LLS, Santos A (2018) Chemical characterization and biological potential of the essential oil of Eucalyptus globulus Labill. J Pharm Pharmacol 6(12). https://doi.org/10.17265/2328-2150/2018.12.001
Deng C, Xie Y, Liu Y, Li Y, Xiao Y (2022) Aromatherapy plus music therapy improve pain intensity and anxiety scores in patients with breast cancer during perioperative periods: a randomized controlled trial. Clin Breast Cancer 22(2):115–120
Dhifi W, Bellili S, Jazi S, Bahloul N, Mnif W (2016) Essential oils' chemical characterization and investigation of some biological activities: a critical review. Medicine 3(4):25–25
Didehdar M, Chegini Z, Shariati A (2022) Eugenol: a novel therapeutic agent for the inhibition of Candida species infection. Front Pharmacol 13:872127
Ebrahimi H, Mardani A, Basirinezhad MH, Hamidzadeh A, Eskandari F (2022) The effects of Lavender and Chamomile essential oil inhalation aromatherapy on depression, anxiety and stress in older community-dwelling people: a randomized controlled trial. Explore 18(3):272–278
El-Kersh DM, Mostafa NM, Fayez S, Al-Warhi T, Abourehab MA, Eldehna WM, Salem MA (2022) GC-MS metabolites profiling of anethole-rich oils by different extraction techniques: antioxidant, cytotoxicity and in-silico enzymes inhibitory insights. J Enzyme Inhib Med Chem 37(1):1974–1986

Elkousy RH, Mostafa NM, Abd-Alkhalek AM, El Hassab MA, Al-Rashood ST, Eldehna WM, Eldahshan OA (2022) GC/MS analysis and potential synergistic effect of mandarin and marjoram oils on Helicobacter pylori. J Enzyme Inhib Med Chem 37(1):1610–1619

El-Nashar HA, Mostafa NM, Abd El-Ghffar EA, Eldahshan OA, Singab ANB (2021) The genus Schinus (Anacardiaceae): a review on phytochemicals and biological aspects. Nat Prod Res 36(18):4833–4851

El-Nashar HA, Sayed AM, El-Sherief HA, Rateb ME, Akil L, Khadra I, Majrashi TA, Al-Rashood ST, Binjubair FA, El Hassab MA (2023) Metabolomic profile, anti-trypanosomal potential and molecular docking studies of Thunbergia grandifolia. J Enzyme Inhib Med Chem 38(1):2199950

Fahmy M, Farghaly A, Hassan E, Hassan E, Hassan Z, Mahmoud K, Omara E (2022) Evaluation of the anti-cancer/anti-mutagenic efficiency of Lavandula officinalis essential oil. Asian Pac J Cancer Prev 23(4):1215–1222

Francisconi RS, Huacho PMM, Tonon CC, Bordini EAF, Correia MF, de Cássia Orlandi Sardi J, Spolidorio DMP (2020) Antibiofilm efficacy of tea tree oil and of its main component terpinen-4-ol against Candida albicans. Braz Oral Res 34:e050

Fung TKH, Lau BWM, Ngai SPC, Tsang HWH (2021) Therapeutic effect and mechanisms of essential oils in mood disorders: interaction between the nervous and respiratory systems. Int J Mol Sci 22(9):4844

Gao S, Liu G, Li J, Chen J, Li L, Li Z, Zhang X, Zhang S, Thorne RF, Zhang S (2020) Antimicrobial activity of Lemongrass essential oil (Cymbopogon flexuosus) and its active component citral against dual-species biofilms of Staphylococcus aureus and Candida species. Front Cell Infect Microbiol 10:603858

Gholijani N, Gharagozloo M, Farjadian S, Amirghofran Z (2016) Modulatory effects of thymol and carvacrol on inflammatory transcription factors in lipopolysaccharide-treated macrophages. J Immunotoxicol 13(2):157–164

Goes TC, Ursulino FRC, Almeida-Souza TH, Alves PB, Teixeira-Silva F (2015) Effect of lemongrass aroma on experimental anxiety in humans. J Altern Complement Med 21(12):766–773

Gonfa YH, Tessema FB, Gelagle AA, Getnet SD, Tadesse MG, Bachheti A, Bachheti RK (2022) Chemical compositions of essential oil from aerial parts of Cyclospermum leptophyllum and its application as antibacterial activity against some food spoilage bacteria. J Chem 2022. https://doi.org/10.1155/2022/5426050

Guo J, Jiang X, Tian Y, Yan S, Liu J, Xie J, Zhang F, Yao C, Hao E (2024) Therapeutic potential of cinnamon oil: chemical composition, pharmacological actions, and applications. Pharmaceuticals 17(12):1700–1700

Hajibagheri A, Babaii A, Adib-Hajbaghery M (2014) Effect of Rosa damascene aromatherapy on sleep quality in cardiac patients: a randomized controlled trial. Complement Ther Clin Pract 20(3):159–163

Halder S, Mehta AK, Mediratta PK, Sharma KK (2012) Acute effect of essential oil of Eugenia caryophyllata on cognition and pain in mice. Naunyn Schmiedeberg's Arch Pharmacol 385(6):587–593

Han X, Gibson J, Eggett DL, Parker TL (2017) Bergamot (Citrus bergamia) essential oil inhalation improves positive feelings in the waiting room of a mental health treatment center: a pilot study. Phytother Res 31(5):812–816

Harada H, Kashiwadani H, Kanmura Y, Kuwaki T (2018) Linalool odor-induced anxiolytic effects in mice. Front Behav Neurosci 12:414763

Hashikawa-Hobara N, Otsuka A, Ishikawa R, Hashikawa N (2019) Roman chamomile inhalation combined with clomipramine treatment improves treatment-resistant depression-like behavior in mice. Biomed Pharmacother 118:109263

Heimesaat MM, Mousavi S, Weschka D, Bereswill S (2021) Anti-pathogenic and immune-modulatory effects of peroral treatment with cardamom essential oil in acute murine campylobacteriosis. Microorganisms 9(1):169–169

Hu Y, Zhang J, Kong W, Zhao G, Yang M (2017) Mechanisms of antifungal and anti-aflatoxigenic properties of essential oil derived from turmeric (Curcuma longa L.) on Aspergillus flavus. Food Chem 220:1–8

Huang X-w, Feng Y-c, Huang Y, Li H-l (2013) Chemical composition, antioxidant and the possible use as skin-care ingredient of clove oil (*Syzygium aromaticum* (L.) Merr. & Perry) and citronella oil (*Cymbopogon goeringii*) from China. J Essent Oil Res 25(4):315–323

Hugar SM, Gokhale N, Uppin C, Kajjari S, Meharwade P, Joshi RS (2022) The effects of lavender essential oil and its clinical implications in dentistry: a review. Int J Clin Pediatr Dent 15(3):385–388

Insuan O, Thongchuai B, Chaiwongsa R, Khamchun S, Insuan W (2021) Antioxidant and anti-inflammatory properties of essential oils from three Eucalyptus species. CMU J Nat Sci 20(4):e2021091

Ishnava KB, Chauhan JB, Barad MB (2013) Anticariogenic and phytochemical evaluation of Eucalyptus globules Labill. Saudi J Biol Sci 20(1):69–74

Jia Y, Zou J, Wang Y, Zhang X, Shi Y, Liang Y, Guo D, Yang M (2021) Action mechanism of Roman chamomile in the treatment of anxiety disorder based on network pharmacology. J Food Biochem 45(1):e13547

Juergens U (2014) Anti-inflammatory properties of the monoterpene 1.8-cineole: current evidence for co-medication in inflammatory airway diseases. Drug Res 64(12):638–646

Júnior J d M d A d M, Damasceno M d BMV, Santos SAAR, Barbosa TM, Araújo JRC, Vieira-Neto AE, Wong DVT, Lima-Júnior RCP, Campos AR (2017) Acute and neuropathic orofacial antinociceptive effect of eucalyptol. Inflammopharmacology 25(2):247–254

Kamatou GPP, Viljoen AM (2010) A review of the application and pharmacological properties of α-bisabolol and α-bisabolol-rich oils. J Am Oil Chem Soc 87(1):1–7

Keefe JR, Mao JJ, Soeller I, Li QS, Amsterdam JD (2016) Short-term open-label chamomile (Matricaria chamomilla L.) therapy of moderate to severe generalized anxiety disorder. Phytomedicine 23(14):1699–1705

Kim D-S, Lee H-J, Jeon Y-D, Han Y-H, Kee J-Y, Kim H-J, Shin H-J, Kang J, Lee BS, Kim S-H, Kim S-J, Park S-H, Choi B-M, Park S-J, Um J-Y, Hong S-H (2015) Alpha-Pinene exhibits anti-inflammatory activity through the suppression of MAPKs and the NF-κB pathway in mouse peritoneal macrophages. Am J Chin Med 43(04):731–742

Kong Y, Wang T, Wang R, Ma Y, Song S, Liu J, Hu W, Li S (2017) Inhalation of Roman chamomile essential oil attenuates depressive-like behaviors in Wistar Kyoto rats. Sci China Life Sci 60(6):647–655

Kong AS-Y, Maran S, Yap PS-X, Lim S-HE, Yang S-K, Cheng W-H, Tan Y-H, Lai K-S (2022) Anti- and pro-oxidant properties of essential oils against antimicrobial resistance. Antioxidants 11(9):1819–1819

Lemmens-Gruber R (2020) Adverse effects and intoxication with essential oils. In: Handbook of essential oils. CRC Press, Boca Raton, pp 517–541

Lenardão EJ, Savegnago L, Jacob RG, Victoria FN, Martinez DM (2015) Antinociceptive effect of essential oils and their constituents: an update review. J Braz Chem Soc. https://doi.org/10.5935/0103-5053.20150332

Leong W, Huang G, Liao W, Xia W, Li X, Su Z, Liu L, Wu Q, Wong VKW, Law BYK, Xia C, Guo X, Khan I, Wendy Hsiao WL (2022) Traditional Patchouli essential oil modulates the host's immune responses and gut microbiota and exhibits potent anti-cancer effects in Apc mice. Pharmacol Res 176:106082–106082

Lizarraga-Valderrama LR (2021) Effects of essential oils on central nervous system: focus on mental health. Phytother Res 35(2):657–679

López V, Nielsen B, Solas M, Ramírez MJ, Jäger AK (2017) Exploring pharmacological mechanisms of lavender (Lavandula angustifolia) essential oil on central nervous system targets. Front Pharmacol 8:280

Maggio F, Rossi C, Serio A, Chaves-Lopez C, Casaccia M, Paparella A (2025) Anti-biofilm mechanisms of action of essential oils by targeting genes involved in quorum sensing, motility, adhesion, and virulence: a review. Int J Food Microbiol 426:110874–110874

Mahmoud OA, Ayoub IM, Eldahshan OA, Singab ANB (2024) The impact of seasonal variation on the composition of the volatile oil of Polyalthia suberosa (Roxb.) Thwaites leaves and evaluation of its acetylcholinesterase inhibitory activity. BMC Complementary Med Ther 24(1):159

Malcolm BJ, Tallian K (2017) Essential oil of lavender in anxiety disorders: ready for prime time? Ment Health Clin 7(4):147–155

Martucci JF, Gende LB, Neira LM, Ruseckaite RA (2015) Oregano and lavender essential oils as antioxidant and antimicrobial additives of biogenic gelatin films. Ind Crop Prod 71:205–213

Mieres-Castro D, Ahmar S, Shabbir R, Mora-Poblete F (2021) Antiviral activities of eucalyptus essential oils: their effectiveness as therapeutic targets against human viruses. Pharmaceuticals 14(12):1210–1210

Moghrovyan A, Sahakyan N (2024) Antimicrobial activity and mechanisms of action of Origanum vulgare L. essential oil: effects on membrane-associated properties. AIMS Biophys 11(4):508–526

Mohammed HA, Qureshi KA, Ali HM, Al-Omar MS, Khan O, Mohammed SAA (2022) Bio-evaluation of the wound healing activity of Artemisia judaica L. as part of the plant's use in traditional medicine; phytochemical, antioxidant, anti-inflammatory, and antibiofilm properties of the plant's essential oils. Antioxidants 11(2):332–332

Moss M, Cook J, Wesnes K, Duckett P (2003) Aromas of rosemary and lavender essential oils differentially affect cognition and mood in healthy adults. Int J Neurosci 113(1):15–38

Naaz S (2022) Antimicrobial and antioxidant activities of fennel oil. Bioinformation 18(9):795–800

Nair A, Mallya R, Suvarna V, Khan TA, Momin M, Omri A (2022) Nanoparticles—attractive carriers of antimicrobial essential oils. Antibiotics 11(1):108–108

Ocaña A, Reglero G (2012) Effects of thyme extract oils (from Thymus vulgaris, Thymus zygis, and Thymus hyemalis) on cytokine production and gene expression of oxLDL-stimulated THP-1-macrophages. J Obes 2012:1–11

Okano S, Honda Y, Kodama T, Kimura M (2019) The effects of frankincense essential oil on stress in rats. J Oleo Sci 68(10):1003–1009

Ou MC, Hsu TF, Lai AC, Lin YT, Lin CC (2012) Pain relief assessment by aromatic essential oil massage on outpatients with primary dysmenorrhea: a randomized, double-blind clinical trial. J Obstet Gynaecol Res 38(5):817–822

Owen L, Laird K (2018) Synchronous application of antibiotics and essential oils: dual mechanisms of action as a potential solution to antibiotic resistance. Crit Rev Microbiol 44(4):414–435

Parys B (1983) Chemical burns resulting from contact with peppermint oil mar: a case report. Burns 9(5):374–375

Paula-Freire LIG, Andersen ML, Molska GR, Köhn DO, Carlini ELA (2013) Evaluation of the antinociceptive activity of Ocimum gratissimum L. (Lamiaceae) essential oil and its isolated active principles in mice. Phytother Res 27(8):1220–1224

Paula-Freire L, Molska G, Andersen M, Carlini E (2015) Ocimum gratissimum essential oil and its isolated compounds (Eugenol and Myrcene) reduce neuropathic pain in mice. Planta Med 82(03):211–216

Pazyar N, Yaghoobi R, Bagherani N, Kazerouni A (2013) A review of applications of tea tree oil in dermatology. Int J Dermatol 52(7):784–790

Pérez-Recalde M, Ruiz Arias IE, Hermida ÉB (2018) Could essential oils enhance biopolymers performance for wound healing? A systematic review. Phytomedicine 38:57–65

Peterfalvi A, Miko E, Nagy T, Reger B, Simon D, Miseta A, Czéh B, Szereday L (2019) Much more than a pleasant scent: a review on essential oils supporting the immune system. Molecules 24(24):4530–4530

Pezantes-Orellana C, German Bermúdez F, Matías De la Cruz C, Montalvo JL, Orellana-Manzano A (2024) Essential oils: a systematic review on revolutionizing health, nutrition, and omics for optimal well-being. Front Med 11:1337785

Pferschy-Wenzig E-M, Pausan MR, Ardjomand-Woelkart K, Röck S, Ammar RM, Kelber O, Moissl-Eichinger C, Bauer R (2022) Medicinal plants and their impact on the gut microbiome in mental health: a systematic review. Nutrients 14(10):2111

Qneibi M, Jaradat N, Hawash M, Zaid AN, Natsheh A-R, Yousef R, AbuHasan Q (2019) The neuroprotective role of Origanum syriacum L. and Lavandula dentata L. essential oils through their effects on AMPA receptors. Biomed Res Int 2019(1):5640173

Rabie O, El-Nashar HA, Majrashi TA, Al-Warhi T, El Hassab MA, Eldehna WM, Mostafa NM (2023) Chemical composition, seasonal variation and antiaging activities of essential oil from Callistemon subulatus leaves growing in Egypt. J Enzyme Inhib Med Chem 38(1):2224944

Rahman MA, Rahman MM, Ahmed N (2012) Essential oil of *Alpinia calcarata* Rosc. Rhizome: heals inflammation and nociception in animal models. J Biol Act Prod Nat 2(6):365–376

Raymundo LJRP, Guilhon CC, Alviano DS, Matheus ME, Antoniolli AR, Cavalcanti SCH, Alves PB, Alviano CS, Fernandes PD (2011) Characterisation of the anti-inflammatory and antinociceptive activities of the Hyptis pectinata (L.) Poit essential oil. J Ethnopharmacol 134(3):725–732

Ribeiro Junior HL (2024) The therapeutic potential of essential oils in cancer treatment: a comprehensive review. Braz J Health Aromather Essent Oil 1(1):bjhae14

Rodrigues de Almeida Costa CA, Kohn DO, de Lima VM, Gargano AC, Flório JC, Costa M (2011) The GABAergic system contributes to the anxiolytic-like effect of essential oil from Cymbopogon citratus (lemongrass). J Ethnopharmacol 137(1):828–836

Rombolà L, Tridico L, Scuteri D, Sakurada T, Sakurada S, Mizoguchi H, Avato P, Corasaniti MT, Bagetta G, Morrone LA (2017) Bergamot essential oil attenuates anxiety-like behaviour in rats. Molecules 22(4):614

Sah SP, Mathela CS, Chopra K (2011) Involvement of nitric oxide (NO) signalling pathway in the antidepressant activity of essential oil of Valeriana wallichii Patchouli alcohol chemotype. Phytomedicine 18(14):1269–1275

Saki M, Seyed-Mohammadi S, Montazeri EA, Siahpoosh A, Moosavian M, Latifi SM (2020) In vitro antibacterial properties of Cinnamomum zeylanicum essential oil against clinical extensively drug-resistant bacteria. Eur J Integr Med 37:101146–101146

Sarmento-Neto J, Do Nascimento L, Felipe C, De Sousa D (2015) Analgesic potential of essential oils. Molecules 21(1):20–20

Sattayakhom A, Wichit S, Koomhin P (2023) The effects of essential oils on the nervous system: a scoping review. Molecules 28(9):3771

Scuteri D, Hamamura K, Sakurada T, Watanabe C, Sakurada S, Morrone LA, Rombolà L, Tonin P, Bagetta G, Corasaniti MT (2021) Efficacy of essential oils in pain: a systematic review and meta-analysis of preclinical evidence. Front Pharmacol 12:640128

Sena G, De Rose E, Crudo M, Filippelli G, Passarino G, Bellizzi D, D'Aquila P (2024) Essential oils from Southern Italian aromatic plants synergize with antibiotics against Escherichia coli, Pseudomonas aeruginosa and Enterococcus faecalis cell growth and biofilm formation. Antibiotics 13(7):605–605

Sentari M, Harahap U, Sapiie TWA, Ritarwan K (2019) Blood cortisol level and blood serotonin level in depression mice with basil leaf essential oil treatment. Open Access Maced J Med Sci 7(16):2652

Severino P, Andreani T, Chaud M, Benites C, Pinho S, Souto E (2015) Essential oils as active ingredients of lipid nanocarriers for chemotherapeutic use. Curr Pharm Biotechnol 16(4):365–370

Shaapan RM, Al-Abodi HR, Alanazi AD, Abdel-Shafy S, Rashidipour M, Shater AF, Mahmoudvand H (2021) Myrtus communis essential oil; anti-parasitic effects and induction of the innate immune system in mice with Toxoplasma gondii infection. Molecules 26(4):819–819

Shahidpour F, Mehrjerdi FZ, Mozayan MR, Marefati N, Hosseini M (2021) The effects of frankincense extract on depression and anxiety-like behaviors induced by lipopolysaccharide in rats. Learn Motiv 73:101708

Sharma N, Roy PSD, Majeed A, Salaria KH, Chalotra R, Padekar SK, Guleria S (2022) Nanoencapsulation-a novel strategy for enhancing the bioactivity of essential oils: a review. Indian J Agric Res 56(3):241–248

Shin S-H, Ye M-K, Chae M-H, Lee D-W (2021) Chamaecyparis obtusa essential oil inhibits house dust mite induced nasal epithelial cell activation and immune responses. J Oleo Sci 70(3):431–438

Sidiropoulou E, Marugán-Hernández V, Skoufos I, Giannenas I, Bonos E, Aguiar-Martins K, Lazari D, Papagrigoriou T, Fotou K, Grigoriadou K, Blake DP, Tzora A (2022) In vitro antioxidant, antimicrobial, anticoccidial, and anti-inflammatory study of essential oils of oregano, thyme, and sage from Epirus, Greece. Life 12(11):1783–1783

Smith A, Matthews O (2022) Aromatic ointments for the common cold: what does the science say? Drugs Context 11:1–9

Stinson RJ, Morice AH, Sadofsky LR (2023) Modulation of transient receptor potential (TRP) channels by plant derived substances used in over-the-counter cough and cold remedies. Respir Res 24(1):45–45

Stojanović NM, Ranđelović PJ, Simonović M, Radić M, Todorović S, Corrigan M, Harkin A, Boylan F (2024) Essential oil constituents as anti-inflammatory and neuroprotective agents: an insight through microglia modulation. Int J Mol Sci 25(10):5168–5168

Sutrisna INGT, Dewi NNA, Wati DK, Linawati NM, Wirasuta IMAG (2025) A systematic review of eucalyptol as an innovative therapeutic agent for respiratory diseases. Int J Appl Pharm 17(3):93–100

Varma S, Blackford S, Statham BN, Blackwell A (2000) Combined contact allergy to tea tree oil and lavender oil complicating chronic vulvovaginitis. Contact Derm 42(5):309–310

Villareal MO, Ikeya A, Sasaki K, Arfa AB, Neffati M, Isoda H (2017) Anti-stress and neuronal cell differentiation induction effects of Rosmarinus officinalis L. essential oil. BMC Complement Altern Med 17(1):549

Vora LK, Gholap AD, Hatvate NT, Naren P, Khan S, Chavda VP, Balar PC, Gandhi J, Khatri DK (2024) Essential oils for clinical aromatherapy: a comprehensive review. J Ethnopharmacol 330:118180

Watanabe E, Kuchta K, Kimura M, Rauwald HW, Kamei T, Imanishi J (2015) Effects of bergamot (Citrus bergamia (Risso) Wright & Arn.) essential oil aromatherapy on mood states, parasympathetic nervous system activity, and salivary cortisol levels in 41 healthy females. Forsch Komplementärmed/Res Complement Med 22(1):43–49

Yadav N, Chandra H (2017) Suppression of inflammatory and infection responses in lung macrophages by eucalyptus oil and its constituent 1,8-cineole: role of pattern recognition receptors TREM-1 and NLRP3, the MAP kinase regulator MKP-1, and NFκB. PLoS One 12(11):e0188232

Yammine J, Chihib N-E, Gharsallaoui A, Dumas E, Ismail A, Karam L (2022) Essential oils and their active components applied as: free, encapsulated and in hurdle technology to fight microbial contaminations. A review. Heliyon 8(12):e12472–e12472

Yanakiev S (2020) Effects of cinnamon (Cinnamomum spp.) in dentistry: a review. Molecules 25(18):4184–4184

Zhang N, Zhang L, Feng L, Yao L (2018) Cananga odorata essential oil reverses the anxiety induced by 1-(3-chlorophenyl) piperazine through regulating the MAPK pathway and serotonin system in mice. J Ethnopharmacol 219:23–30

Zhang L-L, Yang Z-Y, Fan G, Ren J-N, Yin K-J, Pan S-Y (2019) Antidepressant-like effect of Citrus sinensis (L.) Osbeck essential oil and its main component limonene on mice. J Agric Food Chem 67(50):13817–13828

Zhao H, Ren S, Yang H, Tang S, Guo C, Liu M, Tao Q, Ming T, Xu H (2022) Peppermint essential oil: its phytochemistry, biological activity, pharmacological effect and application. Biomed Pharmacother 154:113559

Zonfrillo M, Andreola F, Krasnowska EK, Sferrazza G, Pierimarchi P, Serafino A (2022) Essential oil from Eucalyptus globulus (Labill.) activates complement receptor-mediated phagocytosis and stimulates podosome formation in human monocyte-derived macrophages. Molecules 27(11):3488–3488

Chapter 12
Plant-Based Bioplastics and Sustainable Materials

Tadesse Bassie Gelaw, Limenew Abate Worku, Marie Yayinie, Teshale Adane, Archana Bachheti, and Rakesh Kumar Bachheti

Abstract This chapter presents plant-based bioplastics and sustainable materials as possible alternatives to traditional (petroleum-based) plastics. With an emphasis on thermal, physical, and mechanical properties, it also explains the definitions, origins, classifications, and manufacturing processes of bioplastics derived from plant sources such as cellulose, starch, polyhydroxyalkanoates, and polylactic acid. Extrusion and moulding (injection and blow moulding) are described, including the raw polymer modification and processing methods. Applications for consumer goods, packaging, agriculture, and biomedicine are examined, along with their drawbacks. The chapter concludes by highlighting technological developments and the use of plant-based bioplastics in the circular economy as a viable approach for manufacturing sustainable, renewable, and environmentally friendly materials.

Keywords Plant-based bioplastics · Sustainable materials · Traditional plastics · Properties · Bioplastics

---

T. B. Gelaw · L. A. Worku (✉) · M. Yayinie
Department of Chemistry, College of Natural and Computational Science, Debre Tabor University, Debre Tabor, Ethiopia

T. Adane
Department of Chemical Engineering, Gafat Institute of Technology, Debre Tabor University, Debre Tabor, Ethiopia

A. Bachheti
Department of Environmental Science, Graphic Era (Deemed to be University), Dehradun, Uttarakhand, India

R. K. Bachheti
Centre of Molecular Medicine and Diagnostics (COMMD), Saveetha Dental College & Hospitals, Saveetha Institute of Medical and Technical Sciences, Saveetha University, Chennai, Tamil Nadu, India

L. A. Worku et al. (eds.), *Plant-Based Solutions for Sustainable Agriculture and Healthcare*, Sustainability Sciences in Asia and Africa,
https://doi.org/10.1007/978-981-95-8910-4_12

12.1 Introduction

12.1.1 *Definition of Bioplastics*

According to European Bioplastics standards, a plastic is considered a bioplastic if it is both bio-based and biodegradable. This means that plastic obtained from full or partial plant-based biomass can be biodegradable or non-biodegradable (Union Directive 2022). Bio-based plastics are plastics containing organic carbon derived from renewable sources, such as agricultural, plant, animal, fungal, and microorganism materials from natural environments in equilibrium with the atmosphere. Degradable plastics are plastics designed to undergo a significant change in their chemical structure under specific environmental conditions, resulting in a loss of certain properties that may vary as measured by standard test methods appropriate to the plastic and its application over a period that determines their classification. These plastics are susceptible to biological degradation, producing mineralized products from carbon dioxide, water, and other smaller molecules (inorganic compounds and biomass) (Afshar et al. 2024).

12.1.2 *Overview of Bioplastics and Their Classification (Biodegradable Versus Non-Biodegradable)*

Bioplastics are polymers derived from natural resources, and most are partially or fully biodegradable, breaking down into CO_2, H_2O, and other harmless compounds without leaving toxic residues. Their production conserves petrochemicals, reduces CO_2 emissions, lowers fossil fuel use, and helps manage non-biodegradable plastic waste. Although the cost depends on raw materials and additives, bioplastics currently used in films hold potential for applications in automobiles and electronics with advancing technology.

12.1.3 *Importance of Sustainable Materials*

Bio-based plastic is a perfect alternative to synthetic plastic, as it can be made from renewable, natural sources such as corn starch, potato starch, inedible food waste, and lignocellulosic crop residues. Such plastics are essential for a sustainable, pollution-free ecology as they rely less on fossil fuels, emit fewer greenhouse gases, and reduce hazardous waste generated by oil-derived plastics (Jeremic et al. 2020).

12.1.4 Environmental Impact of Traditional Plastics and the Need for Sustainable Alternatives

Synthetic plastics, derived from petroleum sources such as coal, natural gas, and crude oil, contain toxic compounds that can harm human health and the environment. They are non-biodegradable and require millions of years to degrade, harming flora, fauna, and abiotic environments (Kumari et al. 2023). They are difficult to recycle through natural processes since they are produced in such large quantities. Although they are the cheapest option compared to plant-based bioplastics, they are not biodegradable, impair soil fertility, increase greenhouse gas emissions, and pollute aquatic environments. In contrast, plant-based bioplastics manufactured from sustainable sources are free of hazardous compounds, decompose organically within 180–300 days, are eco-friendly, and do not have a detrimental impact on the environment or organisms.

Additionally, they are non-toxic, easier to recycle, increase soil fertility, and can be disposed of without polluting the environment through landfilling or composting. However, their annual production is insufficient and more costly than that of traditional plastics. Therefore, it is becoming increasingly important to make environmentally friendly materials accessible to mitigate the detrimental impacts of conventional petroleum-based plastics (Liwarska-Bizukojc 2021).

12.1.5 Scope of the Chapter

This chapter examines plant-based bioplastics as sustainable alternatives to petroleum-derived plastics. It highlights the source, classification, and synthesis of plant-based bioplastics, examining their properties, manufacturing methods, and environmental impacts relative to conventional plastics. It also encompasses the life cycles of plant-based and petroleum-based plastics, from sourcing raw materials and polymer synthesis to industrial plastic manufacturing, consumption, and the end-of-life consequences for the environment, as well as possible measures to alleviate the negative impact. This study places particular emphasis on the beneficial application of plant-based bioplastics for eradicating the invasion of petroleum-based bioplastics. It proposes possible solutions for the impediments of plant-based bioplastics, as well as review the current progress on bioplastic processing for replacing petrol-based non-renewable plastics applied.

12.2 Overview of Plant-Based Bioplastics

12.2.1 Sources of Plant-Based Bioplastics

Plant-based bioplastics are processed from renewable natural sources, such as corn, sugarcane, starch, food and agricultural wastes, or lignocellulose crop residues, all of which provide environmentally benign alternatives for reducing carbon emissions and lowering the use of petroleum-based plastics (Mori 2023; Shahar et al. 2025).

12.2.2 Types of Plant-Based Bioplastics

There are different types of bioplastic materials, grouped based on thermal properties (thermoplastics and thermosetting), production methods (natural and synthetic), polymer type (polyesters, polyamides, polycarbonates, vinyl polymers, and polysaccharides), breakdown by natural biological processes (biodegradable and non-biodegradable), and source of raw materials (cellulose, starch, and microbe-based) (Kumari et al. 2023; Dutta and Sit 2024).

12.2.3 Starch-Based Bioplastics

Starch, one of the most abundant and widely studied biopolymers, is a promising alternative to petroleum-based plastics due to its biodegradability, biocompatibility, and applications in edible food coatings. It is also a versatile, inexpensive, and primary energy reserve polysaccharide in plants. It can be extracted from various botanical sources, and differences in amylose and amylopectin content influence its properties and processing behaviour. However, its industrial use is limited by challenges such as its high hydrophilicity, caused by multiple hydroxyl groups, and the inherent brittleness of starch films, which often lack the mechanical strength needed to maintain structural integrity during processing (Mori 2023). Such problems can be easily overcome by blending with a stable, synthetic thermoplastic polymer. On the other hand, starch enhances the biodegradability of synthetic polymers, leading to the production of more affordable biomaterials by significantly reducing the high costs associated with these products. Such hybrid biomaterials become practical through thermoplasticization, crosslinking, esterification, and blending with different polymers (Jeremic et al. 2020).

12.2.4 Polylactic Acid (PLA)

Poly(lactic acid) is a bio-based and biodegradable plastic produced from lactic acid (2-hydroxypropanoic acid), a basic monomer produced either by bacterial fermentation or by chemical synthesis. The industrial production of PLA (lactic acid) monomers is sustainable and renewable, as the carbohydrate sources (corn starch, cassava roots, sugar cane, and potatoes) are renewable (Msuya 2018). It is one of the most efficient biopolymers, as 1 kg of it is produced from 1.6 kg of the fermented feedstock, whereas others require two to three times more carbohydrate feedstock for production. This polymer and its blend are manufactured as granules with various properties and used to produce foil, cups, moulds, bottles, and mulch films. Most PLA-based materials cannot withstand hot liquids due to PLA's low glass transition temperature (Jeremic et al. 2020).

The three production routes of polylactic acid are direct condensation polymerization (production of less desired low-density molecular weight polylactic acid by esterification reaction of lactic acid and free water removal using progressive vacuum and high temperature), polymerization through lactide formation (production of high molecular weight polylactic acid using lactide as an intermediate state in the presence of a metal catalyst), and thirdly direct polymerization in azeotropic solution (continuous removal of condensation water by azeotropic distillation) (Yang et al. 2024).

It is a thermoplastic with high strength and modulus, and it also exhibits high tolerance to various processing techniques and conditions. It also has a glossy appearance and good transparency. Amorphous and crystalline PLA show brittle behaviour at room temperature. Due to its similar mechanical properties, PLA is a potential alternative to conventional polymers such as polyethylene terephthalate (PET), polystyrene (PS), and polycarbonate (PC), for applications in the automotive, electronics, and packaging industries (Viana et al. 2008).

12.2.5 Polyhydroxyalkanoates (PHA)

Polyhydroxyalkanoates (PHAs) are natural biodegradable polyesters, primarily produced by prokaryotes as cytoplasmic water-insoluble storage compounds of carbon and energy (Koller and Mukherjee 2022). Due to their characteristics, PHAs are known to be better oxygen barriers than polypropylene and poly(ethylene terephthalate), as well as better water vapour and fat/odour barriers. It has had limited market success due to high production costs and instability in thermo-mechanical properties, which are affected by fluctuations in molecular weight and synthase activity. However, its mechanical properties and biocompatibility can be enhanced by blending, surface modification, or combining it with other polymers, enzymes, and inorganic materials, thereby expanding its application potential (Kumar et al. 2021). Commercial production of PHA is achieved through microbial fermentation, such as that of *Alcaligenes latus*, *Bacillus megaterium*, Cupriavidus necator, and Pseudomonas oleovorans, which convert various carbon sources (including glucose, fructose, and fatty acids like lauric acid) into PHA (Jeremic et al. 2020).

12.2.6 Cellulose-Based Bioplastics

Cellulose is the most abundant renewable resource in plant cell walls. It can be produced by either plant photosynthesis or microbial synthesis (Carpita 2011; Chawla et al. 2009). It is the first plant-based thermoplastic polymer, represented as $(C_6H_{10}O_5)_n$. The widespread availability, biodegradability, biocompatibility, non-toxicity, and good structural, chemical, physical, and mechanical characteristics of cellulose make it a flexible biopolymer used in various applications like heavy metal absorption, tissue engineering, bone implants, drug delivery, enhancing soil fertility in agriculture, and facilitating controlled release of fertilizers (Dutta and Sit 2024). In addition, its nanocomposites are used in tissue and ligament engineering, wound healing, and bioethanol production. Due to the presence of reactive hydroxyl groups, cellulose can be chemically modified, mainly into esters, ethers, and graft copolymers for advanced material applications (Chen et al. 2022).

12.3 Production Processes

12.3.1 Extraction and Processing of Raw Materials

Obtaining renewable feedstocks, such as starch-rich crops (corn, cassava), sugar sources (sugarcane, beets), cellulose from wood or agricultural leftovers, plant oils, algae, or microbial biomass, is the first step in the extraction and processing of bioplastic raw materials. Depending on their composition, these materials are extracted using different techniques: starch is separated by centrifugation and grinding; sugars are released by enzymatic hydrolysis; cellulose is obtained by pulping and chemical delignification; plant oils are extracted by pressing or solvents; and microbial biopolymers such as PHAs are recovered after fermentation and cell disruption. Every technique separates the essential elements—starch, sugars, cellulose, oils, or microbial polymers—that are used as the building blocks to produce bioplastics (Marzo-Gago et al. 2023).

Methods such as fermentation, chemical modification, and polymerization are used to convert plant extracts into functional biopolymers. For example, dark fermentation of renewable feedstocks followed by ring-opening polymerization has been used to produce high-molecular-weight poly(lactic acid). Additionally, the rapid production of biomass-derived butyrolactones into their corresponding polymers and the transformation of cellulose into cellulose acetate have been reported in plant-based bioplastic polymer synthesis. The functionalized biopolymers are further processed using standard methods, including injection moulding, extrusion, casting, and film blowing. Other supporting materials, such as additives (natural fillers) and plasticizers, are used to improve physical and mechanical properties, which in turn affect cost, performance, and sustainability (Ahmad et al. 2024; Mohanty et al. 2003; Shafqat et al. 2021).

12.3.2 Manufacturing Techniques

The processing of plant-based bioplastics involves various methods, including polymerization (condensation or addition polymerization), polymer modification, and shaping (extrusion and moulding), which depend on the characteristics of the raw material. The polymers can be processed using three methods. In the first method (microbial fermentation), monomers are produced from plant-derived sugars, which are then polymerized into various polymers, such as biobased polyethylene, polyhydroxyalkanoates, bacterial cellulose, poly(lactic acid), and hyaluronic acid, all of which have functional properties competitive with those of traditional bioplastics. The second method commonly employed in polymer manufacturing is a slight modification of biological polymers to improve their processability and mechanical properties. Starch is a typical example, blended with synthetic and other bio-based polymers to reduce the degradability of traditional polymers, lower costs, and improve mechanical properties (Jayarathna et al. 2022). Chitosan modified with poly(ε-caprolactone) is another example used to increase the shelf life of food packaging materials (Joseph et al. 2011). The last method is based on the direct involvement of microorganisms in producing and storing bioplastics within their cells before releasing and processing them into bioplastic products. The conversion of microbial or plant-based raw materials into valuable, sustainable bioplastic products is based on three production routes (Jeremic et al. 2020; Sidek et al. 2019).

12.3.2.1 Extrusion

Extrusion is a crucial method for producing bioplastics in the desired shape by melting and forcing them through an extruder (or die). It is a more practical method for packaging thermo-bioplastics, such as poly(lactic acid), polyhydroxyalkanoates, and ethyl cellulose films, due to its high mixing and moulding efficiency (Zarate-ramírez et al. 2019). Protein-based and other bioplastics with high softening temperatures above their decomposition temperatures can be easily manufactured by extrusion using plasticizers. The optimum amount of these materials lowers the softening temperature, enhances processability, and improves some mechanical properties (Verbeek and Van Den Berg 2010).

12.3.2.2 Injection Moulding

The injection method is widely used for the manufacturing of traditional plastics and is also adaptable for bioplastic production, including biodegradable and compostable plastics. It covers one-third of the polymer product fabricated by this method. This technology is fascinating the world due to its significant contribution to producing plastics with various shapes, such as a net, and sizes, with a three-dimensional structure. It can be done using either an electric injection device or a hydraulic injection machine. The three basic processes involved in moulding are melting the polymer, forcing the melted material to flow into the mould, and cooling and opening the mould to collect the finished product. It is also advantageous due to its design, which features a short cycle, high precision, and efficiency, and because the moulders are automated using advanced technology. Such advanced designs help to monitor various parameters like temperature, pressure, flow rate, and other crucial mould factors. Information technology advancements that enable more intelligent data collection, process optimization, and predictive control have been closely linked to this progress (Fu et al. 2020; Alonso-González et al. 2025; Zhao et al. 2020).

12.3.2.3 Blow Moulding

The blow moulding manufacturing process is used to produce hollow plastic parts, such as bottles and toys, by inflating a parison (a thermoplastic tube) within a moulding machine. In the manufacturing process, there are three basic thermoplastic processes: extrusion, injection moulding, and stretch blow moulding. Among the three plastic blow moulding processes, extrusion blow moulding is the most widely practiced, followed by injection moulding and stretch blow moulding, respectively. Bioplastics can be processed using this method; for example, poly(lactic acid) and polybutylene succinate can be blended to produce bioplastic bottles. In the process, the two polymer blends, additives, and lubricants are mixed in the reactive twin-screw extrusion, then subjected to injection moulding, and finally shaped into bottles through stretch blow moulding (Aversa et al. 2021). A similar procedure is employed to produce lightweight, complex-shaped objects in 3D printing applications. For example, blends of polylactic acid, poly(3-hydroxybutyrate), and cellulose nanocrystals, as well as blends of rice bran (a by-product of rice milling) and poly(lactic acid), have been successfully engineered for the desired application (Frone et al. 2020; Felix et al. 2025) (Fig. 12.1).

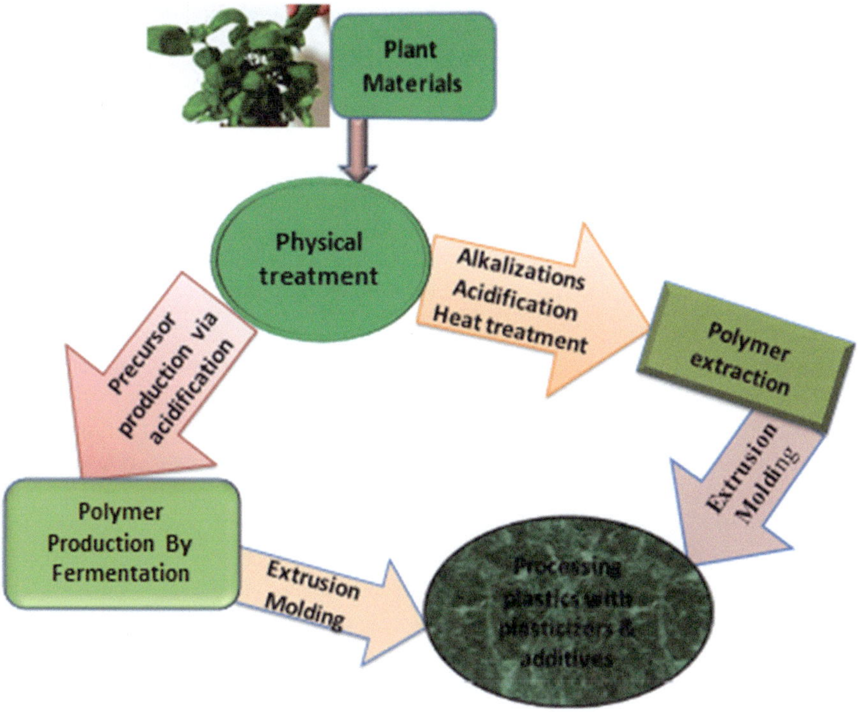

Fig. 12.1 Flow diagram of the production process for plant-based bioplastics (George et al. 2021)

12.3.3 Mechanical Properties

The mechanical properties of plastics are evaluated based on their flexibility, strength, and durability. For instance, conventional plastics like polypropylene, polyethylene, and polyethylene terephthalate can withstand heavy loads and prolonged exposure to the environment without changing their physical or chemical structure, thanks to their high endurance and tensile strength. However, plant-based bioplastics, such as polylactic acid, polyhydroxyalkanoates, polyamide, polyhydroxybutyrate, starch, and cellulose-based bioplastics, exhibit thermal instability, brittleness, poor mechanical strength, and limited processability, which restrict their applications and require high mechanical power (Kumari et al. 2023). Thus, plant-based bioplastics can compete with traditional plastics in terms of mechanical properties by adjusting various parameters, employing different synthetic methods, and blending with additives and fillers. Thermal and ultrasound treatments are processing conditions and post-treatment methods that also improve the mechanical properties and the formation of small-pore structures in soy protein-based bioplastics, respectively (Bouroudian et al. 2020). In addition, the mechanical properties of plant-based bioplastics, including flexibility, can be modified by adding reinforcing agents such as plasticizers, inorganic fillers, and natural fibres. For example, bioplastics with fillers like fibres increase the plastic's ability to withstand heavy loads due to their high strength and modulus. Plasticizers, on the other hand, make the bioplastic less rigid and more flexible (Boey and Lee 2022; Tummala et al. 2006).

12.3.4 *Biodegradability and Environmental Impact*

The life cycle of plant-based bioplastics varies depending on the physical and chemical structures of their constituent materials. In addition to the biopolymer's chemical structure, composition, and physical properties, degradation depends on external factors such as temperature, humidity, oxygen concentration, time, pH, sunlight, nutrient availability, and the presence of microorganisms. The biodegradation of plant-based bioplastics occurs naturally and industrially in both aerobic and anaerobic environments (Folino et al. 2020). The bioplastics degrade biologically into carbon dioxide, ammonium, water, nitrogen, hydrogen, and other biomass by microorganisms (bacteria, fungi, and algae). Most of these compounds, produced by the biodegradation of plant-derived polymers in the natural environment, are also used as nutrients by living organisms. Hence, the biodegradation products are environmentally benign and safe for living systems compared with traditional plastics. Poly(lactic acid), polyhydroxyalkanoates, polycaprolactone, starch-based plastic, cellulose-based plastic, as well as blends of polybutylene succinate and polybutylene adipate terephthalate, are biodegradable. In contrast, biopolyethylene, polyvinyl chloride, bioterephthalate, and polypropylene are not biodegradable, even though they are derived from renewable feedstocks (biomass) (Rujni and Pilipovi 2017). In addition to biodegradation, plant-based bioplastics offer several advantages, including reduced greenhouse gas emissions, decreased landfill waste, reduced dependency on fossil-based plastics, and increased soil fertility for sustainable agriculture (Rajendran et al. 2025).

Conventional plastics are suitable for long-term applications due to their exceptional mechanical strength, flexibility, and durability, as shown in Table 12.1. However, they are not biodegradable and are chemically inert, which can cause environmental pollution. The strong carbon–carbon bonds in their structure make such plastics resistant to degradation. Bioplastics, on the other hand, exhibit greater flexibility and higher biodegradability but lower mechanical strength and durability (Shafqat et al. 2021). They are more environmentally friendly than conventional plastics because they undergo biodegradation and are converted into natural molecules, such as CO_2, H_2O, and biomass (Folino et al. 2020).

Table 12.1 Conventional and bioplastic properties

Property	Bioplastics	Conventional plastics	References
Tensile Strength	30–60 MP	20–80 MPa	McKay et al. (2024) and Gabriel et al. (2021)
Flexibility	Less to moderate	Highly flexible	Jayarathna et al. (2022)
Durability	Degrades under moisture, microbial activity, and UV light over time	It has been resistant to environmental and biological degradation for decades.	Pellicer et al. (2017)
Thermal stability	Softening point (50–70 °C)	Softening points (100–250 °C)	Megha et al. (2024)
Processing temperature	150–190 °C (e.g. PLA and PHB)	160–280 °C (e.g. PP, PET, PS)	Sabalina et al. (2025)
Chemical resistance	Sensitive to hydrolysis and some organic solvents	Highly resistant to acids, bases, and solvents	Nizamuddin et al. (2024) and Gadaleta et al. (2025)
Environmental impact	Renewable, biodegradable, and compostable under proper conditions; lower CO_2 emissions	Non-biodegradable; made from fossil fuels and persistent in the environment	Muhammad Shamsuddin (2017)

12.3.5 *Analysis of Degradation Rates and Environmental Benefits*

Plant-based bioplastics degrade more quickly than petroleum-based plastics; however, the rate of degradation varies with polymer type (i.e. chemical nature and structural composition) and environmental conditions. This faster decomposition minimizes long-term accumulation in landfills and ecosystems, thereby reducing plastic pollution. Materials such as poly(lactic acid), polyhydroxyalkanoates, cellulose, and starch-based plastics can degrade within months under industrial composting conditions. In contrast, traditional plastics, such as polyethylene and polypropylene, can tolerate degradation for decades or more in normal environmental conditions (Shafqat et al. 2021). Furthermore, plant-based bioplastics offer ecological benefits, including reduced greenhouse gas emissions during manufacturing, decreased reliance on non-renewable fossil fuels, and potential circularity through composting or recycling (Rajendran et al. 2025). These benefits, however, are dependent on proper waste management systems, as some bioplastics require specific conditions, such as temperature, humidity, and pH, to break down effectively. Their total ecological impact is also influenced by factors such as feedstock production and energy inputs (Islam et al. 2024).

12.4 Applications of Plant-Based Bioplastics

Due to their renewable source, adaptability, compatibility, recyclability, benign behaviour, and potential for biodegradation, plant-based bioplastics have extensive applications in consumer goods, the textile industry, packaging, agriculture, and medicine (Singh et al. 2024). They are commonly used in packaging materials, such as bottles, films, trays, and bags, where starch-based and polylactic acid plastics offer lightweight, biodegradable options to traditional plastics. They are also used in farming as seed coverings, plant pots, and biodegradable mulch films that reduce plastic waste in the soil. The biocompatibility and lower environmental impact of bioplastics make them a preferred choice for use in the pharmaceutical and medical industries for disposable products, including syringes, medication capsules, and surgical tools. Furthermore, plant-based bioplastics are increasingly being investigated for use in textiles, 3D printing, and consumer goods, including electronic casings, cutlery, and containers, indicating their growing importance as sustainable alternatives across various manufacturing industries (Hariharan et al. 2025) (Table 12.2).

12.4.1 Packaging Materials

In the packaging industry, plant-based bioplastics have gained significant attention as environmentally friendly, high-functionality, and sustainable alternatives to traditional petroleum-based plastics. Most of these materials are biodegradable and compostable under natural conditions because they are derived from renewable plant sources, such as cellulose, starch, and plant biomass. The transparency, lightweight nature, and excellent printability of plant-based bioplastics, such as

Table 12.2 Production processes and applications of plant-based bioplastics

Plant-based bioplastic	Production processes	Application	References
Polylactic acid	Condensation polymerization	Packaging films, disposable cups	Dybka-Stępień et al. (2021)
Polyhydroxyalkanoates	Extraction and purification	Packaging, drug delivery	Prakash et al. (2022)
Starch-based bioplastics	Plasticization using sorbitol	Bags and food packaging	Chan et al. (2021)
Cellulose-based bioplastics	Esterification	Coating and packaging	Ren et al. (2025)
Protein-based bioplastics	Blending	Packaging	Felix et al. (2017)
PHA-starch blends	Blending	Disposable containers	Shahar et al. (2025)
Lignin-based bioplastics	Blending	Coating and packaging	Yang et al. (2019)
Bio-polyethene (bio-PE)	Polymerization	Different types of containers	Shahar et al. (2025)

polylactic acid and starch-based polymers, make them preferable for packaging applications, including the production of bottles, films, trays, and food containers (Singh et al. 2024). They preserve vital packaging qualities, including strength, flexibility, and food product protection, while successfully lowering their carbon footprint and reducing environmental pollution. Furthermore, such bioplastics offer a viable and sustainable alternative to contemporary packaging needs, despite challenges such as poor heat resistance, durability issues, processing challenges, and high production costs. However, ongoing advancements in polymer mixing and processing technologies are improving their processability and performance (Shi et al. 2025).

12.4.2 Agricultural Uses

As sustainable alternatives to traditional plastics, plant-based bioplastics are used in agriculture to improve soil respiration, stimulate microbial activity, and reduce biological carbon turnover. Composting plant-based bioplastic waste helps reduce landfill waste and increase soil nutrient extraction rates, eliminating the need for other fertilizers (Islam et al. 2024). For example, bioplastics derived from maize, sugarcane, starch, cellulose, and potato starch are biodegradable and compostable, meaning they naturally decompose in soil without producing hazardous by-products. Hence, it is not essential to collect the waste of such bioplastics, since they have a positive impact on soil aeration, moisture retention, and fertility. Therefore, plant-based bioplastics offer a feasible, environmentally benign alternative for sustainable agricultural practices and environmental safety (Polunin et al. 2025). Furthermore, plant-based bioplastics, such as poly(butylene succinate), poly-3-hydroxybutyrate, and poly(hexamethylene succinate), form double-sided coating biodegradable mulch films that can be used for encapsulation and controlled release of fertilizers, which is important in accelerating the growth of plants' blossoms and roots (Bi et al. 2021).

12.4.3 Consumer Products

Bioplastics derived from maize, potato, sugarcane, sugar beet, soybean, palm, castor oil, rice husk, wheat straw, crustacean shells, algal biomass, wood chips, fruit peels, and spent grain are used for making food packaging, films, biodegradable containers, coatings, medical implants, antimicrobial films, disposable dinnerware, cosmetic containers, and so on (Kumar et al. 2025). These and other uses of plant-based bioplastics make them ideal for consumer applications due to their low carbon footprint and environmentally friendly properties (Kumar et al. 2025). Among the many plant-based bioplastics, poly(lactic acid) is well-known for its use in manufacturing bottles, personal care packaging, and food containers. Such facts demonstrate that the various applications of plant-based bioplastics encourage manufacturers to shift towards more sustainable and environmentally friendly consumer products (Jamshidian et al. 2010).

12.4.4 Other Applications

Aside from agricultural, packaging, and consumer items, plant-based bioplastics are also used in the textile, medical, and automotive industries. Polyhydroxyalkanoates and poly-3-hydroxybutyrate are among the bioplastics used in medicine, including antibacterial agents, biocidal agents, biodegradable implants, biomedical devices, drug carriers (scaffolds for drug delivery), anticancer agents, and tissue engineering, due to their biocompatibility and non-toxicity (Nanda et al. 2022). In the textile industry, plant-based polymers such as cellulose, poly(lactic acid), and starch are commonly used to manufacture sustainable fibres that increase fabric resistance to harmful UV radiation and replace traditional 3D printing and coatings for various fabrics, thereby reducing environmental impact and dependency on synthetic materials (Patti and Acierno 2022). Furthermore, in the automotive industry, bioplastics derived from plant oils and cellulose are used to replace several components traditionally made from metal and metal alloys, such as panels, seat cushions, and trims, thereby reducing vehicle weight and environmental impact (Pellicer et al. 2017).

12.5 Environmental Impact and Sustainability

The negative impacts of bioplastics on human health and environmental pollution extend not only to their use and disposal but also to their entire life cycle, from the cultivation of raw materials to the end of their use. During this process, they produce pollutants and pose health risks. It depends on the choice of feedstock, cultivation practices, processing methods, and disposal methods. For example, large-scale land use affects biodiversity through habitat destruction. Excessive use of fertilizers and pesticides during cultivation can lead to eutrophication and acidification, contributing to climate change throughout their life cycle. Additionally, improper biodegradation can lead to the fragmentation of bioplastics into micro- and nanoplastics, resulting in ecotoxicity and their entry into the food chain (Islam et al. 2024; Shand et al. 2025). Although bioplastics can harm human health and the environment, they have not yet been compared with traditional bioplastics in terms of sustaining a healthy life and a clean environment (Ferreira-Filipe et al. 2021).

12.5.1 Biodegradability and Compostability

Both biodegradable and compostable materials are types of degradable plastics that degrade through the action of naturally occurring microorganisms, such as bacteria, fungi, and algae. Biodegradation activity, primarily caused by enzyme action, leads to significant changes in the chemical structure of the plastic material (Rana 2019). The plastic material can be considered biodegradable and compostable if it has a biodegradation level of >90% compared to cellulose (positive standard) in 180 days under controlled composting conditions measured through respirometric methods, a disintegration level of >90% in 3 months, ecotoxicity: toxicity tests (germination and plant tests), and chemical testing according to product safety (Nizamuddin et al. 2024; Gadaleta et al. 2025).

12.5.2 Life Cycle Assessment of Bioplastics

A life cycle assessment (LCA) of bioplastics evaluates their environmental impact from raw material extraction to production, use, and disposal, providing a comprehensive comparison to traditional petroleum-based plastics (Ali et al. 2023). LCAs generally indicate that bioplastics produced from renewable feedstocks, such as corn, sugarcane, or cellulose, can reduce greenhouse gas emissions and fossil fuel consumption during production; however, these advantages are heavily dependent on agricultural practices, land use, fertilizer inputs, and energy sources (Daniel Posen et al. 2017). The extraction of raw materials for plant-based bioplastics has a lower environmental impact than that of traditional plastics. Their production process consumes less energy, particularly when efficient technology and renewable sources are used. Traditional plastics take more than 100 years to decompose through composting, whereas plant-based bioplastics are compostable and easily break down into organic matter under specific conditions. In the recycling process, traditional plastics can be recycled multiple times; however, they are often disposed of in landfills or incinerated, resulting in long-term environmental issues. In contrast, the recycling of bioplastics is less developed due to inadequate infrastructure and varies by type. Their decomposition in the natural environment depends on their composition and the conditions, which may require an industrial facility to fully degrade them. Thus, the environmentally benign properties of plant-based bioplastics depend on the feedstock type, farming methods, energy sources, composition, structure, functional groups, and waste management systems (Atiwesh et al. 2021) (Fig. 12.2).

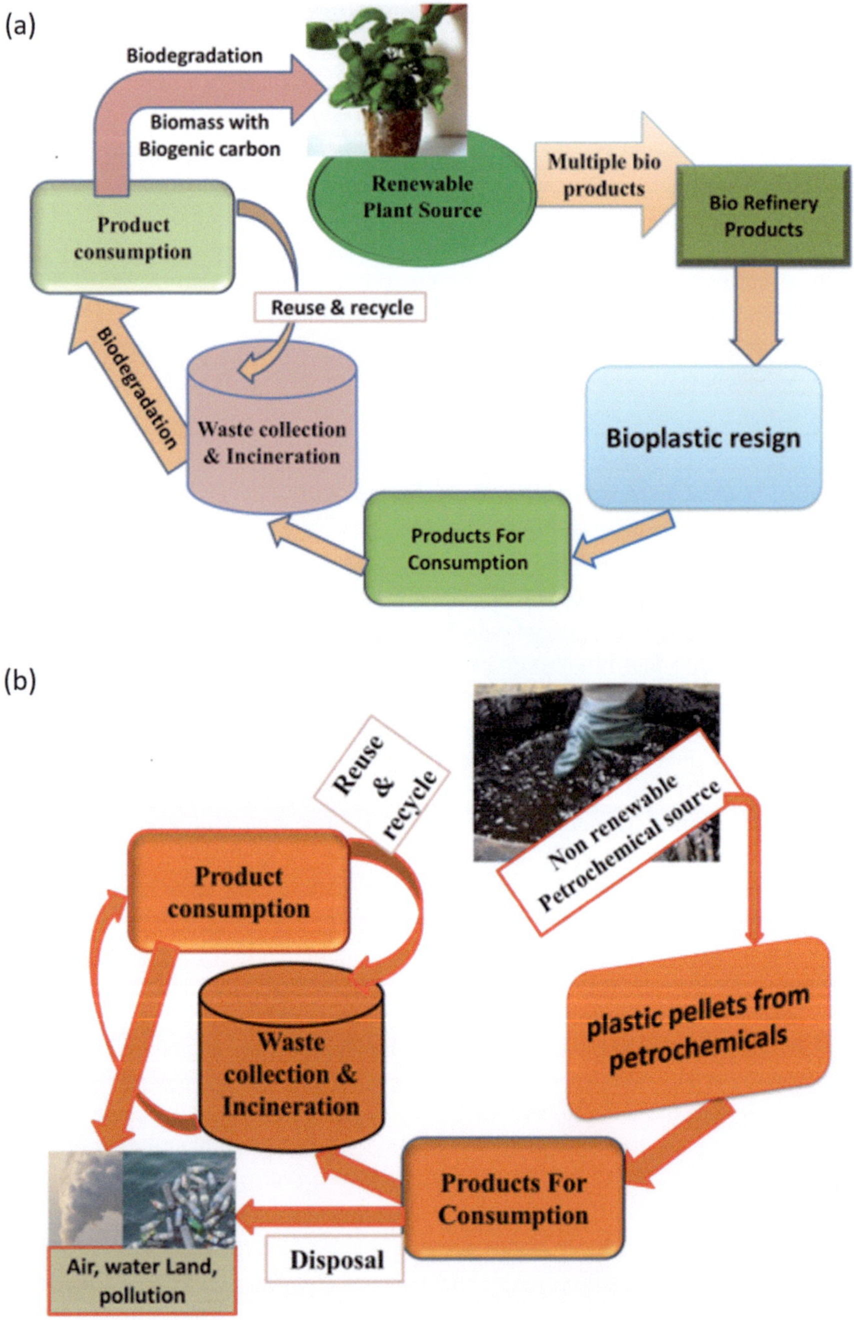

Fig. 12.2 The life cycle assessment of (**a**) plant-based bioplastics and (**b**) traditional plastics

Table 12.3 Comparison of plant-based and fossil-based plastics (Rajendran et al. 2025; Atiwesh et al. 2021)

Plant-based bioplastics	Fossil-based bioplastics
Renewable source	Non-renewable source
Rising global warming	Comparatively little effect on global warming
Relatively expensive	Cheaper
Compostable	Remain a virgin for more than 100 years
Most of them are biodegradable.	Non-biodegradable
Environmentally benign	Toxic to the environment
Composed of organic carbon (plant nutrient)	Disposed of in landfills/incinerations
Sustainable	Ceased as the sources are thoroughly mined
Most of them have improved durability and heat resistance	High durability and heat resistance
Recyclable	Widely recyclable

12.5.3 Comparison with Fossil Fuel-Based Plastics

Plant-based bioplastics (polymers, resins, and composites) are potential alternatives to traditional petroleum-based plastics, offering functional and environmentally friendly substitutes. Because traditional plastics are derived from non-renewable sources, their production ultimately ends when the raw materials are completely depleted, in addition to their toxicity and non-biodegradability. In contrast, plant-based plastics remain sustainable because the raw materials are renewable, non-toxic, and compostable, breaking down into natural nutrients (Rajendran et al. 2025) (Table 12.3).

12.6 Challenges and Limitations

12.6.1 Economic Factors

Various economic factors pose challenges to the widespread use of plant-based bioplastics. The retail price of plant-based bioplastics is commonly higher than that of traditional plastics due to higher manufacturing costs. These costs depend on chemical additives, raw material extraction, production processes, and infrastructure limitations (Arikan and Ozsoy 2015). The limited availability of resources such as chitosan, cellulose, lignin, and starch, which rely on agricultural products, can drive up market prices. In addition, improving manufacturing technology for infrastructure and processing techniques and obtaining certification for the newly developed bioplastics to ensure the sustainability of the product require additional investment, which makes the overall process challenging for the product's economic feasibility. Thus, the plant-based bioplastic becomes competitive only if there are economic incentives, efficient processing technology, and government involvement in policy changes and implementation (Negrete-Bolagay and Guerrero 2024).

12.6.2 Performance Issues

Plant-based bioplastics are emerging as a potential resource for various applications, including packaging, consumer goods, agriculture, and industries such as automotive, medical, and electronic device manufacturing. However, some challenges limit the widespread implementation of plant-based bioplastics. For example, low mechanical strength, thermal resistance, and barrier properties against moisture and gases limit their practicality in packaging, biomedical applications, and energy storage devices. In addition, processing challenges, functional limitations, incompatible recycling processes, environmental concerns, and chemical stability are factors affecting their performance and broad application (Dutta and Sit 2024). The performance issues of these bioplastics can be addressed and made sustainable using various methods, including physical, chemical, and surface treatments. For example, blending with other polymers increases interactions and compatibility within the composites, resulting in higher tensile strength, modulus, and breaking elongation. Similarly, reinforcing biopolymers with bionanofillers or fibres produces low-cost biopolymer composites that are competitive with traditional plastics (McKay et al. 2024).

12.6.3 Public Perception and Market Acceptance

Plant-based bioplastics, compared to conventional plastics, are promising for the circular economy as they reduce dependency on fossil-based plastics and their negative impact on the environment and living systems. The replacement of conventional plastics with plant-based bioplastics depends on public acceptance and market availability. Consumers often prefer these materials due to their renewable source, biodegradability, and promise to reduce environmental pollution. However, a lack of public awareness, reluctance to unfamiliar technology, inconsistent product labelling, unclear policies, misconceptions, limited knowledge, and negative attitudes towards achieving personal goals and values can act as barriers to public acceptance and market demand (Domingos et al. 2025). Consumers prefer products that offer price, quality, visual appeal, and safety over those that are environmentally benign. Hence, possible solutions to promote public awareness and market acceptance that address negative perceptions include consistent product labelling, increasing the availability of low-cost bioplastics, government incentives, and implementing relevant policies. Furthermore, producing and commercializing plant-based bioplastics requires providing consistent, low-cost products and collaborating with governments, industries, and public-private partnerships (de Jong et al. 2025).

12.7 Future Trends and Innovations

12.7.1 Advances in Bioplastic Technology

Advancements in plant-based bioplastics are attracting researchers due to their low environmental impact, renewability, reduced safety risks, and the potential for low-cost industrial-scale production. For example, improvements in the quality of cellulose-based bioplastics can be achieved by incorporating nanofillers, resulting in composites with enhanced thermal, mechanical, and barrier properties suitable for industrial and technological applications. Poly(lactic acid) with nanocellulose is a promising material for industrial applications due to its compatibility, renewability, and improved rheological, mechanical, and thermal properties. Nanocellulose can also serve as a suitable green filler for polyhydroxyalkanoates, providing enhanced properties and potential as a material replacement for traditional plastics in food packaging and medical applications (Shaghaleh et al. 2018). Advancements in processing technologies, such as microbial mixing culture, anaerobic digestion, and pretreatment methods, have led to significant cost reductions and improvements in the efficiency of bioplastic manufacturing, including the production of poly(lactic acid) and polyhydroxyalkanoates. Thus, the sustainable development of plant-based bioplastics is driven by technological advancements and improved processes, from raw material pretreatment to the production of finished products (Asim et al. 2025). Advances in biotechnology, including synthetic biology and microbial engineering, offer new opportunities for the production of cost-effective, scalable bio-based products (Anon 2024).

12.7.2 Potential for Circular Economy Integration

The need for sustainable development worldwide has led to a shift from a conventional linear economic model to a more sustainable and efficient circular economy. Bio-based innovation plays a crucial role in addressing environmental challenges, enhancing resource efficiency, and promoting a circular economy by offering sustainable alternatives to traditional fossil-based materials (Anon 2024). These innovative technologies include the production of biofuels, bioplastics, and biofertilizers through bioconversion, fermentation, and anaerobic digestion. All of these processes use agricultural and other waste to transform into valuable products, making a considerable contribution to the circular economy by minimizing waste, recovering resources, and promoting the use of renewable materials. Even though some environmental risks are associated with the cultivation of raw materials and the intrusion of micro- and nanoparticles into the ecosystem, plant-based bioplastics can still be considered one of the main contributors to the circular economy due to their safe production process, environmental friendliness, material recoverability, and recyclability, which extends the life cycle of the product (Di Bartolo and Infurna 2021).

12.8 Conclusion

Plant-based bioplastics are promising materials for replacing petroleum-based plastics, as they are derived from renewable sources and offer environmentally benign alternatives to reduce carbon emissions and pollutants. The ongoing advancements in innovation, biotechnology, and processing technologies offer new opportunities to enhance their quality, scalability, and economic viability; however, they currently face challenges with cost, mechanical performance, and widespread adoption. For instance, recent studies on reinforcing biopolymers with bionanofillers or fibres have shown that they can produce low-cost biopolymer composites that are competitive with traditional plastics. They can significantly reduce dependency on fossil-based raw materials and alleviate the environmental impacts of traditional plastics by providing eco-friendly alternatives for consumer goods, packaging, agriculture, and biomedical applications. Their sustainability potential is further increased by incorporating non-food and agricultural waste biomass and by improving their recycling and composting capabilities. Generally, producing and commercializing plant-based bioplastics requires providing consistent, low-cost products and collaborating with governments, industries, and public-private partnerships.

References

Afshar SV, Boldrin A, Astrup TF, Daugaard AE, Hartmann NB (2024) Degradation of biodegradable plastics in waste management systems and the open environment: a critical review. J Clean Prod J 434:1–24

Ahmad A, Banat F, Alsafar H, Hasan SW (2024) An overview of biodegradable poly (lactic acid) production from fermentative lactic acid for biomedical and bioplastic applications. Biomass Convers Biorefinery 14(3):3057–3076

Ali SS, Abdelkarim EA, Elsamahy T, Al-Tohamy R, Li F, Kornaros M, Zuorro A, Zhu D, Sun J (2023) Bioplastic production in terms of life cycle assessment: a state-of-the-art review. Environ Sci Ecotechnol 15:100254

Alonso-González M, Felix M, Romero A, Sergi C, Bavasso I, Sarasini F (2025) Optimization of processing conditions for rice bran-based bioplastics through extrusion and injection molding. J Polym Environ 33(1):512–527

Anon (2024) The role of bio-based innovations in circular economy: a biochemical and economic perspective. J Med Sci Biol Chem 1(1):21–27

Arikan EB, Ozsoy HD (2015) A review: investigation of bioplastics. J Civ Eng Archit 9(2):188–192

Asim Z, Muhammad H, Shahzad A, Ghodake G, Mahmoud KA, Almomani F, Rasool K (2025) Transforming agricultural food waste into bioplastics: methods, potential, and technological advances. Adv Sustain Syst 9(3):2400864

Atiwesh G, Mikhael A, Parrish CC, Banoub J, Le TAT (2021) Environmental impact of bioplastic use: a review. Heliyon 7(9):e07918

Aversa C, Barletta M, Vesco S (2021) Injection-stretch blow molding of poly (lactic acid)/polybutylene succinate blends for the manufacturing of bottles. J Appl Polym Sci 139(4):1–16

Bi S, Pan H, Barinelli V, Eriksen B, Ruiz S, Sobkowicz MJ (2021) Biodegradable polyester coated mulch paper for controlled release of fertilizer. J Clean Prod 294:126348

Boey JY, Lee CK (2022) Factors affecting mechanical properties of reinforced bioplastics: a review. Polymers 14:3737

Bouroudian E, Perez-puyana V, Guerrero A, Romero A, Jim M (2020) Evaluation of different strengthening methods in the mechanical and functional properties of soy protein-based bioplastics. J Clean Prod 262:121517

Carpita NC (2011) Update on mechanisms of plant cell wall biosynthesis: how plants make cellulose and other (1→4)-β-D-glycans. Plant Physiol. American Society of Plant Biologists 155:171–184

Chan JX, Wong JF, Hassan A, Zakaria Z (2021) Bioplastics from agricultural waste. In: Deans M (ed) Biopolymers and biocomposites from agro-waste for packaging applications, pp 141–169. https://doi.org/10.1016/B978-0-12-819953-4.00005-7

Chawla P, Bajaj I, Survase S, Singhal RS, Biological E. Limited (2009) Microbial cellulose: fermentative production and applications. Food Technol Biotechnol 47:107–124

Chen C, Xi Y, Weng Y (2022) Recent advances in cellulose-based hydrogels for tissue engineering applications. Polymers 14:1–27

Daniel Posen I, Jaramillo P, Landis AE, Griffin WM (2017) Greenhouse gas mitigation for U.S. plastics production: energy first, feedstocks later. Environ Res Lett 12(3). https://doi.org/10.1088/1748-9326/aa60a7

de Jong E, Goumans I, Visser R, Puente Á, Gruter GJ (2025) The opportunities and challenges of biobased packaging solutions. Polymers 17(16):1–33

Di Bartolo A, Infurna G (2021) A review of bioplastics and their adoption in the circular economy. Polymers 13:1–26

Domingos S, Farias AR, Possidónio C, Luís S, Loureiro A, Cruz B, Portugal E (2025) Consumer demand for bioplastics: an umbrella review of barriers and recommendations. J Risk Res:1–18. https://doi.org/10.1080/13669877.2025.2553084

Dutta D, Sit N (2024) A comprehensive review on types and properties of biopolymers as sustainable bio-based alternatives for packaging. Food Biomacromol 1(2):58–87

Dybka-Stępień K, Antolak H, Kmiotek M, Piechota D, Koziróg A (2021) Disposable food packaging and serving materials—trends and biodegradability. Polymers 13(20):3606

Felix M, Perez-Puyana V, Romero A, Guerrero A (2017) Development of protein-based bioplastics modified with different additives. J Appl Polym Sci 134(42):1–8

Felix M, Romero A, Aliotta L, Alonso-gonz M, Gigante V, Sergi C, Bavasso I, Sarasini F (2025) Innovative approaches to bioplastic development: rice bran/PLA blends via extrusion combined with injection molding and 3D printing. J Environ Manag 389:126081

Ferreira-Filipe DA, Paço A, Duarte AC, Rocha-Santos T, Patríci Silva AL (2021) Are biobased plastics green alternatives? a critical review. Int J Environ Res Public Health 18(15):7729

Folino A, Karageorgiou A, Calabrò PS, Komilis D (2020) Biodegradation of wasted bioplastics in natural and industrial environments: a review. Sustainability (Switzerland) 12(15):1–37

Frone AN, Batalu D, Chiulan I, Oprea M, Gabor AR, Nicolae C-a, Raditoiu V, Trusca R, Panaitescu DM (2020) Morpho-structural, thermal and mechanical properties of PLA/PHB/cellulose biodegradable nanocomposites obtained by compression molding, extrusion, and 3D printing. Nanomaterials 10(51):1–30

Fu H, Xu H, Liu Y, Yang Z, Kormakov S, Wu D, Sun J (2020) Overview of injection molding technology for processing polymers and their composites. ES Mater Manuf 8:3–23

Gabriel AA, Solikhah AF, Rahmawati AY (2021) Tensile strength and elongation testing for starch-based bioplastics using melt intercalation method: a review. J Phys Conf Ser 1858(1):012028

Gadaleta G, Andrade-Chapal JC, López-Ibáñez S, Mozo-Toledo M, Navarro-Calderón Á (2025) Biodegradability of bioplastics in managed and unmanaged environments: a comprehensive review. Materials 18(10):2382

George N, Debroy A, Bhat S, Singh S, Bindal S (2021) Biowaste to bioplastics: an ecofriendly approach for a sustainable future. J Appl Biotechnol Rep 8(3):221–233

Hariharan P, Shri Sathyan GS, Sathyaprakash SV, Rinchen K, Lokesh RS, Shiju A, Agilandeswari K (2025) Sustainability in packaging: from traditional packaging to biodegradable alternatives. J Packag Technol Res 9(3). https://doi.org/10.1007/s41783-025-00188-3

Islam M, Xayachak T, Haque N, Lau D, Bhuiyan M, Pramanik BK (2024) Impact of bioplastics on environment from its production to end-of-life. Process Saf Environ Prot 188:151–166
Jamshidian M, Tehrany EA, Imran M, Jacquot M, Desobry S (2010) Poly-Lactic Acid: production, applications, nanocomposites, and release studies. Compr Rev Food Sci Food Saf 9(5):552–571
Jayarathna S, Andersson M, Andersson R (2022) Recent advances in starch-based blends and composites for bioplastics applications. Polymers 14(21):1–24
Jeremic S, Milovanovic J, Mojicevic M, Skaro-bogojevic S, Nikodinovic-runic J (2020) Understanding bioplastic materials – current state and trends. J Serb Chem Soc 85(12):1507–1538
Joseph CS, Prashanth KVH, Rastogi NK, Indiramma AR, Reddy SY, Raghavarao KSMS (2011) Optimum blend of chitosan and poly-(ε-caprolactone) for fabrication of films for food packaging applications. Food Bioprocess Technol 4(7):1179–1185
Koller M, Mukherjee A (2022) Polyhydroxyalkanoates (PHAs) – production, properties, and biodegradation. In: Biodegradable polymers in the circular plastics economy. Wiley-VCH GmbH, pp 145–204. https://doi.org/10.1002/9783527827589.ch6
Kumar V, Sehgal R, Gupta R (2021) Blends and composites of polyhydroxyalkanoates (PHAs) and their applications. Eur Polym J 161:110824
Kumar K, Khatri A, Thakur IS (2025) Existing scenario and environmental significance of biodegradable plastics: a review for a sustainable future. Green Energy Environ Technol 4:1–35
Kumari S, Rao A, Kaur M, Dhania G (2023) Petroleum-based plastics versus bio-based plastics: a review. Nat Environ Pollut Technol 22(3):1111–1124
Liwarska-bizukojc E (2021) Effect of (bio) plastics on soil environment: a review. Sci Total Environ 795:1–18
Marzo-Gago C, Díaz AB, Blandino A (2023) Sugar beet pulp as raw material for the production of bioplastics. Fermentation 9(7):1–20
McKay I, Vargas J, Yang L, Felfel RM (2024) A review of natural fibres and biopolymer composites: progress, limitations, and enhancement strategies. Materials 17(19):4878
Megha M, Kamaraj M, Nithya TG, GokilaLakshmi S, Santhosh P, Balavaishnavi B (2024) Biodegradable polymers research and applications. Phys Sci Rev 9(2):949–972
Mohanty AK, Wibowo A, Misra M, Drzal LT (2003) Development of renewable resource-based cellulose acetate bioplastic: effect of process engineering on the performance of cellulosic plastics. Polym Eng Sci 43(5):1151–1161
Mori R (2023) Replacing all petroleum-based chemical products with natural biomass-based chemical products: a tutorial review. RSC Sustain 1(2):179–212
Msuya N (2018) Poly (lactic-acid) production – from monomer to polymer: a review. SF J Polym Sci 1(1):1–15
Muhammad Shamsuddin I (2017) Bioplastics as better alternative to petroplastics and their role in national sustainability: a review. Adv Biosci Bioeng 5(4):63
Nanda S, Patra BR, Patel R, Bakos J, Dalai AK (2022) Innovations in applications and prospects of bioplastics and biopolymers: a review. Environ Chem Lett 20(1):379–395
Negrete-Bolagay D, Guerrero VH (2024) Opportunities and challenges in the application of bioplastics: perspectives from formulation, processing, and performance. Polymers 16(18):1–38
Nizamuddin S, Baloch AJ, Chen C, Arif M, Mubarak NM (2024) Bio-based plastics, biodegradable plastics, and compostable plastics: biodegradation mechanism, biodegradability standards and environmental stratagem. Int Biodeterior Biodegrad 195:105887
Patti A, Acierno D (2022) Towards the sustainability of the plastic industry through biopolymers: properties and potential applications to the textiles world. Polymers 14(4):692
Pellicer E, Nikolic D, Sort J, Baró MD, Zivic F, Grujovic N, Grujic R, Pelemis S (2017) Advances in applications of industrial biomaterials. Springer, Cham, pp 1–214
Polunin Y, Domnich B, Setien KP, Voronov A (2025) Bioplastics from natural renewable polymeric resources: a review. Chem Chem Technol 19(1):91–107
Prakash P, Lee WH, Ching-Yeeloo HS, Wong J, Parumasivam T (2022) Advances in polyhydroxyalkanoate nanocarriers for effective drug delivery: an overview and challenges. Nanomaterials 12(1):1–20

Rajendran S, Al-Samydai A, Palani G, Trilaksana H, Sathish T, Giri J, Saravanan R, Isaac Joshua Ramesh Lalvani J, Nasri F (2025) Replacement of petroleum based products with plant-based materials, green and sustainable energy—a review. Eng Rep 7(4). https://doi.org/10.1002/eng2.70108

Rana KI (2019) Usage of potential micro-organisms for degradation of plastics. Open J Environ Biol 4:007–015

Ren H, Lan A, Moy B, Ali C, Lee D, Burés J, Dumanli AG (2025) Solvent-free crosslinking of Hydroxypropyl cellulose via esterification: towards green bioplastics. Carbohydr Polym 371:124479

Rujni M, Pilipovi A (2017) Challenges and opportunities of biodegradable plastics: a mini review. Waste Manag Res 35:132–140

Sabalina A, Platnieks O, Gaidukova G, Aunins A, Eiduks TV, Gaidukovs S (2025) Thermomechanical and mechanical analysis of polylactic acid/polyhydroxyalkanoate/poly(butylene succinate-co-adipate) binary and ternary blends. RSC Adv 15(1):501–512

Shafqat A, Al-Zaqri N, Tahir A, Alsalme A (2021) Synthesis and characterization of starch based bioplatics using varying plant-based ingredients, plasticizers and natural fillers. Saudi J Biol Sci 28(3):1739–1749

Shaghaleh H, Xu X, Wang S (2018) Current progress in production of biopolymeric materials based on cellulose, cellulose nanofibers, and cellulose derivatives. RSC Adv 8(2):825–842

Shahar FS, Balakrishnan TS, Sultan MTH (2025) The evolution and environmental prospects of renewable bioplastics: types, production methods, and sustainability. J Renew Mater 13(6):1071–1101

Shand H, Patra S, Mondal R, Mandal AK, Husen A (2025) Invisible contaminants: environmental distribution and biological consequences of microplastics and nanoplastics. Next Res. https://doi.org/10.1016/j.nexres.2025.101032

Shi X, Cui L, Chao X, Shuping W (2025) Next-generation bioplastics for food packaging: sustainable materials and applications. Materials 18(12):1–48

Sidek IS, Draman SFS, Abdullah SRS, Anuar N (2019) Current development on bioplastics and its future prospects: an introductory review. INWASCON Tech Mag 1:03–08

Singh P, Sharma A, Rathee A, Puri V, Chopra H, Nagpal M, Kaur M, Malik T (2024) Revolutionizing packaging: bioplastics for superior food and pharmaceutical solutions. Polym Polym Compos 32:1–15

Tummala P, Liu W, Drzal LT, Mohanty AK, Misra M (2006) Influence of plasticizers on thermal and mechanical properties and morphology of soy-based bioplastics. Ind Eng Chem Res 45(22):7491–7496

Union, European, Single-Use Plastics Directive (2022) Bioplastic remains plastic: bio-based, biodegradable and compostable plastics. Oceana Europe 2050:1–8

Verbeek CJR, Van Den Berg LE (2010) Extrusion processing and properties of protein-based thermoplastics. Macromol Mater Eng 295(1):10–21

Viana JC, Cunha AM, Simo CL (2008) Mechanical properties of poly (e-caprolactone) and poly (lactic acid) blends. J Appl Polym Sci 112:345–352

Yang J, Ching YC, Chuah CH (2019) Applications of lignocellulosic fibers and lignin in bioplastics: a review. Polymers 11(5):1–26

Yang Z, Yin G, Sun S, Ping X (2024) Medical applications and prospects of polylactic acid materials. iScience 27(12):1–18

Zarate-ramírez LS, Romero A, Bengoechea C, Partal P, Jim M (2019) Bioplastics based on wheat gluten processed by extrusion. J Clean Prod J 239:1–8

Zhao P, Zhang J, Dong Z, Huang J, Zhou H, Jianzhong F, Turng L-S (2020) Intelligent injection molding on sensing, optimization, and control. Adv Polym Technol 2020:1–22

Chapter 13
Circular Bioeconomy: Plants as Renewable Resources

Teshale Adane, Limenew Abate Worku, Marie Yayinie, Tadesse Bassie Gelaw, Sheetal Tyagi, Rakesh Kumar Bachheti, and Archana Bachheti

Abstract Climate change is affecting the whole world these days, as many greenhouse gases are being released into the atmosphere. The fundamental culprit is the economy's reliance on nonrenewable resources, such as energy and products derived from fossil fuels. Therefore, it is necessary to seek other techniques to address these difficulties. Applying the concepts of the circular bioeconomy offers a sustainable way to eliminate negative consequences by boosting the exploitation of bio-based resources. By substituting fossil-fuel-derived energy and goods with plant-based alternatives, the consequences of climate change can be mitigated throughout the product life cycle. In addition, implementing circular bioeconomy principles enables efficient management of biomass-based resources by integrating circular economy concepts into the bioeconomy framework. The key objectives of a circular bioeconomy include ensuring food security, increasing resource-efficient use, reducing greenhouse gas emissions, reducing dependence on fossil fuels, and creating employment opportunities. This research aims to examine the role of bio-based resources within the circular bioeconomy framework and explore their potential to

T. Adane
Department of Chemical Engineering, Gafat Institute of Technology, Debre Tabor University, Debre Tabor, Ethiopia

L. A. Worku (✉) · M. Yayinie · T. B. Gelaw
Department of Chemistry, College of Natural and Computational Science, Debre Tabor University, Debre Tabor, Ethiopia

S. Tyagi
Department of Chemistry, School of Basic and Applied Sciences, Shri Guru Ram Rai University, Dehradun, Uttarakhand, India

R. K. Bachheti
Centre of Molecular Medicine and Diagnostics, Saveetha Dental College & Hospitals, Saveetha Institute of Medical and Technical Sciences, Saveetha University, Chennai, Tamil Nadu, India

A. Bachheti
Department of Environment Science, Graphic Era (Deemed to be University), Dehradun, Uttarakhand, India

L. A. Worku et al. (eds.), *Plant-Based Solutions for Sustainable Agriculture and Healthcare*, Sustainability Sciences in Asia and Africa,
https://doi.org/10.1007/978-981-95-8910-4_13

enhance economic and environmental sustainability. Mainly, key areas such as the conceptual framework of circular bioeconomy, suitable technologies for plant conversion, circularity strategies in plant-based bioeconomy systems, environmental and socioeconomic impacts, policy and regulatory frameworks, and future perspectives and research directions were critically analyzed.

Keywords Bioeconomy · Climate change · Renewable resources · Plant · Biomass

13.1 Introduction

The circular bioeconomy is guided by the principles of the circular economy and focuses primarily on plant biomass as the foundation for sustainable production and consumption (Tan and Lamers 2021). Moreover, energy and other products are produced from biomass rather than fossil fuels (Tan and Lamers 2021). Similarly, the concept of a circular bioeconomy involves all energy and other materials derived from plant biomass rather than petroleum (Sadhukhan et al. 2024).

Over the last few decades, awareness of the circular bioeconomy has increased significantly due to the growing threat of climate change posed by the use of fossil fuels as energy sources (Hassegawa et al. 2022). Therefore, the world economy has begun to transform into a circular bioeconomy that emphasizes the production of biomass-based products and energy, the conservation of resources, and the recycling of waste (Lang 2022; Schipfer et al. 2024). All of these activities must be grounded in scientific understanding, creativity, and sustainable practices. Moreover, the circular bioeconomy aims to achieve both environmental and economic sustainability by transitioning from a linear to a circular model (Lang 2022). Figure 13.1 presents the concepts of a circular bioeconomy, which aims to reduce reliance on fossil fuel resources, mitigate and adapt to climate change, create jobs, ensure food security, and conserve natural resources. However, in a circular bioeconomy, addressing issues such as land use competition and the fuel vs. food debate is just as crucial for food security.

The circular bioeconomy mainly depends on plant biomass, from smaller biomass to large timber. These resources provide a variety of products and ecosystem services, such as raw materials for different industries, climate regulation, carbon sequestration, and biodiversity conservation (Salzman et al. 2018). Additionally, plants can absorb CO_2 during photosynthesis, thereby reducing atmospheric greenhouse gas concentrations (Salzman et al. 2018; Sathre and O'Connor 2010).

Numerous biomass-based products have been developed recently as sustainable replacements for nonrenewable products, including food packaging, textile fibers, composites, and chemicals (Bessaha et al. 2024; Hassegawa et al. 2022). Energy derived from biomass is economically and environmentally sustainable (Hsiao and Hu 2024). For instance, using biomass-based energy sources is a significant factor

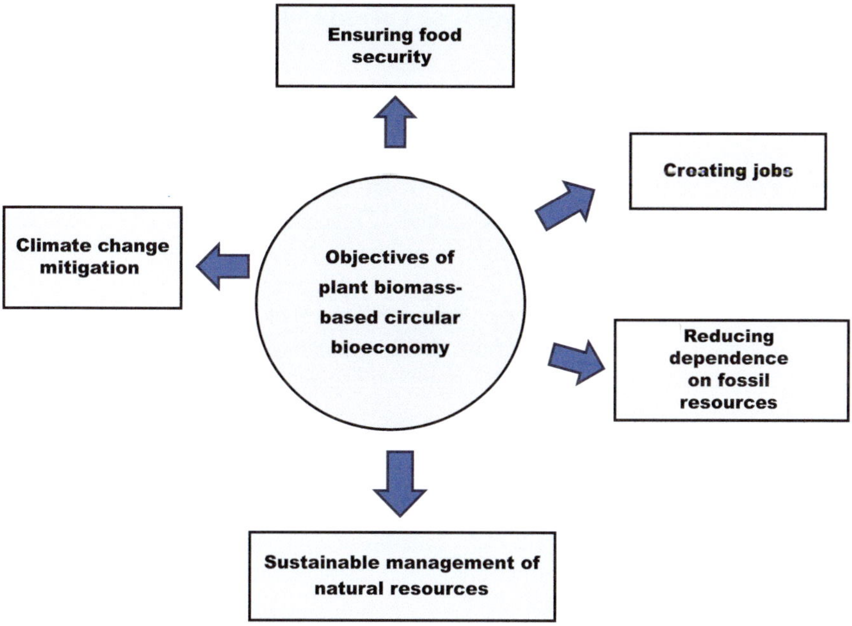

Fig. 13.1 Schematic diagram of the objectives of a plant biomass-based circular bioeconomy

in achieving the global net-zero carbon target by 2050 (Ruiz-Gonzalez et al. 2023). Plants absorb CO_2 during photosynthesis to balance the CO_2 released during combustion. Plant biomass-based circular bioeconomy is typically regarded as carbon neutral (Ruiz-Gonzalez et al. 2023; Siddiqi et al. 2021). As a result, the plant biomass-based circular bioeconomy is independent of fossil-based resources (Basso et al. 2021; Gray 2025; Liu and Song 2022).

Regarding job creation, plant biomass-based circular bioeconomy plays a significant role in strengthening the agriculture, forestry, and biotechnology sectors. This system can promote rural development and support the shift to a more resilient and sustainable global economy by enhancing the use of biomass resources and innovations (Gawel et al. 2019).

Despite the growing popularity of a plant biomass-based circular bioeconomy, there is still a lack of comprehensive review studies. To address this gap, this review examines the critical role of plant biomass in advancing the circular bioeconomy. Specifically, it provides an in-depth analysis of the conceptual framework of a circular bioeconomy; plant biomass as a renewable resource; technologies for converting plant biomass; applications of plant-based renewable resources; circularity strategies in a plant-based bioeconomy; environmental and socioeconomic impacts; policy and regulatory frameworks; and future perspectives and research directions.

13.2 Conceptual Framework of Circular Bioeconomy

In recent years, policymakers and scientists have increasingly focused on promoting economic and environmental sustainability. One of the promotion systems is utilizing circular bioeconomy, which is based on plant biomass resources (Kardung et al. 2021). For the sustainable development of the economy and the environment, a circular bioeconomy framework is crucial (Casau et al. 2022). The conceptual framework of the circular bioeconomy emphasizes reduction, reuse, and recycling (3Rs) (Vural Gursel et al. 2022). This conceptual framework focuses on a closed-loop system that aims to achieve zero waste (Naidoo et al. 2021). Figure 13.2 represents the conceptual framework of the circular bioeconomy. Product optimization, resource-efficient utilization, and environmental sustainability are pillars of the circular bioeconomy framework.

A circular bioeconomy is a conceptual framework in which all products and resources derived from biomass are used, recovered, and regenerated within a closed-loop system (Hsiao and Hu 2024; Tardy et al. 2023). In this system, materials derived from biomass remain recyclable throughout their lifecycles. For example, biogas can be produced from agricultural residues and considered a renewable energy source, while the slurry can be used as a soil amendment to improve soil fertility.

The bioeconomy model can generally be categorized into linear and circular forms, both of which involve the utilization of biomass resources (Didenko et al. 2018). As shown in Fig. 13.2, the linear bioeconomy model follows a traditional production-consumption process that does not incorporate a waste-recycling

Fig. 13.2 The conceptual framework of the linear bioeconomy model

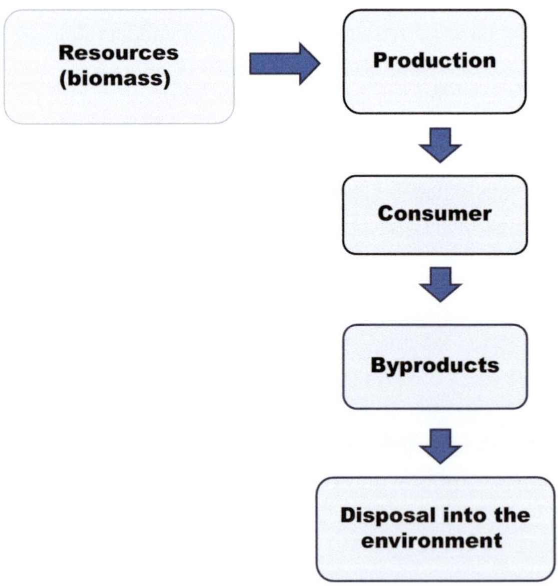

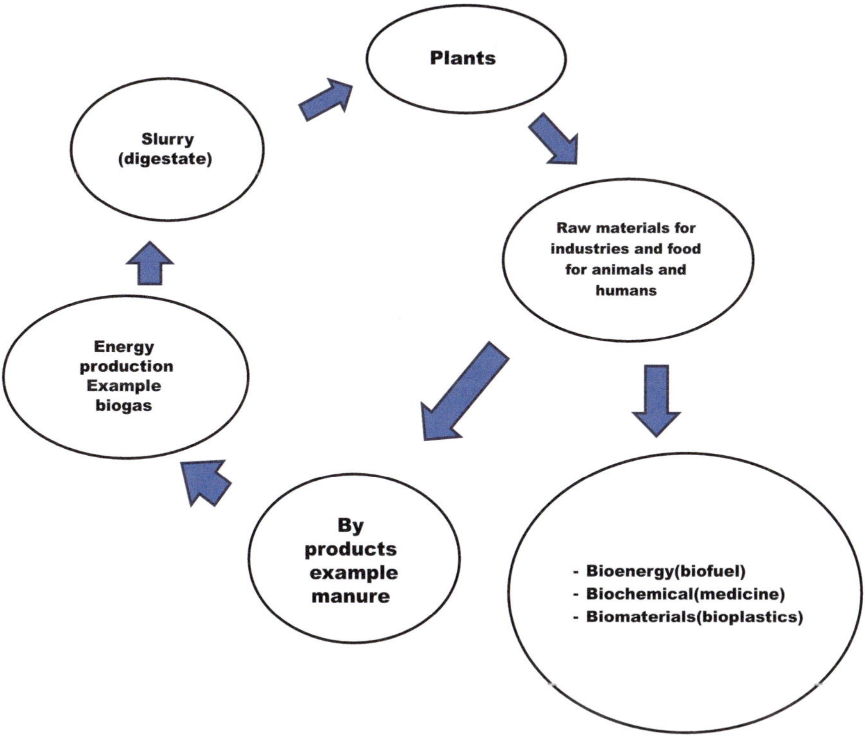

Fig. 13.3 The conceptual framework of the circular bioeconomy model

system. Consequently, this model contributes to excessive greenhouse gas emissions, resource depletion, and environmental degradation.

In contrast, the circular bioeconomy model emphasizes a closed-loop practice system in which resources are used, recovered, and recycled throughout the value chain (Fig. 13.3) (Bai et al. 2019). Additionally, the circular bioeconomy model prioritizes waste reduction, resource conservation, and advancing environmental sustainability. In general, a circular bioeconomy model is more advanced than a linear one.

13.3 Plant Biomass as a Renewable Resource

Plant biomass is considered a renewable resource because it originates from living organisms that can be regenerated through natural photosynthesis. We rely on plant biomass daily for the production of energy, food, chemicals, and various other materials. Owing to its continuous regeneration, plant biomass is widely recognized as a renewable resource (Perea-Moreno et al. 2019; Zardzewiały et al. 2023). Plant biomass can be categorized into different groups, including woody, herbaceous, agricultural residues, and energy crops (Fenta 2024; Popp et al. 2021). Woody biomass is often used for timber production, agricultural residues are suitable for producing biogas, and energy crops such as corn are commonly used for biofuel production.

Plants have unique physical and chemical characteristics that influence their potential use in material and energy production (Popp et al. 2021; Tolessa 2023). The physical characteristics of plants include bulk density, mechanical strength, and moisture content (Wu et al. 2011). Their chemical characteristics include cellulose, hemicellulose, lignin, ash, starch, and protein (Zong et al. 2020).

Cellulose is a polysaccharide composed of D-glucose units connected by β-glucoside bonds (Mazumder and Zhang 2023; Worku et al. 2024). Cellulose provides rigidity and strength to the cell wall (Mazumder and Zhang 2023). Additionally, it contributes to the structural strength of plant tissues due to its long, unbranched chains, which form microfibrils (Khodayari et al. 2021; Worku et al. 2023). $(C_6H_{10}O_5)_n$ is the molecular formula of cellulose, where n refers to the degree of polymerization, which ranges from 1000 to 10,000.

Another essential component of plants is hemicellulose. Hemicellulose is composed of 5-carbon sugars and 6-carbon sugars (Ahmed et al. 2023; Periyasamy et al. 2022). Compared with cellulose, it has a lower degree of order and a lower molecular weight.

Lignin is a three-dimensional aromatic polymer that forms a complex network within the plant cell wall (Periyasamy et al. 2022). It creates an irregular matrix interwoven with hemicellulose and cellulose, offering additional structural support to the plant. However, due to its complex and heterogeneous nature, it is difficult to break down, thereby posing significant problems for various industrial applications (Periyasamy et al. 2022; Abate et al. 2023).

Plant biomass also contains various elemental components, including carbon, hydrogen, nitrogen, and oxygen, as well as trace amounts of sulfur, potassium, calcium, magnesium, and phosphorus. Among these elements, approximately 45% of plant biomass is carbon (Esteves et al. 2023), and about 6% is hydrogen (Rosa and Mazzotti 2022). Nitrogen typically ranges from 0.2% to 1% (Maltsev et al. 2023). As a result, plants with lower nitrogen content generally release less NO_x during combustion. Similarly, most plants contain less than 0.2% sulfur, although some species may contain 0.5–0.7% (Olivo et al. 2024). Chloride is also present in plants (Han et al. 2023; Soltanloo et al. 2022).

13.4 Technologies for Plant Biomass Conversion

Plant biomass conversion technologies involve converting biomass into energy and other products. Physical, biological, chemical, and thermochemical technologies have been developed for the conversion of biomass (Ahmed et al. 2021; Velvizhi et al. 2023; Vickram et al. 2023). The different plant biomass conversion technologies are shown in Fig. 13.4. The type of biomass, its moisture content, and the

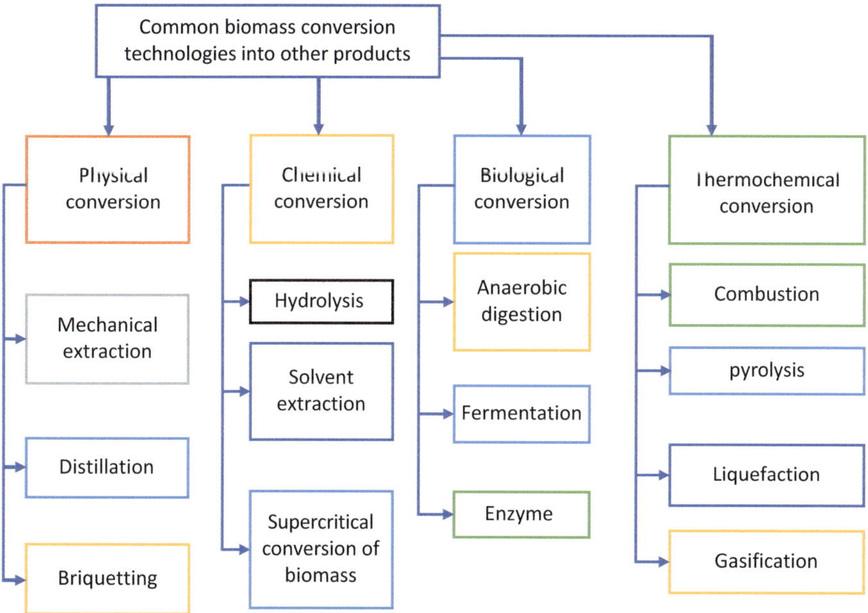

Fig. 13.4 Common biomass conversion technologies

intended final products are among the variables that determine which technology is best (Garba and Garba 2020).

Physical conversion technology can be categorized into three types: briquetting, distillation, and mechanical extraction (Ahmed et al. 2021). In mechanical extraction, valuable products such as oils, fibers, and juices are obtained from plant biomass by applying mechanical force. Grinding and pressing are also physical methods used to extract essential products from biomass, such as oilseed extraction. Distillation, which relies on the difference in boiling point, is another a physical conversion method used to separate valuable components from plant materials. Briquetting is also physical conversion technique where biomass is compressed into dense, solid fuel forms.

Biomass is converted into either a liquid or gas form through biological conversion technologies, such as enzymatic and microbial processes (Paul et al. 2024). This technology includes anaerobic digestion, fermentation, and composting (Kataya et al. 2023).

In the anaerobic digestion process, biomass is converted into valuable products through microbial activities in the absence of oxygen. This process consists of four main phases: hydrolysis, fermentation, acetogenesis, and methanogenesis (Kumar and Pandey 2019; Sangeetha et al. 2020). During anaerobic digestion, approximately 40–65% of the resulting biogas is methane, while about 35–55% consists of other gases such as CO_2, hydrogen, and hydrogen sulfide (Sangeetha et al. 2020).

The remaining slurry or digestate is considered a byproduct, but it is commonly used as a soil amendment due to its nutrient-rich properties (Ghosh et al. 2020).

In the fermentation process, biomass is converted into alcohols and other valuable products through the action of bacteria and yeast, either in the presence or absence of oxygen (Patra et al. 2022). Through fermentation, biomass can be transformed into products such as CO_2, hydrogen, ethanol, butanol, iso-butanol, lipids, and various organic acids. However, materials that contain cellulose, hemicellulose, and lignin should first be decomposed into a simpler form, as it can be challenging during the fermentation process (Patra et al. 2022). Thus, the pretreatment process is essential to enhance the fermentation process (Chopra et al. 2024).

Biomass is transformed into simple humus substrates through microbial activity, either in the presence of oxygen (aerobic environment) or in its absence (anaerobic environment), a process collectively referred to as composting (Zhan et al. 2025). Humus substrate is a crucial, nutrient-rich material commonly used to improve soil fertility. Moreover, it is an essential technique to manage organic solid waste and is utilized to treat about 9% of organic solid wastes in worldwide waste management practices (Mazzà et al. 2020).

Biomass is converted into other valuable products through chemical reactions or catalysis, a process collectively known as chemical conversion technology. This technology includes hydrolysis, solvent extraction, and supercritical conversion (Martins-Vieira et al. 2023). In these processes, materials such as lignin, hemicellulose, and cellulose are decomposed into simpler forms, which can then be used to produce biofuels, biochemicals, and other products.

Thermochemical conversion is another crucial method in which biomass is converted into energy and other chemicals through the utilization of heat. This technology offers faster reaction times and the ability to handle a variety of biomass types, making it more efficient in several applications compared with the biochemical conversion technique (Casau et al. 2022). Combustion, torrefaction, pyrolysis, liquefaction, and gasification are commonly used as thermochemical conversion methods (Pang 2019).

Biomass is directly converted into heat through combustion (Lohri et al. 2017; Zhang et al. 2010). Densified biomass combustion, boiler combustion, and stove combustion are commonly used (Khan et al. 2009). Biomass is converted into charcoal, biochar, bio-oil, and syngas through a limited-oxygen environment referred to as the pyrolysis process (Uddin et al. 2018). Similarly, in the torrefaction process, biomass is transformed into valuable products under conditions of limited oxygen (Mamvura and Danha 2020; Nunes 2020).

Biomass is converted into a liquid bio-crude through the utilization of high pressure and temperature in a water condition, a process collectively referred to as the liquefaction method (Gollakota et al. 2018). Liquefaction has a high conversion performance and is particularly useful for wet feedstocks (Gollakota et al. 2018).

In the gasification process, biomass is transformed into gas mixture that includes CO_2, light hydrocarbons, carbon monoxide, water vapor, and a small quantity of char, ash, and tar (Gu et al. 2019). Widely applied gasifier equipment includes fixed-bed and fluidized-bed designs (Lourinho et al. 2023). The gasification method operates more effectively at higher temperatures (Yang et al. 2021).

13.5 Applications of Plant Biomass-Based Renewable Resources

Plant biomass is considered a renewable resource and serves as a significant source of food, energy, and raw materials. It also plays an important role in supporting environmental sustainability. For instance, various cereal crops, vegetables, agricultural residues, and algae are widely used as sources of food and bioenergy (Rajendran et al. 2025). Figure 13.5 represents the different applications of plant biomass resources.

Plant biomass is widely used as a source of bioelectricity and biofuel. Bioelectricity is derived from biomass through combustion, gasification, and microbial fuel cells. Biofuels, including bioethanol, biodiesel, and bio-oil, are derived from plant biomass through a fermentation process. These fuels are widely used for transportation purposes.

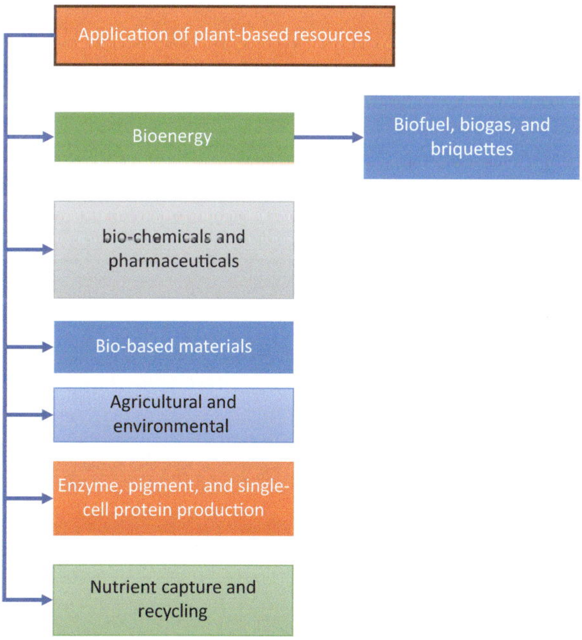

Fig. 13.5 Application of plant-based resources

Textile fibers, composites, and bioplastics are bio-based materials that serve as feedstock for several industries. For instance, textile fibers and composites are commonly derived from cotton, flax, hemp, jute, and bamboo trees. Sugarcane and cornstarch are frequently used to produce bioplastics. Biochemicals, such as drugs, can also be obtained from plant biomass. In general, plant biomass is better suited for producing these materials due to its renewability and environmental friendliness compared to petroleum-based resources.

Plant biomass plays a significant role in improving agricultural productivity and conserving biodiversity. It enhances agricultural productivity by improving soil structure, increasing micronutrient and macronutrient levels, and boosting water retention capacity. These factors lead to greater soil fertility and reduce dependence on synthetic fertilizers. Additionally, plant biomass can absorb CO_2, thereby lowering greenhouse gas emissions and helping to mitigate the effects of climate change.

Table 13.1 provides an overview of the application of plant biomass in various sectors, including agriculture, environmental management, pharmaceuticals, bioenergy, bio-based materials, and biochemicals.

13.6 Circularity Strategies in Plant Biomass-Based Bioeconomy

Circularity strategies in the bioeconomy involve designing production, consumption, and recycling systems that apply closed-loop approaches (Khanna et al. 2024). These strategies aim to eliminate waste and promote the efficient use of biomass resources (Tsakalova et al. 2024). Figure 13.6 presents circularity strategies within the plant biomass resources.

In the circular bioeconomy, circularity strategies prioritize converting agricultural byproducts into valuable resources rather than treating them as waste. Plant biomass residues, in particular, are increasingly utilized for producing value-added products such as biochar, animal feed, bioplastics, textile fibers, and compost (Lin et al. 2014; Ma and Liu 2019; Salazar Sandoval et al. 2024; Zhu et al. 2020).

Agricultural residues can be transformed into bioenergy, such as biogas, thereby reducing reliance on petroleum-based resources (Arias et al. 2021). Moreover, this strategy enhances resource efficiency and contributes to mitigating the impacts of climate change (Salazar Sandoval et al. 2024).

Other circularity strategies within the circular bioeconomy emphasize converting biomass-based byproducts into new products at the end of the life cycle, thereby preventing their accumulation in landfills (Dallaev et al. 2025). In addition, these strategies apply the 3Rs: reduce, reuse, and recycle to enhance both economic and environmental sustainability (Mestre and Cooper 2017). Another key circularity strategy in the biomass-based bioeconomy is the integration of actors across the value chain, such as linking agriculture systems with industry sectors (Murele et al. 2020). Such integration supports the valorization of waste throughout the entire system (Murele et al. 2020).

Table 13.1 Application of plant biomass-based resources

Type plants	Products	Applications	References
Madder	Natural dyes	Leather dying	Mandal and Venkatramani (2023)
Enset	Natural fibers	For food packaging	Punia Bangar et al. (2023)
Onion (*Allium cepa*)		Nonload-bearing	Goyal et al. (2017)
Arjun bark (Terminalia arjuna)	Natural dyes	Cotton dyeing	Habib et al. (2021)
Hemp, kenaf, and bamboo	Natural fiber	Cement concrete composites	Abdalla et al. (2022)
Olives, fruits, grapes, vegetables, spices, herbs, and algae	Phenolic molecules	Antimicrobial and antioxidant	Beya et al. (2021)
Moringa oleifera	Natural coagulant	Treatment of pharmaceutical industry wastewater	Singh et al. (2025)
Corn	Corn fiber	Packaging	Zubair et al. (2022)
Cactus	Cactus juice	Wastewater remediation	Maache et al. (2017)
Mango	Mango seeds	Wastewater remediation	Duque-Acevedo et al. (2020)
Oat	Oat plant roots	Construction, packaging	Zhou et al. (2021)
Orange	Orange leaf	Packaging	Ouakarrouch et al. (2020)
Pineapple	Pineapple leaf	Construction	Le et al. (2020)
Rice	Rice husk	Construction	Busquets-Ferrer et al. (2021)
Sanseviera	*Sansevieria* fiber	Construction	Dénes et al. (2019)
Thespesia Populnea	*Thespesia Populnea* Fiber	Packaging	Vinod et al. (2019)
Leucas Aspera	*Leucas Aspera* fiber	Textile	Zhou et al. (2021)
Pea	Pea seed	Packaging	Mehyar et al. (2012)
Cocos	Cocos Nucifera fiber	Construction	Yaradoddi et al. (2022)

13.7 Environmental and Socioeconomic Impacts

The circular bioeconomy has positive environmental impacts, including reduced greenhouse gas emissions, decreased dependence on petroleum resources, enhanced biodiversity, improved nutrient cycling, and better soil and water retention. These factors play a significant role in both environmental and socioeconomic impacts (Ferreira et al. 2022). Biomass-based resources are renewable, biodegradable, and carbon-sequestering (Su et al. 2024). Moreover, a biomass-based circular

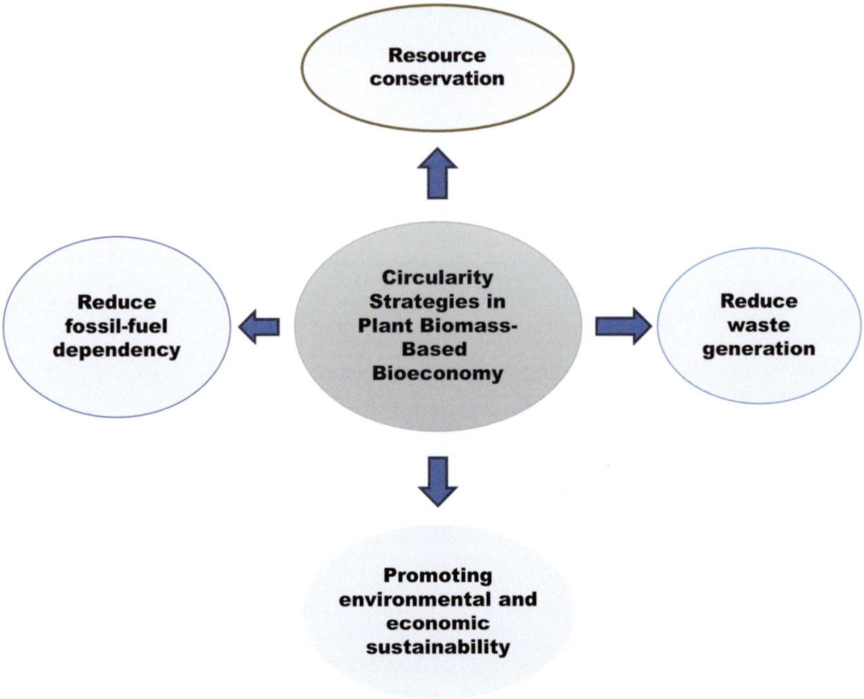

Fig. 13.6 Conceptual framework of circularity strategies in the circular bioeconomy

bioeconomy can significantly reduce greenhouse gas emissions and the overall carbon footprint (Imran 2025).

Additionally, a circular bioeconomy centered on plants can help maintain biodiversity and ecosystem services (Lemoine et al. 2024). These methods seek to preserve the vital ecological processes that sustain the life cycle, restore ecosystems, and conserve biodiversity (Paleari 2024).

Significant job opportunities could be created by the biomass-based economy, especially in rural areas (Gatto and Re 2021). By encouraging value addition and expanding revenue streams, it can also act as a center for green innovation (Gatto and Re 2021).

A circular bioeconomy based on plant biomass, on the other hand, has several drawbacks, such as competition for land use between food and fuel, and difficulties with social acceptance (de Sousa et al. 2024). A successful transformation depends on addressing these ecological and social trade-offs. Circular bioeconomy strategies should prioritize community involvement, land equity, and food sovereignty (de Sousa et al. 2024).

13.8 Policy and Regulatory Frameworks

Sustaining a circular bioeconomy requires the establishment of efficient legislative and regulatory frameworks (Brandão and Santos 2024). These frameworks include the rules, incentives, and safeguards required to promote innovation, protect natural resources, and guarantee equitable benefit distribution (Brandão and Santos 2024).

Climate change, biodiversity loss, and ongoing sustainability challenges have posed a threat to the world in recent years (Dietz et al. 2023). A circular bioeconomy based on plant biomass has been supported and sustained by a number of actors, including national governments international organizations and legislators (Dietz et al. 2023).

To increase the adoption of biomass-based resources within a circular bioeconomy, governments and international organizations have implemented incentives and subsidies (Vasileiadis et al. 2025). These organizations provide financial support to plant-based innovators and the agricultural sector to promote circular bioeconomy initiatives (Gottinger et al. 2020). By boosting market demand and investment in green technologies, such initiatives improve the environment to support a sustainable economic shift. A crucial component of maintaining the circular economy based on plant biomass is preparing certification for investors (Ramirez-Contreras and Faaij 2018).

It is crucial to ensure that the production and use of plant-based biomass do not harm ecosystems, uproot communities, or jeopardize food security as demand for it rises (Ramirez-Contreras and Faaij 2018). In this sense, certification programs and sustainability standards are essential. They ensure that plant-based materials are sourced ethically and support climate mitigation and circular economy goals by establishing standards for the production, harvesting, processing, and trading of biomass (Ramirez-Contreras and Faaij 2018).

Finally public-private partnerships and inclusive stakeholder engagement are critical to the development of the circular bioeconomy. The transition to a bio-based economy based on renewable plant resources requires cooperation across multiple sectors. Public-private partnerships help a range of actors, such as government agencies, private companies, research institutions farmers, and civil society organizations mobilize resources, share technical and financial expertise, and align goals (Szarka et al. 2023). Active stakeholder engagement encourages transparency, innovation, and a persistent commitment to sustainable goals (Leal Filho et al. 2025).

13.9 Future Perspectives and Research Directions

Scientists and policymakers should prioritize strategies that convert nonrenewable resources into renewable alternatives, with a particular emphasis on biomass-based resources. Future research should focus on developing and putting into practice innovative effective and environmentally friendly biomass conversion technologies.

Additionally, research projects should adhere to zero-waste principles ensuring that effluents or byproducts from one industry can be used as inputs for another creating a circular and resource-efficient system. Future research on feedstock diversification and sustainability assessment must look at a range of resources, including food waste, forestry byproducts, algae, and agricultural residues. Techno-economic analysis and lifecycle assessment are also necessary to evaluate the social, environmental, and financial sustainability of these resources and the supply chains that support them. The development of bio-based materials and advancements in green production is another important area of study for advancing sustainable development. Bio-based materials such as biopolymers nanocellulose composites, and bio-based packaging materials have the potential to significantly replace fossil-derived products (Kiran et al. 2024).

The world is currently undergoing a dramatic transition away from conventional systems and toward digital technologies such as artificial intelligence (AI) and the Internet of Things (IoT). These technologies need to be integrated for the circular bioeconomy to progress more swiftly. For instance, tasks that previously required human intelligence are now completed by AI (Bastos et al. 2024). In the plant biomass sector remote sensing data can be fed into machine-learning algorithms to support better planning strategies and predict plant biomass productivity (Javaid et al. 2022). Furthermore, IoT technologies offer significant advantages, such as enhanced connectivity and instantaneous information exchange (Dolgui and Ivanov 2022). Thus, the convergence of AI and IoT offers enormous potential for the development and optimization of the circular bioeconomy.

The study will focus on strategies that plant biomass-based systems enable for climate change adaptation and mitigation. It will specifically look into integrated strategies, such as converting biomass into biochar and using that biochar for long-term carbon sequestration within circular bioeconomy frameworks in order to achieve carbon-neutral goals. According to previous research (Lynch 2016; Trujillo-Cayado et al. 2025), the study will give priority to the development of new technologies, the sustainable use of materials derived from biomass, and more general environmental sustainability conditions. Furthermore, just as crucial to the widespread acceptance of the circular bioeconomy concept as technological advancement will be the creation of innovative business models and implementation strategies that enable large-scale practical deployment at the global level (Donner and de Vries 2021).

In the context of future circular bioeconomy development implementing green chemistry principles is essential for reducing chemical hazards, protecting biodiversity, and fostering broader environmental sustainability (Chen et al. 2020). These concepts have numerous advantages, including reduced waste, enhanced energy and resource efficiency and enhanced safety (Chen et al. 2020). Furthermore, the circular bioeconomy is strengthened by the application of green chemistry frameworks and the sustainable resource utilization principle, which facilitate the production of new goods from agricultural byproducts (Sah et al. 2025).

To sustain the circular bioeconomy future studies must prioritize cross-sectoral collaboration. In actuality, the fields of agriculture, energy, finance, technology, and policy must work closely together to create a circular bioeconomy based on plant biomass. This type of collaboration accelerates innovation, reduces risk, and promotes long-term sustainability by strengthening the integration of infrastructure resources and knowledge (Szarka et al. 2023).

13.10 Conclusions

Scientists and policymakers have focused more on the transition from nonrenewable to renewable energy sources, particularly those derived from plant biomass, over the past few decades as part of the development of a circular bioeconomy. This shift is primarily due to the world's reliance on finite resources, such as fossil fuels. An environmentally friendly approach that can reduce environmental harm while replacing materials derived from fossil fuels is a circular bioeconomy based on plant biomass. This review evaluates how plant biomass-based resources might enhance economic resilience and environmental sustainability and looks at how they fit into the circular bioeconomy framework. Some of the main thematic areas include the circular bioeconomy's conceptual foundations, technologies for converting plant biomass into high-value products, ways to achieve circularity within biomass-based systems, effects on society and the environment, relevant legal and regulatory frameworks, and possible directions for future research. When all is said and done, biomass-based resources in a circular bioeconomy provide a practical and reliable path to long-term economic and environmental sustainability.

References

Abate L, Tadesse MG, Bachheti A, Husen A, Bachheti RK (2023) Medicinal plant-based lignin and its role in nanoparticles synthesis and applications. In: Secondary metabolites from medicinal plants. CRC Press, Boca Raton, pp 87–110

Abdalla JA, Thomas BS, Hawileh RA (2022) Use of hemp, kenaf and bamboo natural fiber in cement-based concrete. Mater Today Proc 65:2070–2072. https://doi.org/10.1016/J.MATPR.2022.06.428

Ahmed I, Zia MA, Afzal H, Ahmed S, Ahmad M, Akram Z, Sher F, Iqbal HMN (2021) Socioeconomic and environmental impacts of biomass valorisation: a strategic drive for sustainable bioeconomy. Sustainability (Switzerland) 13(8):1–32. https://doi.org/10.3390/su13084200

Ahmed I, Asgher M, Nazish N, Hussain SM, Bilal M, Parra-Saldívar R, Iqbal HMN (2023) Biotransformation of lignocellulosic-based biomass waste into value-added energy products. In: Handbook of environmental chemistry, vol 131, pp 437–451. https://doi.org/10.1007/698_2022_963

Arias A, González-Rodríguez S, Vetroni Barros M, Salvador R, de Francisco AC, Moro Piekarski C, Moreira MT (2021) Recent developments in bio-based adhesives from renewable natural resources. J Clean Prod 314:127892. https://doi.org/10.1016/J.JCLEPRO.2021.127892

Bai Y, Ochuodho TO, Yang J (2019) Impact of land use and climate change on water-related ecosystem services in Kentucky, USA. Ecol Indic 102:51–64. https://doi.org/10.1016/j.ecolind.2019.01.079

Basso TP, Basso TO, Basso LC (2021) Biotechnological applications of biomass. https://books.google.com/books/about/Biotechnological_Applications_of_Biomass.html?id=-TpbEAAAQBAJ

Bastos T, Teixeira LC, Nunes LJR (2024) Forest 4.0: technologies and digitalization to create the residual biomass supply chain of the future. J Clean Prod 467:143041. https://doi.org/10.1016/J.JCLEPRO.2024.143041

Bessaha F, Bessaha G, Benhouria A, Benalioua B, Bendahma F, Boucif F, Mahrez N, Ziane S, Çoruh A, Khelifa A (2024) Highly efficient batch adsorption of anionic dye in wastewater using nanocomposite: experimental and theoretical studies. Desalin Water Treat 317:100292. https://doi.org/10.1016/J.DWT.2024.100292

Beya MM, Netzel ME, Sultanbawa Y, Smyth H, Hoffman LC (2021) Plant-based phenolic molecules as natural preservatives in comminuted meats: a review. Antioxidants 10(2):263. https://doi.org/10.3390/ANTIOX10020263

Brandão AS, Santos JMRCA (2024) Sustainability from policy to practice: assessing the impact of European research and innovation frameworks on circular bioeconomy. Sustainability (Switzerland) 16(6):2355. https://doi.org/10.3390/SU16062355/S1

Busquets-Ferrer M, Czabany I, Vay O, Gindl-Altmutter W, Hansmann C (2021) Alkali-extracted tree bark for efficient bio-based thermal insulation. Constr Build Mater 271:121577. https://doi.org/10.1016/J.CONBUILDMAT.2020.121577

Casau M, Dias MF, Matias JCO, Nunes LJR (2022) Residual biomass: a comprehensive review on the importance, uses and potential in a circular bioeconomy approach. Resources 11(4):1–16. https://doi.org/10.3390/resources11040035

Chen TL, Kim H, Pan SY, Tseng PC, Lin YP, Chiang PC (2020) Implementation of green chemistry principles in circular economy system towards sustainable development goals: challenges and perspectives. Sci Total Environ 716:136998. https://doi.org/10.1016/J.SCITOTENV.2020.136998

Chopra J, Rangarajan V, Rathnasamy S, Dey P (2024) Life cycle assessment as a key decision tool for emerging pretreatment technologies of biomass-to-biofuel: unveiling challenges, advances, and future potential. Bioenergy Res 17(2):857–876. https://doi.org/10.1007/S12155-024-10741-8/METRICS

Dallaev R, Papež N, Allaham MM, Holcman V (2025) Biodegradable polymers: properties, applications, and environmental impact. Polymers 17(14):1981. https://doi.org/10.3390/POLYM17141981

de Sousa R, Bragança L, da Silva MV, Oliveira RS (2024) Challenges and solutions for sustainable food systems: the potential of home hydroponics. Sustainability 16(2):817. https://doi.org/10.3390/SU16020817

Dénes O, Florea I, Manea DL (2019) Utilization of sheep wool as a building material. Procedia Manuf 32:236–241. https://doi.org/10.1016/J.PROMFG.2019.02.208

Didenko NI, Klochkov YS, Skripnuk DF (2018) Ecological criteria for comparing linear and circular economies. Resources 7(3):48. https://doi.org/10.3390/RESOURCES7030048

Dietz T, Rubio Jovel K, Deciancio M, Boldt C, Börner J (2023) Towards effective national and international governance for a sustainable bioeconomy: a global expert perspective. EFB Bioecon J 3:100058. https://doi.org/10.1016/J.BIOECO.2023.100058

Dolgui A, Ivanov D (2022) 5G in digital supply chain and operations management: fostering flexibility, end-to-end connectivity and real-time visibility through internet-of-everything. Int J Prod Res 60(2):442–451. https://doi.org/10.1080/00207543.2021.2002969;SUBPAGE:STRING:ABSTRACT;ISSUE:ISSUE:DOI

Donner M, de Vries H (2021) How to innovate business models for a circular bio-economy? Bus Strateg Environ 30(4):1932–1947. https://doi.org/10.1002/BSE.2725;REQUESTEDJOURNAL:JOURNAL:10990836;WGROUP:STRING:PUBLICATION

Duque-Acevedo M, Belmonte-Ureña LJ, Cortés-García FJ, Camacho-Ferre F (2020) Agricultural waste: review of the evolution, approaches and perspectives on alternative uses. Glob Ecol Conserv 22:e00902. https://doi.org/10.1016/J.GECCO.2020.E00902

Esteves B, Sen U, Pereira H (2023) Influence of chemical composition on heating value of biomass: a review and bibliometric analysis. Energies 16(10):4226. https://doi.org/10.3390/EN16104226

Fenta AA (2024) State of the art of biochar in Ethiopia. A review. Heliyon 10(3):e24934. https://doi.org/10.1016/j.heliyon.2024.e24934

Ferreira V, Fabregat-Aibar L, Pié L, Terceño A (2022) Research trends and hotspots in bioeconomy impact analysis: a study of economic, social and environmental impacts. Environ Impact Assess Rev 96. https://doi.org/10.1016/j.eiar.2022.106842

Garba A, Garba A (2020) Biomass conversion technologies for bioenergy generation: an introduction. In: Biotechnological applications of biomass. https://doi.org/10.5772/INTECHOPEN.93669

Gatto F, Re I (2021) Circular bioeconomy business models to overcome the valley of death. A systematic statistical analysis of studies and projects in emerging bio-based technologies and trends linked to the SME instrument support. Sustainability 13(4):1899. https://doi.org/10.3390/SU13041899

Gawel E, Pannicke N, Hagemann N (2019) A path transition towards a bioeconomy—the crucial role of sustainability. Sustainability 11(11):3005. https://doi.org/10.3390/SU11113005

Ghosh P, Sengupta S, Singh L, Sahay A (2020) Life cycle assessment of waste-to-bioenergy processes: a review. In: Bioreactors: sustainable design and industrial applications in mitigation of GHG emissions, pp 105–122. https://doi.org/10.1016/B978-0-12-821264-6.00008-5

Gollakota ARK, Kishore N, Gu S (2018) A review on hydrothermal liquefaction of biomass. Renew Sust Energ Rev 81:1378–1392. https://doi.org/10.1016/J.RSER.2017.05.178

Gottinger A, Ladu L, Quitzow R (2020) Studying the transition towards a circular bioeconomy—a systematic literature review on transition studies and existing barriers. Sustainability 12(21):8990. https://doi.org/10.3390/SU12218990

Goyal R, Guvendiren M, Freeman O, Mao Y, Kohn J (2017) Optimization of polymer-ECM composite scaffolds for tissue engineering: effect of cells and culture conditions on polymeric nanofiber mats. J Funct Biomater 8(1):1. https://doi.org/10.3390/JFB8010001

Gray RS (2025) Including harvested grain biogenic CO2 to address a critical flaw in climate accounting. Aust J Agric Resour Econ 0:1–13. https://doi.org/10.1111/1467-8489.70044

Gu H, Tang Y, Yao J, Chen F (2019) Study on biomass gasification under various operating conditions. J Energy Inst 92(5):1329–1336. https://doi.org/10.1016/J.JOEI.2018.10.002

Habib N, Adeel S, Ali F, Amin N, Khan SR (2021) Environmental friendly sustainable application of plant-based mordants for cotton dyeing using Arjun bark-based natural colorant. Environ Sci Pollut Res 28(38):54041–54047. https://doi.org/10.1007/S11356-021-14536-8/METRICS

Han J, Yu D, Wu J, Yu X, Liu F, Xu M (2023) Effects of torrefaction on ash-related issues during biomass combustion and co-combustion with coal. Part 2: ash fouling behavior. Fuel 334:126777. https://doi.org/10.1016/J.FUEL.2022.126777

Hassegawa M, Van Brusselen J, Cramm M, Verkerk PJ (2022) Wood-based products in the circular bioeconomy: status and opportunities towards environmental sustainability. Land 11(12). https://doi.org/10.3390/land11122131

Hsiao CJ, Hu JL (2024) Biomass and circular economy: now and the future. Biomass 4(3):720–739. https://doi.org/10.3390/BIOMASS4030040

Imran (2025) Carbon cultivation for sustainable agriculture, ecosystem resilience, and climate change mitigation. Commun Soil Sci Plant Anal 56(9):1430–1456. https://doi.org/10.1080/00103624.2025.2453996;ISSUE:ISSUE:DOI

Javaid M, Haleem A, Singh RP, Suman R (2022) Enhancing smart farming through the applications of Agriculture 4.0 technologies. Int J Intell Netw 3:150–164. https://doi.org/10.1016/J.IJIN.2022.09.004

Kardung M, Cingiz K, Costenoble O, Delahaye R, Heijman W, Lovrić M, van Leeuwen M, M'barek R, van Meijl H, Piotrowski S, Ronzon T, Sauer J, Verhoog D, Verkerk PJ, Vrachioli M, Wesseler JHH, Zhu BX (2021) Development of the circular bioeconomy: drivers and indicators. Sustainability (Switzerland) 13(1):1–24. https://doi.org/10.3390/su13010413

Kataya G, Cornu D, Bechelany M, Hijazi A, Issa M (2023) Biomass waste conversion technologies and its application for sustainable environmental development—a review. Agronomy 13(11). https://doi.org/10.3390/agronomy13112833

Khan AA, de Jong W, Jansens PJ, Spliethoff H (2009) Biomass combustion in fluidized bed boilers: potential problems and remedies. Fuel Process Technol 90(1):21–50. https://doi.org/10.1016/J.FUPROC.2008.07.012

Khanna M, Zilberman D, Hochman G, Basso B (2024) An economic perspective of the circular bioeconomy in the food and agricultural sector. Commun Earth Environ 5(1):1–9. https://doi.org/10.1038/S43247-024-01663-6;SUBJMETA

Khodayari A, Thielemans W, Hirn U, Van Vuure AW, Seveno D (2021) Cellulose-hemicellulose interactions – a nanoscale view. Carbohydr Polym 270:118364. https://doi.org/10.1016/J.CARBPOL.2021.118364

Kiran BR, Prasad MNV, Mohan SV (2024) Synthetic biology: an emerging field for developing economies. In: Bioremediation and bioeconomy: a circular economy approach, 2nd edn, pp 767–787. https://doi.org/10.1016/B978-0-443-16120-9.00013-3

Kumar S, Pandey A (2019) Current developments in biotechnology and bioengineering and waste treatment processes for energy generation: an introduction. In: Current developments in biotechnology and bioengineering: waste treatment processes for energy generation, pp 1–9. https://doi.org/10.1016/B978-0-444-64083-3.00001-4

Lang C (2022) Bioeconomy – from the Cologne paper to concepts for a global strategy. EFB Bioecon J 2:100038. https://doi.org/10.1016/J.BIOECO.2022.100038

Le OTH, Tran LN, Doan VT, Van Pham Q, Van Ngo A, Nguyen HH (2020) Mucilage extracted from dragon fruit peel (Hylocereus undatus) as flocculant for treatment of dye wastewater by coagulation and flocculation process. Int J Polym Sci 2020(1):7468343. https://doi.org/10.1155/2020/7468343

Leal Filho W, Sigahi TFAC, Anholon R, Rebelatto BG, Schmidt-Ross I, Hensel-Börner S, Franco D, Treacy T, Brandli LL (2025) Promoting sustainable development via stakeholder engagement in higher education. Environ Sci Europe 37(1). https://doi.org/10.1186/s12302-025-01101-0

Lemoine G, Prasad MNV, Dubois P (2024) Bioeconomy and circular economy approaches for the greening of urban wastelands: focus on biodiversity to achieve sustainability. In: Bioremediation and bioeconomy: a circular economy approach, 2nd edn, pp 475–498. https://doi.org/10.1016/B978-0-443-16120-9.00019-4

Lin CSK, Koutinas AA, Stamatelatou K, Mubofu EB, Matharu AS, Kopsahelis N, Pfaltzgraff LA, Clark JH, Papanikolaou S, Kwan TH, Luque R (2014) Current and future trends in food waste valorization for the production of chemicals, materials and fuels: a global perspective. Biofuels Bioprod Biorefin 8(5):686–715. https://doi.org/10.1002/BBB.1506

Liu Y, Song W (2022) Mapping human appropriation of net primary production in agroecosystems in the Heihe River Basin, China. Agric Ecosyst Environ 335:107996. https://doi.org/10.1016/J.AGEE.2022.107996

Lohri CR, Diener S, Zabaleta I, Mertenat A, Zurbrügg C (2017) Treatment technologies for urban solid biowaste to create value products: a review with focus on low- and middle-income settings. Rev Environ Sci Biotechnol 16(1):81–130. https://doi.org/10.1007/s11157-017-9422-5

Lourinho G, Alves O, Garcia B, Rijo B, Brito P, Nobre C (2023) Costs of gasification technologies for energy and fuel production: overview, analysis, and numerical estimation. Recycling 8(3):49. https://doi.org/10.3390/RECYCLING8030049/S1

Lynch T (2016) United nations sustainable development goals: promoting health and well-being through physical education partnerships. Cogent Educ 3(1). https://doi.org/10.1080/2331186X.2016.1188469;JOURNAL:JOURNAL:OAED20;REQUESTEDJOURNAL:JOURNAL:OAED20;PAGE:STRING:ARTICLE/CHAPTER

Ma Y, Liu Y (2019) Turning food waste to energy and resources towards a great environmental and economic sustainability: an innovative integrated biological approach. Biotechnol Adv 37(7):107414. https://doi.org/10.1016/J.BIOTECHADV.2019.06.013

Maache M, Bezazi A, Amroune S, Scarpa F, Dufresne A (2017) Characterization of a novel natural cellulosic fiber from Juncus effusus L. Carbohydr Polym 171:163–172. https://doi.org/10.1016/J.CARBPOL.2017.04.096

Maltsev Y, Kulikovskiy M, Maltseva S (2023) Nitrogen and phosphorus stress as a tool to induce lipid production in microalgae. Microb Cell Fact 22(1):1–22. https://doi.org/10.1186/S12934-023-02244-6/METRICS

Mamvura TA, Danha G (2020) Biomass torrefaction as an emerging technology to aid in energy production. Heliyon 6(3):e03531. https://doi.org/10.1016/j.heliyon.2020.e03531

Mandal S, Venkatramani J (2023) A review of plant-based natural dyes in leather application with a special focus on color fastness characteristics. Environ Sci Pollut Res 30(17):48769–48777. https://doi.org/10.1007/S11356-023-26281-1/METRICS

Martins-Vieira JC, Torres-Mayanga PC, Lachos-Perez D (2023) Hydrothermal processing of lignocellulosic biomass: an overview of subcritical and supercritical water hydrolysis. Bioenergy Res 16(3):1296–1317. https://doi.org/10.1007/S12155-022-10553-8/METRICS

Mazumder S, Zhang N (2023) Cellulose–hemicellulose–lignin interaction in the secondary cell wall of coconut endocarp. Biomimetics 8(2):188. https://doi.org/10.3390/BIOMIMETICS8020188

Mazzà G, Malek R, Raguzzoni K, Ragazzi M, Ciolli M (2020) The usage of compost-like outputs in Lebanon: regulatory framework, opportunities and threats. A survey. https://www.microfinanzaesviluppo.it/wp-content/uploads/2021/02/Compost_Survey_Lebanon.pdf

Mehyar GF, Al-Ismail K, Han JH, Chee GW (2012) Characterization of edible coatings consisting of pea starch, whey protein isolate, and carnauba wax and their effects on oil rancidity and sensory properties of walnuts and pine nuts. J Food Sci 77(2):E52–E59. https://doi.org/10.1111/J.1750-3841.2011.02559.X

Mestre A, Cooper T (2017) Circular product design. A multiple loops life cycle design approach for the circular economy. Design Journal 20:S1620–S1635. https://doi.org/10.1080/14606925.2017.1352686;WGROUP:STRING:PUBLICATION

Murele OC, Zulkafli NI, Kopanos G, Hart P, Hanak DP (2020) Integrating biomass into energy supply chain networks. J Clean Prod 248:119246. https://doi.org/10.1016/J.JCLEPRO.2019.119246

Naidoo D, Nhamo L, Lottering S, Mpandeli S, Liphadzi S, Modi AT, Trois C, Mabhaudhi T (2021) Transitional pathways towards achieving a circular economy in the water, energy, and food sectors. Sustainability 13(17):9978. https://doi.org/10.3390/SU13179978

Nunes LJR (2020) A case study about biomass torrefaction on an industrial scale: solutions to problems related to self-heating, difficulties in pelletizing, and excessive wear of production equipment. Appl Sci 10(7):2546. https://doi.org/10.3390/APP10072546

Olivo EF, Zaccaron A, Acordi J, Ribeiro MJ, Fernandes ÉMR, Zocche JJ, Raupp-Pereira F, Olivo EF, Zaccaron A, Acordi J, Ribeiro MJ, Maria É, Fernandes R, Zocche JJ, Raupp-Pereira F (2024) Technosol development based on residual fraction of coal tailings processing, agroindustrial waste, and paper industry waste. Sustainability 16(17):7471. https://doi.org/10.3390/SU16177471

Ouakarrouch M, El Azhary K, Laaroussi N, Garoum M, Kifani-Sahban F (2020) Thermal performances and environmental analysis of a new composite building material based on gypsum plaster and chicken feathers waste. Therm Sci Eng Prog 19:100642. https://doi.org/10.1016/J.TSEP.2020.100642

Paleari S (2024) The EU policy on climate change, biodiversity and circular economy: moving towards a Nexus approach. Environ Sci Pol 151:103603. https://doi.org/10.1016/J.ENVSCI.2023.103603

Pang S (2019) Advances in thermochemical conversion of woody biomass to energy, fuels and chemicals. Biotechnol Adv 37(4):589–597. https://doi.org/10.1016/J.BIOTECHADV.2018.11.004

Patra D, Patra BR, Pattnaik F, Hans N, Kushwaha A (2022) Recent evolution in green technologies for effective valorization of food and agricultural wastes. In: Emerging trends to approaching zero waste: environmental and social perspectives, pp 103–132. https://doi.org/10.1016/B978-0-323-85403-0.00001-3

Paul AK, Prasad A, Mahto B, Kumar A, Singh P (2024) Introduction to bioenergy and waste to energy. In: Biodegradable waste processing for sustainable developments, pp 1–33. https://doi.org/10.1201/9781003502012-1/INTRODUCTION-BIOENERGY-WASTE-ENERGY-ATANU-KUMAR-PAUL-ARBIND-PRASAD-BIDYANAND-MAHTO-ASHWANI-KUMAR-PAVITRA-SINGH

Perea-Moreno MA, Samerón-Manzano E, Perea-Moreno AJ (2019) Biomass as renewable energy: worldwide research trends. Sustainability 11(3):863. https://doi.org/10.3390/SU11030863

Periyasamy S, Karthik V, Senthil Kumar P, Isabel JB, Temesgen T, Hunegnaw BM, Melese BB, Mohamed BA, Vo DVN (2022) Chemical, physical and biological methods to convert lignocellulosic waste into value-added products. A review. Environ Chem Lett 20(2):1129–1152. https://doi.org/10.1007/S10311-021-01374-W/METRICS

Popp J, Kovács S, Oláh J, Divéki Z, Balázs E (2021) Bioeconomy: biomass and biomass-based energy supply and demand. New Biotechnol 60:76–84. https://doi.org/10.1016/J.NBT.2020.10.004

Punia Bangar S, Ilyas RA, Chaudhary N, Dhull SB, Chowdhury A, Lorenzo JM (2023) Plant-based natural fibers for food packaging: a green approach to the reinforcement of biopolymers. J Polym Environ 31(12):5029–5049. https://doi.org/10.1007/S10924-023-02849-3/METRICS

Rajendran S, Al-Samydai A, Palani G, Trilaksana H, Sathish T, Giri J, Saravanan R, Lalvani JIJR, Nasri F (2025) Replacement of petroleum based products with plant-based materials, green and sustainable energy—a review. Eng Rep 7(4):e70108. https://doi.org/10.1002/ENG2.70108

Ramirez-Contreras NE, Faaij APC (2018) A review of key international biomass and bioenergy sustainability frameworks and certification systems and their application and implications in Colombia. Renew Sust Energ Rev 96:460–478. https://doi.org/10.1016/J.RSER.2018.08.001

Rosa L, Mazzotti M (2022) Potential for hydrogen production from sustainable biomass with carbon capture and storage. Renew Sust Energ Rev 157:112123. https://doi.org/10.1016/J.RSER.2022.112123

Ruiz-Gonzalez A, Wang M, Haseloff J (2023) Sustainable approaches to incorporate plant-based biomaterials in power generation. Solids 4(2):133–155. https://doi.org/10.3390/solids4020009

Sadhukhan J, Martinez-Hernandez E, Amezcua Allieri MA, Zermeño Eguía-Lis JA, Castillo A, Dominguillo D, Torres-García E, Aburto J (2024) Strategic navigation of world-leading biorefineries and Mexico's policy landscape: a gateway to a sustainable circular bioeconomy. J Clean Prod 434. https://doi.org/10.1016/j.jclepro.2023.140386

Sah MK, Ettarhouni ZO, Pathak R, Gawad J, Bonde C, Arya SP, Bhattarai A (2025) Green chemistry: strategies and sustainable approaches for bridging UN SDGS. ChemistrySelect 10(25):e00847. https://doi.org/10.1002/SLCT.202500847

Salazar Sandoval S, Amenábar A, Toledo I, Silva N, Contreras P (2024) Advances in the sustainable development of biobased materials using plant and animal waste as raw materials: a review. Sustainability 16(3):1073. https://doi.org/10.3390/SU16031073

Salzman J, Bennett G, Carroll N, Goldstein A, Jenkins M (2018) The global status and trends of Payments for Ecosystem Services. Nat Sustainability 1(3):136–144. https://doi.org/10.1038/S41893-018-0033-0;SUBJMETA

Sangeetha T, Rajneesh CP, Yan WM (2020) Integration of microbial electrolysis cells with anaerobic digestion to treat beer industry wastewater. In: Integrated microbial fuel cells for wastewater treatment, pp 313–346. https://doi.org/10.1016/B978-0-12-817493-7.00015-1

Sathre R, O'Connor J (2010) Meta-analysis of greenhouse gas displacement factors of wood product substitution. Environ Sci Pol 13(2):104–114. https://doi.org/10.1016/J.ENVSCI.2009.12.005

Schipfer F, Burli P, Fritsche U, Hennig C, Stricker F, Wirth M, Proskurina S, Serna-Loaiza S (2024) The circular bioeconomy: a driver for system integration. Energy Sustain Soc 14(1):1–19. https://doi.org/10.1186/s13705-024-00461-4

Siddiqi H, Mishra A, Kumari U, Maiti P, Meikap BC (2021) Utilizing agricultural residue for the cleaner biofuel production and simultaneous air pollution mitigation due to stubble burning: a net energy balance and total emission assessment. ACS Sustain Chem Eng 9(47):15963–15972. https://doi.org/10.1021/ACSSUSCHEMENG.1C06202

Singh S, Agarwal P, Verma A, Saini G (2025) Application of sustainable plant-based materials for treatment of pharmaceutical industry wastewater. J Phys Conf Ser 3076(1):012002. https://doi.org/10.1088/1742-6596/3076/1/012002

Soltanloo M, Babaee MH, Hosseini Yeganeh SE, Shafiei A, Akbari AH (2022) Root cause failure investigation of a boiler waterwall tube employed in a 325 MW thermal power plant: caustic corrosion phenomenon and its effects. Eng Fail Anal 133:105974. https://doi.org/10.1016/J.ENGFAILANAL.2021.105974

Su J, Long Q, Lin S, Hu Z, Zeng Y (2024) Carbon neutralization in Yunnan: harnessing the power of forests to mitigate carbon emissions and promote sustainable development in the Southwest forest area of China. Mitig Adapt Strateg Glob Chang 29(8):1–36. https://doi.org/10.1007/S11027-024-10185-7

Szarka N, García Laverde L, Thrän D, Kiyko O, Ilkiv M, Moravčíková D, Cudlínová E, Lapka M, Hatvani N, Koós Á, Luks A, Martín Jimenez I (2023) Stakeholder engagement in the co-design of regional bioeconomy strategies. Sustainability (Switzerland) 15(8):6967. https://doi.org/10.3390/SU15086967/S1

Tan ECD, Lamers P (2021) Circular bioeconomy concepts—a perspective. Front Sustainability 2:1–8. https://doi.org/10.3389/frsus.2021.701509

Tardy BL, Lizundia E, Guizani C, Hakkarainen M, Sipponen MH (2023) Prospects for the integration of lignin materials into the circular economy. Mater Today 65:122–132. https://doi.org/10.1016/J.MATTOD.2023.04.001

Tolessa A (2023) Bioenergy potential from crop residue biomass resources in Ethiopia. Heliyon 9(2):e13572. https://doi.org/10.1016/j.heliyon.2023.e13572

Trujillo-Cayado LA, Sánchez-García RM, García-Domínguez I, Rodríguez-Luna A, Hurtado-Fernández E, Santos J (2025) Emerging trends in sustainable biological resources and bioeconomy for food production. Appl Sci 15(12):6555. https://doi.org/10.3390/APP15126555

Tsakalova M, Onorato C, Karathanasis AZ, Deligkiozi I (2024) A circular economy framework for the assessment of bio-based value chains. Sustain Chem Environ 6:100099. https://doi.org/10.1016/J.SCENV.2024.100099

Uddin MN, Techato K, Taweekun J, Rahman MM, Rasul MG, Mahlia TMI, Ashrafur SM (2018) An overview of recent developments in biomass pyrolysis technologies. Energies 11(11):3115. https://doi.org/10.3390/EN11113115

Vasileiadis V, Samara E, Zafeiriou, Douvartzides S, Tsanaktsidis C (2025) Advancing the bioeconomy through biodiesel: a comparative evaluation of plant-based, animal-based, and mixed feedstocks for sustainable use of heating diesel. Biofuels. https://doi.org/10.1080/17597269.2025.2509046;CSUBTYPE:STRING:AHEAD

Velvizhi G, Jacqueline PJ, Shetti NP, K L, Mohanakrishna G, Aminabhavi TM (2023) Emerging trends and advances in valorization of lignocellulosic biomass to biofuels. J Environ Manag 345:118527. https://doi.org/10.1016/J.JENVMAN.2023.118527

Vickram AS, Saravanan A, Senthil Kumar P, Thamarai P, Yasodha S, Jamuna G, Rangasamy G (2023) An integrated approach to the sustainable development and production of biofuel from biopolymers and algal biomass derived from wastewater. Fuel 349:128691. https://doi.org/10.1016/J.FUEL.2023.128691

Vinod A, Vijay R, Singaravelu DL, Sanjay MR, Siengchin S, Moure MM (2019) Characterization of untreated and alkali treated natural fibers extracted from the stem of Catharanthus roseus. Mater Res Express 6(8):085406. https://doi.org/10.1088/2053-1591/AB22D9

Vural Gursel I, Elbersen B, Meesters KPH, van Leeuwen M (2022) Defining circular economy principles for biobased products. Sustainability 14(19):12780. https://doi.org/10.3390/SU141912780

Worku LA, Bachheti RK, Tadesse MG (2023) Preparation and characterization of carboxylated cellulose nanocrystals from *Oxytenanthera abyssinica* (Ethiopian lowland bamboo) cellulose via citric acid anhydrous hydrolysis catalyzed by sulfuric acid. Biomass Conv Bioref. https://doi.org/10.1007/s13399-022-03718-0

Worku LA, Tadesse MG, Bachheti RK, Bachheti A, Husen A (2024) Synthesis of carboxylated cellulose nanocrystal/ZnO nanohybrids using *Oxytenanthera abyssinica* cellulose and zinc nitrate hexahydrate for radical scavenging, photocatalytic, and antibacterial activities. Int J Biol Macromol 267:131228. https://doi.org/10.1016/j.ijbiomac.2024.131228

Wu MR, Schott DL, Lodewijks G (2011) Physical properties of solid biomass. Biomass Bioenergy 35(5):2093–2105. https://doi.org/10.1016/J.BIOMBIOE.2011.02.020

Yang Y, Liew RK, Tamothran AM, Foong SY, Yek PNY, Chia PW, Van Tran T, Peng W, Lam SS (2021) Gasification of refuse-derived fuel from municipal solid waste for energy production: a review. Environ Chem Lett 19(3):2127–2140. https://doi.org/10.1007/s10311-020-01177-5

Yaradoddi JS, Banapurmath NR, Ganachari SV, Soudagar MEM, Sajjan AM, Kamat S, Mujtaba MA, Shettar AS, Anqi AE, Safaei MR, Elfasakhany A, Haque Siddiqui MI, Ali MA (2022) Bio-based material from fruit waste of orange peel for industrial applications. J Mater Res Technol 17:3186–3197. https://doi.org/10.1016/j.jmrt.2021.09.016

Zardzewiały M, Bajcar M, Puchalski C, Gorzelany J (2023) The possibility of using waste biomass from selected plants cultivated for industrial purposes to produce a renewable and sustainable source of energy. Appl Sci 13(5):3195. https://doi.org/10.3390/APP13053195

Zhan Y, Li R, Chen W, Chen Y, Yang L, Liu B, Tao X, Chen P, Wang Z, Zhang H, Li J, Chang Y, Wei Y, Li J (2025) Deciphering the carbon and nitrogen component conversion in humification process mediated by distinct microbial mechanisms in composting from different domestic organic wastes. Environ Sci Pollut Res 32(22):13474–13486. https://doi.org/10.1007/S11356-024-32224-1/METRICS

Zhang L, Xu C(C), Champagne P (2010) Overview of recent advances in thermo-chemical conversion of biomass. Energy Convers Manag 51(5):969–982. https://doi.org/10.1016/J.ENCONMAN.2009.11.038

Zhou J, Barati B, Wu J, Scherer D, Karana E (2021) Digital biofabrication to realize the potentials of plant roots for product design. Bio-Des Manuf 4(1):111–122. https://doi.org/10.1007/S42242-020-00088-2/FIGURES/13

Zhu Z, Gavahian M, Barba FJ, Roselló-Soto E, Bursać Kovačević D, Putnik P, Denoya GI (2020) Valorization of waste and byproducts from food industries through the use of innovative technologies. In: Agri-food industry strategies for healthy diets and sustainability: new challenges in nutrition and public health, pp 249–266. https://doi.org/10.1016/B978-0-12-817226-1.00011-4

Zong P, Jiang Y, Tian Y, Li J, Yuan M, Ji Y, Chen M, Li D, Qiao Y (2020) Pyrolysis behavior and product distributions of biomass six group components: starch, cellulose, hemicellulose, lignin, protein and oil. Energy Convers Manag 216:112777. https://doi.org/10.1016/J.ENCONMAN.2020.112777

Zubair M, Roopesh MS, Ullah A (2022) Nano-modified feather keratin derived green and sustainable biosorbents for the remediation of heavy metals from synthetic wastewater. Chemosphere 308:136339. https://doi.org/10.1016/J.CHEMOSPHERE.2022.136339

Chapter 14
Ethical and Environmental Challenges in Plant-Based Solutions

L. Srimathi Priya, Manikandan Ayyar, V. P. Santhi, K. Indhumathi, A. Dinesh, N. Subasree, K. Radhakrishnan, and A. Muthukrishnaraj

Abstract Global initiatives to develop plant-based alternatives across industries, particularly food, energy, materials, and healthcare, open the way to new sustainable opportunities but also raise a series of ethical and environmental dilemmas. The chapter examines the complexities surrounding the development and introduction of plant-based systems in the context of biopiracy, safeguarding of indigenous knowledge, access to knowledge, and maintenance of biodiversity. It looks not only at the environmental trade-offs in land use, monoculture, and water consumption and carbon emissions, but also at the socio-economic consequences for food security, rural livelihoods, and cultural values. There is a gap in regulation and governance, and the focus is on certification, policy coherence, and corporate responsibility. The chapter demonstrates the tangled relationship between innovation and ethical

L. S. Priya
Department of Natural Resource Management, Horticultural College and Research Institute (HC & RI), Tamil Nadu Agricultural University (TNAU), Periyakulam, Theni, Tamil Nadu, India

M. Ayyar (✉)
Department of Chemistry, Karpagam Academy of Higher Education (Deemed to be University), Coimbatore, Tamil Nadu, India

Centre for Material Chemistry, Karpagam Academy of Higher Education (Deemed to be University), Coimbatore, Tamil Nadu, India

V. P. Santhi
Department of of Horticulture, Anbil Dharmalingam Agricultural College and Research Institute (ADAC & RI), Tamil Nadu Agricultural University (TNAU), Trichy, Tamil Nadu, India

K. Indhumathi
Krishi Vigyan Kendra, Tamil Nadu Agricultural University, Papparapatty, Dharmapuri, Tamil Nadu, India

A. Dinesh
Department of Chemistry, K. Ramakrishnan College of Engineering (Autonomous), Affiliated to Anna University, Samayapuram, Trichy, Tamil Nadu, India

L. A. Worku et al. (eds.), *Plant-Based Solutions for Sustainable Agriculture and Healthcare*, Sustainability Sciences in Asia and Africa,
https://doi.org/10.1007/978-981-95-8910-4_14

stewardship, using real-world case studies of everything at hand, including soy production in the Amazon, plant-based meat substitutes, and bioplastics. Lastly, it proposes actionable and integrative solutions for designing and creating global plant-based diversity initiatives that are locally situated, equitable, transparent, ecologically sound, and just to all living people.

Keywords Plant-based diversity · Environmental · Food, Energy, and Healthcare industries

14.1 Introduction

As the world population grows, estimated to reach 9.2 billion by 2100 and even 12 billion by 2050, pressure on natural resources has increased. Although developments in technology have led to high food production like never before, these developments have not benefited people equally. It is now understood that inequalities in food systems are deep-rooted and that undernutrition and obesity are equally acute health risks in the population today. At the same time, food production, especially industrial animal agriculture practices, is a significant factor in climate change, biodiversity loss, water contamination, and land degradation. To address these related issues, plant-based diets have become a trending approach toward anthropocentric sustainable development and ethical behavior. A shift from less resource-intensive animal-source products to more plant-based foods has the potential to go a long way toward curtailing the environmental menace and, in turn, address issues of animal welfare, food health, and food parity in global food distribution.

Nevertheless, the switch to plant systems does not come without its environmental and ethical risks. Ethically, the issues discussed relate to moral responsibilities toward the animal world, the freedom of future generations to live on the planet, and the equal right to healthy food. The view of considering the morality of food choices based on the utilitarian, deontological, and virtue theories of ethics is quite diverse, yet complementary. However, for us humans, the fact that plants can reduce the

N. Subasree
Department of Chemistry, School of Basic Science, Vels Institute of Science, Technology and Advanced Studies (VISTAS), Pallavaram, Tamil Nadu, India

K. Radhakrishnan
Department of Chemistry, Karpagam Academy of Higher Education (Deemed to be University), Coimbatore, Tamil Nadu, India

Centre for Material Science, Karpagam Academy of Higher Education (Deemed to be University), Coimbatore, Tamil Nadu, India

A. Muthukrishnaraj
Amrita School of Artificial Intelligence, Amrita Vishwa Vidyapeetham, Coimbatore, Tamil Nadu, India

suffering of animals and humans is another strong reason to follow a plant-based diet, as a utilitarian would see it. Deontological theories focus on justice, non--maleficence, and beneficence duties as they apply to the construction of food systems. At the same time, virtue ethics notably emphasizes the cultivation of compassion, temperance, and ecological acumen.

Environmentally, plant-based systems are usually safer than animal-based systems, but they also have limits and risks. By failing to take proper care, monoculture cropping, overuse of water, reliance on pesticides, and long-distance transportation of certain plant products may jeopardize the aim of sustainability. Additionally, life cycle assessments (LCAs) indicate that the environmental impact of plant-based foods is not always lower, and that, as the theoretical diet model suggests, not all consumers behave the same way. Environmental assessments are further complicated by the lack of region-specific data and variations in agricultural practices (Pandian et al. 2024).

This chapter uses the lenses of ethics and environmental science to illuminate the promise and threats of plant-based solutions. It seeks to feature the complexity of achieving a sustainable transition toward sustainable diets in a world whose ecology has been destroyed, social structures are unbalanced, and morality is uncertain. Combining philosophical knowledge with empirical evidence on the environment, this chapter presents a complex picture in which the application of plant-based interventions can facilitate a more sustainable and equitable food system, while also highlighting the limitations that must be overcome to achieve the desired aims of such interventions.

14.2 Ethical Considerations

The ethics of plant-based solutions overlap significantly with ongoing discussions about societal values in food choices, cultural norms, animal safety, health equity, and sustainability. In sustainable food systems, pressure from global customers is increasing. However, ethical dimensions provide a structural window for individual and collective decision-making. Traditionally, meat-eating has been a core component of the human diet, with compelling cultural, social, and symbolic value. To most communities, meat is not just a source of nourishment: it is a tradition, an identity, and a social bond. Research indicates that most people still justify meat consumption because they believe meat is natural, normal, necessary, and friendly. These reasons run deep in social norms and make it difficult to change the way people consume plant-based foods, even when they see the positive environmental and health outcomes of using less meat.

However, these conventional justifications are disputed by ethical frameworks, which cast doubt on whether practices that cause suffering, environmental degradation, and inequity are ethical, especially when it is possible to act differently. An example is utilitarianism, which judges an action by its consequences. In this view, the morality of reducing animal suffering, minimizing greenhouse gas emissions, and preserving biodiversity through vegan diets is part of the ethical requirements. The big thinkers in utilitarianism maintain that the unhappiness inflicted on animals in an industrial farming system cannot be offset by the comparatively insignificant happiness people get from consuming meat.

On the other hand, deontological ethics emphasizes duty and moral principles rather than outcomes. Such an approach dwells on the fact that some actions, like torturing sentient animals or working under exploitative working conditions, are morally wrong because they violate inherent moral obligations. In this perspective, it would be inhumane and irresponsible, as far as the cultural heritage and consumer demands are concerned, to support food systems that are founded on cruelty, exploitation, or environmental degradation. Virtue ethics is a third dimension of the discussion, emphasizing character and moral growth. By adopting plant-based solutions, one can be identified as fulfilling the requirements of compassion, temperance, and environmental stewardship, which may be part of an ethical, moral lifestyle. A person who behaves in accordance with these virtues not only benefits society but also builds personal integrity and ecological wisdom (Ford et al. 2023).

Even though awareness is on the rise, most ethical obstacles remain, hindering the shift to plant-based diets. There are stereotypes about a vegetarian or vegan diet that seem to be always inconvenient, less tasty, and more costly. Moreover, the understanding of meat substitutes, particularly highly processed ones such as plant-based burgers, has sparked debates over healthfulness and authenticity. Although some consider these products environmentally and ethically superior, others argue that they do not justify health claims because they are highly processed, just like other traditional processed foods, and thus not any better. In addition, inclusivity and equity should be included in ethical criticisms of plant-based solutions. The consumption of plant-based diets should not be the privilege of the privileged; culturally appropriate, healthy, and affordable plant-based diets should be widely available across socio-economic groups. Ethical solutions do not exist in a vacuum, without reference to the very facts surrounding food accessibility, affordability, and cultural acceptability. Overall, plant-based solutions have an ethical dimension that encompasses personal responsibility, systemic reform, and cultural adjustment. The transition to plant-based systems is not just a dietary choice but an ethical one, one that requires reconsidering our connections to animals, the planet, and each other. Ethics in this complicated transition should be based not only on philosophies of harm reduction and justice but also on empathy, humility, and a readiness to question old assumptions about what we consume and why (Giacalone et al. 2022).

14.2.1 Moral Imperatives in Shifting to Plant-Based Systems

Contemporary diets, and in particular the one that essentially defines the typical Western diet and consists of a high load of ultra-processed and animal-centered meals, are ethically and publicly concerning. Not only are these patterns nutritionally unsound, but they also result in excessive damage at the individual, societal, and environmental levels. A powerful moral consideration lies at the heart of this problem: there is a growing imperative to ensure we transition to whole-food, plant--based diets that are ethical and promote our health.

The ethical objection to a diet that relies on animal products has its starting point in the fact that factory farming is a well-documented reality and that farmed animals are exposed to gross levels of confinement, systematic cruelty, and untimely death. Although the moral case for animal suffering is well established, it overlaps with the health consequences for humans. The chronic diseases associated with red and processed meat include cardiovascular diseases, obesity, type 2 diabetes, and some types of cancer, and these diets have been tied to these diseases. On the contrary, a whole-food, plant-based diet has proven not only to prevent but also to reverse most of these conditions, thereby offering a strong health-based case for adopting it.

Nevertheless, the problem does not narrow down to personal health conditions. The factory farming industry itself is a major contributor to global climate change, emitting considerable amounts of greenhouse gases (GHG) such as methane, nitrous oxide, and carbon dioxide. It also contributes to deforestation, water contamination, biodiversity loss, and the emergence of antibiotic-resistant pathogens, problems that worsen the overall public health hazard globally. What is more, animal agriculture is associated with zoonotic spillover events, such as pandemics, further accelerating the demand to restructure diets.

The interconnectivity of these harms leads to a greater societal breakdown of the moral dispensation of resources. In high-income nations, subsidies for agricultural production are overrepresented in livestock feed crops rather than in healthful plant foods directly consumed by humans. Not only does this deepen food inequalities, but it also makes it harder for lower-income groups to access healthy, affordable food sources. Consequently, nutrition equity must be regarded as a moral concern grounded in the principles of justice, autonomy, and beneficence.

From a medical ethics perspective, medical professionals, especially doctors, have an exclusive moral obligation. The strong correlation between dietary patterns and health outcomes, as well as environmental degradation, justifies physicians advocating a plant-based diet not only as permitted but also as ethically mandated. Each clinical contact is an opportunity to teach the patient not only about the risks associated with their health condition but also about the risks to society in general related to eating habits. An individual emphasis on health and personal well-being and a societal emphasis on sustainability are not a hoax; they are a necessity. Openness regarding these two motivations empowers informed judgments in promoting public interests and health aims.

This is a moral obligation that is not confined to an individual. Individuals should not be the only ones burdened by this change process; institutions, policymakers, industry stakeholders, and educators should be too. Just as climate action is not the job of specialists working in the environmental sphere, dietary change cannot be the responsibility of a single sector of the population.

In sum, not only for the welfare of animals and the environment, but also for human health and justice, conversion to a plant-based food system model is ethically required. It calls for a redefinition of nutrition advice, resource distribution,

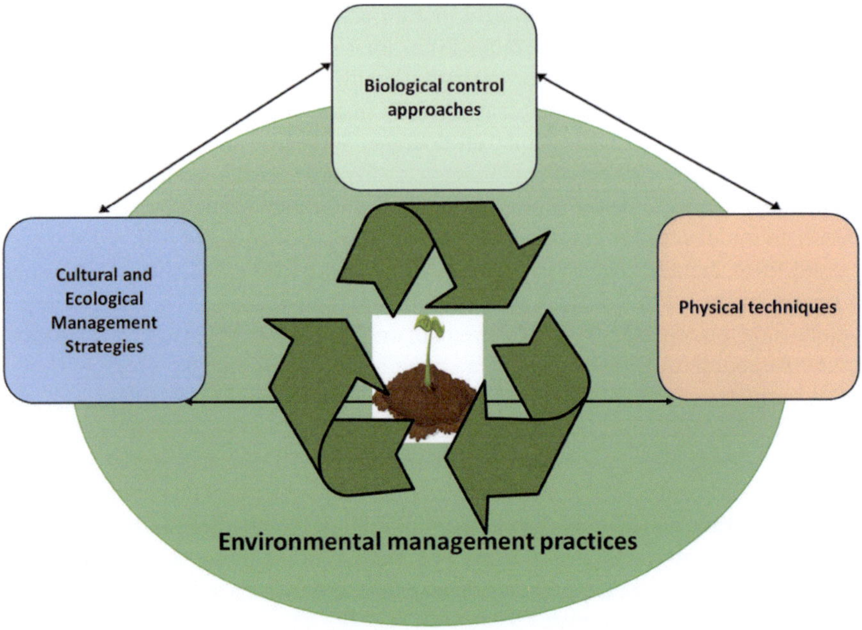

Fig. 14.1 Integrated environmental management practices involving biological control, physical techniques, and cultural-ecological strategies for sustainable agriculture (Cucu et al. 2025)

and health lobbying, in line with the need for immediate system change, as shown in Fig. 14.1. The convergence of ethical reasoning, scientific evidence, and sustainability principles supports a unified call to action, one that makes plant-based solutions not only desirable but morally imperative (Cucu et al. 2025).

14.2.2 *Biopiracy and Intellectual Property Rights*

Herbs and food have been used as the basis for traditional medicine and local cuisine. These are plants that burn with centuries of empirical knowledge built and passed along by indigenous and local communities. However, as global markets and science expanded, another scurrilous ethical issue has arisen: the seizure and patenting of traditional knowledge and biological resources without adequate recognition or reward, commonly called biopiracy.

Biopiracy, in general terms, is the misuse or theft of traditional knowledge (TK) and genetic resources by corporations or research facilities, often for profit. This is commonly seen when bioactive compounds are isolated by pharmaceutical or biotechnology companies, patented by them, and sold, making huge profits without giving a share back to the communities that had discovered and developed these plants in the first place and preserved their age-old knowledge. There are significant ethical issues concerning injustice, consent, and exploitation that the practice evokes.

The IPRs are meant to safeguard innovations and creativity, ensuring that inventors or creators have exclusive rights over their inventions. These are patents, plant variety protection, geographical indications, trade secrets, and trademarks. Hypothetically, IPRs promote innovation by rewarding inventors with the right to monopolize their products under the law. However, the current model of IPRs does not necessarily appreciate that biological materials and traditional knowledge are usually the work of many people over a long time. Traditional knowledge may not meet the patent law requirements for novelty, non-obviousness, and utility, whereas the legal definition of a patent only values this kind of knowledge as valuable in terms of science and culture, not at law.

Several international legal instruments have been developed to fill gaps and ensure free access and fair sharing of benefits. The Convention on Biological Diversity (CBD), the Nagoya Protocol, and the work of the Intergovernmental Committee on Intellectual Property and Genetic Resources (IGC) within the World Intellectual Property Organization (WIPO) are examples of efforts to shape the framework for the rights of indigenous peoples. These treaties reiterate the necessity of prior informed consent, the agreements being concluded in good faith, and the fair distribution of benefits derived from the use of genetic resources and associated knowledge.

Nevertheless, there are a few difficulties. The patent systems used by many countries, especially high-income countries, are often not aligned with the values and practices of indigenous knowledge systems. The murkiness of the patent registration process, insufficient engagement in the community, and legal illiteracy with oppressed groups compound the issues. In addition, biopiracy tends to happen in a setting of power inequality so that communities with financial problems can do little to prevent the actions of multinational companies.

Critics like activist scholar Vandana Shiva have opposed this relationship as a contemporary extension of colonial exploitation, with the affluence of biodiversity-rich areas stripped away, based on legal frameworks developed without the involvement of those who will be most adversely affected. The current trend of the commodification of nature in the form of IPRs, which is done without proper protection, threatens to undermine the cultural heritage of indigenous people and their sovereignty.

Ethnobotany is the study of the interaction between people and plant diversity, which can serve as an example of the possibilities and dangers of working with traditional knowledge. Such ethically conducted research should acknowledge and safeguard the input of local communities and encourage biodiversity conservation. There is a solution that involves incorporating IPRs into ethnobotanical practice to empower people, build non-exploitative, collaborative relationships, and share the benefits. This, however, needs to be done through real community participation, capacity development, and a system that recognizes that knowledge is held collectively, not in an individualistic manner.

When it comes to the use of plants to address environmental and human health issues, intellectual property rights and the prohibition of biopiracy are matters of ethics as much as of law. By safeguarding traditional knowledge, the shift toward new food systems becomes inclusive, respectful, and just. As innovations using plants become increasingly common, it is important to ensure that the production of these innovations does not marginalize the people who have been custodians of biodiversity (Caballero-Serrano et al. 2019).

14.2.3 *Indigenous Knowledge and Benefit Sharing*

Historically and across a broad geographical area, indigenous and local peoples have established a complex, culturally refined relationship with the environments in which they live, which has resulted in ethnobotany, the creation of rich bodies of ethnobotanical knowledge that inform the use of plants in food, medicine, and rituals, and inform ecological management. It is not just culturally relevant but also scientifically important, as it has led to significant pharmaceutical discoveries and the development of plant-based solutions with diverse applications worldwide. Nonetheless, as valuable as they may seem, indigenous knowledge systems are exposed to exploitation, biopiracy, and a lack of gains from their innovations.

These communities have been heatedly debating fairness, consent, and ownership, highlighted by the growing commercial and scientific interests in genetic resources and traditional plant knowledge. Also put forward as a solution to this are Access and Benefit Sharing (ABS) systems, which are proposed as a way to recognize and recompense the communities from which knowledge or resources are drawn. International instruments such as the Convention on Biological Diversity (CBD) and the Nagoya Protocol have been instrumental in establishing these principles and reinforcing countries' sovereignty over their genetic resources, as well as the need for prior informed consent before they are used.

Nonetheless, excellent benefit-sharing mechanisms are yet to be implemented despite these safeguards. A fair negotiation can be problematic due to the power and knowledge asymmetry between the science and indigenous people's institutions. The legal or technical ability of many indigenous groups to participate in ABS negotiations to the full extent is lacking, and benefits in such cases do not reach those who possess knowledge. Such relationships are often complicated by mistrust, differences in cultural understanding, and a lack of clarity in the legal context (Kumar et al. 2021).

Moreover, traditional knowledge does not necessarily fit within the traditional model of intellectual property rights (IPRs), as it is more communal and intergenerational. One such system is patenting, which generally aims to provide incentives to individual inventors by rewarding inventions that are both new and temporal. Traditional knowledge, in turn, is knowledge that has evolved continuously across generations and is usually entwined with the cultural and religious fabric of a given group of people. Such a mismatch has given rise to situations where organizations have patented medicinal plants or formulations founded on indigenous knowledge without authorization, acts that have been disparaged far and wide as biopiracy.

The example of Artemisia annua and the discovery of a life-saving antimalarial compound, artemisinin, is a prime example of how traditional knowledge can be instrumental in advancing science today. Nonetheless, there are also legal and ethical issues regarding ownership, credit, and the distribution of benefits in such cases. There are substantial economic implications: the international market for herbal medicines and natural health products is measured in billions of dollars, with very little or no economic compensation to the communities where the basic knowledge arises.

To address such problems, more inclusive, motivating, and transparent research and development practices should be adopted. Some of the new ways that have emerged to ensure the community is protected and benefits from projects include community-based research agreements, benefit-sharing funds, and the development of traditional knowledge databases. These frameworks must not only help ensure that benefits are distributed equitably, whether in monetary or other terms, but also that communities have the right not to be involved or to dictate how their knowledge is adopted.

Ethical benefit sharing is not only about legal writing; it is also about justice. Appreciating indigenous knowledge is a way of recognizing that communities that have long been custodians of biodiversity make an intellectual contribution to the issue. It also demands that both the scientific and policy-making communities be involved in the decolonization of research practice and in the rebuilding of how value is realized and distributed. To achieve that, plant-based innovations have the potential to be more than a means of exploitation that replicates historical exploitative trends, but also tools of ecological sustainability and social justice. Considering sharing the benefits of such plants with indigenous people to develop appropriate global environmental and health solutions is one of the core elements for developing inclusive and ethical food and health systems. This will require the creation of trust, capacity, and win-win relationships, and will form a foundation for innovation driven by respect for cultural heritage and shared responsibility (Andrzejewski 2017).

14.2.4 Equity and Accessibility of Plant-Based Technologies

The shift to plant-based technologies is an avenue to solve multiple issues affecting the Earth, such as climate change, chronic ailments, and food security. There is, however, a dark side to the transition, involving important issues of equity, access, and justice. As plant-based solutions continue to go mainstream, it is necessary to explore their benefits without excluding the ethical and social considerations of their use and distribution. Increased consumption of alternative plant-based products would significantly disrupt current value chains, labor requirements, and capital investments in food and agricultural supply chains worldwide. There is some moral responsibility attached to these systemic changes, since they affect human health, their jobs, cultural practices, and the rights of both people and animals. From a human rights perspective, the likelihood of such changes should be assessed by their effects on the rights to food, adequate nutrition, fair wages, and safe working conditions. At the same time, the environmental benefits, including reduced greenhouse gas emissions, lower water consumption, and the preservation of biodiversity, could also be viewed in light of intergenerational justice and the value of the natural world in itself.

In addition, the application of plant-based technologies requires ethical pluralism. Varying stakeholders—scientists, policymakers, consumers, and indigenous people—have differing priorities for ethical frameworks, including human rights, animal welfare, environmental ethics, and economic justice. It is also the best moral justification for adopting such technologies, not based on a particular value structure but on a general, inclusive approach that recognizes the validity of different perspectives. The multidimensional nature of such ethical assessment makes decisions more balanced and socially acceptable to a greater extent.

Although the health benefits of a plant-based diet have been reported to include reduced risk of cardiovascular disease, diabetes, and other cancers, these benefits are unevenly distributed. It has been found that high fiber, antioxidant, and unsaturated fat intakes and low consumption of saturated fat and cholesterol characterize plant-based diets. Nutrient-dense, minimally processed plant foods, though, are more limited in most territories, especially in less well-endowed and poorer country locales. Conversely, there is a universality in the availability of ultra-processed plant-based foods, which on many occasions lead to ill health effects, including beverages with added sugar and salty snacks.

The cultural acceptability, availability, and affordability of plant-based diets remain highly pertinent to their consumption. Widespread access to healthy plant-based foods is blocked by barriers such as food deserts, the absence of culturally specific alternatives, and economic disparities. Such differences highlight one of the key ethical issues in plant-based transformations: the potential for them to entrench existing inequalities unless concrete efforts are made to make them inclusive. Food systems equity should include input costs, clear labeling on food products, and policies that facilitate the local production and distribution of healthy plant-based cuisines with cultural significance (Dwivedi et al. 2022).

Plant-based meat alternatives (PBMAs), including soy- or pea-based burgers, have attracted considerable consumer interest in their development and marketing. PBMAs are intended to mimic meat's sensory qualities, making them appealing to omnivores who want to eat sustainably. The products have the potential to reduce meat consumption and its environmental impacts. Nonetheless, there are concerns about their nutritional quality, heavy processing, and low availability. Although PBMAs have been promoted as beneficial to the environment, compared with conventional meat, they tend to be far more costly, target customers in wealthier urban centers, and are not available in most low- and middle-income areas.

According to this report, factors that affect consumer acceptance of plant-based technologies include taste, texture, price, convenience, cultural attitudes, and dietary norms. Such factors might be the only way plant-based alternatives could gain traction in many cases. Moreover, structural impediments, including poor food education, the absence of food sovereignty, and corporate-controlled foodscapes, must be dismantled to empower everyone to make fair food choices, as shown in Fig. 14.2.

To realize their full potential, plant-based solutions must be embedded in policies driven by justice, equity, and sustainability. Governments, civil society, and the private sector will have to collaborate to establish an environment in which the default diet consists of healthy, sustainable, and culturally acceptable plant-based foods. This is achieved through investing in agricultural diversification, local food systems, food literacy programs, and subsidies for nutritious plant-based foods.

Finally, the transition to plant-based technologies should also achieve fairness between the environment and the balance of inclusion and health equity. It must enable societies to acknowledge cultural and human origins and to know that bringing the benefits of the innovation to everyone is possible, no matter where they live, how much money they have, or which identity they hold. Accessibility and equity should not be add-on considerations; they lie at the core of the ethical development of sustainable food systems (Andreani et al. 2023).

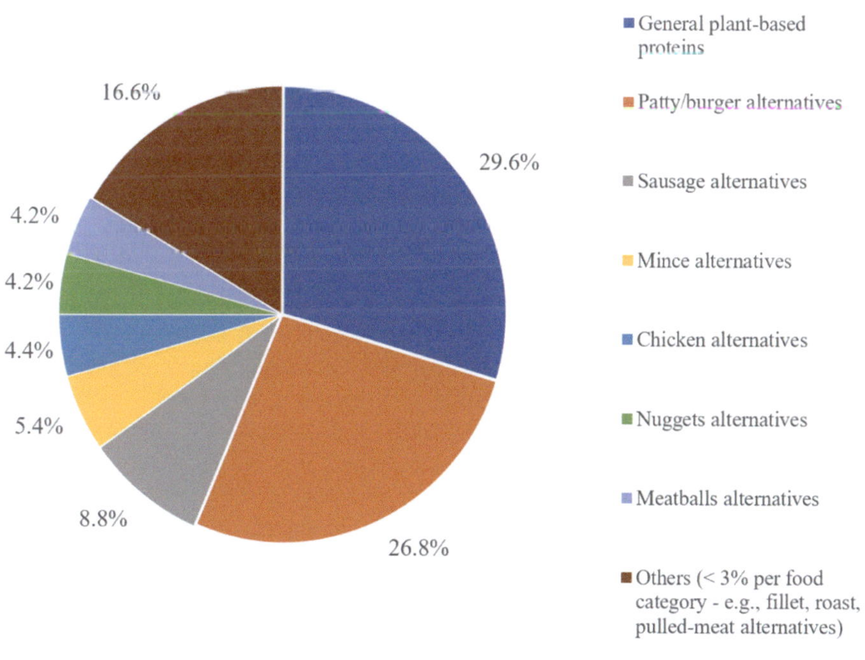

Fig. 14.2 Market distribution of plant-based meat alternatives across various product (Andreani et al. 2023)

14.2.5 Animal Ethics Versus Ecosystem Integrity

Mitigating climate change is critical to the sustainability of human societies and the natural environment. Conventional beef production in countries like the United States continues to contribute to greenhouse gas emissions, land degradation, and water use. There are also animal welfare issues that are of great concern to these systems. They have, however, substantial advantages in terms of the supply of nutrient-rich food, job creation for millions of people, and regional cultural support and economic development. New plant-based meat alternatives (PBMAs) have been touted as a revolutionary innovation that can address numerous environmental/animal welfare, and human health issues arising from meat production. Despite their potential, PBMAs entail a series of ethical trade-offs that should be carefully considered. According to the critics, these alternatives, particularly in their corporate-dominated, food-form, pose a danger of perpetuating an unfair and unsustainable food system. On the other hand, advocates claim that such technologies have the potential to significantly improve ethics by reducing the harms of conventional animal farming, provided their product development and distribution incorporate sustainable practices. Moving to plant-based substitutes is bound to disrupt existing value chains, alter the use of natural resources, and change the labor and capital intensity of agriculture. Such systemic changes are a morally relevant issue because they affect diverse stakeholders, including workers, communities, animals, and ecosystems. It is critical to evaluate such outcomes through various ethical perspectives, including human rights, economic justice, environmental integrity, and animal rights.

On the human rights front, access to healthy food, fair wages, and decent working conditions are essential. Concerns about justice emphasize that equitable health outcomes and equality of benefits and burdens across the population are important. Environmental ethics also raises awareness of the intrinsic value of biodiversity and ecosystems, as well as the need to prevent intergenerational harm from climate change. Similarly, animal ethics also focuses on ethical duties to minimize animal suffering and to acknowledge the duties of non-human animals to live without callousness or manipulation (Rojas-Downing et al. 2017). In addition, discussions of ethics linked to science have flooded the press with news about the human gut microbiota and its health effects. High-fiber diets containing significant amounts of plant fiber increase favorable bacterial fermentation, which raises resistance, reduces inflammation, and boosts metabolic and neurological function. These results favor the health benefits of plant-based diets and provide a biological rationale for changes in dietary habits.

Besides the biomedical and environmental arguments, social neuroscience has investigated the psychological and moral inclinations that inform decisions about eating plant-based foods. Neuroimaging studies have shown that vegetarians exhibit increased activation in brain regions associated with empathy, social mirroring, and self-representation. These discoveries are promising and suggest that ethical reasons behind dietary behavior may be embedded in the neurological front, such as compassion and harm aversion. Despite increased transparency about the ethical and ecological benefits of plant-based nutrition, there is a lack of policy integration. Addressing food system transformation requires a collaborative approach that encompasses scientific breakthroughs, lawful controls, and market-based improvements. Governments are being urged to redesign agricultural, health, and environmental policies to align with sustainable development objectives, ensuring that future food systems will be more ethically and ecologically resilient. Sustainable food systems should not just help reduce environmental loads, but also be clear, culturally appropriate, and beneficial to human and pet health. Therefore, the overall evaluation and implementation of innovations in the plant-based food sector must be grounded in a pluralistic ethic that respects conflicting values and seeks solutions to produce just and inclusive outcomes (Naliyadhara et al. 2023).

14.3 Environmental Impacts

The health benefits of plant-based diets have gained traction in recent years, as they support both human and environmental health. Large-scale animal farming is a major source of environmental degradation, particularly through greenhouse gas emissions, land loss, biodiversity loss, and water loss. Plant-based dietary shifts can be a solution to climate change and a way to reduce stress on ecosystems by decreasing consumption of animal products, including red meat. However, the environmental benefits of plant-based food systems need to be considered in a balanced way that accounts for the resource intensity of the plant-based food produced, as well as its nutritional value.

Approaches at multiple levels of governance in the United States have included a range of proposed policy measures aimed at transforming the paradigm for obtaining nutrition toward plant-based eating. The federal government finances cost-effective activities like the Supplemental Nutrition Assistance Program and the Dietary Guidelines for Americans that, to a great extent, promote healthy vegetable foods to motivate people. However, these programs do not place significant emphasis on diet-related sustainability. Municipal policies have introduced plant-based foods in schools, and complementary state-level efforts have encouraged schools to purchase local, sustainable produce, thereby incentivizing greater procurement. Urban agriculture and community-supported agriculture are other grassroots trends that seek to redirect food systems toward place-based, low-impact modes of production.

Eating habits have significant implications for environmental indicators. High consumption of animal-derived foods and highly processed foods contributes to very high levels of greenhouse gas emissions, water consumption, and land use. Plant-based diets (vegetarian and vegan) show lower environmental footprints across all indicators. Aleksandrowicz et al. (2016) estimated reductions of 22% in GHG emissions, 28% in land use, and 18% in water use across all plant-based diets. Vegan models of diet demonstrated the most significant reductions in emissions and land use, whereas vegetarian diets showed the best efficiency in water use. However, studies assessing the water implications of vegan diets yielded mixed outcomes, suggesting that agricultural practices, crop types, and geographic contexts can greatly influence water demand (Silverwood et al. 2025). Optimization modeling has also been used to guide dietary shifts toward more environmentally sustainable and nutritionally adequate changes. The models take concepts such as planetary limits on climate, land, and water use and combine them with nutrient and health requirements. For example, Chaudhary et al. (2018) examined three dietary scenarios worldwide, inter alia, adherence to global dietary guidelines, vegetarianism, and veganism. Findings revealed large declines in GHG emissions and freshwater use in North America and Europe under plant-based scenarios.

Nevertheless, in other parts of the world, such as South Asia, whose diets are already plant-based, augmenting nutritional sufficiency has only necessitated slight modifications in the use of environmental resources, thus revealing that nutritional and environmental outcomes are not homogeneous under all conditions, as seen in Fig. 14.3. Another dimension that has been under consideration in the definition of sustainable diets is the consideration of such diets, not only based on environmental and health standards, but also on the availability of the diet, its affordability, and cultural acceptability. It requires nutrient-rich foods with a limited environmental impact to be available to all groups of society and adaptable to regional food cultures. Such an intersectional strategy is important to the success of public health and environmental policy as well as to the popularization of sustainable dietary behaviors. To conclude, plant-based diets hold a potential sustainable change in the green agenda, though such a move comes with its own trade-offs. Based on the successful and fair results, approaches would have to be identified as diverse, with elements of environmental science, nutrition, food economics, and traditional society. Policymakers should aim to produce interventions that uphold human well-being and ecological resilience, given the intricacies of global and local food systems (Sahoo et al. 2024).

14.3.1 Land Use Change and Biodiversity Loss

Despite accounting for less than 20% of global calorie intake, the environmental footprint of animal-source foods (ASFs) is disproportionately large, accounting for a significant share of land-use change, biodiversity loss, greenhouse gas emissions (GHGEs), and water consumption. The conversion of pastures to grow crops is by far one of the biggest causes of ecosystem destruction through massive deforestation. Another example is that about the same percentage of the Amazon rainforest has been degraded by cattle farming, leading to significant loss of biological diversity and the release of carbon previously locked up in the rainforest. According to the Food and Agriculture Organization (FAO) report, Livestock Long Shadow, in 2006, it was noted that livestock used 26% of the Earth's surface area, 33% of the world's arable land for feed production, and generated 18% of the world's GHG

Fig. 14.3 Pillars of sustainable environmental technology encompassing energy, waste management, emissions reduction, AI applications, and agricultural safety (Sahoo et al. 2024)

emissions, even though later assessments like the 2009 Worldwatch Institute report had projected that, in fact, "livestock-related emissions" were, perhaps, higher than 51%.

The conversion of land to agricultural use is particularly significant in places with high biodiversity, such as the Amazon, where the landscape supports more than 40,000 plant species, thousands of animal species, and 200,000 native peoples. Such destruction of biodiverse systems not only endangers species but also undermines indigenous cultures and potential pharmaceutical discoveries. This leads to morally questioning the extent to which environmental destruction and cultural destruction can be accepted to fulfill rising ASF consumption, especially among the rich.

Dietary patterns have a significant impact on land use and biodiversity. The adoption of plant-based diets is also linked to decreased demand for water, land, and chemical inputs. For instance, vegetarian diets are 2.9-fold less water-demanding, 2.5-fold less energy-demanding, 13-fold less fertilizer-demanding, and 1.4-fold less pesticide-demanding than omnivorous diets. Besides, plant-based meat and dairy alternatives can be essential tools for reducing land-use impacts. An international model that forecasts that half of the main ASF (beef, pork, chicken, and milk) will be replaced with plant-based equivalents by 2050 suggests that the world terrorist threat will end and that agricultural GHGEs will decrease by 31% compared to 2020 levels. The possibility of carbon capture also exists with this substitution through reforestation of abandoned farmlands, which could deliver twofold climate-mitigation benefits and account for up to 25% of global land restoration goals under the Kunming-Montreal Global Biodiversity Framework.

Nevertheless, the environmental benefits of ASF replacement are non-linear. At over 50% substitution, diminishing saturation effects emerge: deforestation and land-use emissions are reduced, conversion ceases, and land is progressively left unused rather than recovered. Unless there is active reforestation or ecosystem restoration, the potential for dietary shifts in response to climate change is not realized. Furthermore, the consequential increase in GHG intensity in the remaining livestock system due to extensification and a decline in feed quality can be unintentionally triggered by high levels of substitution. The impacts are regionally distributed. China would witness the most transformation in its use of agricultural inputs, with Sub-Saharan Africa and South America poised to record the greatest improvement in their environmental conditions. The greatest benefits of beef substitution, with the most significant reductions in ASF sources, occur when multiple ASF sources are substituted, creating a synergistic effect. Finally, the problem is closely linked to changes in land use and biodiversity loss, given that diets centered around ASF are so. Although popular diets constitute a feasible pathway toward environmental sustainability and ethical food systems, the success of such transitions depends on an integrated set of measures aligned with them, including dietary shifts, technological improvements, policy support, and dynamic ecosystem recovery (Henry et al. 2019).

14.3.2 Water Footprint and Agrochemical Use

The co-occurrence of keywords extracted from environmental studies datasets can provide insight into existing trends, especially in sustainability-related studies. The recurring issues of greenhouse gases, climate change, and carbon footprints indicate that people are increasingly eager to study the effects of human activities on the environment. This tendency demonstrates the rapid development of research on the processes of understanding and mitigating climate change. The emergence of concepts such as life cycle assessment and environmental impact demonstrates that more integrated approaches to assessing the environmental impacts of products and services can be adopted. The approaches align with broader, widely accepted objectives of sustainable development.

Controlled environment agricultural systems, such as hydroponics and aquaponics, perform better than conventional agriculture in terms of efficiency in land, water, and nutrient use, which is necessary to meet the growing demand for food security worldwide. These systems, however, have the disadvantage of consuming much energy, and there is controversy over whether they qualify as organic. A life cycle assessment of the United States found that the practice of aquaponics, which produces fish and vegetables, yielded almost twice as much economic value per month as hydroponics, which produces vegetables only. Moreover, aquaponics displayed an endpoint environmental impact that was 45% lower, despite its total output value being greater. Heat and light electricity play a significant role in environmental costs in both systems, and other costs are associated with water management, fish feed, and fertilizer production. By switching to renewable energy sources, such as solar or wind power, these systems could be less harmful to the environment.

Once alternative protein sources are evaluated, a range of criteria should be considered, especially environmental and nutritional factors and resource efficiency. Their environmental impact is much lower than that of plant-based proteins, especially ruminant meats such as beef and lamb. Insect proteins are considered unconventional, but they are comparable to traditional meats in terms of digestibility and tend to have higher raw protein content than plant-based proteins. Animal proteins are nutritionally complete in the sense that they supply a full amino acid profile compared to some plant proteins that do not supply all nine essential amino acids. This is because protein quality is one of the most important parameters in determining dietary sustainability. The extensive use of agrochemicals has posed one of the significant environmental challenges of modern agricultural practices. This synthesis of farming began with the Green Revolution, which demanded heavy use of synthetic fertilizers and pesticides, greatly boosting yields and also resulting in unexpected ramifications for ecosystems. Agrochemicals have been necessary to increase world food production; yet, excessive use of these agrochemicals is endangering flora and fauna, soil health, and long-term sustainability (Berger et al. 2025).

To counter this, new and environmentally friendly methods are increasingly being embraced. One such strategy is promoting agrobiodiversity. This involves increasing the area under locally adapted landraces, crop rotations, intercropping, trap cropping, and creating habitats with natural predators for pests. Dependency on synthetic pesticides can be minimized through field engineering and plant traits that selectively promote the survival of natural enemies or reduce pest survival. Another way to reduce pesticide use is host plant resistance, which can be achieved through introgression breeding (introducing resistant alleles from wild relatives into elite crop lines). Agro-biodiversity-based ecological pest management is therefore a promising future of agriculture in terms of resilience and sustainability. It is consistent with global trends to reduce chemical use and increase the ecological sustainability of agricultural systems.

Adding credence to this is the book Agrochemicals in Soil and Environment: Impacts and Remediation, which features a comprehensive analysis of the effects of agrochemical use on the soil ecosystem. The book underscores the damage that the overuse of chemicals causes to soil's biological, chemical, and physical properties, including microbial diversity and nutrient retention. This type of degradation reduces crop productivity, putting food security at risk. The text goes on to outline that measurable indicators are required to determine the effects of agricultural contaminants. It examines how plants respond to various soil pollutants, such as heavy metals, synthetic chemicals, and agricultural runoff, and suggests sustainable approaches to soil cleanup, including bioremediation, plant-based technologies, and nanotechnology. These observations are important not only to the researchers and agronomists but also to policymakers concerned with sustainable agriculture and environmental health (Nyaupane et al. 2023).

14.3.3 Carbon Sequestration Potential Versus Emission Sources

Environmental sustainability of agricultural systems is closely linked to their carbon dynamics. In the wake of the mounting seriousness of climate change, the role of produce-based farming systems in reducing greenhouse gas (GHG) emissions and increasing carbon sequestration is attracting growing interest as a possible solution. These benefits cannot be automatically reaped; however, they hinge heavily on the farming practices used, the nature of the crop cultivated, and the overall supply chain actions involved in delivering food to consumers.

Industrial conventional agriculture, especially intensive tillage, monocultures, synthetic fertilizers, and intensive livestock farming, is a significant source of GHGs. These can be mistaken for being produced solely in farms through the use of fossil fuels, the application of fertilizers, and digestion (methanogenesis) by animals, but they are also produced through factory upstream and downstream activities, such as the construction of farms, processing and packaging of the foods, transporting, and selling them. It has been found that as much as 50% of GHG emissions related to food production occur during the production phase, with ruminant livestock accounting for a large share of these emissions, primarily methane (CH_4) and nitrous oxide (N_2O). These gases are much more potent in their global warming potential than carbon dioxide (CO_2), with methane estimated to be 28 times and nitrous oxide 265 times stronger than carbon dioxide over a 100-year interval.

Regenerative and climate-smart food production systems (including no-till farming, cover cropping, crop diversification, and the introduction of livestock to crop or rotational areas) offer an alternative set of approaches that can reduce emissions and sequester carbon, helping achieve both emissions cuts and increased soil carbon storage capacity. Nature-based solutions (NBS) in agriculture are emerging practices grounded in this foundation, which do not focus solely on productivity and income but also on climate resilience and mitigation. The practices enhance the storage of atmospheric carbon in soil by building soil organic matter, thereby improving fertility and long-term productivity while minimizing the impacts of erosion and land degradation.

One of the most significant social and environmental issues is the relative emissions intensity of animal- and plant-based food systems. Empirical research has repeatedly demonstrated that meat and dairy products have much greater GHG emissions in their production per kilogram than products based on plants. As another example, carbon footprints above 10 kg CO_2-equivalents per kg of product are very nearly (by far) limited to meat and dairy, specifically beef and lamb. Comparatively, plant-based protein sources, like soy, produce a fraction of these emissions and have much-reduced effects on acidification, eutrophication, and land use. In addition, switching to plant-based diets has been shown to produce significant reductions in national and individual GHG emissions. Research conducted in the United Kingdom, Finland, and the United States indicates that implementing vegan dietary cultures would reduce food-related emissions by 26–48% and total national emissions by 8% or more. These results highlight the mitigation capacity of dietary shifts, especially in the context of sustainable agriculture.

However, it is important to note that plant-based systems do not have zero emissions. Even the cultivation of some crops, particularly in industrial monoculture, can still involve intensive irrigation, fertilizer use, and long travel distances, all sources of carbon expenditure. Therefore, the environmental benefits of plant-based solutions must be viewed critically within an integrated framework that encompasses production processes, land-use efficiency, supply chain management, and consumer attitudes. To sum up, even though plant-based solutions themselves have huge potential for this end, as they would lower agricultural emissions and lead to more carbon sequestration, the real scope of their influence depends on moral decisions regarding farming practices, fair land policies, and policies that encourage agroecological integrity. The issue is not only the change in diet but also the need to redesign the structure of the systems that are involved in producing, processing, and distributing food in the form of plant products (Wu et al. 2024).

14.3.4 Monoculture Practices and Soil Degradation

It has also led to serious soil degradation and environmental issues through current monoculture farming methods, as many are calling for a more diversified farming model better suited to environmental conditions. UNCTAD (2013) highlights that the world would move toward diversified cropping and reduced dependence on chemical inputs in the production of agricultural crops, thereby benefiting both developed and developing countries. Food production must rise amid limited arable land, water, and energy, worsened by climate change, as the world population is expected to reach nine billion by 2050. The strongholds of genetically homogeneous crops, heavy dependence on fertilizers and pesticides, and intensive monoculture cultivation of land without alternatives have resulted in the loss of biodiversity, water pollution, and degradation of soil organic matter, thus leading to a long-term decline in soil productivity. Genetically modified crops also pose a threat to biodiversity, especially when engineered to be resistant to herbicides and pesticides. Impact assessments confirm that adopting agrobiodiversity and regenerative systems to compensate farmers for ecosystem services, such as clean water and biodiversity conservation, is important (UNCTAD, 2013). Meanwhile, sustainable food systems have come to the fore as people turn to plant-based diets to address the environmental burden of animal farming, including greenhouse gas emissions, land and water use, and pollution. Agriculture accounts for 30% of global greenhouse gas emissions, consumes 70% of usable freshwater, and covers more than a third of cultivable land. Foods of animal origin, especially meat and milk products, contribute so much to land degradation, as the production of livestock feed is the cause of deforestation and loss of habitats. The author gives the following example: in the U.S., intensive monoculture has resulted in serious soil erosion, with cropland eroding 13 times faster than it can withstand, and 60% of the pastureland is overgrazed. The burden on the environment underscores the need to reevaluate monoculture agricultural practices and to facilitate more ecologically stable, diversified systems that ensure long-term food security and healthy soil (Belete and Yadete 2023).

14.4 Socio-economic and Cultural Implications

14.4.1 Food Security and Agricultural Transition

The current movement toward plant-based alternatives as a solution to climate change and animal ethics challenges associated with industrial animal production has a complicated food security dimension. In its real sense, food security is a variance beyond mere dietary availability (e.g., food as a source of calories); it requires reliable access to healthy, safe, culturally acceptable, and nutritious food that can satisfy the nutritional needs of an active and healthy life. However, a narrative about plant-based food systems is that they can support environmental and health outcomes without adequately considering how agricultural shifts will affect food equity, access, and provision across social and economic groups and geographic locations. Such a disparity begs the question of who has won and who has lost amid the trend of adopting a diet based on plant-derived foods.

The protection of food security is often discussed within a limited paradigm that centers on calorie availability and the reduction of hunger. Nevertheless, the nutrition problems facing the world are not linear but multifaceted and involve undernutrition, micronutrient deficiencies, and burgeoning rates of overweight and obesity. Sometimes they occur in the same societies. Although plant-based diets can help support a healthier population and alleviate the burden of chronic conditions, given broader nutrition issues, they are not inherently the solution to all these problems. In low-income areas, where interventions against malnutrition and stunting remain prevalent, meat and dairy products can be replaced with plant-based alternatives, provided they are fortified with iron, vitamin B12, and essential amino acids. Thus, the shift to plant-based systems cannot afford to compromise nutrition needs on the altar of sustainability, per se.

The mismatch between the objectives of food security policies and the practice on the ground is one such ethical issue. Although definitions of food security may include nutritional well-being as an important objective, in reality, agricultural policies are strongly biased toward large-scale production and the efficient delivery of food to markets, rather than diversified food systems that produce nutrient-rich food. This disjuncture leaves a moral gap: Are policies intended to keep encouraging monocultures of large-scale plant crops for export and processing (e.g., soy and corn), or are they intended to favor small, more diverse, and ecologically integrated farming methods that serve local needs?

Some governments have supported plant-based agriculture due to pressures from environmental and ethical imperatives. As an example of such a policy development, the California Assembly Bill 1289 proposes funding and technical assistance to farmers shifting toward crop-based agricultural systems that do not include producing crops for animal feed. This bill points to a trend in which climate resilience has been equated with agricultural reform. However, this kind of initiative also needs to make sure that smaller and mid-sized farmers are not economically marginalized. There should be fair remuneration and infrastructure alongside the provision of new supply chains to ensure that farmers can pursue livelihoods and, in the process, serve environmental agendas in the context of ethical transitions.

There is also the fact that the current global nutrition transition (a shift from traditional diets to highly processed Western diets) compounds this situation of food insecurity. The mass-produced plant-based products aimed at replacing meat, including imitation meat and highly processed vegan snack foods, can be healthier for the environment but may be less nutritious than nutrient-rich whole plant foods. The new products are frequently transported to low- and middle-income nations via globalized food chains and displace traditional food diets that contain legumes, vegetables, and grains. Consequently, the existing populations that are at risk of malnutrition are susceptible to increased risks of non-communicable ailments like obesity, cardiovascular disease, and diabetes. The ethical duality is that, on the one hand, it should encourage sustainable diets, while, on the other hand, ensuring they do not reproduce the same health inequalities associated with industrial food systems.

On the producer side, livestock farmers face specific transition problems due to the integration of these activities into their culture and the economy, particularly in regions where meat production is deeply rooted. A study in Wales found that although some livestock farmers perceived long-term sustainability in plant-based markets, several limitations, such as land constraints, crop cultivation expertise, and rigid social culture, are restraining them. The transformational change in such an environment cannot be based on consumer demand or regulatory pressure. It is necessary to adopt participatory methods that entail inclusivity and embrace the identities, community orientations, and generational attachments of farmers to livestock practices. Seeking input from these stakeholders in the co-design of new plant-based farming approaches helps create a fairer transition that maintains the resilience of rural livelihoods.

In addition, cultural inclinations and food traditions have to be acknowledged in the exchange of ethical food. Sustainable land lifestyles are practiced by people in many indigenous and local communities, who have long sustained themselves with plant-based or semi-vegetarian diets. Nevertheless, dominant plant-based discourse, which is often consumed in Western consumer economies, may neglect these activities or even discriminate against them. On the contrary, plant-based ethical solutions ought to strengthen traditional knowledge systems and enable local communities to control their food futures, thereby improving food sovereignty and ecological stewardship.

Finally, the interface between food security and plant-based agricultural transitions also needs an overall, ethical, health-based approach. Policymakers, scientists, and activists should not only focus on the environmental justification for reducing meat intake but also on the economic, cultural, and nutritional outcomes of such action. The ongoing environmentally unfriendly practices risk reinforcement unless plant-based solutions are accompanied by inclusive planning and equal support systems. Ethical frameworks that prioritize the voices of farmers, as well as vulnerable communities and diverse cultural traditions, will be paramount to the making-over of a food system that is not merely vegetable but also sufficient and just (Bouteska et al. 2024).

14.4.2 Impact on Traditional Farming and Rural Livelihoods

The increased market pressure to adopt plant-based diets and alternative proteins, including lab-grown meat analogues and plant-based meat, presents both opportunities and threats to the continued existence of traditional farming methods. Although environmental and ethical arguments demonstrate that it is time to reduce consumption of animal-sourced foods (ASF) and turn to science and policy, the socioeconomic implications of changing lifestyles in rural communities and among farmers must also be discussed. The sudden transformation of the global food system can destabilize traditional livestock-based livelihoods that are integrally linked to the economic, cultural, and social environments of rural regions. Therefore, the moral aspects of this change extend beyond consumption behaviors to include equity, justice, and sustainability in agricultural livelihoods.

One of the most important issues arising in the literature and in interviews with stakeholders is the threat of corporate monopolization of the alternative protein industry. With the technological complexity of producing alt-meat products and the need to use bioreactors, precision fermentation, and synthetic biology, there is increasing anxiety that very few large corporations will have the capital and expertise to scale these technologies into profitable, sensible operations. Such a monopolistic trend may push out small-scale farmers and ranchers, essentially replicating the injustices of the current industrial food system under the guise of sustainable production. Unless access to infrastructure, intellectual property, and markets is provided, plant-based or cell-based alternatives will be commercialized, and the advantages of the transition may become confined to a few players, leaving rural communities behind.

More recent studies have highlighted the role of farmers and ranchers in the future of agriculture and have thus included them in discussions about it. Inclusive governance models should not view rural producers as passive objects of change or recipients of change, as they are the main stakeholders in the decision-making process. Their standpoints are key not only to the design of socially fair transition plans but also to the identification of feasible avenues of adjustment, including, for example, a shift toward climate-resistant crops, engagement in regenerative agriculture, and the redemption into new value chains focusing on sustainable plant-based foods. In the absence of such participatory strategies, the transition is at risk of deepening past country-urban inequalities and eroding confidence in food system change.

There is also a new social discourse shift that may further marginalize the usual animal farming. As plant-based alternatives continue to rise in popularity and moral standing in popular media, livestock inquiry might shift not only as a matter of environmental degradation but also as an ethical atrocity. This swift shift in social values can stigmatize livestock farmers, making them seem like an impediment to progress rather than heritage and land caretakers. This may not only diminish the rural culture of people whose history and economy are based on animal farming but also cause psychological pressure and result in a loss of sense of purpose among farmers.

Ecologically and nutritionally, the absorption of ASF leaves room for bioengineered plant-based foods, with the abundance of synthetic vitamins, segregated proteins, laboratory-created macromolecules, and devised nutrients raising doubts about the authenticity of this transition and its desirability. Although these innovations theoretically have the potential to offer a similar macronutrient and micronutrient profile to animal products, they also break with traditional foodways and involve centralized, high-tech production, which further disempowers local food producers. Also, as foods reach urban markets, knowledge gaps stemming from ignorance, technological impediments, or a lack of policy support might leave rural communities unable to access or adapt to the economic opportunities these foods offer (Gil et al. 2024).

The shift to plant-based, climate-friendly agriculture also affects the complex ecological and economic interrelationships in mixed farming. Livestock and food production go hand in hand in most parts of the world, but they are also used as a source of manure to fertilize the soil, support soil nutrient cycling, and produce a variety of foods. An improper break in this balance without other approaches to soil fertility may degrade land productivity and pose other threats to farmers' livelihoods. Also, reductions in commodity prices caused by fluctuations in global demand might have a detrimental impact on producers of livestock feed and other agricultural products. Therefore, the environmental benefits of ASF reduction need to be balanced against the economic harm that may be incurred, particularly by entities in the worst position to absorb the cost of transition.

It is critical to note that these developments occur within the broader context of threats to global food security, including climate change, biodiversity loss, and recent geopolitical shocks. In this regard, it is not merely ethically necessary to minimize environmental degradation but to plan transitions that build the resilience of the rural world and promote food sovereignty. The transition can be facilitated by maintaining a policy mix that supports livelihood diversification, investment in local food processing, and the establishment of policy tools such as payments for ecosystem services (PES). These methodologies incentivize farmers to store carbon, preserve biodiversity, and other attributes that are profitable for the environment as a whole and support the optimization of environmental stewardship and rural economic health.

To summarize, the shift toward plant-based food systems should be accompanied by comprehensive measures to support rural populations, with particular attention to those who have traditionally focused on livestock farming. This demands that policy interventions be oriented toward justice, economic inclusion, and adaptive capacity. A socially just transition does not simply mean replacing the source of food with another but reimagining the work of agriculture with respect to livelihoods, cultural traditions, and ecological health. Plant-based solutions will be sustainable and ethically acceptable only when they are transformed to concentrate the voices of rural people and share the benefits of the innovation (Boix-Fayos and De Vente 2023).

14.4.3 Cultural Acceptance and Dietary Shifts

The worldwide movement toward embracing plant-based diets is supported by environmental sustainability and better human health. However, the demonstration is that the shift away from meat-oriented food cultures and toward a vegetable-only diet is neither linear nor universal. Culture, religion, local food systems, and social identity are key factors that shape people's food habits and can be a serious limitation or facilitator of the transition to plant-based systems. In contrast to either nutritional or ecological arguments, what cultural evidence requires is knowledge of deeply rooted values, societal norms, and food symbolism. Eating habits are not just biological necessities; they are also cultural roles and the use of rituals, festivities, and family traditions.

Cultural identity resistance is one of the main hurdles to promoting dietary modification toward plant-based diets. Meat hunting and eating were especially linked with masculinity, wealth, and the intactness of traditional foodways in most societies, especially in the Western world. On the other hand, vegetarian or vegan diets can be marginal or extremist, especially in places where meat is relevant socially or economically. On the same note, some cultures dictate the eating of particular foods, or rather have taboos against certain foods, which adds to the complexity. In another example, vegetarianism might be compatible with Hindu and Buddhist practice but be regarded as the opposite of the cultural norms of other communities or as a form of identifying with the community.

The socio-cultural context also shapes the kind of support or opposition one can receive if he/she opts to use a plant-based lifestyle. Research on factors influencing the decision to transition to a plant-based diet has shown that people are more likely to adopt a plant-based diet when they receive familial and peer support. By contrast, those immersed in cultures where meat consumption has become normalized or even glorified might find it difficult to continue making this change, even though they are aware that it makes them healthier or more environmentally friendly. The media's exuberance in coverage, social discussions, and the presence of plant-based role models are compelling factors in helping people acclimate to new eating habits.

Besides, the availability of plant foods and their affordability are critical structural prerequisites for dietary change. Food behavior can be strongly influenced by the food environment, including the availability of plant-based foods in markets, restaurants, and public institutions. When clients from low-income groups are motivated to consume sustainably, the price difference between animal- and plant-origin products can hinder action. There is, therefore, a combination of cultural acceptability and food justice and equity. The just transition involves ensuring that culturally acceptable, affordable, and nutritionally balanced plant-based alternatives are accessible to the rest of the socio-economic group.

International institutions have recognized that a more comprehensive framework is required that incorporates cultural relevance alongside environmental concerns and health. The Food and Agriculture Organization (FAO) coined the term "sustainable diets," which focuses on the need to have diets that are environmentally sustainable, nutritious, economically sustainable, and culturally acceptable. Such a definition supports the notion that the dietary transition cannot be introduced in a cultural and structural vacuum. Sustainable diets should respect and uphold biodiversity and traditional foodways and should focus on the twofold burden of undernutrition and obesity (Pingali et al. 2023).

Different plant-based dietary patterns represent this. Although some people adhere to rather radical vegan diets that exclude all animal products, others focus on the Mediterranean, New Nordic, or Dietary Approaches to Stop Hypertension (DASH) diet. These diets also focus on plant foods but include some fish, dairy products, or lean meat. All these models have demonstrated different levels of cultural integration, health benefits, and environmental efficiency. For example, the Mediterranean diet, which is based on the traditional foods of Mediterranean countries, has already been well-researched for its beneficial effects on cardiovascular health, lifespan, and biodiversity preservation.

Medical studies promote the favorable health effects of plant-based diets, including lower risk of heart disease, better weight control, and reduced cases of type 2 diabetes. Randomized controlled trials and meta-analyses consistently indicate that people on vegetarian or vegan diets tend to have slightly lower body mass index

(BMI) and superior lipid profiles and glycemic control. Moreover, it has been estimated that plant-based diets are far less burdensome on the environment than omnivorous diets in terms of greenhouse gas emissions, land use, and water consumption. Nonetheless, improperly structured plant diets can result in deficiencies in micronutrients, such as vitamin B12, iron, and omega-3 fatty acids, which is why appropriate nutritional training is important, with a culturally adaptable focus.

Although personal health reasons are highly persuasive, the social and environmental ramifications for planetary health further underscore the need to switch to sustainable food models, making it even more urgent. Existing dieting patterns contribute to the global syndemic, a synergistic relationship between non-communicable diseases, undernutrition, and climate alteration. Therefore, vegan diets may provide an intervention target with strategic value that supports a variety of public health and ecological objectives simultaneously. The shift, however, should be introduced carefully to avoid imposing dietary models that conflict with accepted local traditions, financial situations, or even social forces.

Policymakers, health professionals, educators, and environmental advocates should work together to develop culturally sensitive policies that promote sustainable diets. This incorporates the need to update national dietary recommendations with the environment in mind, not only subsidizing the sale and production of vegetable foods but also incorporating food cultural choices into sustainability models. Education should emphasize the equivalence of plant-based diets with regional cuisines and build on this by aligning with social norms to change community-based strategies. Notably, people should not be made to feel that dietary change is the only responsibility. A decent and efficient change depends on structural support, alongside community empowerment and resource distribution.

To sum up, the moral and environmental reasons to move toward plant-based eating are compelling, but cultural acceptance should be a central issue to address, not rejecting people but supporting them on a systemic level. To achieve effective change in diet, it is not only necessary to understand that individuals should change their attitudes and behaviors but also to redesign the principles of society, the structures of institutions, and the culture of food. With cultural plurality legitimized and incorporated, animal-free solutions no longer need to be niche alternatives but can become more amenable, pluralistically bridging the world (Hess et al. 2025).

14.4.4 Ethical Supply Chains in Plant-Based Industries

Plant-based solutions have emerged rapidly as alternatives to traditional animal-derived food products and are restructuring global food supply chains. As consumer preferences shift toward more ethical, sustainable, and health-focused plant-based product purchases, attention is increasingly focused on ensuring that supply chains go beyond efficiency to be ethically responsible and environmentally friendly. Although the agriculture sector accounts for less than 2% of U.S. employment, the overall food system, which comprises food manufacturing and distribution, employs roughly 11% of the U.S. workforce and occupies over 50% of U.S. land and 80% of consumptive freshwater use. Such figures illustrate the enormity of the current position and the potential transformation of the food industry, especially the consequences of the proliferation of plant-based products.

To assess the potential effects of the widespread substitution of meat products on the economy and the environment, researchers have modified computationally intensive modeling techniques, such as the USAGE-Food model, an extended form of the computable general equilibrium USAGE model. This model represents the reallocation of the economy across 392 sectors and 27 related industries in the food sector and provides information on how resources such as labor, land, and water can be reallocated. The shift in plant-based value chains occurs through the use of inputs released by a beef-related area under several alternative protein scenarios (e.g., ALTP10 to ALTP60). It is important to note that the grain and oilseed processing industries are expected to show significant increases in output and employment and that they serve as the basis for the production of the core ingredients of plant-based food products.

Nevertheless, the transformation of agricultural inputs and overall economic resources toward plant-based production does not automatically address ethical or environmental dilemmas. Although plant-based proteins are widely understood as more sustainable and healthier, including pea protein, soy protein, legumes, and algae, ethical concerns should go beyond environmental considerations in animal agriculture. The issues include, e.g., labor rights, fair trade, the loss of biodiversity through monoculture farming, and the carbon footprint of plant ingredients imported worldwide. Ethical supply chains should be transparent, pay decent wages, have minimal environmental impact, and provide equal access to raw materials and outputs (Rogers et al. 2024).

The example of pea proteins demonstrates both opportunities and complexity. The prominence of peas is shaping their complete amino acid profile and nutritional value, offering a constructive alternative to soy and other protein sources. With increased demand for pea-based proteins, stakeholders should evaluate the structural soundness of supply lines to assess the feasibility, sustainability, and resilience of the pea-based protein market. A recent survey of stakeholders in the pea protein supply chain identified various strengths, such as nutritional superiority and consumerization, as well as weaknesses, including unpredictable supply of raw materials, limited processing capacity, and reliance on international trade channels. The Strengths, Weaknesses, Opportunities, Threats (SWOT) analysis of the pea protein industry identifies areas for improvement and investment. This practice highlighted the necessity of a local production network, infrastructure magnification, and open-governance mechanisms to ensure the ethical feasibility of the industry.

Moreover, despite all this, plant-based industries are commonly marketed as more ethical in practice; the spread of such markets should not repeat the exploitative patterns so closely associated with industrial farming. Ethical issues may arise when cash crops displace native landholders or when large agricultural farms control the food supply chain to the disadvantage of smallholders. Plant-based sustainable development, therefore, involves multi-stakeholders, multi-interventions in policy-making, and observance of international standards in labor and environmental policies. This also involves certification of organic practices, fair trade, and sustainable water management, as well as applying the principles of the circular economy to minimize waste and improve resource efficiency.

Given the 21–37% share of food system emissions in global emissions and the 30–35% of agricultural emissions caused by cattle alone, it is not hard to see that a shift to a plant-based food system is important for meeting climate goals. However, it is not only necessary to reduce emissions. At the same time, it is important to ensure that a sustainable approach to maintaining ethical integrity throughout the whole supply chain can be traced back to the farm and followed to the fork. Therefore, the vegan market will need to grow not only as an environmentally necessary force but also as an ethically innovative one. To sum up, a systems-based approach that combines economic viability, good stewardship, and social fairness in plant-based industries is desirable, as it offers a balanced approach that transcends ethical supply chain issues. By having macro-level foresight tools, including the USAGE-Food framework and case studies of pea protein supply chains, among others, that reveal on-the-ground realities, policymakers and industry leaders could better point the way to grow the plant-based industries sustainably. The development of ethical supply chains is not purely a logistical or economic endeavor; it is, in an ethical turn, a moral requirement that will prove that plant-based solutions really make a difference toward a more just and sustainable food future (Jafarzadeh et al. 2024).

14.5 Regulatory and Governance Challenges

14.5.1 International Frameworks and Protocols

The shift toward plant-based diets is increasingly recognized as a foundation for change across all structures of the global food system as it turns to sustainability. Nonetheless, the regulatory and governance structure is not keeping pace with the rate of innovation and consumer interest. These systems, which include global dietary recommendations, farm strategies, environmental conventions, and trade standards, play a vital role in defining food production, marketing, and consumption patterns across different countries. However, long-held socio-cultural structures, traditions of historical diets, and the lack of unifying international regulatory systems pose a significant obstacle to widespread change. The Lewin (1997) field theory is one of the major theoretical perspectives that support the complexity of a dietary change. He assumed that behavior is situated in a field or the world in which desires, goals, values, and anxieties define a person's life. The frames uphold the two areas, or schemata of interpretation, in the sense that the way people think and act in their world is shaped by them. When applied in food systems, the frames may serve as unconscious barriers to change, even when the evidence of a harmful environmental state or probable risks of health deterioration from meat consumption is irrefutable. Burnes (2004) explain that this typically results in structural and psychological resistance and that dietary behavior is especially obstinate.

For example, meat consumption is a significant part of most cultures and has therefore been seen as a way to have a proper meal. The studies indicate that, despite knowing the moral and environmental consequences of an animal-based diet, customers find it hard to switch to a vegetarian or plant-based diet because it lacks the culture of a full, complete, satisfying, and socially acceptable meal. Although middle- and upper-class groups can afford the cultural capital to explore plant-based diets, poorer groups usually have limited access to the so-called cultural repertoire that would help them embrace these changes (Abbaspour 2024).

Globally, the FAO sustainable diet definition and the UN Sustainable Development Goals (SDGs) are examples of frameworks that focus on the combination of nutrition, environmental sustainability, and social and economic equity. A popular working definition of sustainable diets is diets with low environmental consequences that promote food and nutrition security and a healthy life for the present and future generations (FAO 2010). Such diets should respect diversity in nature and foster self-respect, be culturally acceptable, accessible, economically equitable, and of sufficient nutritional value. Although plant-based diets meet these standards perfectly, the processes by which they could be disseminated in state policies are divided.

A multidimensional conceptual framework has been formulated to inform sustainable food transitions, with four major domains that address the complexity of ethical and environmental issues in plant-based solutions. The focus of the agricultural sector is on reducing the yield gap and achieving food security without further environmental degradation and on the efficient use of land and sustainable agricultural techniques, as shown in Fig. 14.4. The health and nutrition focus area aims to address the nutrient gap by encouraging a variety of nutrient-rich plant-based foods that can support human well-being across populations. The environmental sphere focuses on reducing greenhouse gas emissions, preserving biodiversity, and decreasing the overall ecological footprint through responsible sourcing and manufacturing practices. Finally, the social justice field addresses the issues of centralizing power in the food governance process, ensuring equal access to nutritious foods, and defending indigenous and local food cultures marginalized by dominant agricultural policy. This systems-level framework of the domains provides a holistic approach for policymakers, stakeholders, and researchers who strive to synchronize plant-based dietary change with broader human and environmental well-being objectives. This comprehensive model shows that making people eat a different type of food is not the only way to ensure sustainability; it is also necessary to alter the systems people use to control food production, distribution, and marketing. Take the example of food-based dietary guidelines (FBDGs) in several countries that have begun to incorporate sustainability principles, as they advise people to eat less meat and more plant-based foods. However, these rules differ significantly in terms of enforcement and cultural appeal. Countries such as Brazil and Canada have

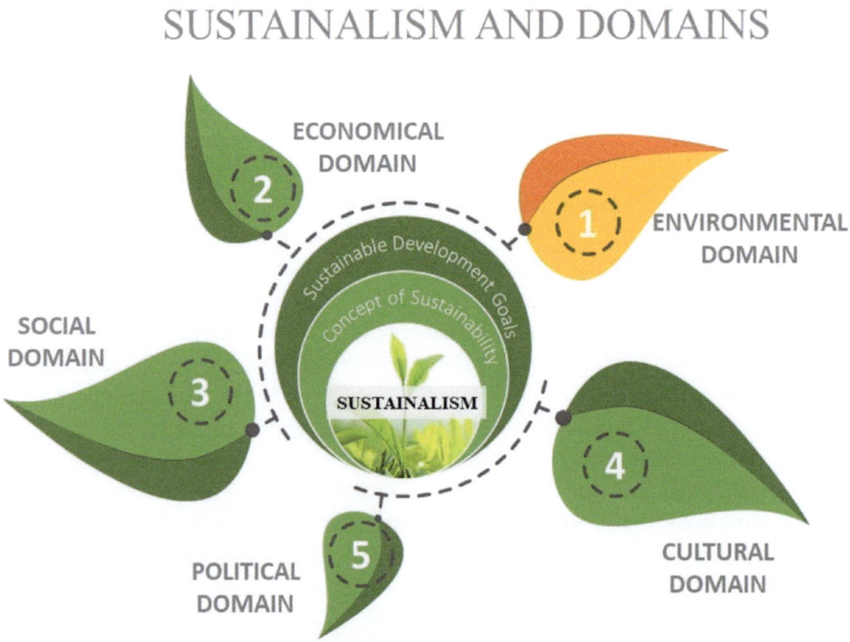

Fig. 14.4 Five key domains of sustainalism (Hariram et al. 2023)

introduced new, comprehensive FBDGs, whereas others still rely on agricultural lobbies and unreasonable nutrition guidelines.

There are also no harmonized global standards in the governance of alternative proteins, such as plant-based meat analogs, fermented proteins, and cultivated meat. This lack creates regulatory uncertainty for both producers and consumers. There are concerns about labeling, food safety, transparency in nutritional information, and assessments of environmental impact. These issues are now considered only in a consolidated manner by international regulatory bodies such as Codex Alimentarius, the WHO, and the World Trade Organization. Moreover, when it comes to the ethical regulation of food systems, the focus should be on research ethics and consumer rights. A recent study approved by the University of the Witwatersrand Ethics Committee (Protocol H22/10/03) followed best practices in data collection, as none of the research procedures coerced or manipulated people to participate, and the use and storage of data were protected by a confidentiality agreement. Although this is an exceptional case, it raises broader concerns about the ethical oversight of research and development in the plant-based sector.

Nevertheless, systemic and structural barriers notwithstanding, increasing international pressure on climate change mitigation, food justice, and sustainable agriculture presents the possibility of recontextualizing plant-based diets beyond individual choices and introducing them as one of the most decisive elements of both national and global policy. It is necessary to develop inclusive, evidence-based, and culturally responsive regulatory frameworks to scale up plant-based solutions to the crisis ethically and productively. Although this study has provided only a brief analysis of strategies to support the shift to plant-based diets and sustainable food systems, it can be concluded that changes in international frameworks and governance procedures are required to support this transition. Such development must take an interdisciplinary approach, with policy innovation grounded in environmental science, public health, sociology, and ethics. Lacking unity and agility in governance, the potential of plant-based solutions will be stopped in its tracks by bureaucratic stagnation, tradition, and a lack of regulatory spine (Hariram et al. 2023).

14.5.2 National Policies on Sustainable Agriculture and Ethics

The rise in living standards, especially in the Western world, over the last 100 years has resulted from innovations, societal changes, and advances in technology. However, these changes have also led to unsustainable practices, notably in contemporary food production systems. Thus, more and more people realize the urgent need to transition to more sustainable farming paradigms, especially by reducing consumption of animal-based foods. National policy systems are now playing a greater role in enacting these sustainability transitions by incorporating ethical considerations into agricultural management. A suggested model to lead this change is the sustainability-based dimensions of transition management through the identification of objectives, obstacles, options, opportunities, and outcomes—the so-called five O. The strategic framework helps to promote the transition to plant-based food systems.

The environmental case against eating animal products is now firmly established, as animal farming has been a key contributor to greenhouse gas emissions, land loss, and water insecurity. These are the systemic pressures on the environment; they concern the regulation of climate, biodiversity, and even equitable access to natural resources. To this end, many nations are taking a global approach, using indicators such as GHG emissions, land-use change, and water use to inform changes to the agricultural sector. However, some of the most basic ethical questions, such as whether plant-based and cultured meat alternatives are morally acceptable and whether their promotion aligns with the values of animal and environmental advocacy circles, are raised by the incentive to promote them. The concerns of individual autonomy, dietary freedom, and nutritional adequacy are also important when discussed in institutional contexts, such as universities, hospitals, and state or publicly available environments.

Researchers such as Ferguson (2025) propose only plant-based catering policies in these institutions as an option that is both ethically acceptable and politically feasible in terms of being a compromise between education and taxation approaches to the issue. The ethical principles in relating to such a situation place being non-stigmatizing about meat consumption, providing accessibility to nutritionally balanced and seemly plant-based systems, and the freedom to choose. Reflective equilibrium is one of the normative ethical methodologies currently used to address conflicts between principled ethical behavior and reality. Together, these emerging policy vehicles, conceptions of ethics, and strategic devices support a new model of advocating for plant-based solutions, one that seeks to balance environmental sustainability, human health, ethical responsibility, and social justice in national food systems (Natarajan et al. 2022).

14.5.3 Certification and Labeling Issues (e.g., Organic, Vegan, Non-GMO)

Consumers' trust and adoption of plant-based products, and the market's adoption of such products, are guided by certification and labeling systems, including organic, vegan, and non-GMO, which also raise complex legal, ethical, and informational issues. Consumers are increasingly interested in plant-based alternatives to traditional animal products, and these consumption orientations, including health, ethics, flavor, environment, and familiarity, usually guide their choices. Because they aim to replicate the existing meat products in terms of their textural and flavor characteristics to attract omnivores, plant-based producers also promote higher performance through sustainability and ethics, creating a paradox that renders regulatory labeling problematic. Such a strain is also present in the same legal discussions of meat-related terms applied to plant-based products, such as whether the terms burger or sausage should be used for a meatless product, with the argument that, when it comes to production differences, that is a reason plant-based products are relevant. In the Americas, regulatory control in the U.S. is split: the FDA regulates most foodstuffs under the Food, Drug, and Cosmetic Act, and the USDA regulates meat, poultry, and eggs.

The Nutrition Labeling and Education Act of 1990 also concentrated power, creating fewer opportunities to interfere with nutrition measures at the state level, and strengthened standardized labeling across the nation. Nevertheless, the practice of food labeling falls under the jurisdiction of both the FDA and the FTC, so other laws against false advertising must also apply to food labeling. Studies indicate that defining inefficiencies, including duplicative certifications (such as USDA Organic and Non-GMO Project Verified, NGPV), can mislead consumers and make them less visible. For example, the USDA Organic seal already excludes GMOs; therefore, the co-label NGPV is redundant but may increase consumer willingness to pay based on a false sense of security. To make all this even more complicated, there is also the case of certification exceptions: under present regulations, USDA Organic can contain up to 5% of non-organic ingredients, and NGPV-certified products can contain substances that are not technically under genetic engineering. These gray zones expose the ethical dilemma of transparency, at least to consumers whose beliefs run counter to anti-GMO positions or to those with the greatest interest in environmental effects. In addition, demographic factors such as age, income, and knowledge of certification schemes have been found to significantly influence buying behavior, according to market research. Due to the growing number of these inconsistencies, the overall effect is a hodgepodge labeling landscape that can lead to poorly informed consumer decisions made in good faith, as well as to the possible unintentional abuse of labeling loopholes by manufacturers. To back the plant-based transition ethically, definitions should be clarified, enforcement strengthened, and consumers educated so that labeling practices are sustainable and morally sound.

14.5.4 Corporate Accountability and Greenwashing

At the same time, though, the existing plant-based meat campaign does not provide severe resistance to the commercial control of agri-food. This section emphasizes that, first, big companies have established dominance in the plant-based meat industry by either partnering with startups or simply investing in them. Second, the solution must be corporate-guided because, in the first place, the current social and environmental issues are largely due to the growth of corporate agri-food, and under such a solution, those who created the problem would provide its solution. Third, there is no progressive opposition to this corporate domination, since the resistance to plant-based production has come from industrial meat producers. Lastly, the business model that Beyond Meat and Impossible Foods use is intertwining technical solutions with political ones. Consequently, although a technical fix exists, the political status quo, the domination of agri-food by corporations, is not only not changed but also not questioned, which continues the same circumstances that created the possibility to develop the current status quo (Kuralt 2021)

This thesis aims to investigate and discuss how Nordic food companies perceive sustainability and to evaluate the initiatives they have undertaken in this regard. In addition, this paper examines the strategic and sustainability checks that Nordic firms face in substituting plant-based product development and sales. With subsequent concerns about the environment, personal health, and ethical issues, people are asking firms to make their products and operations more sustainable. It has created tremendous demand for plant-based products, whose environmental and ethical footprints are also very low. Consequently, the food business is seeing Nordic companies show surprising interest in producing plant-based products to capitalize on this boom in demand. Moreover, Nordic companies are becoming more interested in sustainability and increasingly practicing sustainable production across their organizations. Works devoted to the growing concern for sustainability and sustainability transitions at the corporate level have become increasingly popular over the past decades. However, the food industry has not received sufficient attention, and this is an apparent fact: it is one of the most significant contributors to climate change. Past research on plant-based products has focused on individual consumers, most of whom have examined their perceptions of plant-based foods or the factors that drive them to change their diets. Nevertheless, limited research has examined the strategic and sustainability reasons that compel companies to develop vegetable-based products at the corporate level. This thesis examines the various reasons and drivers compelling companies to create plant-based products, based on nine interviews with experts spanning managers, directors, and even the CEOs of four non-plant-based and five plant-based food companies. Based on the results of this research, organizations focus on sustainability through three central pillars: environmental, social, and ethical. These have proved to be the primary drivers of greater adoption of more sustainable models. Conversely, the key drivers of plant-based product creation are the growing consumer demand for these foods, the significant business opportunities they offer, and their reduced environmental impact and health benefits compared to animal-based alternatives.

The plant-based companies, however, were the only ones to cite ethical considerations as a driver of plant-based products. Non-plant-based companies did not address the ethical considerations of plant-based products. In conclusion, despite the plant-based market being relatively small, the industry is experiencing very high growth. This leads to the conclusion that Nordic food companies should consider the role they want to take on in the field and explore various ways to be more sustainable. In most cases, this thesis creates an avenue for numerous future research studies evaluating the sustainability of the translons applied at the corporate level and for formulating plant-based products. The concept of sustainability has emerged as one of the most talked-about and reported issues in society. Since the climate is changing catastrophically on Earth, we must all change our habits, and in this case, businesses, particularly food companies, should commit to this. They can instead take advantage of the fact that consumers are now more interested in sustainability and engage in some form of greenwashing. Greenwashing is the practice of a firm withholding the truth by claiming it is more sustainable or promoting a product as more sustainable than it actually is. One study (thesis) aimed to examine the phenomenon of greenwashing, more precisely, the aspect of its use in the food industry, and examine the level of awareness of consumers of this phenomenon. The thesis discussed the most common forms of greenwashing and examined how authorities, particularly in the EU, are combating the issue. To gain an understanding of consumers' awareness of and attitudes toward this phenomenon, a survey was conducted as part of this thesis. The vast majority of respondents were Finns, and although much of the study focused on the broader EU markets, this further narrowed the study's focus to the Finnish markets.

The survey findings and theoretical framework were among the inputs used to develop important recommendations for companies in the food manufacturing industry on what they need to consider when selling their products, as well as what they can do to make this process more sustainable. These recommendations were aimed at influencing companies and providing managerial reasons for why they need to conduct their marketing more sustainably. The research found that greenwashing is highly common across the food sector. Things are slowly improving as more and more people become aware of such practices, and greenwashing is increasingly legalized. The study revealed that the negative effects of greenwashing can be very detrimental to companies, and that is why it is also helpful to companies, consumers, and the environment to switch to more environmentally friendly business and marketing strategies (Harwatt et al. 2024).

14.6 Case Studies

14.6.1 Soy Cultivation in the Amazon Basin

Plant-based soy proteins have become a pillar of the global effort to promote sustainable eating, as meat substitutes surge in popularity. In Brazil, a world leader in soy production, this expansion has sparked significant controversy over the environmental, social, and health effects of soy production and processing. Although the marketing for soy-based patty products tends to position them as eco-friendly and health-savvy substitutes for meat, the truth is more complicated and even runs counter to those messages. Currently, the vast area of Brazil is mainly under soybean production, accounting for around 4% of the country, leading to production that increased by more than 300% between 1990 and 2017. The Amazon biome, as a center of deforestation attention, may, however, obscure the fact that most soy farming occurs in the Cerrado biome, especially in Mato Grosso, a zone that, in addition to deforestation, biodiversity loss, and the emission of greenhouse gases (GHG), portrays land change due to land-use change. In addition, soy production is highly dependent on chemicals such as pesticides and fertilizers, leading to soil erosion, pollution of surface and underground water, and posing health hazards to neighboring populations. In meat alternatives made with soy proteins, such as soy burgers, extensive chemical processing is used to impart desirable sensory properties and a nutritional profile. An example is the extraction of soybean oil; hexane, a petroleum-based solvent, is normally used to obtain the oil, leaving behind defatted flakes, with the resulting product mainly used as animal feed. This characteristic of soy served as both a direct and an indirect food source when fed to animals, raising the question of whether the product of such dubious origins is indeed sustainable. Their highly processed character, the intake of admixtures and mixes, including methylcellulose, and the lack of transparent labeling further indicate that soy patties are sometimes not truly sustainable or really healthy. Although soy is undoubtedly a valuable source of both dietary and nutritional proteins, its environmental impact and the industrialization of its production for meat substitutes should be closely questioned.

The Soy Moratorium and the Roundtable on Responsible Soy (RTRS) are collaborative initiatives designed to reduce deforestation, especially in the Amazon, and increased non-governmental organisations (NGO) pressure has played a role in both. Nevertheless, policy debates continue, such as whether to extend protections to the Cerrado and how to balance private sector and government regulation. These efforts are tunnel-visioned on concerns about biodiversity, without taking a wider view of ecology and consumption patterns. The idea of distancing, which Princen presents, can be applied to describe the under-addressed issues with soy, including its environmental impact through a second-order connection to the industry (the use of soy in livestock feed chains). To sum up, despite soy-based meat alternatives seeming like a greener option, the production pathways reveal systemic sustainability issues. Those concerns require a multifaceted solution that includes putting agricultural processes on a sustainable path, enhancing transparency in processing and labeling, and reconsidering soy across direct and indirect food systems (Donoso et al. 2024).

14.6.2 Palm Oil Substitutes and Their Trade-Offs

Given the scale of production and its economic importance, palm oil production is also accompanied by numerous environmental and ethical issues that raise concerns worldwide and drive change. A prime example is the transformation of biodiversity forests into monoculture plantations, which has received substantial empirical support from researchers but has not been spared by distortion promoters and would-be politicians. When even the news media can manipulate public opinion, it is essential to publish clear, precise information about the environmental impact and societal repercussions of palm oil, leaving no room for doubt. It would help policymakers, companies, consumers, and civil society to make informed decisions about palm oil and alternatives because it is clear.

To counter the effects of the unsustainable practice, the world has given rise to certification schemes like the Roundtable on Sustainable Palm Oil (RSPO), which was kick-started and launched in 2004 by a collaboration between Unilever and the World Wildlife Fund. The RSPO and other certification schemes, such as the Indonesian ISPO, the Malaysian MSPO, and the Brazilian SPOPP, aim to control and refine their practices regarding deforestation, biodiversity loss, workers' rights, and communities' welfare. Despite this, less than 1% of the palm oil traded worldwide is RSPO-certified, and a higher proportion is produced under "No Deforestation" pledges. Also, palm oil's complex production system, driven by large-scale industrial plantations and smallholder operations, makes it challenging to draw blanket ethical conclusions.

A large percentage of palm oil producers, such as smallholders, operate in diverse socio-economic, political, and cultural settings, including organized outgrower schemes in Southeast Asia, traditional cultivation in Central Africa, and native areas of Borneo. Moreover, these nuances require contextualized reform and policy strategies. To mitigate environmental destruction and improve lives, some producers have begun experimenting more with sustainable options, such as mixed cropping systems and animal integration. The mere history of palm oil ethical trade-offs cannot be seen in a vacuum. However, it necessitates multidisciplinary cooperation and a more in-depth contact with the communities involved to make the vegetable oil production in the future less harmful to the sustainability agenda of the world and of the entirety of the UN Sustainable Development Goals (SDGs) (Naidu and Moorthy 2021).

14.6.3 Ethical Challenges in Plant-Based Meat Alternatives

These developments in plant-based and cultured meat products have reshaped global discussions about food ethics, sustainability, and people's health. These technologies offer a unique opportunity to untangle meat consumption, animal exploitation, deforestation, and heavy resource consumption. Meat grown from plants like soy, legumes, grains, and vegetables is called plant-based meat, whereas animal meat grown in bioreactors without animal breeding or slaughter is called cultured meat. The two aim to replicate the sensory experience of conventional meat, thereby offering a sustainable solution for the unconventional livestock sector. However, there are ancient roots to plant-based meat since it has long been a staple in Eastern communities. Foods like tofu and seitan have been around for several centuries in Eastern societies, but in the West, the research and development conducted on the plant-based food industry since the 1960s (with veggie burgers) and business in the 1980s set the stage for the advanced meat alternatives that are currently on the market today (Smith AF 2006).

Two decades ago, consumer demand for meat substitutes changed dramatically. Food, which was distinguishable from animal meat, is now highly engineered and almost indistinguishable. Startups like Beyond Meat and Impossible Foods already use high-tech food science to disassemble plant material into its basic components of amino acids, lipids, and minerals, and then recombine them to reproduce both the composition and flavor profile of meat. For example, Beyond Meat prioritizes molecular mimicry, namely recreating plant-based materials into a fibrous, juicy structure that mimics animal tissues (Jang et al. 2024). Such advances have not just led to greater consumer uptake; they have also prompted philosophical and moral arguments about the place of meat in human culture, food, and identity.

Plant-based meat has many ethical implications. First of all, the technology proposes a non-violent alternative to animal farming, thereby considering animal issues such as cruelty, animal welfare, and the mass commodification of sentient creatures. The factory farms are a major concern because there is a lot of overcrowding and inhumane treatment of animals in the farms. Plant-based alternatives may replace conventional meat, thereby eliminating the need to breed, raise, and kill billions of animals per year. This ethical incentive aligns with the interests of vegan and vegetarian philosophies, as well as the growing mainstream consumer concern for cruelty-free, compassion-oriented foodstuffs. Plant-based and cultured meats have a much smaller ecological footprint than food production, according to environmental ethics. Cultured meat has the potential to use 1% of the farmland, 4–18% of the water, and create 4–22% of the greenhouse gas emissions compared to conventional meat production (Tuomisto and de Mattos 2011). These numbers can show how plant-based meat could play a positive role in alleviating climate change, regenerating land, and improving water management. Second, infusing plant-based production may arise and promote the de-selective demand on natural habitats and deforestation and can also address biodiversity loss, all of which are directly connected to the ethical management of ecosystems (Profeta et al. 2021). Even though these outcomes are very positive, there are still some ethical issues. Nutritional integrity is a significant concern. Plant-based meats can mimic the taste and texture of meat. However, some critics suggest they are unlikely to fully substitute for the highly complex nutritional profile of animal meat, including essential amino acids, heme iron, and vitamin B12. The health effects of processed ingredients and additives to give food more flavor, color, and texture are also debated (Gastaldello et al. 2022). An ethical food system ought to minimize harm to animals and the environment while ensuring that human consumers are well nourished. This creates a conflict with what food engineering offers in terms of realism versus the historical wholesomeness of traditional plant-based diets.

What is more, sustainable eating trends raised ethical issues around corporate monopolies, food sovereignty, and equity, as the dominance of plant-based meat at the forefront of the market raised concerns. Big multinational food companies are moving into plant-based food production, potentially marginalizing small-scale producers and culturally entrenched food systems. Commercialization of plant-based meat also threatens to leave the economically weak sections of the population out if the products are not dispensed at comparatively low prices compared to conventional meat. In this way, even when these innovations facilitate ethical realities at the macro level, they create or even expand existing inequalities in access to food, cultural presence, and market dominance. Another ethical aspect of the plant-based meat production is its scale. Although high-tech offers solutions such as cultured meat and molecular reassembly that are more sustainable, the production processes may be resource-intensive, proprietary, and highly dependent on centralized infrastructure. This raises issues of energy use, reliance on technology, and the resilience of food systems. A truly ethical and sustainable change means the focus must not only be on the final product but also on the full lifecycle of production, from sourcing raw materials to distribution and waste disposal.

As an agrarian nation with biodiversity and a largely vegetarian population, India has a unique ability to take on the role of an ethical plant-based innovation leader worldwide. Local crops such as millets, pulses, and legumes can serve as the basis for locally adapted meat alternatives. By integrating this biodiversity, conventional food knowledge, and new food technologies, it can help create culturally acceptable food and sustainable products to cater to the needs of both local and international communities. Nonetheless, community relations, shared decision-making, and investment, especially by the government and private sector, are a must to achieve this potential. Plant-based meat is a significant contributor to achieving the UN Sustainable Development Goals (SDGs). These products directly contribute to SDG 2 (Zero Hunger) in providing alternative sources of protein which have lower resource intensity, SDG 3 (Good Health and Well-being) because of the reduced risks caused by red and processed meat consumption, SDG 12 (Responsible Consumption and Production), SDG 13 (Climate Action), and SDG 15 (Life on Land) due to the less environmental impact. Therefore, by incorporating plant-based meat into international food plans, the sector can advance in health, sustainability, and equity. Alternative meats are a landmark in establishing a new ethical framework for food production. They provide practical solutions to some of the current global threats, such as climate change, animal suffering, and food security. However, their potential should be accompanied by vigilant consideration of emergent ethical issues: nutritional adequacy, social equity, trade-offs between costs and environmental impacts, and cultural inclusivity. As long as research, innovation, and policy continue to support plant-based meat, it has the potential not only to be a technical innovation but also a paradigm shift that is truly ethical in the global food system (Xavier et al. 2025).

14.6.4 Bioplastics and Bioenergy Crops: A Double-Edged Sword

That is why the worsening of environmental degradation driven by the usual petrochemical-based industrial production requires a shift toward more sustainable, eco-friendlier alternatives. Among the most topical plant-based solutions arising in this context are bioplastics and bioenergy crops. Such innovations, which are biomass-based and derived from renewable sources such as starches, lignocellulose, vegetable oils, and agricultural residues, are even more praised for their properties of reducing reliance on fossil fuels, reducing greenhouse gas emissions, and contributing to a circular bioeconomy. Nonetheless, there is controversy surrounding their use. Although they are praised for being environmentally friendly and economically creative, there lies an intricate ethical-puritan and natural-resources balance involved in bioplastics and energy crop production that should be criticized diligently.

Bioplastics such as polylactic acid (PLA), polyhydroxyalkanoates (PHA), and starch-based polymers are touted as biodegradable and having a more minor environmental impact. Contrary to conventional plastics, which can take hundreds or even thousands of years to decompose, bioplastics degrade more quickly under certain conditions. They can be converted into agricultural by-products or into special crops like corn, sugarcane, or cassava. This has led to increased use in packaging, agriculture, medicine, and even textiles, due to its biodegradability. Likewise, bioenergy crops, e.g., sugarcane, switchgrass, jatropha, and miscanthus, are grown as a source of biofuels (ethanol, biodiesel, and biogas) as a renewable energy source that is cleaner and environmentally better than fossil fuels, to help achieve a global objective of renewable energy and a mitigation strategy to climate change. Still, the increasing production and commercial use of even these bio-based solutions pose several ethical and environmental questions. One of the most significant issues is the rivalry between bio-based crop production and food security. Depending on the level of bioplastic production and bioenergy crop development, diverting arable land and rivers to produce feedstock can jeopardize food supplies, particularly in areas already struggling with malnutrition and land shortages. This is usually known as the food vs. fuel or food vs. plastic debate, which has been recorded in nations whose staple products, such as maize or sugarcane, have been industrialized at the expense of local consumption, resulting in higher food prices and social and economic disparity (Ng et al. 2024).

Loss of biodiversity and interference with ecosystems are other important issues. In growing bioenergy crops or using plant materials as feedstock for bioplastic production, monoculture plantations resulting from such practices can turn natural habitats, such as forests, wetlands, or grasslands, into intensive agricultural areas. This conversion of the land is not only leading to the loss of biodiversity but also destabilizing ecosystem services such as carbon sequestration, water regulation, and pollination. Additionally, the intensive application of agrochemicals, water, and mechanized cultivation in these plantations can further degrade soil, pollute water, and, in turn, increase greenhouse gas emissions, thereby defeating the aforementioned sustainability outcomes.

An interesting case study can be presented with quinoa farming. Although quinoa is valued as a nutritious food and can be grown in arid areas where other crops cannot be cultivated, its increasing popularity across countries has had both favorable and unfavorable impacts. On the one hand, the production of quinoa has sustained local economies in the Andean region, enhanced food security, and facilitated integration of land use with that of llamas, thereby enabling natural fertilization and minimizing soil depletion. In other cases, however, intensification of quinoa production to meet market demands has led to the overexploitation of marginal lands, the expansion of monocultures, and the loss of biodiversity in quinoa production. Nevertheless, the new bioeconomic uses of quinoa, such as converting its saponins and farm waste into bio-based chemicals, textiles, and even bioplastics, raise the prospect of sustainable value addition, provided such activities are informed by fair and ecologically friendly policies.

This is a similar story with açai, a superfood that has been actively cultivated in the Amazon. Although the pulp of acai fruit has high market value, its seed, which accounts for more than 80% of the fruit, is often wasted, making it a significant source of organic waste. Critics say this biomass is a waste material that poses an environmental liability due to its high volume of disposal. However, recent developments aim to convert cai seeds into bioenergy or cement substitutes, thereby establishing a circular cai bioeconomy. These types of innovations demonstrate that it is possible to convert the by-products of the agro-industrial complex into valuable products, reduce waste, and make their processes more sustainable. Nonetheless, issues of benefit sharing with indigenous peoples, cultural sustainability, and ecological boundaries need to be part of such transitions.

In the broadest perspective of ethics, bioplastics and bioenergy plants are a sobering reminder of the two-edged sword of plant-based solutions. There is no discounting their advantages: lower carbon dioxide output, less plastic pollution, and the use of renewable resources. Their unintended consequences, however, such as displacement of indigenous communities, externalities on the environment, and social-economic inequalities, cannot be neglected. It is important not to evaluate only the cycle of emissions and biodegradability of these materials but also the justice of their production and use. The question is, who owns the land? Who are the benefactors of the innovations? Who is carrying the cost of the environment?

The solution to such complexities is a holistic form of governance, i.e., one that incorporates environmental science, social justice, indigenous rights, and economic policy. Participatory land-use planning, life cycle assessments (LCAs), and certification systems, e.g., FairBio or sustainable biomass labeling, are relevant for minimizing harms and providing good ethical choices. Also, the interdisciplinary nature of studies and multi-stakeholder discussions has to be actively promoted, since a shift toward plant-based materials cannot replicate the extractive logic of fossil capitalism in a green disguise. In short, bioplastics and bioenergy crops have significant potential to shepherd mankind toward a more sustainable tomorrow. However, to realize this potential, it is necessary to abandon technological optimism and pursue a critically reflective, ethically minded, socially inclined course of action. This is the only way the innovations will make a substantial contribution to ecological resilience and socio-economic equity during the era of a climate crisis (Chowdhury et al. 2025).

14.7 Innovations and Ethical Frameworks for Sustainability

14.7.1 Integrating Circular Economy in Plant-Based Systems

Bringing a plant-based solution to the overall vision of the radical shift in the food system needs a drastic shift in the economic approach of that economic model, where nature remains a resource to be depleted, a linear stressed model of taking, making, and waste, where culture can sit on a transition of the overall economic model to the regenerative, circular economic model. With the world facing the challenges of resource scarcity, population growth, urbanization, and climate change, a circular economy model can provide environmental benefits by promoting sustainability, resource efficiency, and ethical food production. The given model can not only reduce environmental impacts but also align with ethical principles, ensuring fair access, reducing waste, and refraining from exceeding planetary boundaries.

Circular food systems are all about maximizing resource use by valorizing unavoidable food waste and using plant-based inputs as efficiently as possible. In the case of plant-based diets, the first form of circularity is sustainable crop choice, which demands less water, land, and carbon footprint. Some legumes and pulses are good examples, as not only do they require fewer inputs for cultivation, but they can also enrich the soil by nitrogen fixation. Numerous studies have shown that plant-based proteins produce far lower greenhouse gas emissions, land use, and water use than animal-based proteins, based on comprehensive life cycle assessments (LCAs). It supports the environmental argument for shifting to plant-based diets and their contribution to climate change mitigation.

However, circularity has to go thoroughly toward food and, consequently, the whole-food value chain. Unaffordable agricultural and food processing waste streams, such as fruit peels, husks, and stems, can be converted into high-value bio-based products. These by-products can now be converted into bioenergy, bioplastics, nutraceuticals, and functional components thanks to new biorefinery systems that have become possible, which promote industrial symbiosis. For example, fruit processing waste can be processed to generate feedstock for regenerative biosurfactants in food, cosmetics, and pharmaceuticals. These types of systems encourage the closed-loop use of biomass, improving the usage of fossil fuels and developing bio-based circular economies.

In addition, when implementing the plant-based systems in a circular economy model, food packaging, distribution, and end-of-life considerations should be reconsidered. Agricultural residues can be used to create environmentally friendly alternatives to petroleum-based plastics, such as biodegradable and compostable materials. The materials reduce environmental pollution caused by plastics and contribute to soil conditioning via composting. Nevertheless, to enable such innovations to scale, it is essential to offer the supporting infrastructure and consumer awareness. Product take-back systems, reuse, and industrial composting facilities should be encouraged through policies focused on resource recovery, while landfilling and incineration are discouraged.

Circular plant-based systems incorporate social justice and ethical aspects. Decentralization of production, the promotion of local value chains, and the embodiment of circular models ensure that smallholder farmers have power in relation to dependence on global supply chains and the protection of indigenous knowledge systems. Also, any reduction in food losses and waste will directly contribute to food security by maximizing the utilization of available nutrients and calories. It is also possible to normalize sustainable consumption by promoting educational campaigns or institutional procurement policies, such as a policy requiring school and hospital foods to be plant-based, without violating the rights of those individuals.

Although the circular economy has many excellent prospects, there are still implementation obstacles, such as unclear policy advances, inadequate industrial preparedness, and a lack of financial support for circular practices. Paradoxically, the supported energy recovery techniques, such as incineration, can take precedence over the priorities of circular strategies, since the government has been more inclined to support them. There must be a clear direction in the policy to align the desired waste valorization with environmental and ethical objectives, with prevention, reduction, and upcycling taking precedence.

To conclude, applying the principles of the circular economy to plant-based systems is a crucial direction toward environmental sustainability, ethical integrity, and resource resilience. Getting to the point, the process of waste valorization and industrial innovation through sustainable dietary transitions and the circular models provides a plan for redesigning food systems in the twenty-first century. The multi-stakeholder approach to such integration requires collaboration on scientific innovation, progressive policy, industrial flexibility, and individual engagement, the success of which depends on all of them. It is only in the broadest, most participatory way that we can develop food systems that are not only ecologically sustainable but also just and ethical (Ungerman and Dědková 2020).

14.7.2 Ethical Design and Life Cycle Assessment Tools

Plant-based solutions should be innovative and responsible, which is why, in the transition to sustainable food systems, ethical design and environmental assessment methods are essential for developing new solutions. The scope of design-driven innovation, especially with respect to user engagement, encourages the creation of value-creating systems that can co-create with consumers, rather than responding to market demands. This user-focused strategy increases ethicality, sustainability, and social inclusivity by aligning strategy with implementation and promoting never-ending innovation in the food systems. Ethical design models play a crucial role in reforming food systems to be inclusive, transparent, and responsive, as organizations further explore the benefits of global networks to develop tailored, value-laden experiences. In tandem with this process, artificial intelligence (AI) is swiftly transforming food production, distribution, and consumption by making the process more efficient while generating lower environmental consequences.

Nonetheless, AI integration also poses serious ethical challenges, including algorithmic discrimination, privacy concerns, unequal access, and environmental harm from AI infrastructure. A collaborative, multi-stakeholder model and robust ethics frameworks are required to make AI a sustainable and socially equitable tool, not the opposite. The concept of sustainable food design is an interdisciplinary framework that deals with the entire lifecycle of food systems, such as their production and processing, consumption, and wastage, in efforts to minimize environmental impacts, enhance the health of the population, safeguard biodiversity, and be locally beneficial and fair to all within the economies. However, most studies have not comprehensively examined the interplay among AI, ethical imperatives, and the sustainability of food systems. The current literature focuses either on AI applications or on sustainable food systems individually, without necessarily discussing the subtle ethical and ecological trade-offs that their coexistence would entail. This section aims to fill this gap by providing a comprehensive review of how AI can be used to aid the design of sustainable food, while introducing new sets of ethical issues to consider. The Environmental Life Cycle Assessment (LCA) can provide insight into the environmental costs of dietary changes. LCA analyses have consistently shown that low-animal-product diets, especially vegetarian and vegan diets, result in significant reductions in greenhouse gas (GHG) emissions, land use, and water use. Although these are undeniable environmental advantages, policy has paid little attention to large-scale dietary change because of socio-cultural resistance and a reluctance to disrupt norms perceived as embedded in values. The industry has tried solutions like plant-based meat, such as Beyond Meat and Impossible Foods, as long as they are and prove to be better in a sustainable way. Several comparative LCA studies carried out in the intervening years 2000–2019 have confirmed the environmental benefits of such substitutes, and 34 studies are identified that meet high inclusion criteria regarding scope, methodology, and quantifiable reductions in environmental impact. Although LCAs can easily encounter obstacles such as variable system boundaries or inconsistent reporting (e.g., failure to report on transportation impacts in specific reports), they are invaluable for quantifying sustainability improvements and for policy-making. Individually, the principles of ethical design and sound LCA approaches can be instrumental in assessing and steering sustainable innovations on the path to plant-based systems. However, when combined, they can serve as a solid basis for assessing and steering sustainable innovations ethically, ensuring they are not only technologically feasible and environmentally positive but also ethically correct and socially just (Marques et al. 2025).

14.7.3 Role of AI and Data Transparency in Ethical Plant-Based Solutions

The global health crisis, the looming climate emergency, and rapidly increasing healthcare expenditure have further underscored the importance of sustainable and effective dietary transitions. The transformation to plant-based diets (PBDs) is among the most evidence-based interventions, with high potential to prevent, manage, and even reverse chronic non-communicable diseases (CNCDs), including coronary artery disease, type 2 diabetes mellitus, high blood pressure, and obesity. Case studies like those at Bethsaida Hospital in Indonesia have shown that patient outcomes using structured PBDs can be transformative, with about a 70% drop in restenosis and medical dependency achieved at this hospital through synergistic integration with refined cardiology. Artificial intelligence (AI) and its contribution to enhancing the effectiveness of PBDs are gaining prominence. AI would enable the development of individualized nutrition plans, monitor metabolism in real time, and implement adaptive feedback control mechanisms that could improve dietary adherence and clinical performance. Wearable devices, predictive modeling, and nutritional algorithms are driving the revolution in dietary medicine, making this practice scalable and accurate. These systems not only empower patients but also provide clinicians with real-world decision-support tools. In addition, concepts like the Deep Connect value provide an economically oriented explanation of the possibility of AI-based innovations, including plant-based foods and biosimilars, to produce multidimensional societal and environmental goodness. This is a new value paradigm, with historical precursors including a dynamic economy of business ethics and economic synergies, in which future AI-driven plant-based innovation, combined with community-based manufacturing formats, may become a major macroeconomic NC construct by 2035.

The functionality and acceptability of plant proteins are also accelerated by integrating next-generation food processing technologies, including high-pressure processing, pulsed electric fields, precision fermentation, and 3D extrusion, as shown in Fig. 14.5. AI is significant in controlling and managing these technologies to achieve greater accuracy when tailoring texture, nutrient profiles, and sensory properties. Though life cycle assessments (LCAs) indicate that meat substitutes offer greater environmental benefits than cultured meat, their popularity depends on consumer behavior and the ability to change. The shift in diet is also psychologically complex, as highlighted through behavioral studies. Emotions such as pride and guilt can promote the use of less meat. Disgust, both physical and moral, can produce mixed effects across individuals depending on their stage of change and moral orientation. The intervention must use customized communication that takes these emotional aspects into account. Notably, applying AI to such behavioral models in the context of real-time diet tracking is beneficial, as it fosters transparency and individualization, aligning with the ethical values of letting people do what they want with transparency and justice, and providing people with accessible, fair nutrition. With the development of AI and digital technologies, ethical data governance,

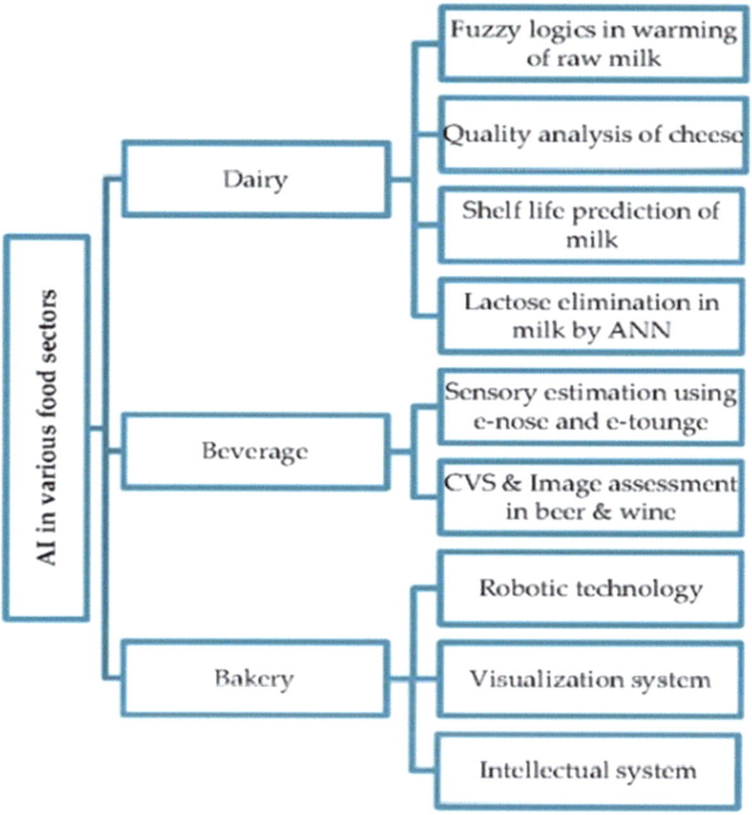

Fig. 14.5 Applications of AI in the Food sector (Taneja et al. 2023)

reduced algorithmic bias, and fair access are now required. By combining AI integration and data transparency, such a solution not only maximizes the efficiency of available plant-based solutions but also assumes the ethical responsibility to create inclusive, sustainable, and health-promoting food systems in the future (Taneja et al. 2023).

14.7.4 Community-Led and Participatory Approaches

Overall, in light of the climate emergency, food system localization and community-based planning and policies have become a highly significant, fast-growing force in the decarbonization of food systems, as they have demanded vastly greater rates and extents of decarbonization than have been achieved primarily through governmental initiatives. Participatory, action-based, community-empowered net-zero food initiatives offer a localizable, populist approach to facilitating plant-based shifts and simultaneously improving food equity and food security. An exemplary case study of Bristol, UK, demonstrates how the co-design of interventions, which can be used to enable the adoption of lower carbon, plant-based diets, can involve the community stakeholders, including policymakers, local businesses, and civil society, using relevant interventions such as public procurement reform, urban community gardens, and food choice architecture. On the one hand, these measures are encouraging, but they face obstacles, including the need for special funding, citizen opposition, and misinformation.

However, enablers like faith centers and community hubs, many of which come with kitchens and land as well as networks, provide viable systems for providing food inclusively, particularly to vulnerable populations. The example of community climate action in Bristol, the "Grow, Cook, and Eat" intervention, provides evidence of a measurable benefit: an estimated 0.8 tons/year reduction in CO_2, with a significant part of the benefit coming from reducing food waste. These forms of hyperlocal focus on holistic planning that incorporates climate, health, and social equity objectives and does not require direct governmental intervention. In the United States, parallel efforts like One Grub Community and Planty for the People in El Paso, Texas, demonstrate how urban gardening, volunteer-based activity, and plant-based meals can help combat food insecurity in underserved border communities.

Developed with assistance from community-based participatory research (CBPR) and urban political ecology approaches, these projects not only capture acts of social entrepreneurship but also make nutritionally rich, culturally suitable food accessible to underprivileged groups. They have employed a multipronged strategy comprising operational activities such as cafes, advocacy practices, and education outreach, which address both the emergency health conditions (e.g., food deserts and disparities) and structural inequalities, making them more resilient in the face of a crisis such as the COVID-19 pandemic. Also, these models prioritize inclusivity, emotional safety, and intersectional outreach and remind us that plant-based solutions cannot just be about a change in the diet but must be transformative tools of healing, empowerment, and resistance to structural change. Collectively, these grassroots movements are a sign of a greater ethical mandate: the need to rebalance our food systems with both planetary and ethical boundaries; to put it in other words, to shift our food systems to values of justice, dignity, and community resilience (Fanzo et al. 2021).

14.8 Future Outlook and Recommendations

14.8.1 Balancing Ethical, Environmental, and Economic Goals

Over the past few years, an increasingly broad assortment of alternative protein (AP) products, comprising plant-based meat and milk substitutes in the past and cell-cultured meat and seafood lately, has commenced occurring in the international markets, especially those in the high-income nations such as the United States. The products mimic the sensory and nutritional characteristics of meat-based foods while also addressing critical environmental, ethical, and health issues. Based on technological affordances and market discourses, the AP industry represents an upward shift in alignment with consumer preferences, market imperatives for sustainability, and ethical standards in animal welfare. But a new generation of these alternatives also introduces its own round of wicked problems, as they are complex, interconnected challenges that are very difficult and slow to solve. The center of the whole thing is the dainty balancing of ethical intent, environmental pragmatism, and economic viability. On the one hand, AP products are a guarantee of fewer greenhouse gas emissions (GHG) and lower land and water consumption, as well as higher animal welfare. In particular, life cycle assessments (LCAs) have estimated that plant-based alternatives such as the Beyond Meat burger can help cut GHG emissions, land use, and water use by up to 90 per cent compared to beef. Nonetheless, cell-cultured meat is resource-efficient compared to livestock in the long term, but due to its high energy requirements, it raises questions about scalability and, therefore, its overall environmental impact. These trade-offs emphasize that the impact of production on the upstream and the downstream effects on food system change should be considered.

At the health level, although there is a positive association between reduced processing of plant-based foods and overall well-being, there is no reason to believe that highly processed AP products can promote better health. Several of these items contain additives, taste enhancers, and emulsifiers to give them a meat-like flavor and texture, hence interfering with nutritional value. In addition, the fast-changing environment of the marketing industry, with competing stories and aggressive advertising, can lead to confusion among consumers, which is fueled by false claims about the real health and sustainability value of AP products. The food marketplace is crowded and confusing, and despite the claims of many food brands, terms used to describe food, such as plant based, natural, and sustainable, are interpreted loosely and are not well-regulated.

The scenario is further complicated by regulatory inertia. Development of standardized labeling criteria, education programs, or comprehensive policies advancing how AP products could be incorporated into a sustainable diet regimen has been a sluggish effort by the federal agencies in most countries, including the U.S. This creates a vacuum of knowledge within the general societal knowledge base and gives unethical information a free run further to embrace healthy and environmentally compatible food choices in the society. Moreover, unequal acceptance and trust in AP innovations are driven by differences in consumer perceptions shaped by cultural, social, and economic contexts, especially among underrepresented communities that have traditionally not been served by the health and food systems (Surya Ulhas et al. 2023). It is a systematic coordination of efforts on many levels to balance these complex dynamics. There is also a need for policymakers to establish science-based standards for labeling AP products and for claiming sustainability. Types of dietary intake should be aware of the distinctions between minimally and ultra-processed plant-based foods and be informed about them by educational institutions and health campaigns. Scientists should also conduct more research on the long-term health effects, as well as on the supply chain emissions and resource use of various protein production models. In the meantime, food producers should be held liable for guaranteed transparency in sourcing, ethical labor standards, and equal distribution systems that guarantee access for everyone.

The implementation of sustainable diets, after all, has to be more than a change in consumption, with meat swapped out in favor of plant-based alternatives; it will require a fundamental redesign of the food system. These are not only initiatives to consume a whole-food, plant-based diet but also to support local and regenerative farming and to promote ethical principles (fairness, animal welfare, and intergenerational justice). Applied separately to AP innovation, the UN Sustainable Development Goals, the Paris Climate Accord, and biodiversity conservation will demand not only technological creativity but also political will and social participation. Ethical, environmental, and economic objectives cannot be competitive priorities in the long-run goal; they are simply indicators of a genuinely sustainable food future (Velenturf and Purnell 2021).

14.8.2 Policy Recommendations for Ethical Transitions

The necessity of developing a way out of the negative consequences of the present global food system cannot be underestimated. The traditional omnivorous diet, high in meat, dairy, and ultra-processed foods, has been cited as one of the primary drivers of greenhouse gas emissions, deforestation, water pollution, and biodiversity loss. To address these adverse consequences, it is necessary to systematically shift current eating patterns toward more sustainable, plant-focused practices. But the transition will not be natural or evenly distributed until it is steered by mutually consistent, encompassing, and evidence-based policies that go beyond focusing solely on environmental outcomes and also address ethical and social equality objectives. Key to this reform is the mobilization of strategic communication as wielded by force multipliers, including governments, the food system, and civil societies, to educate, motivate, and encourage substantial-scale change in diets.

Governments are central in shaping society's behavior and food environments. Representing one of the most effective interventions, revising national food-based dietary guidelines to incorporate the existing scientific evidence on the health-promoting effects of plant-based diets is highly valuable. The new guidelines must clearly recommend reducing meat and dairy intake and integrating various whole plant foods. In addition, these policy frameworks must not only be scientifically sound but also culturally sensitive and affordable. They should be accompanied by simplified yet effective, actionable messages for the general public and stakeholders in the food value chain, including producers, retailers, educators, and health care providers (Leroy et al. 2022).

Along with updating dietary guidelines, it is also essential that governments engage with the population through so-called public communication to legitimize plant-based diets as a beneficial and mainstream choice. These include incorporating sustainability into school learning, mass health campaigns, procurement policies, and media campaigns. The government should be at the forefront, setting a good example by adopting a plant-forward menu and minimizing reliance on animal products, as seen in schools, hospitals, and government cafeterias. At the same time, government subsidies and tax policies, including those targeted at high-emission products or that reward sustainable agricultural practices, can shift market dynamics, making plant-based decisions more cost-efficient.

Another potential tool of change is the food industry. Food manufacturers, distributors, and marketers wield significant power over consumer behavior, as they are the gatekeepers of what appears on store shelves and restaurant menus. The industry must focus on the transparency of product labeling and clearly provide details about the environmental and health outcomes of the presentations they portray. Consumer adoption can also be driven by investing in research and development to make plant-based alternatives tastier, with better texture, and more affordable. Greenwashing should not be used as a marketing approach; marketing strategies should support the verifiable advantages of plant-based products, such as reduced emissions, improved animal welfare, and health benefits. Retailers and brands can also influence default choices, e.g., by prominently displaying plant-based products and selling them concurrently at competitive prices.

Last but not least, citizen contributions, as consumers of food and as activists, are critical to influencing ethical food system transformation. People can promote vegetable foods at school and their place of employment, engage with local governments to support community-supported agriculture, and use digital sources to share information. There is already a change in social norms regarding sustainability, animal ethics, and the consumption of health-friendly foods, initiated by citizens. By giving these voices an equal footing via participatory policy-making, community empowerment, and grassroots education, one will enhance their voices. In addition, online communication and social media-based campaigns, particularly those created by influencers, doctors, and environmental activists, can help demystify the process of adopting plant-based diets and create a sense of communal destiny and empowerment. In conclusion, a multi-stakeholder communication approach must aim to change norms, systems, and incentives, in addition to raising awareness, to facilitate the necessary ethical and environmentally sound transition in diet. At the leadership level, governments should also be at the forefront, with policies and messages for the people. The food industry must align its products and narratives with sustainability values. It is not enough to allow citizens to make decisions; we must also advocate for plant-based solutions. An evidence informed, ethically sound, and coordinated communication structure is needed to guide the food system toward a sustainable, fair, and healthier future (Ullah et al. 2020).

14.8.3 Stakeholder Collaboration and Global Responsibility

A growing reliance on ultra-processed and animal products has characterized the global dietary pattern shift over the last few decades. This has also been coupled with increasing environmental degradation, increased cases of non-communicable diseases, and the issues of food system sustainability. A shift toward more plant-based diets is a potential way to avoid such negative effects by promoting more ecologically sustainable food production and reducing health disparities among populations. However, this transition is intricate, requiring not only changes in consumer behavior but also the harmonization and concerted effort of various stakeholders at the national and international levels.

A good example can be traced to Spain, which was considered the epicenter of the Mediterranean diet, which emphasizes the consumption of many vegetables. The demand for animal products over plant products has led to a rise in interest in Spain in recent years. Nevertheless, as is the case in the food industry, a grade stakeholder analysis of the Spanish agri-food sector shows that the introduction of new plant-derived food products is met with varying levels of reception across stakeholder categories. Although consumers largely perceive these alternatives as more sustainable, healthier, and economically more promising, environmental non-governmental organisations (NGOs), policymakers, and researchers, as well as stakeholders in the core chain of production, such as producers, processors, and distributors, remain skeptical. They usually have apprehensions about factors such as taste, texture, complex processing, and excessive expenditure associated with plant-based innovations (Kearney 2010).

This difference highlights the core issue of a mismatch between traditional forms of food system support (the stakeholders who prioritize production) and the broader social need to shift diets toward environmental sustainability. This divide can only be bridged through encompassing discourse, the exchange of ideas, collaboration, and policy guidelines. Stakeholder cooperation is not only desirable but also crucial for reversing the multidimensional challenges inherent in food system change. Moreover, there is still a significant research gap in assessing the attitudes of non-consumer stakeholders toward the plant-based food system. Though many studies have been conducted on the sensory acceptability, nutritive quality, sustainability, and economic feasibility of plant-based alternatives, the system-related implications, especially the socio-economic, technological, and organizational implications, are under-investigated. The majority of scientific questions focus on consumers, thereby obscuring the opinions and interests of key players in the agri-food chain, which directly affect production, distribution, and policy implementation.

In trying to overcome this gap, superior assessment techniques such as Multicriteria Analysis (MCA) and the Strengths, Weaknesses, Opportunities and Threats (SWOT) model and frameworks have come in handy. The methods provide systematic, open-minded grounds for assessing multidimensional and complex problems. Among them, Multicriteria Mapping (MCM) has proved to be an interesting approach, particularly in regions with high levels of uncertainty and competing values, such as in food system transitions. MCM enables a variety of stakeholder groups to think systematically about the relative merits and trade-offs between traditional and novel food products across common and disputed value spheres, including economic profitability, health, environmental sustainability, sensory appeal, and technologies.

When applied to the Spanish context, a fragmentation of attitudes worth noting emerges from combining an MCM and a SWOT analysis. Lack of familiarity, insufficient investment in innovation, and poor integration with existing agro-industrial systems are among the criticisms producers and processors have made of plant-based alternatives. Conversely, policymakers and researchers cited new opportunities related to sustainable global goals, changing consumer tastes, and the potential for economic diversification. The results show the need for a model of collaborative governance that fosters mutual understanding and collective problem-solving among diverse stakeholder groups.

Moreover, the aspect of global responsibility should also be examined from the context of food system ethics, especially regarding technological solutions such as biofuels. Despite efforts to present biofuels as a way to address fossil fuel dependence, their emergence has sparked ethical controversies, especially over food versus fuel. As witnessed during the 2008 food crisis, diverting agricultural crops to fuel production led to higher food prices, which in turn led to social unrest in many regions across the world. This highlights the limited scope of technological fixes, which are not ethically reviewed and designed to include all stakeholders and, therefore, increase inequality and food insecurity.

Besides, though developments in biotechnology, such as the second generation of biofuels and the use of plants as catalyst sources in fermentation processes, are promising, they are not ethically neutral. This necessitates interdisciplinary work to develop these technologies, which require knowledge bases in genomics, chemistry, nanotechnology, and sustainable farming. Nonetheless, such activities tend to occur without the context of ethics, public relations, or social-political responsibility. The transparency with which these technologies are presented and propagated, more than the large scientific institutions can, obscures vital discussions about their real implications for society. In context, responsible innovation is concerned not only with technological progress but also with ethical foresight, democratic governance, and international coordination. To ensure that plant-based and bio-based solutions support food justice, environmental sustainability, and equitable development, it is important that the voices of diverse individuals, particularly long-term marginalized groups, are heard in shaping research agendas, policies, and implementation. In conclusion, global accountability, interdisciplinary innovation, and stakeholder collaboration in the food system are the priority areas that would enable ethical and sustainable transitions in this global system. The future of food does not depend solely on how consumers behave, but also on how they will improve their relationships, power dynamics, and institutional orderings that guide the production, distribution, and consumption of food. We will only achieve a sustainable and inclusive transformation of the food system by working together, grounded in respect, ethical motivation, and responsibility (Liu et al. 2025).

14.9 Conclusion

The shift toward plant-based solutions is a strong possibility for addressing one of the most urgent ethical and environmental concerns of today. Plant-based innovations have the potential to curb climate change, reduce carbon footprints, and prevent ecological destruction by minimizing the use of animal products and systems that rely heavily on fossil fuels. There are also some complexities associated with this promising shift. Ethically, there are concerns about the high consumption of alternatives as plant-based products rise. These can be unfair intellectual property rights, commercialization of genetic materials, biopiracy, and the ghettoization of the indigenous people and their conventional knowledge base. To make the transition just and inclusive, it is mandatory to pursue equitable access to the development and deployment of plant-based technologies while ensuring cultural sensitivity and respect for community-based innovation. Environmentally, the fact that a food is plant-based does not necessarily make it sustainable. The ecological integrity of plant-based production systems can be impaired by monoculture farming, land-use conversion, excessive water consumption, and the use of chemical fertilizers and pesticides. Those complications make it obvious that the world cannot be truly sustainable without implementing context-relevant agricultural methods, regenerative land management, and sustained control over the environment.

Also, it is not possible to deny the economic and socio-cultural consequences of moving to plant-based systems. A vast proportion of smallholders and rural societies rely on the current agricultural systems. A shift that is not reckoned with these dynamics can lead to the emergence of new forms of social inequality and economic displacement. Community-based economies that empower local farmers, maintain traditional crop expertise, and create ethical transparency and responsibly sourced supply chains must be central to the vision of an authentically responsible transformation of food systems. Ethical frameworks and collaborative, transparent governance mechanisms should be used to shed light on the future of plant-based solutions. Responsible innovation should not be confined solely to product development but also involve a commitment to engaging constructively with stakeholders, including farmers, consumers, policymakers, and marginalized communities. The main reason systemic thinking is essential is to prevent unintended consequences and to allow holistic solutions to emerge by taking into account all aspects of the environment, namely, environmental, social, economic, and cultural. All in all, it is possible to make plant-based systems fundamentally rooted in sustainable development and ethical advancement on one condition: that they are done with purpose, integrity, and inclusivity. Leveraging the potential of plant-based solutions is possible alongside technological development, and a justice, equity, and ecological stewardship approach is the way to build the kind of resilience, ethics, and environmental soundness for our future.

References

Abbaspour N (2024) Fermentation's pivotal role in shaping the future of plant-based foods: an integrative review of fermentation processes and their impact on sensory and health benefits. Appl Food Res 4:100468. https://doi.org/10.1016/j.afres.2024.100468

Andreani G, Sogari G, Marti A, Froldi F, Dagevos H, Martini D (2023) Plant based meat alternatives: technological, nutritional, environmental, market, and social challenges and opportunities. Nutrients 15:452. https://doi.org/10.3390/nu15020452

Andrzejewski A (2017) Traditional knowledge and patent protection: conflicting views on international patent standards. Dermatol Peru 13:93–125. https://doi.org/10.17159/1727-3781/2010/v13i4a2701

Aleksandrowicz L, Green R, Joy EJM, Smith P & Haines A (2016) 'The Impacts of Dietary Change on Greenhouse Gas Emissions, Land Use, Water Use, and Health: A Systematic Review', A.S. Wiley (ed.),PLOS ONE, 11(11):e0165797

Belete T, Yadete E (2023) Effect of mono cropping on soil health and fertility management for sustainable agriculture practices: a review. J Plant Sci 11:192–197. https://doi.org/10.11648/j.jps.20231106.13

Berger M, Gerbens-Leenes W, Karandish F, Aldaya MM, Boulay A-M, Hogeboom RJ, Link A, Manzardo A, Mialyk O, Motoshita M, Nuñez M, Pfister S, Rosenbaum RK, Scherer L, Su H, Wöhler L (2025) Complementary strengths of water footprint and life cycle assessments in analyzing global freshwater appropriation and its local impacts – recommendations from an Interdisciplinary discussion series. Ecol Indic 174:113458. https://doi.org/10.1016/j.ecolind.2025.113458

Boix-Fayos C, De Vente J (2023) Challenges and potential pathways towards sustainable agriculture within the European Green Deal. Agric Syst 207:103634. https://doi.org/10.1016/j.agsy.2023.103634

Bouteska A, Sharif T, Bhuiyan F, Abedin MZ (2024) Impacts of the changing climate on agricultural productivity and food security: evidence from Ethiopia. J Clean Prod 449:141793. https://doi.org/10.1016/j.jclepro.2024.141793

Burnes B (2004) 'Kurt Lewin and the Planned Approach to Change: A Re-appraisal', J Manag Stud 41(6):977–1002

Caballero-Serrano V, McLaren B, Carrasco JC, Alday JG, Fiallos L, Amigo J, Onaindia M (2019) Traditional ecological knowledge and medicinal plant diversity in Ecuadorian Amazon home gardens. Glob Ecol Conserv 17:e00524. https://doi.org/10.1016/j.gecco.2019.e00524

Chowdhury P, Mahi NA, Yeassin R, Chowdhury N-U-R, Farrok O (2025) Biomass to biofuel: impacts and mitigation of environmental, health, and socioeconomic challenges. Energy Convers Manag: X 25:100889. https://doi.org/10.1016/j.ecmx.2025.100889

Cucu MA, Choudhary R, Trkulja V, Garg S, Matić S (2025) Utilizing environmentally friendly techniques for the sustainable control of plant pathogens: a review. Agronomy 15:1551. https://doi.org/10.3390/agronomy15071551

Chaudhary A, Gustafson D & Mathys A (2018) 'Multi-indicator sustainability assessment of global food systems', Nat Commun 9(1):848

Donoso VG, Hirye MCM, Gerwenat C, Reicher C (2024) Amazon deforestation and global meat consumption trends: an assessment of land use change and market data from Rondônia that shows why we should consider changing our diets. Sustainability 16:4526. https://doi.org/10.3390/su16114526

Dwivedi YK, Hughes L, Kar AK, Baabdullah AM, Grover P, Abbas R, Andreini D, Abumoghli I, Barlette Y, Bunker D, Kruse LC, Constantiou I, Davison RM, De' R, Dubey R, Fenby-Taylor H, Gupta B, He W, Kodama M, Mäntymäki M, Metri B, Michael K, Olaisen J, Panteli N, Pekkola S, Nishant R, Raman R, Rana NP, Rowe F, Sarker S, Scholtz B, Sein M, Shah JD, Teo TSH, Tiwari MK, Vendelø MT, Wade M (2022) Climate change and COP26: are digital technologies and information management part of the problem or the solution? An editorial reflection and call to action. Int J Inf Manag 63:102456. https://doi.org/10.1016/j.ijinfomgt.2021.102456

Fanzo J, Haddad L, Schneider KR, Béné C, Covic NM, Guarin A, Herforth AW, Herrero M, Sumaila UR, Aburto NJ, Amuyunzu-Nyamongo M, Barquera S, Battersby J, Beal T, Bizzotto Molina P, Brusset E, Cafiero C, Campeau C, Caron P, Cattaneo A, Conforti P, Davis C, DeClerck FAJ, Elouafi I, Fabi C, Gephart JA, Golden CD, Hendriks SL, Huang J, Laar A, Lal R, Lidder P, Loken B, Marshall Q, Masuda YJ, McLaren R, Neufeld LM, Nordhagen S, Remans R, Resnick D, Silverberg M, Torero Cullen M, Tubiello FN, Vivero-Pol J-L, Wei S, Rosero Moncayo J (2021) Viewpoint: rigorous monitoring is necessary to guide food system transformation in the countdown to the 2030 global goals. Food Policy 104:102163. https://doi.org/10.1016/j.foodpol.2021.102163

Ford H, Gould J, Danner L, Bastian SEP, Yang Q (2023) "I guess it's quite trendy": a qualitative insight into young meat-eaters' sustainable food consumption habits and perceptions towards current and future protein alternatives. Appetite 190:107025. https://doi.org/10.1016/j.appet.2023.107025

Food and Agriculture Organization of the United Nations (FAO) (2010) The State of World Fisheries and Aquaculture 2010. Rome: FAO

Ferguson Z (2025) 'Why you shouldn't serve meat at your next catered event', Politics, Philos & Econ, 24(1):3–24

Giacalone D, Clausen MP, Jaeger SR (2022) Understanding barriers to consumption of plant-based foods and beverages: insights from sensory and consumer science. Curr Opin Food Sci 48:100919. https://doi.org/10.1016/j.cofs.2022.100919

Gil M, Rudy M, Duma-Kocan P, Stanisławczyk R, Krajewska A, Dziki D, Hassoon WH (2024) Sustainability of alternatives to animal protein sources, a comprehensive review. Sustainability 16:7701. https://doi.org/10.3390/su16177701

Gastaldello A, Giampieri F, De Giuseppe R, Grosso G, Baroni L & Battino M (2022) 'The rise of processed meat alternatives: A narrative review of the manufacturing, composition, nutritional profile and health effects of newer sources of protein, and their place in healthier diets', Trends in Food Science & Trends Food Sci & Technol 127:263–71

Hariram NP, Mekha KB, Suganthan V, Sudhakar K (2023) Sustainalism: an integrated socio-economic-environmental model to address sustainable development and sustainability. Sustainability 15:10682. https://doi.org/10.3390/su151310682

Harwatt H, Benton TG, Bengtsson J, Birgisdóttir BE, Brown KA, van Dooren C, Erkkola M, Graversgaard M, Halldorsson T, Hauschild M, Høyer-Lund A, Meinilä J, van Oort B, Saarinen M, Tuomisto HL, Trolle E, Ögmundarson O, Blomhoff R (2024) Environmental sustainability of food production and consumption in the Nordic and Baltic region – a scoping review for Nordic Nutrition Recommendations 2023. Food Nutr Res. https://doi.org/10.29219/fnr.v68.10539

Henry RC, Alexander P, Rabin S, Anthoni P, Rounsevell MDA, Arneth A (2019) The role of global dietary transitions for safeguarding biodiversity. Glob Environ Chang 58:101956. https://doi.org/10.1016/j.gloenvcha.2019.101956

Hess JM, Robinson K, Scheett AJ (2025) Vegetarianísh—how "flexitarian" eating patterns are defined and their role in global food-based dietary guidance. Nutrients 17:2369. https://doi.org/10.3390/nu17142369

Jafarzadeh S, Qazanfarzadeh Z, Majzoobi M, Sheiband S, Oladzadabbasabad N, Esmaeili Y, Barrow CJ, Timms W (2024) Alternative proteins; a path to sustainable diets and environment. Curr Res Food Sci 9:100882. https://doi.org/10.1016/j.crfs.2024.100882

Jang J & Lee D-W (2024) 'Advancements in plant based meat analogs enhancing sensory and nutritional attributes', npj Sci of Food 8(1):50

Kearney J (2010) Food consumption trends and drivers. Philos Trans R Soc B 365:2793–2807. https://doi.org/10.1098/rstb.2010.0149

Kumar A, Kumar S, Komal, Ramchiary N, Singh P (2021) Role of traditional ethnobotanical knowledge and indigenous communities in achieving sustainable development goals. Sustainability 13:3062. https://doi.org/10.3390/su13063062

Kuralt KM (2021) The importance of labeling products with a GMO or non-GMO label. Med Law Soc 14:43–76. https://doi.org/10.18690/mls.14.1.43-76.2021

Leroy F, Abraini F, Beal T, Dominguez-Salas P, Gregorini P, Manzano P, Rowntree J, Van Vliet S (2022) Animal board invited review: animal source foods in healthy, sustainable, and ethical diets – an argument against drastic limitation of livestock in the food system. Animal 16:100457. https://doi.org/10.1016/j.animal.2022.100457

Liu B, Chen M, Kharbeche M, Kerbache L, Haouari M, Gu X, Wang W, Guo W (2025) A systematic review of food accessibility in the food supply chains. Supply Chain Anal 11:100149. https://doi.org/10.1016/j.sca.2025.100149

Lewin K (1997) Resolving social conflicts and field theory in social science. Am Psychol Assoc

Marques C, Güneş S, Vilela A, Gomes R (2025) Life-cycle assessment in agri-food systems and the wine industry—a circular economy perspective. Foods 14:1553. https://doi.org/10.3390/foods14091553

Naidu L, Moorthy R (2021) A review of key sustainability issues in Malaysian palm oil industry. Sustainability 13:10839. https://doi.org/10.3390/su131910839

Naliyadhara N, Kumar A, Kumar Gangwar S, Nair Devanarayanan T, Hegde M, Alqahtani MS, Abbas M, Sethi G, Kunnumakkara A (2023) Interplay of dietary antioxidants and gut microbiome in human health: what has been learnt thus far? J Funct Foods 100:105365. https://doi.org/10.1016/j.jff.2022.105365

Natarajan N, Newsham A, Rigg J, Suhardiman D (2022) A sustainable livelihoods framework for the 21st century. World Dev 155:105898. https://doi.org/10.1016/j.worlddev.2022.105898

Ng KMP, Tan IS, Foo HCY, Lau JSY, Lam MK, Wong MK (2024) Durable polylactic acid bionanocomposites with biomass-derived nanocellulose additives: recent advances in production. J Taiwan Inst Chem Eng:105774. https://doi.org/10.1016/j.jtice.2024.105774

Nyaupane S, Mainali RP, Joshi T, Duwal R (2023) Plant-based agro-biodiversity solutions for reducing agrochemical use and effects. In: Ogwu MC, Izah SC (eds) One health implications of agrochemicals and their sustainable alternatives. Springer Nature, Singapore, pp 545–563. https://doi.org/10.1007/978-981-99-3439-3_20

Pandian K, Mustaffa MRAF, Mahalingam G, Paramasivam A, John Prince A, Gajendiren M, Rafiqi Mohammad AR, Varanasi ST (2024) Synergistic conservation approaches for nurturing soil, food security and human health towards sustainable development goals. J Hazard Mater Adv 16:100479. https://doi.org/10.1016/j.hazadv.2024.100479

Pingali P, Boiteau J, Choudhry A, Hall A (2023) Making meat and milk from plants: a review of plant-based food for human and planetary health. World Dev 170:106316. https://doi.org/10.1016/j.worlddev.2023.106316

Profeta A, Baune M-C, Smetana S, Broucke K, Van Royen G, Weiss J, Hieke S, Heinz V, Terjung N (2021) Consumer preferences for meat blended with plant proteins – empirical findings from Belgium. Future Foods 4:100088. https://doi.org/10.1016/j.fufo.2021.100088

Rogers H, Dora M, Tsolakis N, Kumar M (2024) Plant-based food supply chains: recognising market opportunities and industry challenges of pea protein. Appl Food Res 4:100440. https://doi.org/10.1016/j.afres.2024.100440

Rojas-Downing MM, Nejadhashemi AP, Harrigan T, Woznicki SA (2017) Climate change and livestock: impacts, adaptation, and mitigation, climate. Risk Manage 16:145–163. https://doi.org/10.1016/j.crm.2017.02.001

Sahoo PK, Datta R, Rahman MM, Sarkar D (2024) Sustainable environmental technologies: recent development, opportunities, and key challenges. Appl Sci 14:10956. https://doi.org/10.3390/app142310956

Silverwood SM, Lichter KE, Kleiner HS, Katsumoto TR (2025) A unified strategy for addressing climate change, improving public health, and mitigating environmental degradation. Agric Commun 3:100090. https://doi.org/10.1016/j.agrcom.2025.100090

Surya Ulhas R, Ravindran R, Malaviya A, Priyadarshini A, Tiwari BK, Rajauria G (2023) A review of alternative proteins for vegan diets: sources, physico-chemical properties, nutritional

equivalency, and consumer acceptance. Food Res Int 173:113479. https://doi.org/10.1016/j.foodres.2023.113479
Shurtleff W (2014) History of Soybeans and Soyfoods in Japan, and in Japanese Cookbooks and Restaurants outside Japan (701 CE to 2014). United States: Soyinfo Center
Smith AF (2006) Encyclopedia of junk food and fast food
Taneja A, Nair G, Joshi M, Sharma S, Sharma S, Jambrak AR, Roselló-Soto E, Barba FJ, Castagnini JM, Leksawasdi N, Phimolsiripol Y (2023) Artificial intelligence: implications for the agri-food sector. Agronomy 13:1397. https://doi.org/10.3390/agronomy13051397
Tuomisto HL & Teixeira De Mattos MJ (2011) 'Environmental Impacts of Cultured Meat Production', Environ Sci & Technol 45(14): 6117–23
Ullah H, Khan RU, Tufarelli V, Laudadio V (2020) Selenium: an essential micronutrient for sustainable dairy cows production. Sustainability 12:10693. https://doi.org/10.3390/su122410693
Ungerman O, Dědková J (2020) Model of the circular economy and its application in business practice. Environ Dev Sustain 22:3407–3432. https://doi.org/10.1007/s10668-019-00351-2
United Nations Conference on Trade and Development (UNCTAD) (2013) Trade and Development Report 2013: Adjusting to the Changing Dynamics of the World Economy. New York and Geneva: United Nations
Velenturf APM, Purnell P (2021) Principles for a sustainable circular economy. Sustain Prod Consum 27:1437–1457. https://doi.org/10.1016/j.spc.2021.02.018
Wu L, Miao H, Liu T (2024) Development in agricultural ecosystems' carbon emissions research: a visual analysis using CiteSpace. Agronomy 14:1288. https://doi.org/10.3390/agronomy14061288
Xavier JR, Shashikumar SH, Vats D, Chauhan OP (2025) Future trends in plant-based meat: consumer perception, market growth and health benefits. Future Foods 11:100551. https://doi.org/10.1016/j.fufo.2025.100551

Chapter 15
Advances in Plant Biotechnology for Sustainable Applications

L. Srimathi Priya, K. Radhakrishnan, V. P. Santhi, K. Indhumathi, A. Dinesh, N. Subasree, Manikandan Ayyar, and A. Muthukrishnaraj

Abstract Plant biotechnology has grown to become a revolutionary technology that holds big potential to mitigate global sustainability issues of agriculture, environment, food security, and bioeconomy. The chapter describes new accomplishments in plant biotechnology and how they are applied to enhance sustainable development. It discusses contemporary genetic technologies like genome editing in the CRISPR (clustered regularly interspaced short palindromic repeats)/Cas system and transgenic crop technologies to enhance crop resilience, nutritional value, and production quantity. The effectiveness of tissue culture and micropropagation in reproductive conservation and in mass reproduction is discussed and how molecular

L. S. Priya
Department of Natural Resource Management, Horticultural College and Research Institute (HC & RI), Tamil Nadu Agricultural University (TNAU), Periyakulam, Theni, Tamil Nadu, India

K. Radhakrishnan
Department of Chemistry, Karpagam Academy of Higher Education (Deemed to be University), Coimbatore, Tamil Nadu, India

Centre for Material Science, Karpagam Academy of Higher Education (Deemed to be University), Coimbatore, Tamil Nadu, India

V. P. Santhi
Department of Horticulture, Anbil Dharmalingam Agricultural College and Research Institute (ADAC & RI), Tamil Nadu Agricultural University (TNAU), Trichy, Tamil Nadu, India

K. Indhumathi
Krishi Vigyan Kendra, Tamil Nadu Agricultural University, Papparapatty, Dharmapuri, Tamil Nadu, India

A. Dinesh
Department of Chemistry, K. Ramakrishnan College of Engineering (Autonomous), Affiliated to Anna University, Samayapuram, Trichy, Tamil Nadu, India

L. A. Worku et al. (eds.), *Plant-Based Solutions for Sustainable Agriculture and Healthcare*, Sustainability Sciences in Asia and Africa,
https://doi.org/10.1007/978-981-95-8910-4_15

marker-assisted breeding can be incorporated to make precision breeding of crops. The use of plant–microbe interactions and its application in the fields of biofertilization, biocontrol, and soil health management is also touched upon in the chapter. The methods of metabolic engineering towards improving the production of high--value phytochemicals and secondary metabolites are presented and how they can be utilized in pharmaceuticals, nutraceuticals, and industrial products. Besides, the use of engineered plants in phytoremediation and the clean-up of the environment is another indication of how biotechnology can aid in the restoration of the ecosystems. Sustainable practices in agriculture and industry are being transformed due to innovations in plant-based bioenergy and biomaterials that are enabled by digital technologies, such as artificial intelligence, remote sensing, and bioinformatics. Although there has been so much progress, issues such as regulatory, ethical, and social acceptance are also very vital. The future of how to combine advances in biotechnology with responsibility in innovation and to create a stronger and more sustainable future is left in the chapter.

Keywords Plant biotechnology · Agriculture · Environment · Food security · Bioeconomy · Sustainability

15.1 Introduction

It is estimated that the world population may increase by 25% by 2050, to around nine billion people demanding more food, fiber, and fuel, the production of which is put to a tremendous strain by the agricultural systems. It is estimated that to guarantee food security, the productivity of agriculture has to boost by at least 50%. The ability to meet this target is, however, hampered by the dwindling supply of arable land, most of which is being swallowed up in urbanization, salinization, desertification, etc., and the even greater dearth of freshwater, thus posing a threat of increased abiotic stress. Upon these stresses are losses because of the biotic stressors, which include pests and pathogens that are likely to break out once the crop is ripely grown,

N. Subasree
Department of Chemistry, School of Basic Science, Vels Institute of Science, Technology and Advanced Studies (VISTAS), Pallavaram, Tamil Nadu, India

M. Ayyar (✉)
Department of Chemistry, Karpagam Academy of Higher Education (Deemed to be University), Coimbatore, Tamil Nadu, India

Centre for Material Chemistry, Karpagam Academy of Higher Education (Deemed to be University), Coimbatore, Tamil Nadu, India

A. Muthukrishnaraj
Amrita School of Artificial Intelligence, Amrita Vishwa Vidyapeetham, Coimbatore, Tamil Nadu, India

by which time a lot of natural and human resources have already been engaged (Food and Agriculture Organization (FAO, 2009).

Against this scenario, there has been the advent of biotechnology as a transformational tool that promises a lot in respect to the sustainable development of agriculture. With the intervention of the new generation of biotechnology, such as genetic engineering and molecular breeding, it is now feasible to obtain crop varieties that are more drought, salinity, high-temperature, pest and pathogenic stress resistant, better enhanced in yield, and nutritional profiles. Besides enhancing agricultural productivity, biotechnology is useful in terms of creating environmental sustainability. A case in point is that genetically engineered plants have been created to clean up heavy metal (HM) contamination of soils and, in order to minimize the use of chemical pesticides and herbicides, hence minimize ecological footprints (Tripathi et al. 2019).

Modern agricultural biotechnology encompasses a broad array of applications—from molecular diagnostics and vaccine development to biofortification and the creation of transgenic crops. It is also important in the enhancement of livestock, fish farming, and production of biodegradable biopolymers and biofuels (Thitisaksakul et al. 2012; Kumar et al. 2020). Consequently, biotechnology is transforming food systems in the world through the optimization of agricultural systems that are more effective, resilient, and sustainable. Methods like CRISPR (clustered regularly interspaced short palindromic repeats)-Cas genome editing and OMICS have gone further in expediting the process by enabling more accurate editing of genetic material in a clear manner and paving the way for a new era of biotechnological advances that are safe and more focused.

Sustainable agriculture is an idea that defines the production systems that are ecologically sound, economically viable, and socially responsible. It demands an integrated process, which should be a trade-off between the need to achieve higher productivity and the preservation of natural resources as well as the rural communities. Amy (2015) defines sustainable agriculture as a complex of plant and animal cultivation techniques and practices which in the long run meets human food and fiber requirements, promotes ecological quality, efficiently uses on-farm resources, enhances the economic sustainability of the farm, and improves the quality of life of the farmers and society in general. The applications of biotechnology directly related to these objectives are more efficient and resilient production systems.

In spite of its huge potential, biotechnology continues to encounter scepticism and regulatory problems, especially with respect to the safety and environmental effect of genetically modified organisms (GMOs). These are related to ethical issues, biodiversity preservation, and socially fair distribution of technology, which are very important aspects that need to be discussed and solved with clear and open communication, safety testing, and policy development (Khan et al. 2017). Overall, a combined approach of sustainable agriculture through biotechnology provides a strategic route towards overcoming the multiple characteristics of global food security and environmental and climate change. This chapter examines the achievements of plant biotechnology and ways it can be used to create viable and efficient sustainable agriculture.

15.2 Genetic Engineering and Genome Editing

The transformation in the process of plant biotechnology by genetic engineering and genome editing platforms has provided efficient and highly specific and precise methods of improving crops in sustainable systems of agricultural practice. These technologies give the opportunity to modify the genomes of plants, allowing them to acquire desirable characteristics, i.e., improved yield, stress resistance, better nutritional value, and pest and disease resistance. Genetic engineering makes it highly accurate in inserting or altering specifications of certain genes in comparison to conventional methods of breeding, which are laborious and not specific to the desired modifications; it proves to save time and other resources to generate better crop breeds.

The past 15 years saw a considerable amount of development in the area of plant genome deciphering and determining of the gene functions, due to development in the areas of genomics and molecular biology. The successful addition of transgenes belonging to other species into plant genomes has been made successful through genetic engineering that has resulted in genetically modified (GM) crops with desirable traits. Take the case of Bt cotton and herbicide-resistant soybeans that have seen a huge success in pest management as well as weed control, thereby enhancing productivity and chemical resistance. Further, with genetic engineering, biofortified crops, including the Golden Rice, which are fortified with provitamin A, are being developed to solve micronutrient deficiencies in the developing nations.

The most recent development in the field of plant biotechnology is genome editing technology, and specifically the CRISPR/Cas9 technology. Its CRISPR-based systems enable highly accurate changes to be made in particular points in the genome without using the insertion of foreign DNA, thus addressing most of the regulatory and public apprehensions of conventional GMOs. Value-added traits using this technology have been achieved in the form of protein knockouts of disease-susceptible genes, improved drought tolerance or salinity tolerance, and better use of nutrient pathways to provide metabolic efficiency. The amenity of planning, cost-effectiveness, and high specificity of CRISPR systems have made them a favorite instrument at the center of basic research and applied plant breeding schemes (Hamdan and Tan 2024). miRNAs and transcription factors have equally become sensitive targets of genetic manipulation as they have been found to regulate gene expression during stress. By regulating the activity of these molecular regulators, it is possible to finetune the plant reactions to abiotic stresses like drought, heat, and salinity. In a similar manner, recent developments in synthetic biology and metabolic engineering now provide an additional opportunity to seek other areas to optimize biosynthetic pathways in plants to produce valuable bioproducts, biofuels, and pharmaceuticals, thus further contributing to sustainability.

The concept of precision breeding was based on traditional genetic engineering techniques, including zinc-finger nucleases and transcription activator-like effector nucleases, which had complicated protein engineering and delivery. Conversely, the CRISPR systems are versatile and multiplexing-able, permitting multiple-gene editing at the same time, which is very useful in polyploid varieties, such as wheat, canola, and sugarcane. Also, the newest methods of genome editing, e.g., base editing and prime editing, will allow going even further and modifying a single nucleotide without damaging the double-strand structure, thus reducing off-target effects and increasing safety.

These tools of genome editing combined with high-throughput phenotyping, bioinformatics, and machine learning (ML) create further acceleration in the discovery of gene–trait associations and the creation of better cultivars. This synergy will facilitate climate-smart agriculture through facilitating a quick response of the crops to new environmental changes as well as the introduction of new crop pests and diseases. Despite all the technological developments, there are still challenges. There are regulatory complications, the view of society, and the low efficiency of change in some crops that prevent the industry-wide integration of genetic and genome editing. Moreover, ethical issues and the preservation of biodiversity have to be considered in order to provide a responsible utilization of such instruments. However, as research progresses and regulations are harmonized, genetic engineering and genome editing are immensely promising in building a sustainable, productive, and resilient agricultural future, even though alone these activities cannot solve world hunger (Li et al. 2024).

15.2.1 CRISPR/Cas and New Breeding Technologies

The current trends in the form of increasing challenges to agriculture through the decrease in arable lands, shortage of water, change in climatic conditions, and the fast-developing pathogens have highlighted that new technology in breeding is needed to create tough, high-yielding, and environmentally friendly breeds. One of these new technologies is genome editing with CRISPR/Cas, which has become one of the most promising tools of precision breeding as it opens up an unprecedented degree of control over the plant genome and makes it possible to improve it in ways that do not involve adding any foreign DNA.

CRISPR and the protein present that makes it, Cas9, are part of an adaptive immune system that was first identified in prokaryotes. In the context of plant genome editing, the CRISPR/Cas9 tool garnered widespread attention because it can create double-stranded breaks (DSBs) in specific genomic locations, which is then repaired by the built-in DNA repair mechanisms, and the outcome is either a mutation, insertion, deletion, or precise base substitution. CRISPR/Cas9 technology has, since its initial use in plants in 2013, revolutionized plant biotechnology quickly, because of its ease of use, low cost, and heritable, targeted nature of alteration. With CRISPR/Cas technologies, crops that are disease resistant have been produced that knock out susceptibility (S) genes or perform editing on promoter sections so that the pathogen will no longer recognize them. As an example, the MLO gene in tomato and wheat is mutated to give powdery mildew resistance, and the promotion of specific SWEET genes in rice has been edited to increase immunity to bacteria blight. In the same work, abiotic stress (tolerance), nutrient, and plant architecture (optimization) improvement through genome editing have been demonstrated (Ndudzo et al. 2024).

New breeding technologies (NBTs) do not confine themselves to traditional transgenic methods and also encompass a family of CRISPR-based technologies, including base editing, prime editing, and multiplex genome editing. Precise modifications to nucleotides can be conducted in these more sophisticated systems without inducing DSBs or the use of donor templates. Such base editors as cytosine base editors and adenine base editors can efficiently convert nucleotides at the single--nucleotide level. Prime editing goes a step further, allowing programmable insertions, deletions, and substitutions with even greater specificity and fewer off-target effects.

The other important strength of CRISPR-based NBTs is that they can generate transgene-free edited plants that are usually no different than plants developed conventionally. This aspect has major regulatory consequences, where it could cause some gene-edited crops to be GMO-exempt in certain places and enhance public acceptability. Nevertheless, the efficient delivery of CRISPR components (especially in intractable crops), high editing precisions, and heritability are obstacles in the application that should be addressed. Further, there are issues on the off-target effects, unintended as well as the genome manipulation ethical issues which remain debatable. To overcome these difficulties, genome editing has to be combined with other molecular technologies which include the use of high-throughput phenotyping, transcriptomics, and ML to predict the gene function and enhance trait selection. Finally, CRISPR/Cas and other NBTs are the paradigm NBTs in plant breeding. They enable breeders and researchers to hasten the evolution of crops, improve stress tolerance, and adapt them to evolving ecological conditions, and progress all these changes with minimum ecological impacts. With the maturity of these technologies and changes in regulatory frameworks, they will play an ever-important role in the realization of sustainable agricultural systems, as these play crucial roles in the food and nutritional security in the 21st century (Wei and Li 2023).

15.2.2 Transgenic Crops for Abiotic and Biotic Stress Resistance

Both abiotic and biotic stresses are major constraints in agricultural productivity and cause huge losses in crop yields on a global scale. Drought, salinity, high temperatures, low temperatures, and water shortage are all stresses in agricultural farming that have been increasing with climate change worldwide, whereas infections by bacteria, fungi, viruses, or pests have been a burden upon the world in terms of food security throughout history. Transgenic technologies have been effective in alleviating the presence of these debilitating factors because they have become a feasible means of establishing stress-resistant crops through the introduction or alteration of genes that are linked to the process of stress tolerance.

One of the breakthroughs in the field of plant biotechnology came with the discovery of regulatory networks that respond to stress and how these networks have been harnessed. As opposed to the earlier-used traditional monogenic methods, the decoding of the polygenic character of abiotic stress tolerance has enabled scientists to be able to target more than one gene at a time. Stable expression of numerous stress-resistance genes within the plant genome can be achieved by genetic engineering on a stable basis by employing transformation techniques, including *Agrobacterium tumefaciens*-mediated transformation and particle bombardment. They have a particularly distinct advantage in that large gene constructs can be instilled into these procedures, and thus, complex traits can be incorporated into a single transformation event (Anwar and Kim 2020).

Agro-mediated is still the most widespread technology of gene transfer because of its efficiency and the possibility to transfer large pieces of T-DNA. Conversely, the process of particle bombardment enables the direct uptake of DNA into plant cells and hence may be applied to a broader family of plant species, including those monocots that are less prone to infection by Agrobacterium. These two methods have enabled the production of transgenic plants with an increased degree of tolerance to drought, salinity, cold, and oxidative stress. Indicatively, fructan--accumulating transgenic plants exhibited an increase in proliferation of up to 55% and a dry weight of up to 59% more than the controls during the state of drought, implying that there was considerable improvement in growth under problematic times. Similarly, transgenic plants engineered to accumulate osmoprotectants such as d-ononitol and sorbitol displayed improved photosynthetic efficiency and reduced physiological damage under salt and drought stress conditions.

Other than abiotic stress, biotic stress is a looming threat to world food production since 20–40% of the crop yield is lost each year due to biotic stress. Existing chemical pesticide-based approaches of pest and disease management exert a significant risk on both the environment and human health. To counter this, transgenic technology comes with a long-term solution by injecting terms of resistance in crop bodies, hence reducing the usage of pesticides. As another example, disease--resistant crops were developed by the incorporation of genes that code antifungal proteins or pathogen-responsive signaling molecules or RNA interference (RNAi)-based silencing mechanism. There is also an interesting situation where transgenic crops that are resistant to pathogen species such as *Phytophthora infestans*, which is the causative agent of potato late blight, could help to avoid devastating losses to crops and famines, which had occurred in the past.

Moreover, the use of nanotechnology to manage plant stress is also being considered as a supplemental practice. One of the recent applications of engineered nanomaterials in the agricultural field is in boosting plant defense mechanisms and targeted delivery of agrochemicals in a more precise manner. Nanoparticles in the form of stress-responsive gene carriers or even elicitors could be used to induce plant immune responses and hence present a green alternative in managing biotic stress. The use of transgenic crops that are more resistant to abiotic and biotic stresses is a great advance in the direction of sustainable farming. Nonetheless, there are still some obstacles in keeping the gene stably expressed in the field, authorizations of the regulatory bodies, and acceptance of GMOs by people. The combined approach such as genomics, transcriptomics, molecular breeding, and nanotechnology will be crucial in circumventing the existing constraints and attaining large-scale acceptance of transgenic stress-resistant crops (Su et al. 2023).

15.2.3 Nutritional Biofortification Through Genetic Engineering

Genetic engineering has presented itself as a revolutionary method that can address the issue of micronutrient malnutrition that has occurred in more than three billion people in the world. As part of an expanding demand to add nutrition to staple food crops, this approach will aim to enrich plants with more of the nutrients people need (iron, zinc, vitamin A, selenium, and folate) directly in the edible tissues of the plant. Conventional breeding and conventional agronomics have been used over decades but are again limited by the genetic variation of the crop species or are not adequate to deliver the nutrient density that is required. Genetic engineering gives the ability, however, to place biosynthetic pathways or regulatory factors exactly where required with the flexibility and precision that proves the leveraging of a now present nutrient level or the achievement of a more desirable nutrient state.

Important developments of crops with enhanced levels of iron, provitamin A (such as Golden Rice), and folate among others are transgenically done successfully. As an example, genes that facilitate iron uptake and storage (that is genes coding ferritin and NAS genes) have been incorporated in staple cereals to increase iron content. On the same note, the biological carotenoid pathway has been manipulated into produce types of rice and maize that contain much higher levels of provitamin A. The advent of functional genomics has also helped in the identification of important gene targets and also, the availability of the tissue-specific promoters has helped in guaranteeing that the nutrients are stored in the edible tissues of the plants without affecting the growth or development.

With the recent advances in genome-editing technologies, especially CRISPR/Cas systems, biofortification has become a revolutionized approach, providing precise, marker-free, and transgene-free modifications, which increase efficiency and public acceptance. These technologies have been in use to boost zinc and iron concentrations, improve the bioavailability of calcium, and even to genetically design pathways of omega-3 fatty acids by oilseed crops, thus spreading the nutritional advantage to more people (Mmbando and Missanga 2024).

Along with enriching micronutrients through biofortification, enhancement of health beneficial phytochemicals and macronutrients such as essential amino acids and dietary fiber content is also done. Engineered crops that contain modified nutritional properties, functional foods, are also becoming accepted as having health promoting and chronic disease prevention properties. The main problem, however, is nutrient bioavailability. It is induced not only by the crop composition but it also makes a dependent issue on human physiological and microbiome relations. Thus, plant biotechnology is currently coming into contact with sciences such as nutrigenomics to make nutrition individualized and gain insights into the consequences of bioengineered crops on metabolic condition.

Although progress has been made, we still face some challenges in the form of regulatory barriers and consumer suspicion on the one hand and unequal market access on the other. Misperceptions concerning GMOs, the absence of a common regulatory policy on biosafety, and inadequate knowledge of the advantages of using biofortified crops are some of the issues that prevent widespread use. However, there are some success stories such as iron-rich beans, zinc-enriched wheat, and vitamin A-rich sweet potatoes, which prove the functionality and scalability of this strategy. Economic studies also show that investment in biofortification programs will be high since it will help cut the healthcare expenditure and increase productivity by overcoming the hidden hunger.

In the future, a multidisciplinary effort that combines genetic engineering, breeding, soil science, and nutrition with favorable policies and popular education will be the key to ensuring that biofortification efforts are maximized. With climate change and continuous population growth exerting continuous pressure on global food systems, environmentally scalable and cost-effective solution to enhance human health and reach nutritional security on a global scale comprises transgenic biofortified crops that will alleviate food systems to ensure that the global population remains well-nourished (Chaudhary et al. 2025).

15.2.4 Regulatory and Ethical Considerations of GM Plants

Over the last decades, there has been a raging debate about the development and usage of (GM plants due to beliefs that there are different regulatory settings, consumer beliefs, and moral values regarding their usage between various regions of the globe. In 1996, the initial big bang transpired via the importation of GM soy into Europe (US made) and sparked off the first controversy. This incident was a turning point, especially in the European Union (EU) where citizens became concerned spurring the formulation of high regulatory standards and resultant labeling requirements. Unlike the United States which was more lenient, the EU came up with process-based regulations in 1997, which required that any food that was cooked using GMOs must be labeled, assuming that any modification in the product could be detected. These regulations have since been adapted to include even cleansed materials such as oils and sugars obtained from GM plants since they are chemically indistinguishable from non-GM original ones. This has made the EU regulatory system one of the most intense in the world that has had the effect of taking excessive time to approve and import biotech crops several times. Such asynchronous approval delays of up to 20 months longer than the duration in similar countries such as the USA, Brazil, and Canada has brought about trade complications, more so when trace levels of unapproved GMOs are found in shipments. In parallel to the regulatory trends, there has been impressive progress in the subject of plant biotechnology, especially the development of NBTs, including intragenesis, cisgenesis, and the industry-leading precise genome-editing tool core CRISPR/Cas9. These new technologies make it possible to edit the genomes of plants so that they can be changed more precisely and lead to fewer off-target effects than earlier methods commonly employing genetic material of the same species. These systems are presenting new formal definitions of genetic modification, and a political and scientific issue continues to grow over whether they should be considered under current GMO regulations. While these NBTs offer a promising avenue for creating crops with improved traits, regulatory uncertainty persists, potentially stifling innovation and delaying market access (Caradus 2023).

The GM plant debate goes beyond safety testing and its governance. The social and philosophical dilemmas, which are based in schools of thought such as consequentialism, deontology, and virtue systems, take into account the ramifications of using such technologies as a whole. When considering consequentialism, it is generally the case that the ethical acceptability of GM crops is determined by opinion based on the outcomes, e.g., improved food security, reduced pesticide use, or drought-resistant crops. Nonetheless, issues of biodiversity loss, environmental hazards, unintended gene transfer, and socio-economic inequality are also included in the ethical calculations. Other opponents point out that contemporary biotechnology can strengthen corporate dominance of seeds and foods, thus pertaining to equity and farmer agency.

Certain new ethical aspects arise in medical practice, as in the case of plant-made vaccines, where ethical concerns may arise over clinical testing, worldwide availability, and the realistic possibility of delivering the vaccines in low-resource environments. The products claim economically viable and heat-resistant solutions to conventional vaccines but introducing them should include a clear definition of how they are viable commercially, clear regulatory standards as well as an approach that satisfies the community to make them ethically sound. Acceptance is still affected by public suspicion which has been perpetuated by controversies and misinformation in the past. Owing to this fact, the levels of science communication and open risk-benefit analyses are paramount to advancing informed debate by people.

In sum, it can be concluded that transformation in plant biotechnology brings revolutionary changes to sustainable agriculture, but its development is closely associated with the ability to recognize the regulatory environment and multifaceted ethical issue-related aspects. Since gene-editing tools are pushing the envelope of what makes an organism a "genetically modified organism," there is an increasing imperative to bring harmonized, science-based policies into place that strike an equilibrium between innovation, values of society, ecological stewardship, and global equity (Gregorowius et al. 2012).

15.3 Plant Tissue Culture and Micropropagation

Plant tissue culture as a technology has become a primary technology in plant biotechnology, providing effective tools in basic research as well as in practical applications in agriculture. It covers a wide scope of in vitro methods to culture and manipulate organs, tissues, and cells of plants in a controlled environment. These methods are critical in the research of plant developmental biology, functional genomics, improvement of crop, and the conservation of biodiversity. Among the most important implications of plant tissue culture, there are the capacity of plant tissues to facilitate large-scale micropropagation so as to facilitate clonal propagation of high-quality, disease-free plant material of uniform quality as shown in Fig. 15.1. Horticultural, medicinal, and ornamental plant industries are particularly useful with regard to this.

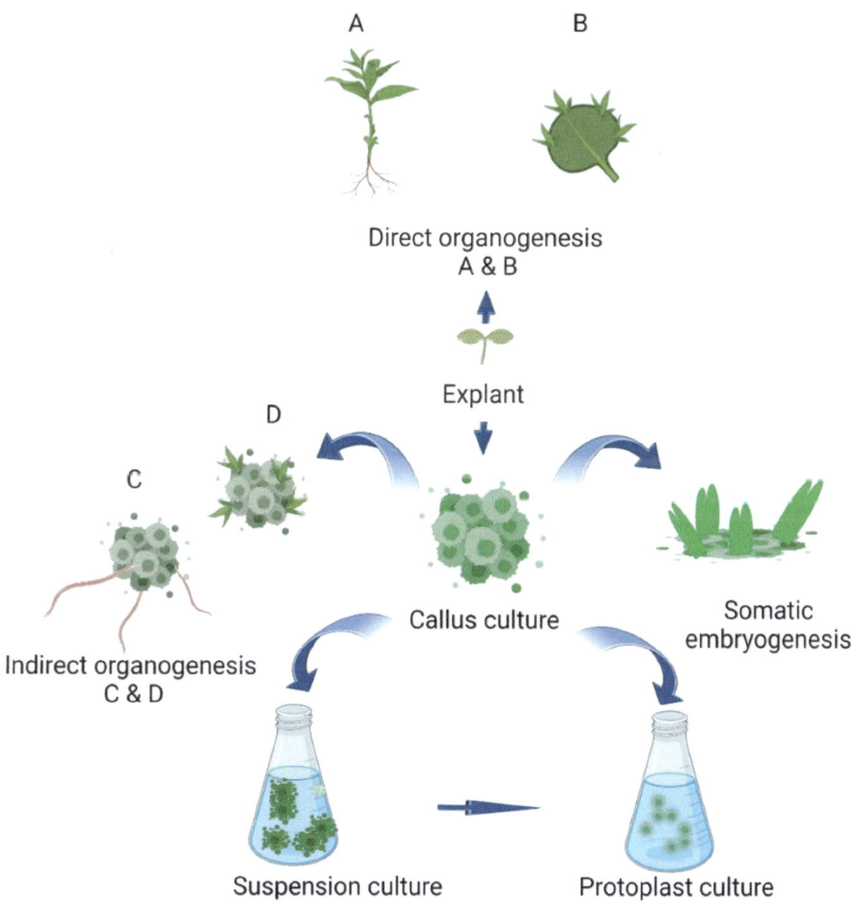

Fig. 15.1 Overview of plant tissue culture pathways, including direct and indirect organogenesis, somatic embryogenesis, and subsequent applications such as suspension and protoplast culture (Wijerathna-Yapa and Hiti-Bandaralage 2023)

Tissue culture has wide range of applications in various fields, such as its use in the removal of viruses using meristem culture, acts as a rescue embryo by relaxing the barrier to interspecific hybridization, the generation of somatic embryos, and regeneration protocol for recalcitrant species. Further, it is part of industrial breeding, and it enables high-velocity homozygous line production with doubled haploidy, somatic hybridization through protoplast fusion, and genetic transformation with *Agrobacterium tumefaciens* or biolistics. More recently, the incorporation of omics technologies, genomics, transcriptomics, proteomics, and metabolomics into tissue culture has provided a great deal of insight into organogenesis and cell differentiation, which is critical in providing an efficient and consistent rate of regeneration across species (Wijerathna-Yapa and Hiti-Bandaralage 2023).

Besides the implications that it has in regard to research and breeding, plant tissue culture ensures the sustainability of plant genetic resources, which is critical. Habitat destruction, climate change, and overexploitation are threats to many wild relatives of modern crops not to mention the rare and endangered species. One cost--effective way of conserving these genetic materials is through in vitro with ex situ conservation techniques, which would help in conservation and future crop tolerance. In order to preserve vegetatively propagated or seed-recalcitrant species, micropropagation, cryopreservation, and slow-growth storage methods are gaining popularity to take care of germplasm collections.

The high-value secondary metabolites produced using plant cell and organ cultures by tissue culture advances also have acquired the ability to be produced on large scales. Bioactive compounds are highly valuable in the industrial and pharmaceutical fields and sustainable substitutes to wild harvesting of medicinal plants. Added to this is the use of elicitors, metabolic engineering, and bioreactor systems which have increased the yields and scalability. The current technological discovery, such as the use of silicon nanoparticles under in vitro conditions, which has shown promise in promoting plant growth rates and minimizing physiological manifestations, is a classic example of the power of nanotechnology integrating with tissue culture to achieve greater results.

Moreover, in vitro nature has been studied in the conservation and multiplication of genetically diverse organisms like *Oryza officinalis,* and the same have the potential to transform cultured rice. A range of rare and endangered medicinal plants has been developed to protocols of callus induction, somatic embryogenesis, and seed germination. It is part of the strategic value of in vitro biotechnologies in the maintenance of plant diversity and the food and ecological security of future generations. To conclude, plant tissue culture and micropropagation are biotechnological methods that cannot be ignored as they contribute both to science discovery and capitalism and environmentally sound environmental management. Even with their continual transformation, particularly in conjunction with elaborate molecular tools, they are still opening up new opportunities for robust, efficient, and sustainable plant systems (Pathirana and Carimi 2022).

15.3.1 Principles and Techniques

The concept of totipotency (the principle that a single cell of a plant can grow into a complete plant under a suitable in vitro environment) is the basis on which plant tissue culture is established. This was originally proposed by Gottlieb Haberlandt in 1902 as the basis of tissue culture today. In spite of the fact that his original experiments failed as he made an incorrect explant choice and poor media selection, his hypothesis and early forecasts became the foundation for further advances in plant biotechnology. Tissue culture has since then evolved to be a mature, accurate, and adaptable process of growing free plant cells, plant tissues, plant organs, and even plant cell organelles under defined physical and chemical climates. Such in vitro methods are used to conduct basic plant science as well as a wide range of commercial purposes.

The most important factors concerning success in plant tissue culture are the use of an actively enriched culture medium, a sterile environment, and accurate regulation of the environmental conditions such as light, temperature, and humidity. Macronutrients, micronutrients, vitamins, a carbon source, e.g., sucrose, and crucially plant growth regulators (PGRs), i.e., auxins and cytochromins, are usually present in the culture media. These PGRs depend on the type and the concentration of the specific PGR to dictate the subsequent developmental pathway of the explant either to form a callus, shoots, or roots, or somatic embryos. Of particular significance was the discovery of plant hormones, in the case of indole acetic acid and kinetin, which demonstrated that exogenous addition of such compounds refined culture procedures and guided morphogenetic events.

Tissue culture directly leads to micropropagation such that plants can be multiplied rapidly and in large numbers, and they are genetically identical, as well as free of diseases. It finds wide application in the high-scale multiplication of the elite germplasms, especially of horticultural, ornamental, and medical species. The micropropagation occurs in several demarcated steps such as (1) selection and sterilization of explants, (2) establishment, (3) in vitro multiplication by shoots, (4) rooting, and (5) acclimatizing the plants to a new environment. Among the widely used techniques are shoot tip culture, nodal culture, and axillary bud proliferation. Another critical technique, called somatic embryogenesis, may take place either directly on the somatic cells or indirectly through a prior culture stage, and its benefit is due to the scalability, labor-saving, and ability to automate the process (Shahzad et al. 2017).

Other sophisticated applications that are linked to the use of the tissue culture process in addition to micropropagation include: These are embryo rescue to allow crossing of barriers to hybridization, virus-free plant production through meristem tissue culture, and somaclonal variation as a source of genetic variation. The technologies that allow the somatic hybrids to be created are the protoplast fusions and culture, which allow the transfer of the genes to jump species. Moreover, suspension cultures and organ cultures are widely applied in the case of industrial-level cultivation of bioactive secondary products, especially in medicinal plants that are rare and/or endangered.

Recent developments in tissue culture incorporate nanomaterials to the culture, silicon nanoparticles to promote growth and minimize somaclonal defects, and metabolomics and transcriptomics to adjust culture conditions and increase efficiency of regeneration. The possibility of tissue culture has been increased further as it has been integrated with genome-editing tools, especially CRISPR/Cas systems, to make targeted gene modifications in crop species. Plant tissue culture and micropropagation techniques, therefore, form key aspects in plant biotechnology, where they facilitate the acquisition of sustainable agriculture by way of propagation of high-quality planting material and preservation of genetic material as well as crop improvement achievements. The principles and procedures are evolving with time due to the advances in molecular biology, materials science, and automation technologies, and hence, plant tissue culture remains on the cutting edge of the current agricultural and technological discoveries (Mehbub et al. 2022).

15.3.2 Conservation of Endangered Plant Species

The dangerous level of biodiversity loss has reached up to 1000- and 10,000-fold higher than the natural loss mark, resulting in what is now referred to as the sixth extinction crisis. Out of these taxa, the taxa whose extinctions are considered to be the most at stake are the plant species due to their exposure to the threats of habitat destruction, pollution, overexploitation, invasive species, and climate change that threaten their existence. This rising crisis highlights the importance of sustainable and effective plans of conservation. In this regard, plant biotechnology provides the effective means of supplementing the conventional conservation methods. In situ and ex situ approaches to preserving rare and endangered species of plants have been found to complement tissue culture-based preserver species most adequately.

Ex situ conservation: Ex situ conservation also includes preservation of the genetic material of plants outside of their natural habitat and is especially important in the circumstances where in situ preservation is not possible because of the excessive degradation of the habitat or the threatened imminent extinction of the plants. Among the in vitro approaches to the ex situ form of conservation are micropropagation, somatic embryogenesis, embryo rescue, photoautotrophic culture, cell suspension cultures, and cryopreservation. The techniques can enable the reproduction, banking, and reproduction of species of plants that are not easy to conserve by other methods like seed banking or growing them in botanical gardens.

Micropropagation is the main method applied in the conservation of endangered plants because it can multiply genetically identical and disease-free plantlets which were developed using small or even single source of explants. It is especially effective in plants such as low-viability seed plants or even vegetatively propagated plants. Also, the benefits of somatic embryogenesis can be brought to bear by the capacity to regenerate plants through somatic tissue, amplifying the scalability and efficacy of protocols, ensuring conservation. The other important technique is the embryo rescue, which involves the rescue of viable plant of those embryos that would have not grown into plant due to premature abortion, which would occur in natural cases (Mamathaba et al. 2022).

Any long-term storage has been revolutionized since cryopreservation freezing of plant materials under ultra-low temperatures (−196 °C), mainly in liquid nitrogen. It allows preserving viable cells, tissues, embryos, or shoot tips indefinitely without genetic mutation. The importance of such a method is especially true of species that have recalcitrant seeds, avoiding the drying and storage of such species in traditional seed banks. Cryopreservation gives genetic stability and can be used to back up field collections and botanic gardens as well as seed banks.

These biotechnological methods are also illustrative in lots of aspects such as in the conservation of wild, crop, medicinal, and ornamental plant species. An example of this is the case of Europe, where the botanical diversity has been successfully conserved through in vitro culture techniques, even though Europe has more than 12,500 species of vascular plants, many of them in an endemic and endangered state. Nations like Greece, Italy, Romania, Bulgaria, and Spain have focused more on the biotechnology-based conservation projects as a result of the endemic flora abundance there.

Although botanical gardens are an older and more conspicuous form of ex situ preservation, with more than a third of flowering plants in the world cultivated in botanical gardens, the botanical gardens have several issues, including limited space where the number of plants is always a trade-off against the area held, high costs of maintenance, and related problems of acclimatization and genetic drift. Comparatively, in vitro conservation offers a lower-cost and space-efficient and scalable solution that has the ability to sustain extensive genetic repositories with limited physiological and genetic erosion.

To sum up, adding plant tissue culture technology to the conservation regime is a sustainable, scientifically sound method to address the problem of biodiversity loss in plants. The techniques are capable of the conservation of plant genetic resources not only to ensure continuity to the future but also to enable re-introduction and habitat restoration projections. With the increasing climate change and anthropogenic impacts, even greater significance will be obtained by biotechnological means of tissue culture and cryopreservation of biological materials worldwide in order to preserve the botanical diversity of the world (Kulus and Zalewska 2014).

15.3.3 Mass Propagation for Commercial Applications

The trees have always played a significant role in the life of human beings and have proved to be a source of their vital needs, which include food, fodder, fuel wood, timber, and fiber as well as other non-timber products. Nevertheless, expedited deforestation, unsustainable use of natural resources, and the growing demand of the world to use forest-derived products have resulted in the loss of forest cover and much threat to genetic diversity of tree species. Although simple propagation techniques have a lot of value, they are not sufficient to address the needs of commerce and conservation, particularly due to the biological complexity of forest trees, including long juvenile periods, high levels of heterozygosity, self-incompatibility, and low growth rates. Against this background, plant tissue culture and micropropagation have been found to be powerful instruments to counter these constraints in order to be able to have rapid and large-scale multiplication of elite tree genotypes.

A massive propagation through the in vitro methods, especially micropropagation, provides the feasible and scalable method of high-quality clonal planting material production. This is essential in not just commercial forestry and agroforestry practices but also the biomass energy and the preservation of elite, endangered, or slow-growing species of trees. Clonal fidelity and uniformity are paramount in the utilization of desirable traits; hence, fast growth, wood quality, disease resistance, and stress tolerance are guaranteed by micropropagation. Genetically identical plantlets may be produced through the protocol of controlled culture conditions, and the selected elite individuals have their advantageous traits maintained across multiple generations.

Although useful, in vitro propagation of woody perennials is a technical challenge because of lack of regeneration of the mature explants in the laboratory. The developmental stages commonly found in trees include embryogenesis, juvenile growth, transition, and mature reproductive stage. During the transition between juvenile stages and mature stages of trees, changes in morphogenic competence of explants greatly influence the necessary in vitro responsiveness. Therefore, it may be hard to propagate mature elites. To curb this, a variety of approaches have been used in an attempt to somehow utilize the tissues of juvento either to rejuvenate the tissues of the adult (Saha and Sundriyal 2012).

Traditional rejuvenation methods involve harsh pruning, hedging, the use of root suckers to grow above it or PGRs to induce growth of epicormic buds. The techniques assist in getting physiologically younger tissue that is more susceptible to regeneration. To enhance the efficiency of regeneration, in vitro methods that include epicormic shoot cultures, repeated subculturing, micrografting to juvenile rootstocks, production of adventitious buds, and regeneration through somatic embryogenesis are also practiced. Somatic embryogenesis especially enables one to grow embryos in somatic cells and has the prospect of being scaled up with bioreactors and thus automatized and is thus very appropriate in batch-wise mass clonal propagation.

Micropropagation has been successfully commercialized in various tree species, such as Eucalyptus, Populus, *Tectona grandis* (teak), and Acacia, among others. Some of the typical stages of these protocols include selection of explants and sterilization of the explants, initiation and multiplication of shoots, induction of roots, and plantlet acclimatization in a greenhouse or nursery. From optimizing culture media to application of PGRs to physiological status of explants, there has been an enhanced success rate.

In addition to this, biotechnological activities have enabled the spreading of species that are economically valuable and have low natural regeneration abilities. Under plantation forestry, clonal propagation is used to result in even growth and dependable yields and better management. With bioenergy crops and short rotation woody biomass species, micropropagation results in very quick establishment in the field and prolonged productivity. Mass propagation within the conservation field assists in rescuing and rehabilitating threatened or rare species and gives an opportunity to traditional seed-based strategies that might be constrained by either dormancy, low vigor, or limited supply.

Drawing a conclusion, it can be stated that mass cloning inventions based on plant tissue culture are a critical component of contemporary plant biotechnology and help solve two major problems of the medicinal plant world: commercial productivity and the survival of biological diversity. Further research with a view to enhancing the regeneration process to become more efficient and culturally successful and, to some extent, combining it with the genetic improvement approaches will make the technologies even more scalable and effective within the framework of sustainable use of forest resources (Shmakov et al. 2024).

15.3.4 In Vitro Stress Tolerance Screening

Various types of biotic and abiotic stresses eternally put plants in jeopardy and retard their growth, development, as well as productivity. Such stresses, such as drought, salinity, pathogens, and HM toxicity, are feared to be tremendous threats to food security in the world, especially in changing climatic conditions. Classical approaches of breeding have been used to produce stress-tolerance genotypes, but they are labor intensive, require a lot of years and resources as well as limited by a slew of plant genomics and environmental fluctuations. Here, in vitro screening and development of in vitro tissue culture of plants have proved to be an extremely efficient, repeatable, and cost-effective method to assess and improve stress tolerance in plants.

The basis of in vitro plant tissue culture is that whether it is an individual plant cells or tissues can be regenerated into complete plants, as a result of cellular totipotency, in a layer under controlled conditions in sterile conditions. This feature can be applied in testing and analyzing the reactions of plants when exposed to different stress agents, such as salinity, drought, toxins, HMs, and pathogens, in a young stage of growth without being interfered by external and unpredictable environmental conditions. Because of the defined culture media, control, to the extent possible, of the physical and chemical parameters, and the capability of applying desired selection pressures, in vitro systems are an attractive candidate as a powerful platform to conduct screening of stress tolerance.

The first important benefit of in vitro screening is that it considerably reduces the length of time it takes to determine stress-tolerant genotypes. Researchers can use test conditions in plant cells, calli, or organogenic tissues to expose them to controllable levels of stress, including media containing NaCl (salinity), polyethylene glycol (PEG), or mannitol (drought) or pathogen-derived toxins, to then select the most tolerant lines on the basis of responses at the physiological and biochemical levels. The regeneration of the selected tolerant cells then forms the whole plants, and they retain the desired stress-resistant characters. Not only does this procedure increase the efficiency of breeding but also it becomes possible to select candidate plants to which there can be an evaluation at the field level (Pinski et al. 2025).

A key role in biotic stress-resistance screening is also carried out by in vitro systems. With the help of pathogen- or toxin-derived elicitors or toxins in the culture medium, resistant somaclones or cells may be identified. Moreover, transgenic solutions in combination with in vitro tissue culture systems allow the creation of genetically engineered plants that can have a stronger resistance to biotic and abiotic stresses. An illustration of this is the fact that plants with genes that encode stress-related proteins (including LEA proteins, heat shock proteins, or antioxidative enzymes) could be regenerated and be screened in vitro under pressure in order to prove successful integration and expression of stress-tolerant characters.

Other traits that have been selected through in vitro selection, besides drought, salinity, and disease resistance, include metal hyperaccumulators, which are plants that are able to absorb and tolerate high levels of HMs, such as cadmium, arsenic, and lead molecules. The plants can be utilized in phytoremediation, which is a green solution to decontaminating the soil and water. The toxicity of such metal-tolerant species to sensitive plants such as crops can be screened at an early stage by simply adding metal salts to the growth media, making it possible to identify hyperaccumulating lines which may then be cultivated and used in the field.

In spite of the stacks of merits, screening in vitro stress tolerance is not problem free. The first critical factor is that well-established and species-specific protocols of tissue culture are required because different species of plants and even genotypes of a species may not respond in a similar manner under in vitro conditions. Furthermore, when stress response is measured on a cellular or tissue scale and when it is measured as a whole plant response, the aspects are not always easy to compare. Transient responses to the tissue culture process, such as somaclonal variation or stress adaptation response, can result in a lack of stable tolerance of staple crops that would be evident in long-term field performance. To lessen these problems, short selection (one step) procedures are advised to lessen the probability of epigenetic interference.

Nevertheless, in vitro screening is an effective supportive approach to the field selection approaches. It is an early platform that can guide and inform any next field testing, thus taking less time and resources involved in different screening programs at large scale. In addition, it can be used in both traditional and biotechnological breeding pipelines, which makes it the central point in contemporary plant biotechnology that wants to develop sustainable, climate-resistant crop systems.

In one word, in vitro stress tolerance soaking is an important tool in the bag of the plant biotechnologist. It provides a rigorous, accurate, and efficient method with which to select and breed genotypes of plants that have increased resistance to a great variety of biotic and abiotic stresses. The mainstreaming of the in vitro methods into overall breeding and conservation activities will therefore play a key role in ensuring sustainable agricultural productivity as the need to produce stress-resilient crops rises ever higher (ŠVábová and Lebeda 2005).

15.4 Molecular Marker-Assisted Breeding

15.4.1 Types of Molecular Markers (SSR, SNP, AFLP, etc.)

Molecular marker-assisted breeding (MAB) is an innovation of plant biotechnology that is groundbreaking in the precision of crop improvement schemes by having the capacity to identify and select the desirable characteristic at the DNA scale. At the core of MAB is a reliance on molecular markers. These are sequences of DNA that can be detected and that are linked to a gene of interest or a region of interest within the genome that enables the genetic variation to be mapped, tracked, and manipulated. The markers are especially useful in rapidly breeding the animals and improving the level of selection as well as surmounting the limitations of phenotyping caused due to the complex nature of traits that are environment-dependent.

Simple sequence repeats (SSRs), or as they are also referred to microsatellites, stand out among the many types of molecular markers because of their high rates of polymorphism, co-dominant inheritance, and ease of amplification using polymerase chain reaction (PCR). SSRs have superior reproducibility and genome distribution, which makes them effective in comparisons of diversity, linkage mapping, and marker-assisted selection (MAS). Due to their multi-allelic nature, they give excellent genetic information, but the development depends on pre-sequencing. The most common variant in the plant genome is the single nucleotide polymorphisms (SNPs), which is composed of single-base changes. SNPs have high-throughput potential genotyping, codominance, and are amenable to automation and standardization, and thus make them very appropriate in genome-wide association study (GWAS), quantitative trait locus (QTL) mapping, and genomic selection (GS). Nevertheless, SNPs are bi-allelic and might need large panels to harvest enough diversity mainly in low-diversity crops (Jiang 2013).

The amplified fragment length polymorphism (AFLP) markers occupy the middle ground between the PCR-based and restriction fragment methods. AFLP has no prerequisite of any information on the sequence and, moreover, enables the scanning of the entire genome by selective amplification of restriction fragments. This marker-based system produces many highly polymorphic and reproducible markers applicable in genetic diversity studies, DNA fingerprinting, and development of saturated genetic maps. Though these are the advantages, AFLP profile interpretation may be difficult because of the dominant inheritance and the allele scoring challenges. Besides these, other marker systems such as random amplified polymorphic DNA (RAPD), inter simple sequence repeat (ISSR), sequence-characterized amplified regions (SCARs), and restriction fragment length polymorphism (RFLP) have been extensively used. RAPD and ISSR are very quick and cost-effective but have reproducibility problems and RFLP, which is highly informative, is also closed-dominant, labor-intensive, and not applicable to large-scale analysis. Next-generation sequencing (NGS) technology has made the creation of genotyping-by-sequencing and ultra-high-density SNP arrays possible, further refining molecular marker analysis, by increasing the level of resolution and throughput. FMMs are even more accurate in selection, constituted using known phenotype-controlling genes. They play a major role in the identification and introgression of genes conferring yield, nutritional quality, and biotic and abiotic stresses. In general, the choice of the right type of marker will vary depending on the breeding goal, complexity of the genome, available resources, and resolution needed. Such combination of strategic uses of multiple marker systems are still continuing to make plant breeding a more strategic, efficient, and sustainable process in the face of global food security and climate resilience (Sharma et al. 2006).

15.4.2 Marker-Assisted Selection for Crop Improvement

MAS is a game-changer in contemporary crop breeding because it is relatively new, and breeders can now locate and narrow down on specific plants that possess the traits they desire faster and more efficiently than when they used phenotypic selection. Through the combination of principles of molecular genetics and traditional breeding, MAS enables the possibility of early and precise detection of genotypes with useful alleles and thereby hastens the production of better crop varieties. The given strategy is particularly useful in combating the worldwide issues of food security, climate change, and the necessity of sustainable agricultural policies.

MAS is based on molecular marker technology, where DNA fragments that are closely associated to genes or QTLs controlling a particular trait, including resistance to diseases, tolerance to abiotic stress, grain quality, and yield-related characteristics, are used. After a faithful correlation has been found between a marker and a trait of interest, plants may be screened according to whether the trait is present or not with simple laboratory techniques without the need for their life cycle or exposure to growth environments in order to reveal the phenotype character corresponding to the trait.

MAS has been associated with high success in the introgression of resistance genes in wild relatives such as landraces into the elite cultivars using a backcross technique to transfer the resistance genes without the undesired genes. In crops such as rice, wheat, maize, soybean, and tomato, MAS has enabled the production of pathogen-resistant varieties of plants to pathogens like Xanthomonas, *Magnaporthe oryzae*, and *Phytophthora infestans*. Equally, it has played a key role in pyramiding several genes on resistance to the biotic stresses, thereby enhancing the resistance durability and lessening the dependency on chemicals (Mangal et al. 2024).

In MAS, the individuals are typically selected by isolating those in a breeding population who possess one or more advantageous marker alleles. This can be done by foreground selection (of target genes), background selection (of recurrent recovery of parental genome), and recombinant selection (breaking linkage drag). Moreover, combining MAS with GS which uses genome-wide data of the markers to predict breeding value is opening doors to make crop improvement more efficient even on the complex traits that are affected by multiple genes and the environment.

The inclusion of high-throughput genotyping platforms, NGS, and phenotyping platforms that can accurately measure traits has extended the applicability of MAS and has allowed breeders to work with greater population sizes and with more traits at the same time. The reducing cost of genotyping as well as the increasingly accessible marker-trait databases are further making MAS less prohibitive in both developing and developed nations.

Conclusively, MAS is an attestation that is considered as part of this tool to modern plant biotechnology, and it is the bridge that brings together molecular biology and field-level breeding. Not to mention that its use not only grows breeding cycles and intensifies genetic gains but also creates climate-resistant, high productivity, and nutritional enhancement in crop strains, which, in turn, contributes to the realization of sustainable agriculture and food security in the twenty-first century (Kim et al. 2021).

15.4.3 Genomic Selection and Predictive Breeding

They are GS and predictive breeding; these are revolutionary techniques in the current plant biotechnology, which provide a paradigm shift from the current phenotype-based selection into genotype-driven improvement procedures. As opposed to the MAS where very few markers were used linked to the major-effect genes, GS is based on the whole-genome marker data to predict the individuals based on the breeding values, both by major and minor genes. With this broad-based technique, the high-quality genotypes can be chosen early in the breeding cycle, thus saving tremendous amounts of time, money, and inputs.

GS is based on the construction of predictive models based on dense genomic marker data, e.g., SNPs with phenotypic data in a training set of individuals. Such models can then be used to predict genomic estimated breeding values of selection candidates without having to subject them to large field tests. This promotion of faster breeding cycles and increases in the overall number of complex traits that can be improved simultaneously with each other by facilitating the use of GS has made it particularly useful when it comes to traits that are polygenic in nature and have a high degree of environmental interaction, such as yield potential, drought tolerance, and nutrient use efficiency.

The fundamental infrastructure to enable efficient GS is the availability of genomic resources, such as high-throughput genotyping platforms, reference genomes, transcriptome profile, and database of gene expression. Further, the GWAS can make the trait-marker links more precise, establishing an even higher precision on the predictive models. Utilization of GS coupled with NGS technologies has augmented selection precision and resolution, which has allowed the utilization of genetic variation that remained previously unexploited and hard-to-exploit favorable alleles that are rare most of the time using conventional methodologies (Abdul Aziz and Masmoudi 2025).

The evolution of artificial intelligence (AI) and ML is taking predictive breeding to a new level as well. Random forests, support vector machines, and deep neural networks (ML algorithms) allow building robust models that have the capacity to solve intricate patterning in large-scale multi-omics data (genomic, phenotypic, environmental, and agronomic datasets). In these models, plant performance is well predicted, and also the ideal combinations of parents to be used in crossbreeding are identified as well as the expression of traits in varied environmental conditions. Particularly, deep learning allows the automation of the analysis of the high-throughput phenotyping data, including drone images and sensor-based IoT devices, to further improve the utility and precision of the trait analysis and selection.

The intersection of data science and plant breeding supports a more complete and system-wide view of crop development, also known as precision breeding. Predictive analytics are established by extrapolating genome, transcriptome, and proteome analyses to promote breeding strategies to suit agroecological areas and adverse situations. As an example, the combination of climatic data and soil parameters with genomic data allows breeders to create crop breeds that survive in a specific climate in the future, thus providing a contribution to the field of climate-resistant agriculture.

There are, however, a lot of challenges even with these advancements. The validity of predictive models is closely linked to the quality and size of training and training populations, marker density and distribution, and representative of phenotypically distributed data. In addition, complex models, especially deep learning algorithms, might lack interpretability, which considers their broader use. The fact that most of the data is found in various platforms poses challenges concerning data harmonization and data standardization. Also, lack of infrastructure and technical expertise may constrain the use of such tools particularly in small-scale breeding programs.

However, the future of plant breeding will be an orchestration of genomic tools, predictive modeling with AI, and high-throughput phenotyping in an effort to hasten creation of accurate and more sustainable crops. Thanks to GS and predictive breeding, there are now opportunities like never before to enhance yield, quality, stress resilience, and resource use efficiency of crops to respond to increasing global pressures in food security and environmental sustainability. With the democratization of predictive breeding, breeders in all parts of the world will be able to realize the potential of plant genetic resources as technologies evolve and data become more accessible (Dublino and Ercolano 2025).

15.4.4 Applications in Biodiversity Conservation

Biodiversity as the complete set of genetic, species, and ecological diversity is the core of ecological health, agricultural health, and human health. Nevertheless, biodiversity is facing a severe threat because of the loss of habitat, climate change, pollution, overexploitation, invasion of non-native species, among other reasons. The International Union for Conservation of Nature says that almost one-third of the global plant species face extinction risk at present. In that regard, molecular marker-assisted breeding (MMAB) offers a novel solution to the evaluation as well as conservation of the genetic variation of plants, supplementing thus more significantly to conservation of biodiversity.

The molecular markers, including SSRs, SNPs, and AFLPs, have come out to be effective in determining the level of genetic diversity, population structure, and evolutionary relationship among and between plant species. Such markers allow the accurate identification of rare alleles and special genotypes that are significant in preserving the capability of adaptation within plant populations. Such molecular understanding will enable formulations of core collections, as well as priority accessions targeted against conservation, to limit the duplicate copy and maximize the genetics in germplasm banks.

Molecular markers are used in ex situ conservation work that entails the establishment of gene banks, seed banks, and in vitro repositories as a way of examining genetic integrity, somaclonal variation, and clonal fidelity of plants propagated in vitro. Plant tissue culture, somatic embryogenesis, and cryopreservation are biotechnology methods that are add-ons to marker-assisted approaches, as they help to multiply and long-term conserve the species that are endangered and recalcitrant.

In the case of in situ conservation, molecular tools can be used to provide continuous investigation of the genetic variation within the natural populations and evaluation of gene flow, as well as detection of bottlenecks caused by the fragmentation of the habitat. Such tools also assist with ecological restoration because they help to select genetically diverse and ecologically appropriate plant material to reintroduce into degraded habitats.

Moreover, by introducing MMAB into conservation and breeding schemes, the conservation and practice are connected with each other. Markers can be used to characterize conserved germplasm for traits such as stress tolerance and disease resistance, and subsequently, the germplasm can be used with both MAS and GS to improve the crop. With this practice, not only is genetic diversity being preserved but also it provides functional value to material that is conserved, making them even more relevant to sustainable agriculture. Conclusively, molecular marker-assisted breeding has gained transformation in the biodiversity conservation strategies such that biodiversity is managed efficiently, sustainably, and on the basis of data. It makes sure that conservation efforts are focused on other longer-term ecological and agricultural objectives, thus being able to make a positive contribution to the global sustainability efforts (Pfenning-Butterworth et al. 2024).

15.5 Plant–Microbe Interactions and Biofertilizers

The interaction between plants and microbes, especially of the rhizosphere, where a thin layer of soil is heavily affected by the interaction of roots and related microbial activity, is a key element in plant growth, health, and vigor as well as environmental stress resistance. These influences are complex and entail the existence of both pathogenic and beneficial microorganisms coveting the pathogenic microbial organisms, but it is the beneficial microbial communities that have recently been a preoccupation to utilizing them in cultivating sustainable agriculture. The replacement of the chemical-based inputs with those derived biologically and the introduction of a safer and more sustainable system of crop production are not only the reaction to the environmental degradation and the loss of soil fertility but also the innovation towards a more sustainable and safer system of crop production.

Recent changes and advances in the field of systems biology and omics technologies—genomics, transcriptomics, proteomics, and metabolomics—have turned around the way we do our research on plant–microbe interactions. The functionalomic studies of model soil microbes, i.e., rhizobacteria and arbuscular mycorrhizal fungi (AMF), have provided very important information in order to analyze the genetic determinants involved in the colonization of the rhizosphere, acquisition of nutrients, and also tolerance to stress. The mapping of microbial traits has also been made possible through whole-genome sequencing and bioinformatics, and newer transcriptomic tools such as microarrays and metatranscriptomics have helped measure in-situ gene expression, providing a more dynamic view of microbial activities in contact with plants.

The most promising discovery of this study is the production of biofertilizers, solutions of living microorganisms, which enrich the soil in nutrients, activate plant growth, and promote the eco-health of the soil. Such microbial inoculants are N-fixing bacteria (e.g., Rhizobium, Azospirillum, Azotobacter, and Azolla), phosphate-solubilizing bacteria, K-mobilizing bacteria, and plant growth-promoting fungi, such as *Piriformospora indica* (formerly *Serendipita indica*) and AMF. Biofertilizers perform different activities such as nitrogen fixation, phosphate solubilization, phytohormone production (e.g., auxins, cytokinins, and gibberellins), production of siderophore to obtain iron, and inducing systemic resistance in host plants (Solomon et al. 2024).

Notably, these microbial communities do not only augment the crop productivity but also make the soil more fertile in the long term by increasing the soil organic matter, the soil physical structure, as well as assist in the removal of toxic compounds, which include HMs. They can be used as a part of the environmentally friendly approach to farming, making the use of the synthetic fertilizers less frequent and environmental pollution less significant. Further, microbial inoculants will also be important in abiotic stress tolerance, which occurs in the plant adaptations to drought, salinity, and metal toxicity via osmolyte accumulation, antioxidant synthesis, and enhanced root system architecture.

There is growing use of biotechnological tools to make and perfect these microbial formulations. As molecular identity methods, 16S rRNA sequencing and molecular fingerprinting help in the proper classification and quality check of a given microbial strain. Genetic engineering also advances the performance characteristics of the said microbes, making them workable under diverse soil and weather conditions. Finally, the plant–microbe relationships form the key element of sustainable farming. These interactions would be exploited through the application of biofertilizers that would enhance the nutrient efficiency and lead to a reduction in the use of agrochemicals and the sustainability of the ecosystems. Combining high biotechnology technology with this line is amplifying the prowess of microbial inoculants and making them highly precise and efficient bio-inputs of future agriculture (Marzouk et al. 2025).

15.5.1 Rhizosphere Engineering for Enhanced Growth

The rhizosphere is a thin slice of soil affected by plant roots that possesses a motile and heterogeneous microbiome, which is crucial in nutrient cycling, plant development, and its abilities to overcome environmental stresses. Rhizosphere engineering is a recent, rapidly developing sphere of agricultural biotechnology which is the engineering of microbial communities within the rhizosphere so as to maintain the best plant–microbe interactions in terms of crop production, resistance to stresses, and sustainability.

The plant circadian rhythm (CR) is one of the most important factors affecting the rhizosphere dynamics, a molecular time-keeping system that changes many of the constantly regulated physiological processes in the plants in response to the environmental conditions. The circadian-rhythm-governing genes are also very sensitive to extrinsic factors like light, temperature, and nutrients. Disturbance or defect of clock genes may therefore strongly affect the time course of root excretion, contributing to a change in the structure and activity of the rhizosphere microbial community, especially rare taxa with key activity on plant beneficial traits. Targeted rhizosphere engineering would be a promising frontier that involves the comprehension of circadian clock genes in the plant and how it impacts root–microbiome signaling.

Rhizosphere engineering approaches are based on the use of plant growth-promoting rhizobacteria (PGPR). PGPR are able to mediate plant growth in a direct way by nitrogen fixation, solubilization of phosphate and potassium, siderophore iron uptake, and the production of phytohormones (auxins, gibberellins, and cytokinins). Indirect pathways are also employed, such as production of antimicrobial substances, pathogen outcompeting, disruption of quorum sensing, prevention of biofilm, and development of systemic resistance. Further, during stress, enzyme production, like the production of ACC deaminase, decreases the ethylene production in plants, thus making it favorable in terms of root development, etc.

The development of growth in the omics technologies, genomics, transcriptomics, proteomics, and metabolomics has made the realization of microbial functions in the rhizosphere much quicker. These tools, when combined with bioinformatics and ML, are used to identify major microbial taxa and functional genes that integrate nutrient mobilization, stress alleviation, and plant immunity. The current engineering efforts are targeted at choosing or developing microbial consortia that have the ability to synergistically carry out these functions under certain soil as well as climatic conditions.

Microbiome-mediated smart agriculture system is an emerging concept that is being discussed about as a futuristic model of finding a rhizosphere engineering solution within the context of sustainable agriculture. The cultural concept of this indication is the creation of tailored-made microbial formulations that are bioinoculants or biofertilizers and done according to the soil, the type of the crop, and the existing biotic and abiotic stresses. Nonetheless, the variable conditions of the soil, flexibility of microorganisms, and interactions among species pose a challenge to successful engineering of microbial biota in the soil. So, it is essential to thoroughly study the soil biotic–abiotic border, on the basis of which the introducing microbes can respond in case of any disruption.

Finally, rhizosphere engineering can be endlessly useful in a modern farming industry to improve the performance of the crops with sustainability for the environment. Taking advantage of the complex mutualism existing between plants and the microbiota, together with the application of biotechnological modules to engineer beneficial microbial consortia, a sustainable and resistant agroecosystem can be achieved, promoting food safety with the preservation of the ecological equilibrium (Ahkami et al. 2017).

15.5.2 *Plant Growth-Promoting Rhizobacteria (PGPR)*

The increasing food demands in the world due to the increasing population and the creation of non-manmade products of agriculture have prompted the mass application of synthetic agrochemicals. These inputs have enhanced production of crops in the short run, but their continued application has drastically affected the soils, biodiversity, and environment in terms of quality. To this effect, biofertilizers, especially the ones that use PGPR, are being found as attractive, sustainable options. PGPR represent a large collection of bacteria that colonize the root and reside in the rhizosphere or the bulk soil and have a positive impact on plant growth and development. They have their beneficial role in the work through various effects that increase the ability to absorb nutrients, mitigate abiotic and biotic stresses, and regulate the functioning of a plant.

The working process of PGPR is both direct and indirect. Directly, they help in gaining necessary nutrients by means of fixing nitrogen that is present in the air in the form of nitrate, making phosphates and potassium soluble, and this occurs by manufacturing siderophores that chelate iron, and by this, iron is made more available to plants. What is more, they produce phytohormones like auxins, gibberellins, and cytokinins that stimulate the elongation of the roots as well as their growth and the whole plant's vigor. The protection of plants directly through PGPR is indirect in that it induces a system of resistance to pathogens, releases antibiotics and enzymes that break down the cell walls of the pathogens, and also competes with the detrimental microbes to gain niche and nutrients.

PGPRs when used as biofertilizers have been described as a sustainable agricultural approach that enhances soil fertility, plant fitness, and productivity in a way that does not have a negative impact on the environment as compared to the use of chemical fertilizers. Such microbial preparations may be live or inactive bacterial strains that come into use when applied on seeds, roots, or soil. In addition to traditional biofertilizers, microbial products like organic extracts and vermicompost teas have also been found to hold some promise in the enhancement of plant health and disease resistance in a manner that is environmentally friendly (Basu et al. 2021).

Even though the positive attitude of PGPR under laboratory and greenhouse conditions has been proved, field performance of PGPR is frequently uneven. This diversity is mainly associated with the complexity of soil ecosystem, environmental heterogeneity, plants–microbe–soil interactions, and the lack of the capacity of certain strain to successfully colonize the rhizosphere in the field environment. Other problems involve the stability of the formulation, the duration of storage, and limitation in the shelf life that limits the uses of the microbial products.

However, the current studies in genetic and functional diversity of PGPR and improvement in microbial genomics and bioinformatics and formulation technology are leading to the emergence of second-generation biofertilizers. Techniques that are under examination to overcome these limitations and to maximize efficacy include strain consortia design, carrier-based delivery, and adaptation to crop and environment. Finally, PGPR can be an excellent candidate for biotechnological application towards sustainable intensification of agriculture. Their possible benefits in maintaining plant growth, enhancing nutrient use, and minimizing their reliance on the applied chemical inputs are also compatible with the objectives of climate adaptation and environmentally friendly agricultural systems. As the areas of microbial screening, field verification, and commercialization channels continue to develop, PGPR-based biofertilizer could take over as a keystone to the increasingly sustainable and more productive agricultural future (Liu et al. 2023).

15.5.3 Mycorrhizal Associations for Nutrient Efficiency

Mycorrhizal fungi, especially the arbuscular mycorrhizal (AM) fungi, are important in promoting the nutrient acquisition of plants and also in enhancing the overall soil fertility status with a symbiotic state with plant roots. This symbiotic association is known to increase much the water and nutrient (mostly phosphorus, nitrogen, and micronutrients) uptake capacities of the plants significantly, which happens in the soil and, in turn, lessens the addition of chemical fertilizers and promotes sustainable agricultural systems. Mycorrhizal biotechnology has demonstrated significant potential in many areas, including agriculture, horticulture, landscaping, and phytoremediation, and in the general use of plants in the market.

The development of new technology in recent years has greatly enhanced the prospect of the use of AM–plant symbiosis. Massive production of AM fungal inoculum has been conventionally performed on large-scale using in vivo cultivation techniques, involving applications of pot cultures or sterilized soils, hydroponics, aeroponics, or greenhouse; however, these techniques have a disadvantage of slowing down inoculum growth as the process continues to sour. Even though the procedures are common, they are often disrupted by the problem of cross-contamination and varied inoculum quality. Improved techniques of production have resulted due to the demand of quality and contamination-free AM fungal propagules. The transformed root organ culture method is one of the most advanced and productive methods of cultivating high-purity AM fungal inoculum. This method enables the growth of AMF under a sterile environment as they are inoculated to transformed root systems, producing propagules without microbial contaminations. This in vitro system has transformed this field by giving consistent, scalable, and quality mycorrhizal products that can be used commercially and in conducting research.

Inoculant products have since been developed and commercialized in a variety of mycorrhizal formulations, such as seed coatings, liquid inoculants, and biostimulants. The uses of such formulations are thus increasingly found due to their ability to create strong symbioses with regard to a wide variety and usually in less-than-ideal soil and climatic conditions. Plant resilience, better nutrient and water productivity, and overall plant health attributes associated with AMF colonization of plant roots enhance the performance of plants generally and under conditions of environmental stress, including drought, salinity, and nutrient deficiency in particular.

The demand for biofertilizers based on mycorrhizal is increasing in the world due to the mounting interest in sustainable agriculture. Through commercialization using particular investment in different inocula production, both our developed and developing nations are making gigantic efforts to reach the bio-inputs market with an aim of finding solutions to the hitches in farming in a bid to commercialize mycorrhizal strains. It still has its problems in the sector, especially on quality assurance and traceability. There is a hitch in ascertaining the actual origin and identity of fungal strains comprising most of the commercial inoculants. Although there exist regulatory frameworks with an aim of ensuring strict quality standards, it has been noted that most of these products available in the market are not uniformly prepared and thus may prove unreliable in the job they are claimed to be doing.

However, existing science in fungal taxonomy, genomics, and production technology is advancing and, as a result, increasing the accuracy and efficacy of mycorrhizal candidacy. These inventions are opening the gates to large-scale and efficient use of mycorrhizal biofertilizers in sustainable crop production systems. Mycorrhizal associations are therefore an invaluable part of the technology toolkit of next-generation plant biotechnology, compatible with the climate-smart and ecologically responsible agriculture goals, namely, by improving the nutrient use efficiency, decreasing the use of agrochemicals, and improving soil health (Farhaoui et al. 2025).

15.5.4 Biocontrol Agents for Sustainable Pest Management

Sustainable pest management has turned out to be one among the key pillars of current farming, particularly after the increasing worries about the ecological and health dangers instigated by conventional synthetic pesticides. An alternative to chemical pesticides is the application of biocontrol agents, including beneficial microorganisms (BCAs), such as bacteria, fungi, and endophytes, which are an adequate option and environmentally friendly. This is because these agents restrain the plant pathogens and pests through the production of antimicrobial compounds, provocation of plant resistance, competition for nutrients or ecological niches, parasitizing them, or attacking the pathogens directly.

In organic farming, the focus is on the use of natural farming methods and production of food, which is free of synthetic agrochemicals. The control of pests, diseases, and weeds in organic farming increasingly uses biocontrol strategies. This strategy is based on naturally occurring antimicrobial agents, whether of microbial origin and/or of botanical origin, including neem (*Azadirachta indica*), garlic (*Allium sativum*), eucalyptus (*Eucalyptus globulus*), turmeric (*Curcuma longa*), and tobacco (*Nicotiana tabacum*). Nonetheless, the organic form of farming is prone to plant disease breakouts because there are no synthetic means of chemical intervention. This has given rise to the increased use of strong nonchemical substitutes like BCAs.

Biocontrol agents inhibit the plant pathogens in a number of ways, with the generation of antibiotics and secondary metabolites being just one of them. Such molecules of low molecular weight are able to block the growth or reproduction of pathogens directly. The mechanism of action is usually interfering with crucial cellular functions, including protein synthesis or cell membrane integrity. As an example, *Bacillus subtilis* AU195's antibiotic is *Bacillomycin D*, *Trichoderma virens* proteins are gliotoxin, and *Pseudomonas fluorescens* F133 excretes 2,4-diacetylphloroglucinol (DAPG), which has been shown to possess broad fungicidal-like antifungal activity. Other BCAs, e.g., *Bacillus cereus* strain UW85, can make more than one antibiotic (Zwittermycin A and Kanosamine), which increases its biocontrol potential towards a variety of phytopathogens (Tyagi et al. 2024).

Besides, genetic engineering has been utilized in increasing the biocontrol of microbial strains. For example, *Pseudomonas putida* WCS358r is a genetically engineered strain to produce phenazine, and DAPG resistance to both of the products was better to target disease in wheat crops. These developments have also underscored the optimism in synthetic biology and microbial engineering in designing the next-generation of biopesticides in targeting the pathogens or target environment-specific biopesticides.

Biocontrol areas that are gaining a lot of attention include the use of endophytes, non-pathogenic microorganisms that are found in the plant tissues. These microorganisms form symbiotic associations with plants and can take part in the defense of the plants against disease-causing organisms. However, the complexity of associated microbial communities with plants, together with their diversity, impedes the isolation and characterization procedures of the functional endophytes with their traits. The majority of them are not utilized due to the limitation of classical culturing techniques and genetic technology.

In order to mitigate such disadvantages, researchers were able to access novel biotechnological technologies, such as CRISPR/Cas9-surgery of genes, nanotechnology, and multi-omics environments (genomics, transcriptomics, proteomics, and metabolomics). These techniques allow studying endophytic associations at the molecular level, which provides the chance of identification of new genes, pathways, and metabolites with the potential to safeguard the plant organism. Of

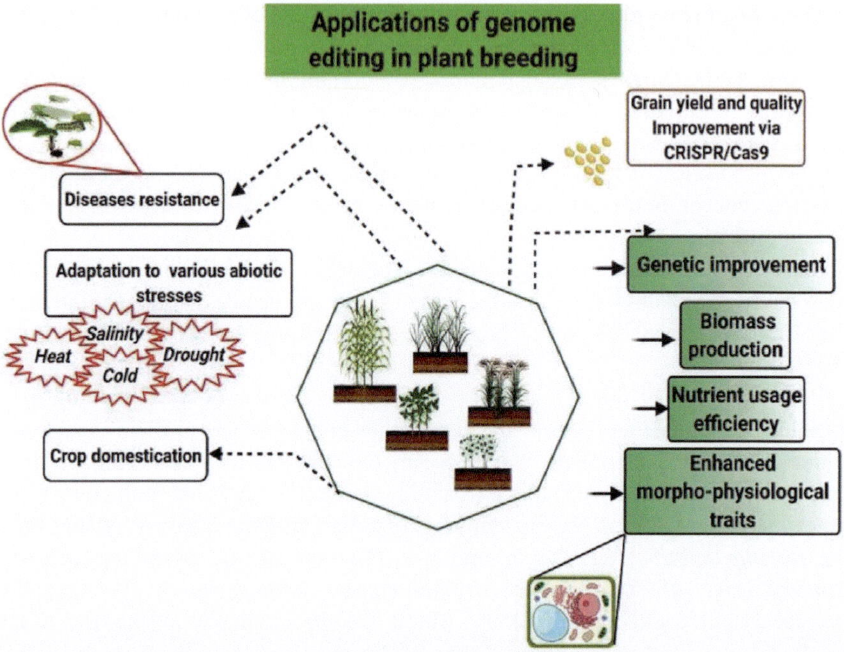

Fig. 15.2 Applications of genome editing (e.g., CRISPR/Cas9) in plant breeding for trait enhancement, stress adaptation, and yield improvement (Rasheed et al. 2021)

particular importance is the CRISPR/Cas9, which will be able to introduce specific editing in plants and microbial genes, including higher stages of enhancing valuable traits and deactivation of diseased features as shown in Fig. 15.2. Nanoparticle research is underway so that improvements in the lining of the microbial inoculants and phytochemical compounds could be obtained as a means of providing security in the field.

The combination of omics technologies would allow a system-level interpretation of the plant–microbe interactions, enabling the researchers to construct model systems and even predictive systems to successfully optimize the biocontrol potential in a predictive manner. These advancements would make this possible through translating lab-scale research into scalable biocontrol solutions that are ready to work on the field.

Pest management should use strong, multipurpose microbial consortia that can be used to protect plants across a host of pathogens simultaneously in a manner that does not harm soil fertility and crop yields. The biocontrol agents not only make the farming less chemically dependent, but they also work to establish a robust agroecosystem. They are being applied in sustainable farming, which is reflective of the philosophy of environmental stewardship, economic feasibility, and biological integrity as being emblematic of amounts of the globe trying to achieve climate-resistant and integrative agricultural systems (Rasheed et al. 2021).

15.6 Secondary Metabolite Production and Phytochemicals

15.6.1 Metabolic Engineering for Enhanced Compound Yield

Plants are unsurpassed chemical factories with an unsurpassed variety of biosynthetic pathways to produce a myriad of secondary metabolites, many of which are nutritionally, pharmaceutically, industrially, and ecologically important. The possibility of deciphering and engineering plant metabolic pathways to increase the synthesis of high-valued products is becoming an increasing possibility with the advent of high-throughput sequencing technologies, computational biology, and synthetic biology tool sets. Metabolic engineering: Metabolic engineering is a way in which genetic manipulation is done to target the metabolic pathways to enhance their potential. This is a highly effective tool to enhance the yield, efficiency, and sustainability of plant-derived phytochemicals.

Identification and functional characterization of major key genes and enzymes involved in the biosynthesis of a particular metabolite is one of the fundamental strategies in metabolic engineering. With the use of transcriptomics, metabolomics, and bioinformatics, researchers can even map such pathways and identify rate-limiting steps or bottlenecks in the regulation. These steps can then be genetically optimized in the native plant or recreated in heterologous hosts like *Escherichia coli* or *Saccharomyces cerevisiae* or *Nicotiana benthamiana* to increase metabolite yields or to make production and purification even easier.

Overproduction of rare or low-abundance phytochemicals can be achieved by reconstitution of the plant biosynthetic pathways in a microbe or another plant. This method has been effective in compounds such as artemisinin (anti-malarial drug), paclitaxel (anti-cancer drug), and many alkaloids, flavonoids, and terpenoids. As an example, the production of plant alkaloid precursors and intermediates in engineered yeast has led to effective bioproduction processes that are arguably more scalable and less environmentally intensive, with the natural sources, than conventional extraction of alkaloids in native media.

Metabolic engineering of native plants can be employed to reroute flux toward certain pathways, augment the abundance of precursors, or delete competitive branches to bring about a rise in the output of desired compounds. Advances in genome editing via means like CRISPR/Cas9 as well as RNA interference (RNAi) and transgene stacking have allowed metabolic networks in plants to be good targets. These modifications can confer agronomic benefits as well, including improved stress tolerance, enhanced flavor or nutritional quality, and better storage characteristics (Elbouzidi et al. 2025).

In addition to enhancing the yield of a compound, either as an industrial or pharmaceutical compound, metabolic engineering is equally promising in possible solutions to major environmental problems. To illustrate, the problem of contaminants of HMs in soils has become one of the major threats to agricultural productivity and the ecology. Phytochelatins are a few specific peptides that are naturally synthesized by the plants and bind and accumulate the harmful HMs like cadmium. Researchers have GM plants to have superior capacity to synthesize phytochelatins through engineering so that they can survive in the polluted areas and act as phytoremediators.

A good example portrays the phytochelatin synthase (PCS) gene of *Arabidopsis thaliana* (AtPCS1); in this case; this was subjected to directed evolution in an attempt to enhance cadmium tolerance of plants like *Brassica juncea*. Interestingly, mutants of AtPCS1 that enhanced tolerance turned out to have less catalytic activity when compared to the wild-type enzyme. They proved to be better at keeping the precursor levels at levels necessary to suffice with phytochelatin synthesis and redox balance when stressed though. Here, this observation highlights the important feature of metabolic engineering that overall performance of an engineered enzyme should be considered within the context of the whole cell and whole metabolism so as to realize a desired phenotype.

In addition, pathway engineering that has been combined with omics technologies gives a systems-perspective look into plant metabolism. Metabolomics, together with flux balance analysis, enables prediction of the pathway phenotype of genetic variation. These observations facilitate a rational design of stable metabolic circuits that are not only productive but will also be stable in changing environmental conditions. To sum up, metabolic engineering holds transformative potential to improve the production of plant secondary metabolites that have a wide range of applications in pharmaceuticals, agriculture, nutrition, and environmental sustainability. The further optimization of genome editing and synthetic and systems biology-based manipulation of plant metabolism is likely to come, and in the long run, this should lead to the sustainable production of valuable phytochemicals using renewable plant resources on a large scale (Vignesh et al. 2024).

15.6.2 Production of Pharmaceuticals and Nutraceuticals

The presence of a huge variety of secondary metabolites of high biological activity tells us those plants have always been a rich source of pharmaceuticals and nutraceuticals. Due to the increased demand for natural therapies and the emergence of preventive medicine, plants have acquired a new spotlight in the medical, nutritional, and therapeutic arena in the context of bioactive compounds. The recent boom of metabolic engineering, genomics, and biotechnology has gone a great way in assisting us to identify, isolate, and develop production of these valuable compounds. Nutraceuticals (a portmanteau of nutrition and pharmaceutical) are functional foods, or dietary ingredients that have a physiologic or health benefit or preventative or therapeutic effect, such as the prevention and treatment of disease. These are vitamins, polyphenols, flavonoids, alkaloids, terpenoids, carotenoids, and peptides, most of which are originally seen in the therapeutic plants, and a growing number of these are being investigated via the scientific substantiation. Epidemiological data also have attributed the intake of phytonutrients to the lower risks of chronic conditions, which include cardiovascular disease, diabetes, obesity, and some cancers. Turmeric (*Curcuma longa*) with curcumin, green tea (*Camellia sinensis*) with catechins, and soybeans with isoflavones are some functional foods with a high proportion of these compounds and have been widely investigated as having a modulatory effect on the immune system, having reduced oxidative stress and having anti-inflammatory effects.

With the aid of biotechnological instruments, there are now possibilities of the further biosynthesis of these compounds both in the native plant of these compounds as well as in heterologous expression. As an example of this, it is possible to overexpress nutraceuticals of interest by engineering their metabolic pathways by altering control factors or by introducing rate-limiting enzymes. Again, synthetic biology methods are being used to allow large-scale production of plant metabolites that are otherwise not rare or are rare, which involves transporting entire complicated biosynthetic gene clusters to microbial hosts such as yeast or even *E. coli*.

There has also been a lot of development with regard to plant biotechnology in pharmaceuticals. Drugs, such as anticancer agents (e.g., paclitaxel from *Taxus brevifolia*), antihypertensives, and analgesics as well as antimicrobial agents, are based on natural products of plants. Improved knowledge in recombinant DNA technology now makes it possible to engineer a therapeutic protein therapeutic and a vaccine as well as shot peptide drugs to be produced on plant-based expression systems (molecular farming). Major breakthroughs have been pioneered in transgenic crops to produce antigenic proteins to pathogens like norovirus and Hepatitis B, paving new opportunities in the development of a cost-effective vaccine (Aghaali et al. 2024).

A major limitation in the production of pharmaceuticals is the production of the source organisms on a large scale. Nevertheless, hairy root culture and plant cell suspension cultures have become effective systems to culture high value-added substances in the confined environment. Besides, the synergists of omics technologies, including genomics, transcriptomics, proteomics, and metabolomics, are applied to identify new ways of biosynthesis and to optimize the conditions in order to achieve the highest yield.

In addition, novel immunogenic peptides and subunit vaccines developed using plant-based systems are being capitalized upon. Plants can be used to produce more pure, safer, and specific antigenic peptides through metabolic and protein engineering. These methods hold enormous potential in the field of low-cost immunization approaches, especially where resources are stretched. To sum up, the intersection of plant biotechnology and the domains of metabolic engineering, synthetic biology, and omics-based technologies presents the emergence of new opportunities related to the production of pharmaceuticals and nutraceuticals. The application and further development of plant metabolism pathways do not only solve the global health problems but also avail a viable trend to meet the challenges in therapeutics and wellness products of the future (Bapat et al. 2023).

15.6.3 Plant Cell and Hairy Root Cultures

Plant cell culture and hairy root technologies have become revolutionary technologies of sustainable plant biotechnology that can provide a controlled and renewable supply of phytochemicals of high value. And as global issues like climate change and the amount of freshwater available and the increasing requirements of trendy, clean, green ingredients persist to increase, these biotechnological applications are providing sustaining farmers to use over conventional farming strategies, since they allow the manufacture of bioactive substances with decreased carbon footprints.

15.6.3.1 Plant Cell Culture Technology

Plant cell cultures consist of undifferentiated callus cells or differentiated organ cultures that are able to produce a great variety of primary and secondary metabolites. These systems have many benefits, which include production throughout the year, standardization of quality of compounds, and do not rely on climate and topography limitations. The other innovation has been exploring how to enhance culture yields and the ability to scale with systems that have bioreactors designed to suit food and cosmetic applications. The trend to find more natural and sustainable ingredients to use in cosmetic product development and formulations has especially fueled the interest in obtaining active compounds via plant cell culture. The biological uses of these extracts are appreciated on their own for diverse compositions, high secondary and primary metabolites, and a multi-use effects, such as antioxidant, anti-aging, and anti-inflammatory effects.

In spite of these encouraging accomplishments, the use of plant cell culture in the food industry can be considered to be fairly minimal, largely because the production of sufficient biomass and cost-effectiveness remains problematic. Such successful pilot cases as Cocovanol 00 TM (cocoa polyphenol extract) and PhytoVanilla 00 TM (vanilla flavor cell-culture-based) have already shown proof-of-concept but suffered in market sustainability due to the cost of production and resulting lack of economies of scale. However, as bioprocess optimization continues, cost-effective and scalable production continues to become a possibility (Rawat et al. 2019).

15.6.3.2 Hairy Root Culture Systems

Agrobacterium rhizogenes-induced hairy root cultures have been proven to be more efficient secondary metabolite production systems due to quick growth, genetic stability, and output of metabolites. These cultures are very similar to the whole plant metabolism and frequently accumulate bioactive substances in greater amounts than those found in traditionally growing plant tissues or in suspension cultures. Hairy roots are especially useful in the production of alkaloids, terpenes, flavonoids, and therapeutic proteins.

Hairy root cultures have enjoyed a great deal of technological improvements with regards to their suitability in industry. A key challenge of all of this, the scale-up, has been tackled and has resulted in the creation of bioreactors that promote the maximization of nutrient and oxygen supply with limited mechanical damage. Reactor systems such as bubble column, airlift, and nutrient mist reactors can provide gentle mixing and good mass transfer, hence retaining the structural integrity and productivity of hairy roots. The combination of synthetic biology and metabolic engineering also enhances biosynthetic efficiency by flexing and redirecting metabolic pathways and creating new ones. The tailor-made production of hard-to-ferment or produce or rare and complex phytochemicals is also possible with these strategies and is hard to achieve in the traditional methodologies.

Hairy root technologies have found application beyond metabolite production to phytoremediation, biosensing, and sustainability of the environment. They have potential in green technologies given that they have the capacity to take in the HMs and to break down organic pollution. The fact that hairy root cultures can be adapted to pose as a pharmaceutical source through the application in drug development, along with the resilient aspect towards the environmental factor, makes them a favorable platform that can be harnessed to offer a multifunctional and sustainable future of biotechnological advancement. Conclusively, the plant cell and hairy root culture technologies are strong, sustainable options applicable to producing high value-added phytochemicals in pharmaceutical, nutraceutical, cosmetic, and environmental management. Further optimization of bioreactor development, metabolic engineering, and regulatory packages should help in wider adoption and commercialization of the said technologies sooner (Tian 2015).

15.6.4 Industrial Applications and Market Potential

Plant biotechnology, particularly through secondary metabolite production and phytochemical synthesis, plays a pivotal role in advancing industrial applications that promote sustainable development. In industrial biotechnology, plant-based systems are used to produce bioactive compounds at a commercial scale in an environmentally friendly way that also reduces carbon intensity compared to chemical methods. This is in addition to pharmaceuticals and nutraceuticals to also include food additives, industrial enzymes, cosmetics, and agricultural inputs. Combined with bio-based industries, plant biotechnology can help bring about the transformation of value chains within the agriculture, healthcare, and manufacturing sectors in developing and least developed countries. By factoring in innovations in biotechnological applications, whether in the aspect of better agricultural mechanization and new varieties of crops with high yield to manufacturing new phytochemicals with health and industry applications, nations can achieve higher production, less wastage, and better competitiveness in the international market.

Fermentation technology continues to form the backbone of industrial biotechnology, and microbial and plant-engineered fermentation systems have evolved to make a great diversity of commercially valuable products, including amino acids, vitamins, alcohols, and polysaccharides. Microbial and plant cell platforms have been further enhanced on the basis of recombinant DNA and protein engineering, as they increase the yield at lowered costs of production. As the industry in industrial fermentation is increasing incredibly with more than USD 127 billion market cap, hairy root systems and plant cell cultures are being promoted as metabolites production ecosystems with sustainability. By virtue of their genetic stability and expandability, through them rare and high-value compounds, which are otherwise challenging to extract as whole plants, may be synthesized.

Another fast-growing part is enzyme biocatalysis, which demonstrates the diversity of plant-based biotechnological uses. Plant and microbial systems are being increasingly developed to produce industrial enzymes, which find their application in the food, textile, paper, and biofuels industries, among others. An example would be enzymes used in dairy processing or baking or enzymes to process textiles, which may either be engineered biological systems, and the industrial enzyme market is worth more than USD 4 billion and is predicted to reach USD 7.1 billion. The rising per capita income and demand for the high-quality food products will make the developing countries significant consumers in this segment.

Plant biotechnology in the medical and healthcare sectors has played a role in advancing the biopharmaceuticals and the diagnostic medical applications. Global healthcare access is also achieved through the establishment of monoclonal antibodies, therapeutic enzymes, and vaccines based on the application of recombinant protein production in plant cell systems. Given that biopharmaceutical sales are more than USD 140 billion and categorize almost all the high-selling drugs worldwide, plant-based systems present scalable, safe, and affordable options to produce therapeutic proteins. To give an example, synthetic biology is trying to produce insulin, interferons, and antimalarial drugs such as semisynthetic artemisinin in plant-based expression systems, marrying synthetic biology and metabolic engineering, to satisfy the global health needs.

Besides, even the idea of sustainable agriculture is enhanced by plant biotechnology through its contribution of decreasing agrochemical reliance. The transgenic crops that are more resistant to pests and have better absorption of nutrients eliminate the extensive use of synthetic fertilizer as well as pesticides in the soil and water pollution. Nevertheless, there are difficulties in adequate accessibility and evading the temptations of biotechnology for the benefit of those that are mainly based in the industrialized countries with the exclusion of the public arena. It should be emphasized that in the future, the development should be based on inclusive science-based applications taking into account the knowledge systems in the process of sustainable environments and most importantly in the regions such as South America, whereby fertilizers and pesticides have been offering excessive exploitation, leading to ecological imbalances.

In sum, the industrial uses and market opportunities of plant-derived secondary metabolites and phytochemicals point to the revolutionary importance of plant biotechnology. With the ever-growing needs of the world to consume natural goods, eco-friendly solutions along with functional ones, the entry of plant biotechnology into industrial processes is a chance at achieving a bioeconomy that is profitable in terms of economy and sustainable in terms of environment (Chandran et al. 2020).

15.7 Phytoremediation and Environmental Applications

15.7.1 Engineered Plants for Heavy Metal and Pollutant Removal

The pollution of the environment due to HMs and organic pollutants can be observed as a continuous challenge to the quality of the soil, the condition of the vegetation, and food security. Traditional remediation systems can be very expensive, intrusive, and disruptive to the environment. Phytoremediation as a method of using plants to remove, stabilize, or even degrade any pollutant in the environment provides an eco-friendly way, and it is a sustainable strategy. Nonetheless, slow growth and biomass production of most plants as well as poor stress tolerance are barriers to the natural phytoremediation capacity. The genetic engineering of plants has been another product of the development of plant biotechnology, where the plants are built with improved abilities to cope with, hold, and filter damage factors, as well as contaminant components in the water and soil surroundings.

Oxygen species (ROS)-mediated oxidative damage of the cell is accompanied by failure of cellular homeostasis and occurs under HM stress. In order to endure such stress, it is natural that plants deploy a number of defense solutions, like production of phytochelatins and metallothioneins, sequestering of metals in vacuoles, induction of antioxidant enzymes, and metal transporter control. Genetic engineering further improves on these mechanisms by the addition or activation of genes mediating metal uptake, chelation, detoxification, and signaling of metal stress.

Some of the plants that have been engineered have posed much increased phytoremediation effectiveness. By way of example, T-DNA-transformed *Arabidopsis thaliana*, which overexpresses PCS, has an increased cadmium accumulation and tolerance. On the same note, mercury volatilization and detoxification improved in *Brassica juncea* with bacterial merA and merB genes. Increased gene expression of metal transporters (e.g., HMA4, NRAMP, and ZIP) has enhanced the uptake and translocation of metals like Zn, Cu, and Pb in different plant models. The species of poplars (*Populus nigra* and *Populus alba*) also have good prospects as phytoremediation trees as they grow rapidly, have deep roots, and are adaptable. The simultaneous and differential expression of metallothionein genes in engineered poplar trees has resulted in a greater tolerance to Cd and Zn as well as a higher sequestration of metals in the cell. These trees also serve as bioindicators in the polluted ecosystem that demonstrate changed physiological reactions and gene expression on exposure to air-borne and soil-borne contaminants.

Phytoremediation is also further boosted by the incorporation of genetic engineering and omics tools and nanotechnology. An example is that nanomaterial-mediated delivery of genes can enhance the transformation process, and transcriptomic profiling can point towards the identification of key regulatory genes in the stress pathways. These combined technologies make it possible to specifically develop lines of a plant based on particular contaminants and environmental conditions. Engineered phytoremediation has its obstacles, despite the potential, such as ecological, regulatory, and the general acceptability of GMOs. However, the limitations can be overcome through further studies and conscientious implementation, leading to the creation of green technology that is reliable and highly efficient to combat pollution on a sustainable level (Sánchez-Castro et al. 2023).

15.7.2 Role of Biotechnology in Soil and Water Cleanup

Addressing contaminated soil and water is an area of great potential where biotechnology is radically changing the way that external contamination is getting cleaned up and presents solutions that are less environmentally degrading as well as equally efficient as traditional means of remediation. With industrialization, intensive farming, and urbanization causing ever-increasing degradation of the natural ecosystems, biotechnological intercession has come to play the important role of ecological restoration and enhancement of the environmental quality. An example of important biotechnological methods would be microbial remediation, where a mixture of different bacteria and fungus, e.g., Pseudomonas, Bacillus, and Arthrobacter, would metabolize and detoxify pollutants such as pesticides, hydrocarbons, and HMs in the environment. These microorganisms that are probably inherent to pollution soils or have been designed to act in an improved manner breakdown all the pollutants into nontoxic substances and contribute to the inherent regeneration of soil systems. In addition to this, the combination of microbial consortia with phytoremediation exploits the synergistic relationship of the plant root and soil microbes to increase the uptake, transformation, and stabilization of the contaminants. Use of organic fertilizers such as compost, horse manure, and biochar further increases microbial activities in the soil, enhances soil structure, and also enhances the biodegradation of the toxic compounds.

Anaerobic digestion is one of the green methods used in water purification because the pollutants undergo biodegradation to produce biofertilizer and biogas fuels in addition to minimizing resource wastage in favor of renewable energy production. MFCs are a two-in-one gaming platform in that they purify the wastewater concomitantly and produce power, particularly with municipal, agricultural, and industrial effluents. The ability of extremophilic microbes to treat saline waste water and even chemically aggressive waste water has even further broadened the scope of microbial use in a variety of environmental conditions. New technologies like biomineralization and biosorption involve the involvement of bioagents to trap or extract the HMs and other forms of contaminants, and they apply to natural and industrial contaminants such as electronic waste. Such biological steps are immediately relevant to the concepts of a circular economy by turning wastes into valuable consumables, e.g., in the case of formulating biodegradable plastics like polyhydroxyalkanoates by processing agro-industrial wastes or biohydrogen through the fermentation of organic matter. Although exposures to biosafety risks and regulatory obstacles to GMOs will continue to affect the progress in the field of synthetic biology, systems biology, and metagenomics, more useful, accurate, and scalable solutions are being created to remediate the environment. In the end, biotechnology presents a more holistic and sustainable method of decontaminating soils and water, affirms ecologic integrity, and increases the resilience of agricultural and industrial landscapes (Oro et al. 2024).

15.7.3 Sustainable Urban Landscaping and Green Infrastructure

Urbanizing and green infrastructure is a melding of the ecological sciences, biotechnology, and urban design that provides groundbreaking innovations to the current and looming environmental and urbanization problems of climate change and rapid population growth. By overcoming the common landscaping methods, putting emphasis on the aesthetics of ornamental plants and exotic species in their usage, sustainable landscaping introduces a different approach that promotes use of the native plant species, effective water usage, repairing the degraded soils and creating the balance of ecosystems. The transition also helps a great deal in preserving the biodiversity, carbon storage, and the control of the local microclimates and local hydrological cycles. As more cities continue to grow and green spaces become smaller, these ecologically driven solutions are becoming even more crucial to help promote the well-being of humans and their resilient cities.

Green infrastructure: A broad set of technologies and design strategies, including vertical gardens, bioswales, green roofs, and manmade wetlands, go into making up green infrastructure that gets incorporated into the built environment to improve ecosystem services. The buildings assist in effects such as the reduction of the urban heat island effect, the quality of the air and water, and how stormwaters are managed. Other than this, they also give landings to pollinators and urban wildlife, hence adding biodiversity to the urban. Notably, such systems also can feature biotechnological improvements, e.g., enhancing the soil with desirable microorganisms (bioaugmentation of soil) or utilizing genetically improved crops that are specifically helpful at the uptake of the undesirable pollutants or that may be actively resistant to drought.

A new aspect of sustainable landscape design, the microbiome-inspired green infrastructure (MIGI) concept, also introduces a new level of consideration of the presented approach as the microbial communities play an important role in ensuring ecosystem functionality and resilience. MIGI helps to advocate for the deliberate design and maintenance of urban green open spaces that encourage healthy environmental microbial compositions, which subsequently allow positive outcomes on human health through the promotion of prized human–environment microbial interactions. Including microbiome science in the design of green infrastructure, cities can create more responsive, adaptable, and health-enabling environments.

The socioeconomic benefits are also provided through sustainable landscaping. Multifunctionality of green spaces in the city means that such spaces are less expensive to maintain, help to socialize the population, can be more aesthetically pleasing, and have increased recreational value, as well as prevent potential damage caused by severe weather events like floods and hot spells. Policywise, sustainable landscaping fits into the wider development sustainability agenda and encourages a circular use of resources, like recycling urban wastewater as an irrigation system or urban organic waste as a compost in order to fertilize the ground.

Biotechnology is very critical in seizing the ability to get these green systems to perform in an optimum way. This involves the use of plant-growth-promoting rhizobacteria, gene-edited drought-tolerant plant varieties, or microbial consortia to clean up soil. The presence of such integrative methods means that beyond making cities look pretty, sustainable landscaping will also act as an infrastructure that can mediate pollution, improve biodiversity in cities, and promote much needed human–nature contact. Hence, the sustainable urban landscaping and green infrastructure are a multifaceted systems that provide an efficient approach to simultaneously tackling the environmental degradation, climate vulnerability, and urban health issues. And with the help of biotechnology, these systems may be enhanced to be more efficient, multifunctional, and up to the long-term sustainability objective, and consequently indispensable part of the future-ready urban ecosystems (Paudel and States 2023).

15.7.4 Case Studies in Environmental Biotechnology

The real-life benefits of environmental biotechnology can best be explained by means of the case studies that point at the groundbreaking role of environmental biotechnology in the restoration of the ecosystems, the cleaning of the pollutants, and the sustainable development. The above real-life scenarios show the potential of plant-based and microbial biotechnologies in addressing some of the most nagging environmental issues arising due to anthropogenic tasks, which include industrial pollution, urbanization, and mismanagement of wastes. Plastic pollution is considered to be one of the most significant environmental issues on the planet nowadays. Plastics, since they are made out of a synthetic polymer, are able to resist natural degradation very well. Over the past couple of years, there are studies that have indicated the use of GM plants and microorganisms which are said to be able to degrade particular plastic polymers. To illustrate, polyethylene terephthalate (PET) plastic remediation has been made possible by means of the discovery of *Ideonella sakaiensis* bacterium which is useful in producing PETase bacterium enzyme. Synbio versions of this enzyme, expressed in a plant expression system or a microbial consortia, could greately speed up the ambient rate of plastics degradation, thereby providing the holy grail of plastic accumulation and solution.

The second case study that is important according to the report is the phytoremediation with GM plants. Such genes include the merA and merB genes of bacteriological origin and have been engineered into poplar trees to gain the property of transforming toxic ionic mercury into elemental mercury, which is less harmful and volatile in nature. Such transgenic trees have been utilized in environmentally contaminated sites with mercury to lower HM poison in soil and water. On the same note, bioengineering of Indian mustard (*Brassica juncea*) has resulted in increased resistance to cadmium and arsenic accumulation and uptake, as a technique with which to clean up polluted soils (Kuppan et al. 2024).

The treatment of wastewater using constructed wetlands employing aquatic plants such as *Phragmites australis* and *Typha latifolia* and, in some instances, rhizospheric bacteria has been found to be highly promising in the field of wastewater management. They are supplemented by inoculating PGPRs and mycorrhizal fungi that make plants more resistive to HMs and organic pollutants and also enhance nutrient cycling and water purification. These combinations of systems not only clean up the pollution of domestic and industrial effluents but also heal the damaged wetland systems.

The use of biofilms comprising *Pseudomonas*, *Bacillus*, and *Shewanella* species at industrial effluent locations can be given as a very appealing instance of microbe-assisted bioremediation. These biofilms make it easy to break up and reduce the HMs like chromium and lead into less harmful compounds. The native microbial community can be sustained or augmented through bioaugmentation by inoculation of a specific series of microbial strains into an affected site in order to increase the effectiveness of a remediation process. Another important example is the successful cleanup of oil spills by the use of hydrocarbon-degrading species, e.g., *Alcanivorax borkumensis*, which is important because of the opportunity that a customized microbial community has regained marine life.

Environmental biotechnology has found effective use in sites of acid mine drainage (AMD). High acidity and dissolved HMs in water are a leading concern in the environment surrounding mining sites, and this is known as AMD. Some acidophilic bacteria, like *Acidithiobacillus ferrooxidans,* have been used in bioreactors in order to precipitate metals such as iron, copper, and zinc in AMD and, in the process, minimize the amount of impact they cause on the environment and, in the process, also recover metals. All these case studies point to the flexibility and scalability of environmental biotechnology. The potential of creating sustainable remediation is vast, whether by phytoremedification with genetically engineered plants to rid the ground of the various substances or microbe consortia specific to the degradation of certain contaminants or by constructed wetlands coordinating the use of plant–microbe systems. Such approaches have further been enhanced by the incorporation of the omics technologies, which include genomics, transcriptomics, and metabolomics, in microbial and plant responses to pollutants, which has enabled us to have a more precise engineering of bioremediation systems. These applications merge traditional ecological knowledge with the newest biotechnology to create scalable, low-cost, and environmentally friendly solutions to pollution and degradation of ecosystems. With environmental pressures only increasing, plant and microbial biotechnologies are getting more entwined in the process of ecological restoration in order to be used in the accomplishment of the long-term sustainability aims (Ali et al. 2020).

15.8 Bioenergy and Biomaterials

15.8.1 Genetic Modification for Enhanced Biomass Yield

Increasing the production of biomass is one of the main objectives in plant biotechnology, especially since it applies to the use of biomass in bioenergy and more broadly in biomaterials. Another advantage of genetic manipulation is that it provides control over major metabolic and hormonal pathways that determine plant growth, development, and biomass accumulation. With the advent of modern biotechnology, the researcher has been able to manipulate genes that will enable carbohydrate metabolism, hormone biosynthesis, and cellular development, which have greatly contributed to the overall biomass production of the different plant species.

One of the most basic approaches will be to alter the metabolic pathways of carbohydrates so as to enhance biomass accumulation. An increase in expression of several genes has been reported to stimulate primary growth; these include UDP-glucose pyrophosphorylase, sucrose synthase, and sucrose phosphate synthase. As an example, the concurrent upregulation of these enzymes in *Nicotiana tabacum* (tobacco) resulted in more than 50% enhancement in the height of the plant. In a like manner, AtPAP2 (purple acid phosphatase) in *Arabidopsis thaliana* was overexpressed leading to marked increase in sucrose phosphate synthase activities, which led to a twofold increase in total biomass than the wild plant. This shows the direct relationship between the sugar metabolism and vegetative growth and the energy storage in plants.

Other manipulations through metabolism involve inhibition of starch phosphorylation. As exemplified, silencing of glucan waterdikinase, a major enzyme involved in starch phosphorylation using the RNA interference (RNAi) method in *Triticum aestivum* (common wheat), resulted in a decrease in starch phosphate and, more importantly, successful and repeatable increases in the vegetative biomass, grain size, and yield of the plant under greenhouse and field environments. These observations prove the viability of energy storage pathway manipulation to achieve improvement in yields. Directed genetic manipulation has been carried out in even the fiber crops like cotton. Expression of a high-quality fiber line, SusA1 gene in *Gossypium hirsutum* (cotton), increased both fiber length and fiber strength when overexpressed. Enhanced GhSusA1 transcript levels in vegetative organs as well as increasing fiber quality also led to a proportionate rise in the overall biomass of the seedling, which is advantageous to industrial waste.

The biosynthesis and pathways of plant hormones also have the most significant role in plant architecture and growth. A type of growth hormone known as gibberellins has attracted great research regarding this aspect. A gene, probably an important enzyme, GA 20-oxidase involved in the gibberellin production pathway, was overexpressed, and this resulted in the hyperactivity of a gibberellin-inactivating enzyme, GA 2-oxidase. This was later reverted by the silencing of GA 2-oxidase in the tobacco and shown to have major gains in the growth metrics as compared to both the wild-type and the GA 20-oxidase-overexpressing plants. Likewise, gene repression of PtGA2ox4 and PtGA2ox5 in Populus (poplar) resulted in enhanced leaf size (50%), stem height (20%), stem diameter (10%), and total biomass (30%), with no influence on the root system, indicating tissue-specific improvements in growth through modification of hormones. Transplastomic tobacco lines that contained Bgl-1 expressed a high level of gibberellin (GA 1 and GA 4) as well as other growth-related hormones like zeatin and indole-3-acetic acid. Such adjustments translated to early flowering, almost a doubling of biomass, and increased leaf size, which further serves to affirm the role of maximizing hormone in improving biomass yield.

Besides regulating the hormones, cell proliferation has also been regulated. Expression of the ARGOS gene found first in Chinese cabbage and Arabidopsis resulted in an oversized organ with an enlarged cell proliferative capacity but without enlargement of the specific cell. In *Nicotiana tabacum*, this translated into visible gains in biomass and plant vigor. Moreover, expression of cytokinin biosynthesis genes like isopentenyl transferase (ipt) has shown to have an enormous effect on plant growth and yield. In narrow-leafed lupin (*Lupinus angustifolius*), transgenic lines expressing the ipt gene under flower-specific promoter demonstrated highly developed branchiness coupled with a high pod set with no reduction in grain size. Equally, tobacco plants having an engineered AOC-ipt gene in them used a 23-fold greater level of cytokinins, which caused them to grow faster, flower longer, produce their seed in bigger quantities, and have a greater chlorophyll content. The amount of total dry biomass of AOC-ipt plants was found to be divided by 1.27 times as much as controls.

Thinking in general, these are biotechnological examples of how specific genetic manipulations can be applied strategically to improve the level of biomass in plants. Targeting metabolic engineering and hormonal control and cell cycle management, plant scientists are coming up with strong bioenergy crops and species high in biomass that can be used in the sustainable production system. Such developments facilitate the ever-increasing call to renewable resources and the emergence of the green bioeconomy (Ahmad and Ming 2024).

15.8.2 Algae- and Plant-Based Biofuel Production

The algae- and plant-based biofuels are up to the mark and can be considered the most promising and sustainable substitute to fossil fuels because they are renewable, carbon-neutral, and have a really enormous potential for large-scale planting and transformation into different forms of fuels. Such biofuels utilize solar energy by harnessing photosynthesis and can be processed into chemical energy stored as lipids, carbohydrates, and other organic molecules and could then be metabolized into high-energy-density fuels like biodiesel, ethanol, biobutanol, and biohydrogen. It has been seen that microalgae have become a feasible and cost-effective feedstock because they can grow very fast and contain lots of lipids, along with the fact that they can grow on non-arable land with wastewater or salty water. *Chlorella*, *Tetraselmis*, *Dunaliella*, *Isochrysis*, *Nannochloris,* and *Nannochloropsis* are good examples of species that contain between 20% and 50% lipids, and thus effective in the production of biofuel. Metabolic engineering strategies are emerging that are able to modify microalgae in order to turbocharge their lipid biosynthesis, and this approach is likely to raise the overall productivity and commercial feasibility of algal biofuels.

Simultaneously, bioethanol and biodiesel have been long produced in higher plants, especially those that have high biomass yield, such as sugarcane, maize, soybean, and switchgrass. The first-generation biofuels majorly depended on starch and sugar crops like corn and sugarcane. Nevertheless, since food security and competition over available land have become an issue, second-generation biofuels, based on lignocellulosic biomass, including crop residues, grasses, and woody materials, have started to take the place of first-generation biofuels. Hemicellulose and cellulose rich in these feedstocks are capable of enzymatic hydrolysis into fermentable sugars; they can be fermented to ethanol or butanol. The biodiesel industry has spread in the world, as there are some sources of feedstock such as soybean, canola, and camelina, waste oil of cooking, and animal fat. It is obtained as a result of the transesterification of triglycerides, where fatty acid methyl esters and glycerol are produced. Algal oils are rich in unsaturated fatty acids and therefore ideal in producing biodiesel with better oxidative stability and lower freeze points.

National and international policies such as the Renewable Fuel Standard II (RFS II) in the United States also play an important role in the development of biofuel industries. This policy requires the gradual growth in production and utilization of renewable fuel up to 36 billion gallons till 2022, 16 billion gallons of which must be produced based on lignocellulosic materials and 1 billion gallons on biodiesel. Those objectives emphasize the significance of the plant and algal biotechnological innovations in making the production of biofuels more efficient. The breakthroughs in genetic engineering, synthetic biology, and systems biology have allowed scientists to design and increase biomass accumulation in plants, alter the composition of their cell walls to be more easily processed, and become more resistant to stressors. In like manner, research in algal biotechnology targets to overexpress those genes that have a role in biosynthesis of lipid, augment carbon assimilation, and improve photosynthetic performance. Systems of biofuel production that are economically viable and environmentally sustainable can be met through integration of these innovations. All in all, algae- and plant-based biofuels are the most vital part of the world transitioning to a low-carbon future. Further research investments, favorable policy, and technologies with flexibility in scale deployments will be necessary to achieve the potential of such bioenergy sources (Anto et al. 2020).

15.8.3 Lignocellulosic Feedstocks for Bioplastics

The natural bioresources that have the greatest potential in the use of sustainable bioplastics relate to lignocellulosic feedstocks, which are composed of cellulose, hemicellulose, and lignin. These resources provide a renewable, biodegradable alternative to plastics made of petroleum products and have enormous potential to make economies less energy-intensive and environmentally damaging. Conventionally, cellulose-based fibers have commonly been used in the reinforcement of bioplastics where lignin may need to be removed in order to enhance the compatibility between the fibers and the matrix. Nevertheless, this delignification process is technology-intensive, requires much energy, and results into wastewater as well as discarding lignin as a byproduct-though lignin is the second most abundant polymer in plant biomass and the most abundant natural aromatic compound. Because of advantages of good mechanical properties, UV resistance, thermal stability, and biodegradability, studies indicate that lignin is valuable with regard to use in bioplastics. Rather than extracratically extract lignin, direct addition of lignin to bioplastics is also being considered as a component of lignocellulosic fibers or lignin on its own as an additive. Not only does such an approach make things more sustainable, but it also makes the processing easier, less costly, and the end properties of bioplastics better.

Hydrophilicity of untreated lignocellulosic fibers and the subsequent significant crosslinking of these materials are one of the key factors limiting the usage of untreated lignocellulosic fibers due to ineffective interfacial adhesion between untreated lignocellulosic fibers and hydrophobic biopolymer matrices in general (PLA, PHB, or starch). Such a lack of compatibility leads to the loss of mechanical strength and dimensional stability in the presence of moisture. In order to surmount this, different techniques of surface modification have been utilized in order to improve bonding between the fibers and matrices. Nevertheless, the majority of such treatments are cellulose-oriented and do not take the benefits of lignin into account. New studies, however, show that it is quite technically possible to retain lignin in the fibers or indeed reinforce the bioplastics directly with industrial lignin that is a by-product of the paper manufacturing industry that aims to yield around 70 million tons of it annually. The potential of lignin to be matched with a variety of biopolymers, as well as its adjustment of a chemical structure by means of functionalization, is one of the reasons why the material may be popular among next-generation composite materials.

On the whole, incorporation of lignocellulosic products, and lignin-bearing fibers, in particular, into bioplastics material science is a breakthrough towards sustainability. This practice will not only lessen the dependence on fossil fuels but will also generate revenues with the extractable compounds of agricultural and industrial by-products. Improved plant biotechnology and processing of materials are leading to the production of lignin-rich feedstocks optimized towards the bioplastic products. Further developments on the design of biocomposites with enhanced interfacial adhesion, fiber dispersion, and the compatibility of the polymeric components will also broaden the scope of use of these materials, making lignocellulosic feedstocks a backbone of future biomaterial sustainability (Kawaguchi et al. 2022).

15.8.4 Sustainability Metrics for Plant-Based Bioeconomy

The dimension of sustainability metrics is also important in the context of the assessment and rendering of advice about the transition to plant-based bioeconomy (pBE) which has been put to the fore as a result of global environmental, climatic, demographic, and resource-related pressures. Societies are dealing with aging populations, food insecurity, and an increasing environmental degradation, and, in this context, pBE is increasingly viewed as a transformative pathway towards the replacement of fossil-based systems with renewable and plant-based systems. Green biotechnology as a form of plant biotechnology is a fundamental driver of this shift through increasing crop yields and improving resource use and the production of value-added bioproducts, including bioenergy, bioplastics, and bio-based chemicals. In order to determine the effectiveness and the future implications of such innovations, detailed sustainability evaluation is required. These indicators include environmental, economic and social aspects which include greenhouse gas emissions, land and water use productivity, carbon footprint, biodiversity conservation, economic feasibility, employment generation and acceptance in the society. In order to compare plant-based bioeconomy systems with alternative modes of production such as the conventional petrochemical or food production process, quantitative models have been used widely, such as life cycle assessment, techno-economic analysis, and sustainability impact assessments as shown in Fig. 15.3.

At the policy and technological development scale, such countries as Germany or programs and initiatives as the Horizon 2020 already recognize the plant-based bioeconomy as a strategic area to transform the economy into a circular and low-carbon. National policies spotlight the need to incorporate the advancements in biotechnology, including the design of better breeding strategies, metabolic engineering, and processes of valorization of biomass to produce sustainable and efficient bio-based production systems. The analysis of patent citations in Germany during the period between 1995 and 2015 indicates that there has been a shift in the technological development of plant biotechnology; not only was the investigative field concerned with basic genetic and molecular research, but more commercial technological advances were being developed with regard towards sustainable energy, food, and medicinal solutions. The trajectory helps to understand the essence of monitoring technological advancements with references to the sustainability indicators that may reveal scalability/replicability of the emerging bio-based technology

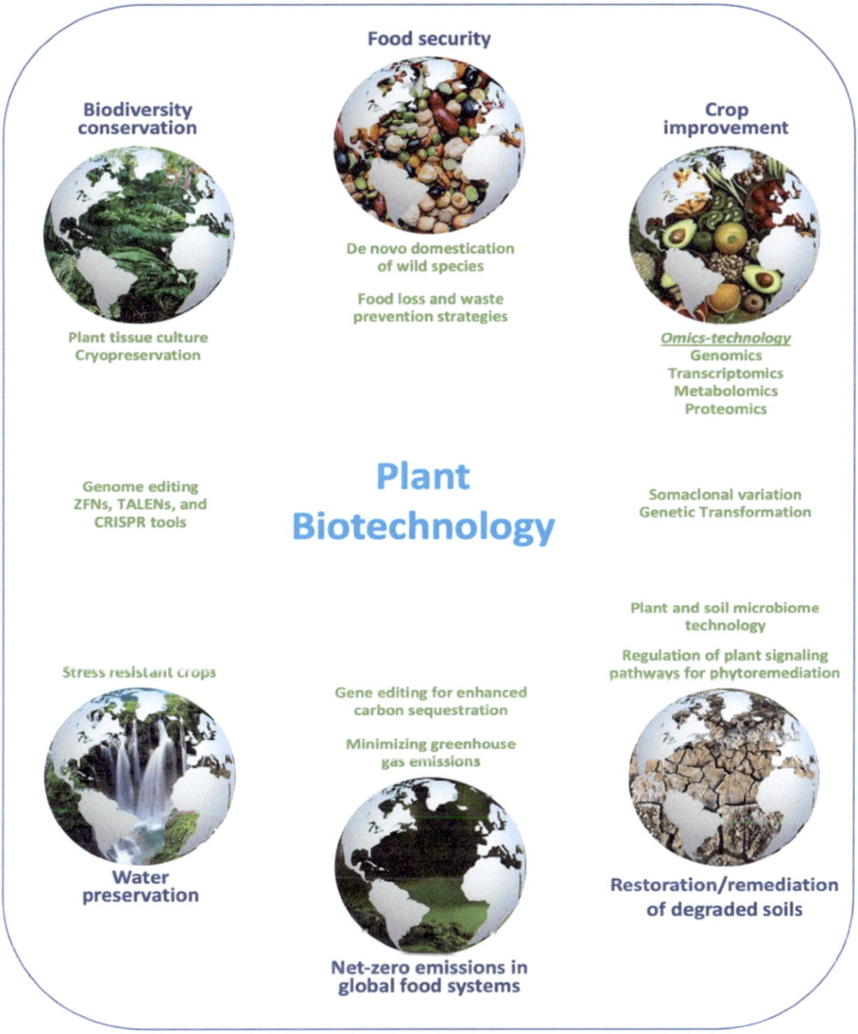

Fig. 15.3 Multifaceted roles of plant biotechnology in food security, biodiversity conservation, crop improvement, and environmental restoration (Pathirana and Carimi 2024)

and environment-related trade-offs. Furthermore, metrics must be adapted to regional agro-climatic contexts, crop diversity, and socio-economic conditions to provide relevant insights for policymakers and stakeholders.

All in all, the integration of strong sustainability indicators in the plans of the plant-based bioeconomy is the indispensable measure to make sure that an exploitation of renewable resources would not mean an exploitation of ecological integrity and social equity. Such measures would allow open measurement of environmental performance, streamline resources, and promote accountability in the route of innovation. With plant-based solutions becoming the backbone of climate actions and sustainable development strategies in all regions, the optimization and the use of these measures will play a crucial role in shaping the bioeconomy transitions to a more resilient, inclusive, and renewed future (Pathirana and Carimi 2024).

15.9 Digital and Precision Tools in Plant Biotechnology

The collider of digital technologies and plant biotechnology creates the era of gifts in a sustainable agricultural system that can provide solutions to the current worldwide issues concerning climate change, food insecurity, and resource constraints through innovations. Precision and digital instruments have made it possible to make real-time, data-based crop decisions in agriculture, which substantially increases levels of accuracy and efficiency of crop supervision, input control, and output forecasting. The utilization of Internet of Things (IoT) sensors, drones, remote sensing (RS), machine-learning algorithms, and data analytics platforms creates the possibility to continuously collect and analyze the environmental and plant physiological parameters such as soil moisture, temperature, humidity, and nutrients. Such developments bring practical information helpful to adaptive crop management, efficient use of resources, and increased productivity with less environmental impact. Precision agriculture also works hand in hand with plant biotechnology to add a greater advantage to such benefits. Genetic engineering, CRISPR-based gene editing, and marker-assisted selection are some of the techniques that have been used to produce crop varieties with superior characteristics like drought tolerance, pest and disease resistance, and better nutritional content. Such biotechnology innovations are able to suitably curb the negative implications of climate variation by assuring that crops may survive well in harsh weather conditions, among other abiotic stresses. It is augmented by other digital technology, including simply the application of biotechnology, such as precision irrigation systems, site-specific fertilizer application, and real-time pest monitoring, to make agriculture sustainable intensity in the least chemical input and help in the sustainability of agriculture with a positive ecological balance.

Although promising, the digitization and precision tool in plant biotechnology undergoes various hindrances in its enactment. There is a cybersecurity and privacy concern in managing and securing large amounts of agricultural data. Also, GMOs continue to be a source of debate among consumers and regulators as far as food labeling and safe consumption of food are concerned. Accessibility risks are also among the disparities between developed and developing regions that access advanced technology, so the benefits of that technology may not be distributed in an even way. The most important aspect to deal with these concerns is to have strong regulatory measures, invest in digital infrastructure, and incorporate policies to make sure that the smallholder farmers and the marginalized communities are not left behind. Going forward, digitized and precision tools in plant biotechnology are towards the acquisition of intelligent self-control with the formation of real-time agricultural systems and even the subsequent predictive models of biology. AI and big data analytics have the potential to improve the accuracy and scale of intervention in agriculture. Match the application of smart farming tools and resilient biotechnology crops to form an effective way of attaining sustainable, climate-resilient, and nutritionally secure food systems. The ethical, environmental, and socioeconomic forward-looking considerations as these technologies approach maturity will be major determinants in whether they will be able to truly exploit the potential of these technologies in efforts to achieve global agricultural sustainability (Ahmad et al. 2024).

15.9.1 Integration of AI and Big Data in Plant Research

The integration of AI and big data analytics into plant biotechnology marks a critical advancement in the pursuit of sustainable agricultural systems. Modern agriculture faces a set of global challenges that are interdependent among one another, and these include climate change, dwindling natural resources, rising populations, and the mandate to achieve food security without compromising on the ecological balance. AI, together with biotechnological methods including genetic engineering, CRISPR-Cas9, and synthetic biology, has the potential to transform ability to fix these problems. The big data systems capture and handle at scale a combination of complex data of wide origin including genomics, phenomics, transcriptomics, field sensors based on IoT, drones, and satellite imaging. These data provide opportunities to perform predictive modeling as well as identify traits and optimize plant-breeding solutions when subjected to AI and ML algorithms.

In plant biotechnology, AI can make genetic modification more precise and efficient, based on target identification of the gene to modify, improvement of gene editing protocols, and prediction of the phenotypic consequence of modifications. Deep learning and neural network analyses analyze genomic sequences and locate those regulatory factors, or positive traits, associated with stress resistance, pest escape, and better nutritional value. AI is also used in in silico modeling and virtual screening of metabolic pathways to help design synthetic biology applications. That is how data integration powered by AI can help develop climate-resilient, high-yielding, and resource-efficient crop varieties, which will be essential to supply the world population of 11 billion or more with food towards 2100.

The adoption of AI also extends to monitoring the health of plants, precision irrigation, disease prediction, and also predicting the yield. Imaging technologies and IoT devices provide real-time data that is processed with the help of AI and results in recommendations related to site-specific management. This increases crop production, reduces resource consumption, and provides greater sustainability because the system becomes less reliant on synthetic fertilizers and pesticides. In addition, AI facilitates decision-making through modeling of environmental relations as well as simulating different agronomic settings, thus enabling dynamic and adaptive management systems as shown in Fig. 15.4. Nevertheless, a number of

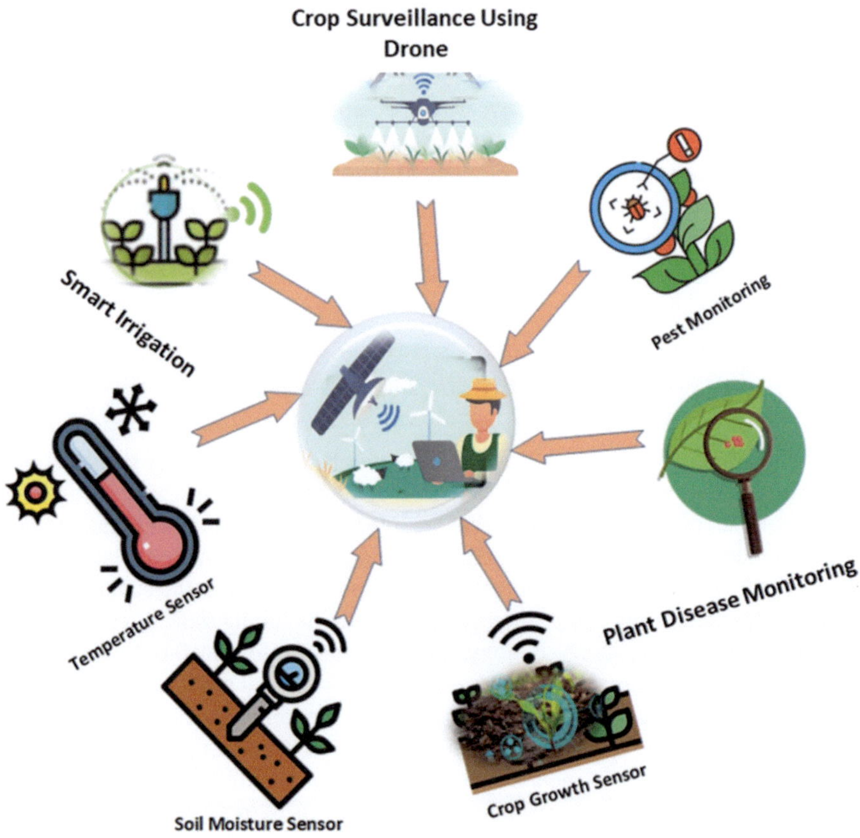

Fig. 15.4 Smart agriculture framework using drone-based crop surveillance and sensors for irrigation, pest, disease, and growth monitoring (Kumari et al. 2025)

obstacles should be addressed to enjoy this potential of AI and big data in relation to plant research. These are the data standardization, heterogeneous dataset integration, information security issues, and the digital divide among the smallholder farmers in the developing world. Ethical concerns as regards AI-based genetic interventions as well as fair access to biotechnology innovations are also very problematic. To solve such issues, multidisciplinary huddling, inclusive policies, and clarity of the regulatory system will allow innovation to flourish while enabling trust. Nevertheless, the synergistic use of AI and big data in plant biotechnology is one of the major branches of Agriculture 4.0, a data-smart, resilient, and sustainable agricultural model. The next changes that we will see are probably going to be around self-adjusting biosystems, the human intelligence augmentation of genetic editing, the independent crop management services, and completely coupled digitalization of the real-time decision-making process in agriculture. Such convergence is not only a revolution towards understanding the plant but also buttresses international initiatives aimed at creating food security, environmental stability, and stability in response to the growing global challenges (Kumari et al. 2025).

15.9.2 Remote Sensing, IoT, and Phenotyping Platforms

In the age of increasing global food demand and climate uncertainty, timely and accurate information on crop production, management, and yield has become necessary to ensure food security and to meet sustainable development goals (SDGs) and more specifically the one that attains zero hunger. RS, the IoT, and automated phenotyping platforms are becoming critical digital technologies in the field of plant biotechnology, as they are the tools that could be used to manage agricultural systems at a high level of resolution and in real-time, as shown in Fig. 15.5. The RS has since then developed tremendously since the deployment of its first satellite in 1957 with the incorporation of satellites and remotely piloted aircraft, which are fitted with the latest sensors able to distinct high-level spectral, spatial, and temporal resolutions. Such systems have made it possible to give proper plant physiological characteristics and environmental status and soil condition monitoring. Wireless sensor technologies also enable real-time data capture, and the storage and manipulation of the large amounts of data they create is made possible by the use of cloud computing.

In order to draw actionable insights, there is an incorporation of AI and big data snapshots to these systems, which provide predictive insights on crop growth, infectious diseases, irrigation programming, and harvest optimization. At the same time, high-throughput plant phenotyping platforms (HT3Ps) have proved to be strong mechanisms to investigate complex phenotypes in large plant populations non-destructively. Such systems include continual growth and growth chambers as well as field facilities, which range in size and utilize both ground-contact and air-based installations. HT3Ps eliminate the drawbacks of traditional phenotyping as they provide a means to collect phenotypic data constantly when using a variety of sensors and imaging technologies. A digital agricultural ecosystem consisting of the integration of HT3Ps with RS and IoT is not only expediting the process of discovering stress-resilient and high-yielding genotypes but also precision farming practices is also improved.

Furthermore, in addition to employing biotechnological solutions such as vertical farms, rain gardens, and bioelectrochemical sensors tied up to a network of IoT in cities, modern urban agriculture is in the process of integrating these digital tools into smart city systems. The applications entail self-powered biosensors and cyber-physical systems (CPS) to monitor environmental variables and optimize urban food production in such systems. The issues that have been seen to impede the progress of these digital innovations include data security, investments in infrastructure, and a healthy supply of talent, although this marked revolution in the field of plant biotechnology has brought in positive gestures. Their planned application plays the crucial role in developing a more resilient, efficient, and sustainable agricultural future (Tiwari and Park 2024).

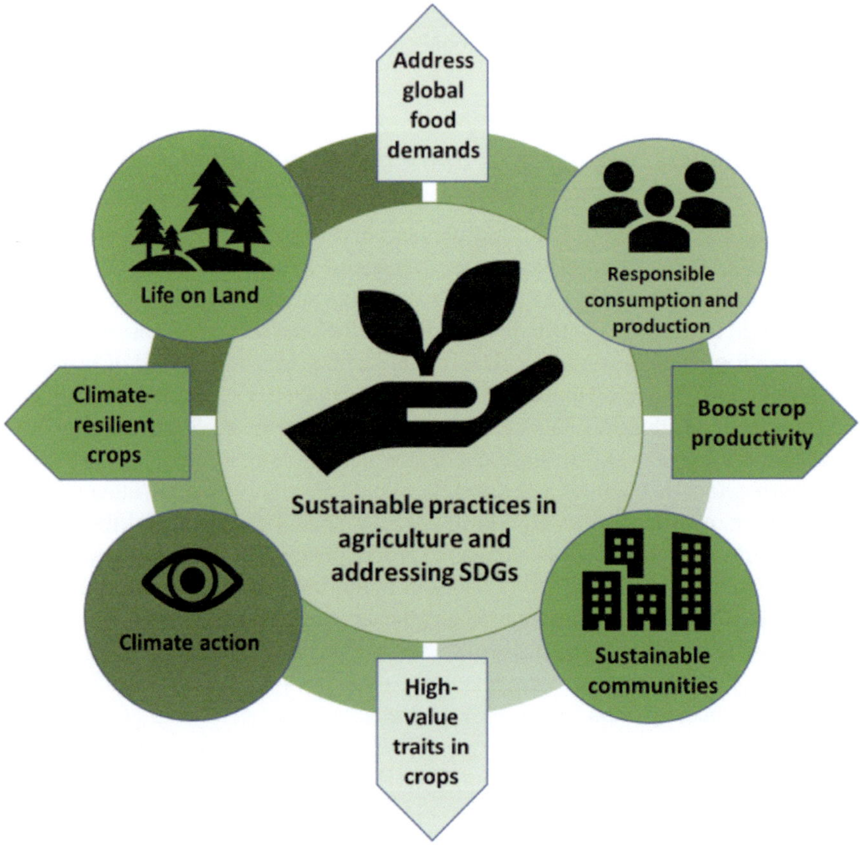

Fig. 15.5 Sustainable agricultural practices aligned with SDGs to enhance crop resilience, productivity, and environmental stewardship (Tiwari and Park 2024)

15.9.3 Bioinformatics and Systems Biology

In the essential foundations of plant biotechnology approaches needed to sustainably take advantage of plant systems, bioinformatics and systems biology have become crucial pillars that have made it substantially easier to comprehend, experiment with, and improve plant systems. The evolution of high-throughput sequencing technologies and the increase of biological data exponentially have caused the appearance of the advanced computational tools so as to store, analyze, and interpret huge amounts of data efficiently. Bioinformatics as the field between biology, mathematics, and computer science helps researchers to organize, analyze, and gather data on genomics, proteomics, and transcriptomics in a systematic manner. The discipline offers the necessary instruments and algorithms to interpret complicated biological data, including nucleotide sequences, protein structure, gene expression patterns, and evolutionary relations. Within the scope of plant biotechnology, bioinformatics is used to speed up the diversion of important genes and regulatory systems that are related to stress tolerance, yield improvement, disease resistance, and metabolic engineering, thereby enhancing crop improvement to a higher and more efficient degree.

Simultaneously, systems biology provides a comprehensive and integrative way to comprehend plant development and functionality. Giving the idea of living organisms a different form, systems biology conceives organisms as dynamic and interrelated networks of molecular components that work across different levels of hierarchies. It explains the difficulty of how protein and gene environment interactions can be realized and understood to be used in coming up with emergent properties that govern plant phenotypes. The method is based on the application of multi-omics technologies, including genomics, transcriptomics, proteomics, and metabolomics, and the integration of these technologies using computational modeling in order to reconstruct whole biological pathways and systems and simulate them in detail. Systems biology offers chances to learn more about multidimensional traits and responses that are otherwise hard to be analyzed utilizing normal genetic models, allowing scientists to represent the way plants grow and develop, as well as react to environmental influences at a system level.

Although they have significant transformative potential, there are issues with the complete utilization of bioinformatics and systems biology in relation to plant biotechnology. These are the relative slowness of genome annotation of plants in the context of animal systems, low immediate accessibility to curated databases specific to plants, and high computing requirements of integrative studies. Also, there is a need to have interdisciplinary experts who can close the gap between the computational model and experimental biology. However, with ongoing improvement in ML and AI, as well as big data analytics, many of these limitations are soon to be suppressed, and these digital tools would be far more predictive and useful. Combined, bioinformatics and systems biology provide the basis of precision in plant bioengineering and the resultant intensive system to improve crops and produce a robust agricultural system relative to the principle of sustainability and food security (Zhang et al. 2025).

15.9.4 Predictive Modeling for Sustainable Crop Performance

The application of predictive modeling has emerged as a very fine tool of digital and precision tools in the biotechnology of crops, and its achievement of sustainable crop performance can be transformational. The historic methods of agronomy and breeding are not enough as agriculture is faced with increasing pressures posed by climate change, population, resource constraints, and necessity of sustainable food systems. Consequently, as a response, predictive analytics is a potent method utilizing ML, AI, and big data to predetermine how crops behave, in addition to optimizing agricultural inputs to determine how to best resolve this issue, responding to this on a real-time level. Through their ability to process large volumes of data containing genomic data, phenotypic evaluation, soil variables, and weather as well as crop management practices, predictive models assist researchers and farmers to predict the behavior of plants in response to stress and disease as well as yield fluctuations with excellent precision.

Predictive modeling has found utility in various areas of crop improvement in relation to plant biotechnology. As an example, it supports the expression of genotype x environment involvements that make it possible to choose stress-tolerant genotypes tailored to agroecological regions. Prediction tools are also used in the creation of precision agriculture systems whereby water, chemicals, and other inputs to crops are used precisely in an optimal fashion in accordance with model predictions so as to create efficiency and less environmental effects. They are also instrumental in disease detection and pest forecasting at an early stage, essential in reducing the loss and the achievement of food security. Predictive modeling has been found valuable in real-life applications geared towards making agricultural resilient and yielding. Satellite imagery, sensor data, and climatic trends are now combined in tools that are used to inform sowing schedules, irrigation regimes, and harvest timings, increased the efficiency of the use of resources as well as the yield. In addition, predictive breeding programs are also used to predict the performance of the new plant varieties in the fields by employing GS models, which helps to considerably shorten the breeding cycle and save time-to-market of the cultivar.

However, broad usage of predictive modeling is not without its difficulties. These amount to scarcity of high-quality and carefully curated datasets, difficulty to bring together data of different types (e.g., genomics, weather, and phenomics), and technical expertise to create and analyze complex models. Moreover, the smallholder farmers in developing countries might be closed away in the usage of these technologies owing to the restrictions on infrastructure. The democratization of prediction analysis is, however, becoming a possibility with the current development of open-access platforms, cloud-based systems, and mobile-based apps. In the future, the current combination of predictive modeling and other digital technologies like IoT devices, RS systems, and systems biology frameworks will result into a more connected and more smart landscape in terms of the agricultural sphere. Such combined systems have the ability to deliver real-world, real-time decision support based on local situations, and ultimately this will result in more sustainable and resilient agricultural systems. Amidst the ever-changing nature of plant biotechnology, predictive modeling will always form the backbone of any adaptive improvements in crop performance, environmental preservation, and food security amidst the modern challenges that the world has never witnessed (Karunathilake et al. 2023).

15.10 Challenges and Future Perspectives

15.10.1 Public Perception and Policy Barriers

The main threats that plant biotechnological innovations can face on their way to successful realization and adoption include the issue of public perception and obstacles in the policy sphere. Even though biotechnology holds great potential in meeting the global agricultural and environmental issues (such as food security, climate resilience, and sustainable resource management), the journey towards its adoption in high frequencies is plagued by sociopolitical activities, lack of regulations, and diverse trust among the general population. The negative history of agricultural biotechnology, especially GMOs, has left an impression on the attitudes of people that has shaped the skepticism and resistance towards the crop science and engineering of new plant breeding techniques in the form of CRISPR and gene editing. Various factors contribute to the public willingness to accept these tools, and they can be risks and benefits of the tools, personal values, how scientifically literate one can be, trust towards the institutions, and information presentation. Cultural values, cultural beliefs, and sociodemographic features, as well as the product-specific features, also influence consumer attitude toward biotech-derived food products. Evaluations reveal that visibility, effective communication plans, plus participation at an early stage of stakeholders can go a long way to improve acceptability and knowledge of the general population.

Besides, there is yet another layer in the form of policy and regulatory frameworks. The regulatory systems, especially within the European Union, have constantly been perceived as too excessive, risk averse, and incompatible with other global regions. Such stringency of regulation may hinder commercialization and approval of products of biotechnological innovation, which will kill innovation and develop uncertainty for the developers. Moreover, regulatory interpretations varying depending on the country raise trade barriers and hinder cooperation between nations. The absence of coordinated policies in the implementation of biotechnology is a setback to the global attempts in implementing biotechnology as a panacea for food security and problems associated with climate change in agricultural production. Moreover, it has been noted that, whereas policymakers focus on risk analysis, they have little regard for the possible benefits, which led to the formulation of policies that do not follow the middle way of managing innovations. Unlike agricultural biotechnology that is more mature and advanced, forestry biotechnology is also relatively younger and less developed but also has the same concerns or even more. The high life cycle, uncertainties on ecology, and loss of biodiversity stir up further concerns of the people. In the forestry literature, the legal, ethical, and governance obstacles are more emphasized, which points in the direction that risk perception within the environment of tree biotechnology is highly intertwined and multi-faceted in relation to the general effects on the environment and society. Although social sciences are increasingly becoming relevant when it comes to solving these problems, an extremely small percentage of literature in forestry and agriculture shows any evidence of taking into account the results of studies in public perception or models of engagement with stakeholders. To deal with these constraints, biotechnology governance has to move to inclusive, upstream methods of participation. This involves an early participation of the rights holders, native communities, farmers, consumers, and environmental activists in research activities and decision-making operations. The policies should be flexible, scientific, and aligned with the needs and values of society with an intent to develop trust in the population through openness and methods of participation in governance. It is not only essential to bridge the gap between science innovation and the popular sentiment so that plant biotechnology can be implemented successfully but also to see to it that the demands of sustainable development can be met in an equitable manner (Moshelion and Altman 2015).

15.10.2 Climate Change Adaptation through Biotech Solutions

Climate change is presenting serious and worsening problems to agricultural systems all over the world where it has a significant impact on plant health, productivity as well as food security at large. Due to rising temperatures and unpredictable rain patterns, frequent droughts, floods, and outbreaks of pests and diseases, yield is going down, soil is getting degraded, and arable land is witnessing a reduction. Such transformations have put huge strains on agriculture, especially in the vulnerable areas like sub-Saharan Africa and South Asia which already face large risks of food insecurity and undernutrition. As an illustration, in the 1980–2015 period, the worldwide yields of wheat and maize dropped by 20.6% and 39.3%, respectively, as a result of climate-related causes. Predictions are that the condition will be even worse given that drought stress alone is estimated to cut crop production by over half in the next 50 years. The area of concern is rice growing in rainfed areas such as South and Southeast Asia with high demands of water, up to 3000–5000 L/kg of seed, and with prolonged droughts, the losses can be more than 40%.

The solutions to these growing challenges are provided by biotechnological innovations as the stocks of armaments. Genetic engineering, genome editing (e.g., CRISPR/Cas systems), RNA interference, NGS, and genome mapping are modern methods that open the way to the development of climate-resilient crops at a defined location with the highest level of precision and speed. The technologies support the development of cultivars that are resistant to abiotic stresses like drought, heat, salinity, and flood scenarios more commonly being experienced as a result of climate change. As an example, salt stress currently impacts more than 20% of the agricultural soils worldwide, and it is estimated that climate change will increase the amount of salt-affected soils twice as much in 2050. Biotechnology offers the capability to create salt-tolerant crops by manipulating the genes that govern stomatal behavior, root system, and also osmotic equilibrium, thus maintaining the growth of the plants under saline conditions.

Moreover, the mentioned technologies improve water and nutrient utilization in crops, which also fits into sustainability. Since farming takes up to 70% of all the freshwater supplies, it is of importance to increase the efficiency of the water use. Through traits derived by biotech methods, water needs can be minimized, and drought and soil moisture shortage effects can be reduced. More so, molecular breeding and transgenic technologies provide the opportunity to combine various traits in crop lines, including pest resistance, temperature adaptation, and nutritional enhancement to optimize crop performance in compound stress conditions. Involving climate-resilient biotechnology in agricultural practices will not just entail technical progress but also the stabilizing related regulations, community participation, and infrastructural investment, especially in the low-middle-income nations. Adaptation using plant biotechnology is not an option anymore as effects of climate change increase. It is a potential solution to stabilizing global food security, conservation of environmental resources, and sustainability of agriculture under a changing and unpredictable climate (Malhi et al. 2021).

15.10.3 Bridging Research and Field-Level Application

The challenge has always been bridging the gap between laboratory research and the real-world, or field-level, application, which has been a thorny topic in plant biotechnology, particularly in the pursuance of sustainable agriculture. Although much has been achieved in genetic engineering, genome editing, synthetic biology, and data-driven precision agriculture, there is a common problem in transferring these findings to systems that can handle them in large-scale environments that operate in the field. The lack of real-world applicability, as well as how often new technologies and innovations fail to deliver the exact same performance they did in the controlled environment of a laboratory or a greenhouse in the complex environment of the field, is frequently attributed to the difference between these two conditions and characterized by the difference in the soil properties, changing climatic conditions, and the makeup of the microbial community. To give an example, plant–microbe interactions in sterile or semicontrolled systems often do not reproduce in the field because of varying diffusion rates, chemical signals, and soil physical properties. Subsequently, more field-based research is urgently needed, especially when it comes to explaining the viability and efficiency of commercial microbial inoculants and biofertilizers. Such biocontrol agents should be tested in long-term studies in order to verify their functionality in various geographies and farming systems.

The combination of the technologies of precision agriculture and smart field platforms is the promising way forward in the attempt to work on these issues. These platforms leverage a set of environmental sensors, digital images, real-time data analytics, and autonomous vehicles to make fine-grained monitoring and control over applied crop inputs possible. As an example, fixed solar-powered soil sensor nodes can provide real-time data on soil respiration, emissions of greenhouse gases, moisture, temperature, salinity, and pH levels, and the mobile ground- and aerial-based systems expand the spatial resolution and the flexibility of monitoring. The information gathered based on these sources can be transferred in a wireless way and examined with the help of AI and ML algorithms to make the decisions that will influence the productive growth of the crops in real-time and optimize the interventions that should take place in the course of the crop. It does not only enhance productivity and resilience but also ensures that there is no unnecessary excessive use of water, fertilizers, and pesticides to maximize the negative impact on the environment.

Moreover, there is a need to engage cross-disciplinary areas of collaboration such as plant scientists, agronomists, microbial ecologists, engineers, and data scientists to develop scalable, affordable, and easily adaptable solutions within particular agricultural ecosystems. Smart fields are a vibrant blend of biotechnology and information technology (IT), not a leap but a concrete path to get the latest research out of the laboratories and into practices in the field. Moreover, blockchain and the IoT add to transparency and traceability in agricultural supply chains, which allows consumers to get confidence in food sustainability. In spite of these developments, challenges are still on the way. The adoption of new biotechnologies may be thwarted by ethical factors, restrictions on regulations, insufficient funding, and a perceived dismissal by the population. Thus, it is necessary to take into consideration policy support, public awareness campaigns, and the further development of translational research in order to overcome these challenges. The future of sustainable agriculture depends on the successful implementation of the innovations in the area of plant biotechnology at the field level. Through the creation of resilient foundations that spearhead real-time on-site management in crops management and creation of effective intersectorial collaborations, the division between research breakthroughs in laboratories and their agricultural application can be effectively closed, which would help ensure food security in the long run and sustainability of the environment (Abiri et al. 2023).

15.10.4 Roadmap for Responsible Innovation

To have sustainable solutions to the complex issues of climate change, shortage of resources, and food security, there is need to have a comprehensive roadmap of responsible innovation of plant biotechnology. With the rise in environmental pressure in the world, the ability to combine advanced IT with biotechnology can therefore be considered to be unprecedented in providing the opportunity and platform to transform the sustainable agricultural processes. Such a synergistic convergence enables the generation of innovative technologies like precision agriculture, genetic engineering, bio-informatics enhanced crop optimization, and smart farming systems. The technologies not only can increase yield and stress tolerance but also enhance the efficiency of resources and lower environmental impact and instigate the sustainability over the long term. Big data analytics combined with AI and the IoT can be used to monitor crop and soil health and real-time climate conditions to enable farmers to use data to optimize their inputs and outputs.

Despite these good opportunities, however, there are a number of obstacles that limit the way forward to its popularization, among which are technical limitations, prohibitive costs of implementation, lack of infrastructure in developing nations, and ethical issues about GMOs and data privacy. Addressing these gaps will require a well-designed roadmap ranging from the short-term goals to the medium-term goals as well as long-term goals. Awareness building, capacity building, and development of the public–private parallels are some of the mechanisms that should be developed in the short term to facilitate knowledge transmission and transfer of technology. The medium-term options should focus on policy reforms, investment in infrastructure, and harmonization of regulatory frameworks that promote good practices in the utilization of biotechnology and IT in agriculture. The establishment of interdisciplinary studies and more inclusive innovation systems in the long term will work to generate scalable and socially acceptable solutions.

The implementation of blockchain-based traceability systems in the supply chains or the use of CRISPR to increase crop yields are successful case studies that attest to the capabilities of such innovations to promote environmental and economic sustainability. The examples show why there is a need to involve cross-sector collaboration of researchers, industry players, policymakers, and civil society to deal with complex challenges in a whole-way approach. Further, the incorporation of the principles behind the circular bioeconomy and the evolution of biodegradable agricultural inputs indicates the evolution of the attempts to comply biotechnology with environmental management.

The facilitation of responsible innovation requires policymakers and industry leaders to implement changes that could affect the regulatory and economic environment positively. This involves providing research in open access and maintaining ethical standards and producing incentives for the development of sustainable technologies. Sustainability and equity will need to be integrated into all the aspects of the innovation lifecycle (research and development, deployment, and monitoring) in order to make plant biotechnology a potent driver of sustainability all over the globe. With the growing issues of environmental and agricultural concerns that the world has to deal with, the world needs an easy roadmap on responsible innovation, which can provide a paradigm shift to capitalizing on the technological convergence to provide a sustainable and resilient future (Sánchez-García et al. 2024).

15.11 Conclusion

Biotechnology in plants has come out as an innovator in fostering sustainable farm activities, renewal of the natural environment, and development of bio-product industries. Plant scientists are also utilizing the innovative use of genetic engineering, genome editing, and molecular breeding tools to help crops respond to altering climatic patterns, increasing population, and diminishing natural resources. The developments do not only raise the output and the tolerance to stress and nutrition but also help to lessen the footprint of agriculture on the environment. Beneficial plant–microbe interactions, plant tissue culture, and micropropagation are some of the technologies that support sustainable propagation, conservation of biodiversity, and development of environmentally spared alternatives of chemical-based inputs. Synthetic biology and metabolic engineering have widened the horizons of plant biotechnology to the production of high (value) products related to pharmaceuticals, nutraceuticals, and bioplastics. Moreover, phytoremediation and the bioenergy trends reveal the possibility of the plant-based approaches to environmental pollution as well as the implementation of circular economy patterns.

The digitization of technologies, including AI, RS, and bioinformatics, also increases the level of precision and scale when applying biotechnology to reach practical applications in the field. Nevertheless, to harness the maximum benefit of plant biotechnology, there are a number of challenges to be solved, such as the regulatory issues, acceptance by the population, accessibility to the technology, and the delivery of biosafety and ethical integrity. We must be prepared to maintain interdisciplinary approaches in the collaborative effort, embrace broad-based innovation, and have well-informed policymaking to ensure that the future of plant biotechnology adds value to the delivery of sustainable goals in the world. When the technological innovation is combined with the stewardship of the environment and societal responsibility, plant biotechnology can be instrumental in providing a strong, healthy, and prosperous future to all future generations.

References

Abdul Aziz M, Masmoudi K (2025) Molecular breakthroughs in modern plant breeding techniques. Hortic Plant J 11:15–41. https://doi.org/10.1016/j.hpj.2024.01.004

Abiri R, Rizan N, Balasundram SK, Shahbazi AB, Abdul-Hamid H (2023) Application of digital technologies for ensuring agricultural productivity. Heliyon 9:e22601. https://doi.org/10.1016/j.heliyon.2023.e22601

Aghaali Z, Naghavi MR, Zargar M (2024) Collaboration of hairy root culture and scale-up strategies for enhancing the biosynthesis of medicinal and defensive alkaloids in Papaver sp. Curr Plant Biol 40:100381. https://doi.org/10.1016/j.cpb.2024.100381

Ahkami AH, Allen White R, Handakumbura PP, Jansson C (2017) Rhizosphere engineering: enhancing sustainable plant ecosystem productivity. Rhizosphere 3:233–243. https://doi.org/10.1016/j.rhisph.2017.04.012

Ahmad K, Ming R (2024) Harnessing genetic tools for sustainable bioenergy: a review of sugarcane biotechnology in biofuel production. Agriculture 14:1312. https://doi.org/10.3390/agriculture14081312

Ahmad Z, Niyazi S, Firdoos A, Wang C, Manzoor MA, Ramakrishnan M, Upadhyay A, Ding Y (2024) Enhancing plant resilience: nanotech solutions for sustainable agriculture. Heliyon 10:e40735. https://doi.org/10.1016/j.heliyon.2024.e40735

Ali S, Abbas Z, Rizwan M, Zaheer I, Yavaş İ, Ünay A, Abdel-DAIM M, Bin-Jumah M, Hasanuzzaman M, Kalderis D (2020) Application of floating aquatic plants in phytoremediation of heavy metals polluted water: a review. Sustainability 12:1927. https://doi.org/10.3390/su12051927

Amy SR, Heard MS, Hartley SE, George CT, Pywell RF & Staley JT (2015) 'Hedgerow rejuvenation management affects invertebrate communities through changes to habitat structure', Basic Appl Ecol 16(5):443–51

Anto S, Mukherjee SS, Muthappa R, Mathimani T, Deviram G, Kumar SS, Verma TN, Pugazhendhi A (2020) Algae as green energy reserve: technological outlook on biofuel production. Chemosphere 242:125079. https://doi.org/10.1016/j.chemosphere.2019.125079

Anwar A, Kim J-K (2020) Transgenic breeding approaches for improving abiotic stress tolerance: recent progress and future perspectives. Int J Mol Sci 21:2695. https://doi.org/10.3390/ijms21082695

Bapat VA, Kavi Kishor PB, Jalaja N, Jain SM, Penna S (2023) Plant cell cultures: biofactories for the production of bioactive compounds. Agronomy 13:858. https://doi.org/10.3390/agronomy13030858

Basu A, Prasad P, Das SN, Kalam S, Sayyed RZ, Reddy MS, El Enshasy H (2021) Plant growth promoting rhizobacteria (PGPR) as green bioinoculants: recent developments, constraints, and prospects. Sustainability 13:1140. https://doi.org/10.3390/su13031140

Caradus JR (2023) Intended and unintended consequences of genetically modified crops – myth, fact and/or manageable outcomes? N Z J Agric Res 66:519–619. https://doi.org/10.1080/00288233.2022.2141273

Chandran H, Meena M, Barupal T, Sharma K (2020) Plant tissue culture as a perpetual source for production of industrially important bioactive compounds. Biotechnol Rep 26:e00450. https://doi.org/10.1016/j.btre.2020.e00450

Chaudhary R, Sharma C, Kumar V, Rajput V, Naik B, Prasad R, Sharma S, Kumar V (2025) Millet biofortification for enhanced iron content: roadmap for combating hidden hunger. J Agric Food Res 19:101654. https://doi.org/10.1016/j.jafr.2025.101654

Dublino R, Ercolano M (2025) Artificial intelligence redefines agricultural genetics by unlocking the enigma of genomic complexity. Crop J 13(5):1350–1362. https://doi.org/10.1016/j.cj.2025.05.008

Elbouzidi A, Haddou M, Baraich A, Taibi M, El Hachlafi N, Pareek A, Mesnard F, Addi M (2025) Biochemical insights into specialized plant metabolites: advancing cosmeceutical applications for skin benefits. J Agric Food Res 19:101651. https://doi.org/10.1016/j.jafr.2025.101651

Farhaoui A, Taoussi M, Laasli S-E, Legrifi I, El Mazouni N, Meddich A, Hijri M, Lahlali R (2025) Arbuscular mycorrhizal fungi and their role in plant disease control: a state-of-the-art. Microbe 8:100438. https://doi.org/10.1016/j.microb.2025.100438

Food and Agriculture Organization (FAO) (2009) Global agriculture towards 2050. High Level Expert Forum: How to Feed the World in 2050, Rome, Italy

Gregorowius D, Lindemann-Matthies P, Huppenbauer M (2012) Ethical discourse on the use of genetically modified crops: a review of academic publications in the fields of ecology and environmental ethics. J Agric Environ Ethics 25:265–293. https://doi.org/10.1007/s10806-011-9330-6

Hamdan MF, Tan BC (2024) Genetic modification techniques in plant breeding: a comparative review of CRISPR/Cas and GM technologies. Hortic Plant J 11(5):1807–1829. https://doi.org/10.1016/j.hpj.2024.02.012

Jiang G-L (2013) Molecular markers and marker-assisted breeding in plants. In: Plant breeding from laboratories to fields. IntechOpen. https://doi.org/10.5772/52583

Karunathilake EMBM, Le AT, Heo S, Chung YS, Mansoor S (2023) The path to smart farming: innovations and opportunities in precision agriculture. Agriculture 13:1593. https://doi.org/10.3390/agriculture13081593

Kawaguchi H, Takada K, Elkasaby T, Pangestu R, Toyoshima M, Kahar P, Ogino C, Kaneko T, Kondo A (2022) Recent advances in lignocellulosic biomass white biotechnology for bioplastics. Bioresour Technol 344:126165. https://doi.org/10.1016/j.biortech.2021.126165

Khan FF, Ahmad K, Ahmed A, Haider S (2017) Applications of biotechnology in agriculture-review article. World J Biol Biotechnol 2:139–142. https://doi.org/10.33865/wjb.002.01.0013

Kim M-S, Yang J-Y, Yu J-K, Lee Y, Park Y-J, Kang K-K, Cho Y-G (2021) Breeding of high cooking and eating quality in rice by marker-assisted backcrossing (MABc) using KASP markers. Plants 10:804. https://doi.org/10.3390/plants10040804

Kulus D, Zalewska M (2014) Cryopreservation as a tool used in long-term storage of ornamental species – a review. Sci Hortic 168:88–107. https://doi.org/10.1016/j.scienta.2014.01.014

Kumar A, Padhee AK & Kumar S (2020) 'How Indian agriculture should change after COVID-19', Food Secur 12(4):837–40

Kumari K, Mirzakhani Nafchi A, Mirzaee S, Abdalla A (2025) AI-driven future farming: achieving climate-smart and sustainable agriculture. AgriEngineering 7:89. https://doi.org/10.3390/agriengineering7030089

Kuppan N, Padman M, Mahadeva M, Srinivasan S, Devarajan R (2024) A comprehensive review of sustainable bioremediation techniques: eco friendly solutions for waste and pollution management. Waste Manag Bull 2:154–171. https://doi.org/10.1016/j.wmb.2024.07.005

Li Z, Yang J, Cai X, Zeng X, Zou J-J, Xing W (2024) A systematic review on the role of miRNAs in plant response to stresses under the changing climatic conditions. Plant Stress 14:100674. https://doi.org/10.1016/j.stress.2024.100674

Liu X, Mei S, Salles JF (2023) Inoculated microbial consortia perform better than single strains in living soil: a meta-analysis. Appl Soil Ecol 190:105011. https://doi.org/10.1016/j.apsoil.2023.105011

Malhi GS, Kaur M, Kaushik P (2021) Impact of climate change on agriculture and its mitigation strategies: a review. Sustainability 13:1318. https://doi.org/10.3390/su13031318

Mamathaba MP, Yessoufou K, Moteetee A (2022) What does it take to further our knowledge of plant diversity in the megadiverse South Africa? Diversity 14:748. https://doi.org/10.3390/d14090748

Mangal V, Verma LK, Singh SK, Saxena K, Roy A, Karn A, Rohit R, Kashyap S, Bhatt A, Sood S (2024) Triumphs of genomic-assisted breeding in crop improvement. Heliyon 10:e35513. https://doi.org/10.1016/j.heliyon.2024.e35513

Marzouk SH, Kwaslema DR, Omar MM, Mohamed SH (2025) "Harnessing the power of soil microbes: their dual impact in integrated nutrient management and mediating climate stress for sustainable rice crop production" A systematic review. Heliyon 11:e41158. https://doi.org/10.1016/j.heliyon.2024.e41158

Mehbub H, Akter A, Akter MA, Mandal MSH, Hoque MA, Tuleja M, Mehraj H (2022) Tissue culture in ornamentals: cultivation factors, propagation techniques, and its application. Plants 11:3208. https://doi.org/10.3390/plants11233208

Mmbando GS, Missanga J (2024) The current status of genetic biofortification in alleviating malnutrition in Africa. J Genet Eng Biotechnol 22:100445. https://doi.org/10.1016/j.jgeb.2024.100445

Moshelion M, Altman A (2015) Current challenges and future perspectives of plant and agricultural biotechnology. Trends Biotechnol 33:337–342. https://doi.org/10.1016/j.tibtech.2015.03.001

Ndudzo A, Sibanda Makuvise A, Moyo S, Bobo ED (2024) CRISPR-Cas9 genome editing in crop breeding for climate change resilience: implications for smallholder farmers in Africa. J Agric Food Res 16:101132. https://doi.org/10.1016/j.jafr.2024.101132

Oro CED, Saorin Puton BM, Venquiaruto LD, Dallago RM, Tres MV (2024) Effective microbial strategies to remediate contaminated agricultural soils and conserve functions. Agronomy 14:2637. https://doi.org/10.3390/agronomy14112637

Pathirana R, Carimi F (2022) Management and utilization of plant genetic resources for a sustainable agriculture. Plants 11:2038. https://doi.org/10.3390/plants11152038

Pathirana R, Carimi F (2024) Plant biotechnology—an indispensable tool for crop improvement. Plants 13:1133. https://doi.org/10.3390/plants13081133

Paudel S, States SL (2023) Urban green spaces and sustainability: exploring the ecosystem services and disservices of grassy lawns versus floral meadows. Urban For Urban Green 84:127932. https://doi.org/10.1016/j.ufug.2023.127932

Pfenning-Butterworth A, Buckley LB, Drake JM, Farner JE, Farrell MJ, Gehman A-LM, Mordecai EA, Stephens PR, Gittleman JL, Davies TJ (2024) Interconnecting global threats: climate change, biodiversity loss, and infectious diseases. Lancet Planet Health 8:e270–e283. https://doi.org/10.1016/S2542-5196(24)00021-4

Pinski A, Mehdi SMM, Betekhtin A (2025) Molecular mechanisms underlying abiotic stress responses in buckwheat. Plant Sci 357:112526. https://doi.org/10.1016/j.plantsci.2025.112526

Rasheed A, Gill RA, Hassan MU, Mahmood A, Qari S, Zaman QU, Ilyas M, Aamer M, Batool M, Li H, Wu Z (2021) A critical review: recent advancements in the use of CRISPR/Cas9 technology to enhance crops and alleviate global food crises. Curr Issues Mol Biol 43:1950–1976. https://doi.org/10.3390/cimb43030135

Rawat JM, Bhandari A, Raturi M, Rawat B (2019) Agrobacterium rhizogenes mediated hairy root cultures: a promising approach for production of useful metabolites. In: New and future developments in microbial biotechnology and bioengineering. Elsevier, pp 103–118. https://doi.org/10.1016/B978-0-444-63504-4.00008-6

Saha D, Sundriyal RC (2012) Utilization of non-timber forest products in humid tropics: implications for management and livelihood. Forest Policy Econ 14:28–40. https://doi.org/10.1016/j.forpol.2011.07.008

Sánchez-Castro I, Molina L, Prieto-Fernández M-Á, Segura A (2023) Past, present and future trends in the remediation of heavy-metal contaminated soil – remediation techniques applied in real soil-contamination events. Heliyon 9:e16692. https://doi.org/10.1016/j.heliyon.2023.e16692

Sánchez-García E, Martínez-Falcó J, Marco-Lajara B, Manresa-Marhuenda E (2024) Revolutionizing the circular economy through new technologies: a new era of sustainable progress. Environ Technol Innovation 33:103509. https://doi.org/10.1016/j.eti.2023.103509

Shahzad A, Sharma S, Parveen S, Saeed T, Shaheen A, Akhtar R, Yadav V, Upadhyay A, Ahmad Z (2017) Historical perspective and basic principles of plant tissue culture. In: Abdin MZ, Kiran U, Kamaluddin, Ali A (eds) Plant biotechnology: principles and applications. Springer, Singapore, pp 1–36. https://doi.org/10.1007/978-981-10-2961-5_1

Sharma BS, Jansen GB, Karrow NA, Kelton D, Jiang Z (2006) Detection and characterization of amplified fragment length polymorphism markers for clinical mastitis in Canadian Holsteins. J Dairy Sci 89:3653–3663. https://doi.org/10.3168/jds.S0022-0302(06)72405-5

Shmakov VN, Belkov VI, Konstantinov YM (2024) Using the rejuvenation phenomenon to obtain vegetative progeny of trees. Russ J Plant Physiol 71:171. https://doi.org/10.1134/S1021443724606682

Solomon W, Janda T, Molnár Z (2024) Unveiling the significance of rhizosphere: implications for plant growth, stress response, and sustainable agriculture. Plant Physiol Biochem 206:108290. https://doi.org/10.1016/j.plaphy.2023.108290

Su W, Xu M, Radani Y, Yang L (2023) Technological development and application of plant genetic transformation. Int J Mol Sci 24:10646. https://doi.org/10.3390/ijms241310646

ŠVábová L, Lebeda A (2005) In vitro selection for improved plant resistance to toxin-producing pathogens. J Phytopathol 153:52–64. https://doi.org/10.1111/j.1439-0434.2004.00928.x

Thitisaksakul M, Jiménez RC, Arias MC & Beckles DM (2012) 'Effects of environmental factors on cereal starch biosynthesis and composition', J CeR Sci 56(1).67–80

Tian L (2015) Using hairy roots for production of valuable plant secondary metabolites. In: Krull R, Bley T (eds) Filaments in bioprocesses. Springer International Publishing, Cham, pp 275–324. https://doi.org/10.1007/10_2014_298

Tiwari P, Park K-I (2024) Seed biotechnologies in practicing sustainable agriculture: insights and achievements in the decade 2014–2024. Appl Sci 14:11620. https://doi.org/10.3390/app142411620

Tripathi AD, Mishra R, Maurya KK, Singh RB, Wilson DW (2019) Estimates for world population and global food availability for global health. In: The role of functional food security in global health. Elsevier, pp 3–24. https://doi.org/10.1016/B978-0-12-813148-0.00001-3

Tyagi A, Lama Tamang T, Kashtoh H, Mir RA, Mir ZA, Manzoor S, Manzar N, Gani G, Vishwakarma SK, Almalki MA, Ali S (2024) A review on biocontrol agents as sustainable approach for crop disease management: applications, production, and future perspectives. Horticulturae 10:805. https://doi.org/10.3390/horticulturae10080805

Vignesh A, Amal TC, Sarvalingam A, Vasanth K (2024) A review on the influence of nutraceuticals and functional foods on health. Food Chem Adv 5:100749. https://doi.org/10.1016/j.focha.2024.100749

Wei J, Li Y (2023) CRISPR-based gene editing technology and its application in microbial engineering. Eng Microbiol 3:100101. https://doi.org/10.1016/j.engmic.2023.100101

Wijerathna-Yapa A, Hiti-Bandaralage J (2023) Tissue culture—a sustainable approach to explore plant stresses. Life 13:780. https://doi.org/10.3390/life13030780

Zhang C, Singh RP, Yadav P, Kumar I, Kaushik A, Roychowdhury R, Mubeen M, Singh SK, Kumar A, Wang J (2025) Recent advances in biotechnology and bioengineering for efficient microalgal biofuel production. Fuel Process Technol 270:108199. https://doi.org/10.1016/j.fuproc.2025.108199

MIX
Papier aus verantwortungsvollen Quellen
Paper from responsible sources
FSC® C105338

If you have any concerns about our products,
you can contact us on
ProductSafety@springernature.com

In case Publisher is established outside the EU,
the EU authorized representative is:
Springer Nature Customer Service Center GmbH
Europaplatz 3, 69115 Heidelberg, Germany

Printed by Libri Plureos GmbH
in Hamburg, Germany